# Research and Development in Intelligent Systems XXIX

## Incorporating Applications and Innovations in Intelligent Systems XX

Max Bramer • Miltos Petridis
Editors

# Research and Development in Intelligent Systems XXIX

## Incorporating Applications and Innovations in Intelligent Systems XX

## Proceedings of AI-2012, The Thirty-second SGAI International Conference on Innovative Techniques and Applications of Artificial Intelligence

*Editors*
Max Bramer
School of Computing
University of Portsmouth
Portsmouth, UK

Miltos Petridis
School of Computing, Engineering
and Mathematics
University of Brighton
Brighton, UK

ISBN 978-1-4471-4738-1 ISBN 978-1-4471-4739-8 (eBook)
DOI 10.1007/978-1-4471-4739-8
Springer London Heidelberg New York Dordrecht

Library of Congress Control Number: 2012951849



Printed on acid-free paper

Springer is part of Springer Science+Business Media (www.springer.com)

# PROGRAMME CHAIRS' INTRODUCTION

M.A.BRAMER, University of Portsmouth, UK
M.PETRIDIS, University of Greenwich, UK

This volume comprises the refereed papers presented at AI-2012, the Thirty-second SGAI International Conference on Innovative Techniques and Applications of Artificial Intelligence, held in Cambridge in December 2012 in both the technical and the application streams. The conference was organised by SGAI, the British Computer Society Specialist Group on Artificial Intelligence.

The technical papers included new and innovative developments in the field, divided into sections on Data Mining, Data Mining and Machine Learning, Planning and Optimisation, and Knowledge Management and Prediction. This year's Donald Michie Memorial Award for the best refereed technical paper was won by a paper entitled "Biologically inspired speaker verification using Spiking Self-Organising Map" by Tariq Tashan, Tony Allen and Lars Nolle (Nottingham Trent University, UK).

The application papers included present innovative applications of AI techniques in a number of subject domains. This year, the papers are divided into sections on Language and Classification, Recommendation, Practical Applications and Systems, and Data Mining and Machine Learning

This year's Rob Milne Memorial Award for the best refereed application paper was won by a paper entitled "Swing Up and Balance Control of the Acrobot Solved by Genetic Programming" by Dimitris C. Dracopoulos and Barry D. Nichols (University of Westminster).

The volume also includes the text of short papers presented as posters at the conference.

On behalf of the conference organising committee we would like to thank all those who contributed to the organisation of this year's programme, in particular the programme committee members, the executive programme committees and our administrators Mandy Bauer and Bryony Bramer.

Max Bramer, Technical Programme Chair, AI-2012
Miltos Petridis, Application Programme Chair, AI-2012

# ACKNOWLEDGEMENTS

## AI-2012 CONFERENCE COMMITTEE

## TECHNICAL EXECUTIVE PROGRAMME COMMITTEE

Prof. Max Bramer, University of Portsmouth (Chair)

Dr. Frans Coenen, University of Liverpool

Dr. John Kingston, Health and Safety Laboratory

Dr. Peter Lucas, University of Nijmegen, The Netherlands

Prof. Daniel Neagu, University of Bradford (Vice-Chair)

Prof. Thomas Roth-Berghofer, University of West London

## APPLICATIONS EXECUTIVE PROGRAMME COMMITTEE

Prof. Miltos Petridis, University of Brighton (Chair)

Mr. Richard Ellis, Helyx SIS Ltd

Ms. Rosemary Gilligan, University of Hertfordshire

Dr Jixin Ma, University of Greenwich (Vice-Chair)

Dr. Richard Wheeler, University of Edinburgh

## TECHNICAL PROGRAMME COMMITTEE

Andreas A Albrecht (Queen's University Belfast)

Ali Orhan Aydin (Istanbul Gelisim University)

Yaxin Bi (University of Ulster)

Mirko Boettcher (University of Magdeburg, Germany)

Max Bramer (University of Portsmouth)

Krysia Broda (Imperial College, University of London)

Ken Brown (University College Cork)

Frans Coenen (University of Liverpool)

Bruno Cremilleux (University of Caen)

Madalina Croitoru (University of Montpellier, France)

Ireneusz Czarnowski (Gdynia Maritime University, Poland)

John Debenham (University of Technology; Sydney)

Nicolas Durand (University of Aix-Marseille)

Frank Eichinger (SAP Research Karlsruhe, Germany)

Adriana Giret (Universidad Politécnica de Valencia)

Nadim Haque (Earthport)

Arjen Hommersom (University of Nijmegen, The Netherlands)

Zina Ibrahim (Kings College, London, UK)

John Kingston (Health & Safety Laboratory)

Konstantinos Kotis (VTT Technical Research Centre of Finland)

Ivan Koychev (Bulgarian Academy of Science)

Fernando Lopes (LNEG-National Research Institute, Portugal)

Peter Lucas (Radboud University Nijmegen)

Stephen G. Matthews (De Montfort University, UK)

Roberto Micalizio (Universita' di Torino)

Dan Neagu (University of Bradford)

Lars Nolle (Nottingham Trent University)

Dan O'Leary (University of Southern California)

Juan Jose Rodriguez (University of Burgos)

Thomas Roth-Berghofer (University of West London)

Fernando Sáenz-Pérez (Universidad Complutense de Madrid)

Miguel A. Salido (Universidad Politécnica de Valencia)

Rainer Schmidt (University of Rostock, Germany)

Sid Shakya (BT Innovate and Design)

Frederic Stahl (Bournemouth University)

Simon Thompson (BT Innovate)

Jon Timmis (University of York)

Andrew Tuson (City University London)

Graham Winstanley (University of Brighton)

Nirmalie Wiratunga (Robert Gordon University)

## APPLICATION PROGRAMME COMMITTEE

Hatem Ahriz (Robert Gordon University)

Tony Allen (Nottingham Trent University)

Ines Arana (Robert Gordon University)

Mercedes Argüello Casteleiro (University of Salford)

Ken Brown (University College Cork)

Sarah Jane Delany (Dublin Institute of Technology)

Richard Ellis (Helyx SIS Ltd)

Roger Evans (University of Brighton)

Lindsay Evett (Nottingham Trent University)

Rosemary Gilligan (University of Hertfordshire)

Adrian Hopgood (De Montfort University)

Stelios Kapetanakis (University of Greenwich)

Alice Kerly (Artificial Solutions UK, Ltd)

Shuliang Li (University of Westminster)

Jixin Ma (University of Greenwich)

Lars Nolle (Nottingham Trent University)

Miltos Petridis (University of Brighton)

Rong Qu (University of Nottingham)

Miguel A. Salido (Universidad Politécnica de Valencia)

Wamberto Vasconcelos (University of Aberdeen)

Richard Wheeler (Edinburgh Scientific)

Simon Coupland (De Montfort University)

John Gordon (AKRI Ltd)

Elizabeth Guest (Leeds Metropolitan University)

Chris Hinde (Loughborough University)

Roger Tait (University of Cambridge)

Patrick Wong (Open University)

# CONTENTS

## Research and Development in Intelligent Systems XXIX

### BEST TECHNICAL PAPER



### DATA MINING



### DATA MINING AND MACHINE LEARNING

## PLANNING AND OPTIMISATION



## KNOWLEDGE MANAGEMENT AND PREDICTION



## SHORT PAPERS

## Applications and Innovations in Intelligent Systems XX

### BEST APPLICATION PAPER



### LANGUAGE AND CLASSIFICATION



### RECOMMENDATION



### PRACTICAL APPLICATIONS AND SYSTEMS

# Research and Development in Intelligent Systems XXIX

# BEST TECHNICAL PAPER

# Biologically inspired speaker verification using Spiking Self-Organising Map

**Tariq Tashan, Tony Allen and Lars Nolle**[1]

**Abstract** This paper presents a speaker verification system that uses a self organising map composed of spiking neurons. The architecture of the system is inspired by the biomechanical mechanism of the human auditory system which converts speech into electrical spikes inside the cochlea. A spike-based rank order coding input feature vector is suggested that is designed to be representative of the real biological spike trains found within the human auditory nerve. The Spiking Self Organising Map (SSOM) updates its winner neuron only when its activity exceeds a specified threshold. The algorithm is evaluated using 50 speakers from the Centre for Spoken Language Understanding (CSLU2002) speaker verification database and shows a speaker verification performance of 90.1%. This compares favorably with previous non-spiking self organising map that used Discrete Fourier Transform (DFT)-based input feature vector with the same dataset.

## 1 Introduction

Speaker verification is the process of proving or denying the identity of a claimed speaker from a test speech sample. This type of application is classified as an open-set problem, as the test sample may belong to the claimed speaker or may belong to an unknown speaker which the network did not experience during the training process. Another speaker recognition problem is speaker identification, where the task is to identify the test speech sample as one of a set of pre-known speakers. Speaker identification is classified as a closed-set problem since the test sample is meant to belong to one of the speakers in the closed-set. Based on the type of classifier used, a speaker recognition system can be categorised as either: Probabilistic, for example Hidden Markov Model (HMM) or Gaussian Mixture Model (GMM) [1], or statistical, for example Support Vector Machine (SVM) [2] or Artificial Neural Networks (ANN). Speaker verification systems using several types of neural network architecture have been investigated: Multi Layer Percep-

1 Nottingham Trent University, NG11 8NS, UK
tariq.tashan@ntu.ac.uk, tony.allen@ntu.ac.uk, lars.nolle@ntu.ac.uk

tron (MLP) [3], Radial Basis Function [4], Neural Tree Network [5], Auto Associative Neural Network [6], Recurrent Neural Network [7], Probabilistic Neural Network [8] and Self Organising Maps (SOM) [9].

Recently, more biologically inspired neural networks methods have been investigated. Spiking neural networks have been employed to solve the speaker recognition problem. George, *et al.* presented a dynamic synapse neural network in speaker recognition application [10]. The network was trained using a genetic algorithms, with gender classification applied first using a rule based method. Two networks were designed for each gender; each network having an input layer of 16 nodes and an output layer of two nodes.

A nonlinear dynamic neural network is presented by Bing, *et al.* for speaker identification task [11]. In that paper, a higher order synapse model for data transfer through the neurons was used. The network contains two dimensions, each representing a space of features. The main concept of the network is to capture the distinctive feature components and magnify their effect.

Timoszczuk, *et al.* use pulse coupled and MLP neural networks for speaker identification [12]. The paper proposes two layers of pulse coupled neural networks for feature extraction followed by an MLP network for classification. Pulse neurons are represented using a Spike response model. The first layer converts the inputs into a pulse modulated sequence while the second layer extracts the features for the MLP network.

Wysoski, *et al.* use spiking neural network in text-independent speaker authentication [13]. The paper investigates two and three layer networks. Each neuron in the spiking neural network is an integrate-and-fire neuron. The first network contains 19 input neurons, two maps of 80 neurons in the first layer and two outputs neurons in the second layer. The two maps and the two output neurons represent the speaker model and the background model respectively. In the second network an additional layer is added to provide normalisation for the score similarity. Feature vector components were encoded into train of spikes using Rank order coding.

In this paper a spiking self organising map is suggested with biologically inspired features. Section two details the physiology of hearing in the human auditory system. Section three then presents a delayed rank order coding scheme as a suggested input feature vector. The proposed spiking SOM algorithm is described in section four, whilst section five presents the algorithm evaluation results with comparison to a non-spiking SOM based algorithm. Section six provides a final conclusion with recommendations for future work.

## 2 Physiology of Hearing

Sound waves are captured in the human auditory system over three stages: outer ear, middle ear and inner ear (cochlea). The outer ear consists of the pinna or the 'concha' which is the external visible part of the ear and the ear canal. The main two tasks of the concha are to collect the sound vibrations and introduce position information into the incoming sound. The ear canal works as a resonator with a peak frequency of 3 kHz [14].

The middle ear consists of the tympanic membrane connected to a combination of three small bones (ossicles) the malleus, the incus and the stapes. The ossicles are supported by two small muscles to the middle ear cavity. The tympanic membrane moves according to the sound vibrations received at the end of the ear canal, transferring the movement freely across the ossicles. This movement reaches the stapes which has a footplate end which connected to the cochlea through an oval window.

The cochlea is the auditory part of the inner ear. It has a snail-shaped bony structure of about two and a half turns with an uncoiled length of 3.1 to 3.3 cm. Along the fluid-filled tube of the cochlea is the 'basilar membrane' which splits the tube into two parallel tubes connected at their end. The front-end of one tube is connected with the oval-shaped footplate of the stapes bone, receiving the vibrations in a piston like movement. This vibratory movement of the stapes is converted into fluid pressure waves that travel along the basilar membrane. The flexible membrane of the rounded window at the end of the cochlea duct allows the pressure wave to propagate easily, following the piston act movement of the stapes. These pressure waves, in turn, displace the basilar membrane at specific positions corresponding to the frequency components of the received signal.

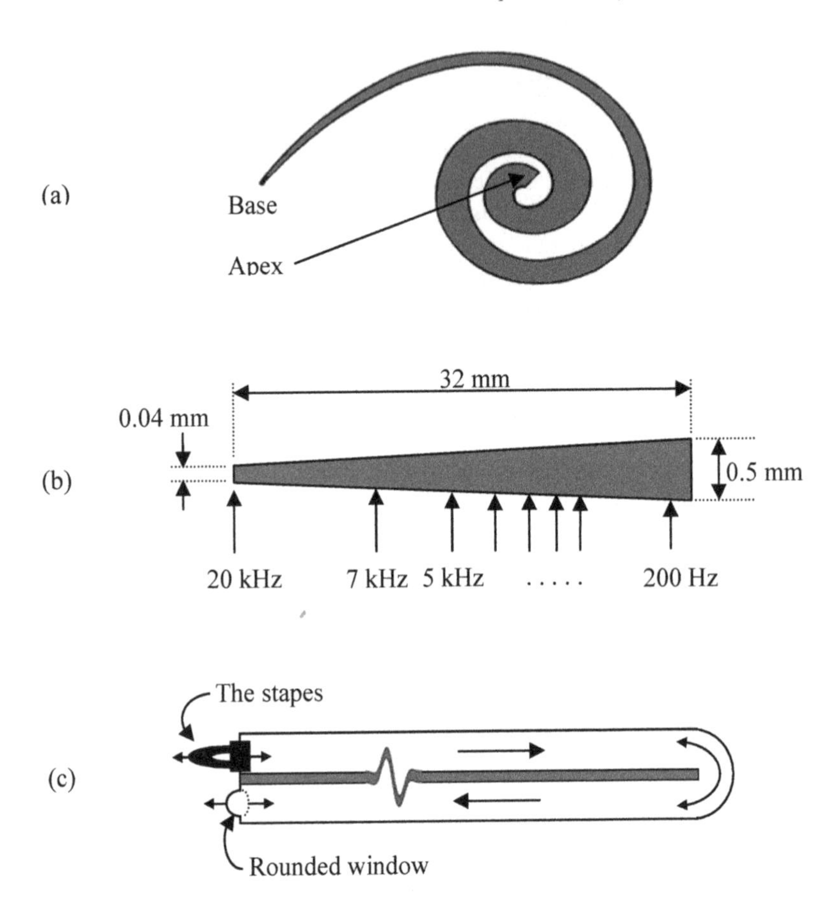


**Figure 1.** Different views of the basilar membrane. (a) Spiralled top view. (b) Unfolded top view, showing dimensions and frequency sensation positions. (c) Side view, showing how the movement of the stapes is propagated as pressure wave inside the cochlea duct.

Along the basilar membrane lies the sensory organ of hearing called the organ of corti. The corti contains the 'hair cells' which are connected directly to the auditory nerve. When the basilar membrane vibrates the hair cells at that position will 'bend' converting this mechanical movement into electrical pulses through a bio-chemical process [14]. As each hair cell is connected to a unique auditory nerve, the position of the active hair cells on the basilar membrane provides frequency information to the cochlear nucleus section of the brain. At low sound levels, the intensity of frequency components are represented by the rate of spikes generated by the hair cells movement [15]. However, at normal conversational speech levels (60-80dB) [16] the dominant hair cells spiking rates are saturated and it is the number of phase locked saturated fibres that indicates the intensity of the central frequency [17]. Figure 2 shows how the captured spectrum by the basilar membrane is transformed into discharged spike trains travelling through the auditory nerve.

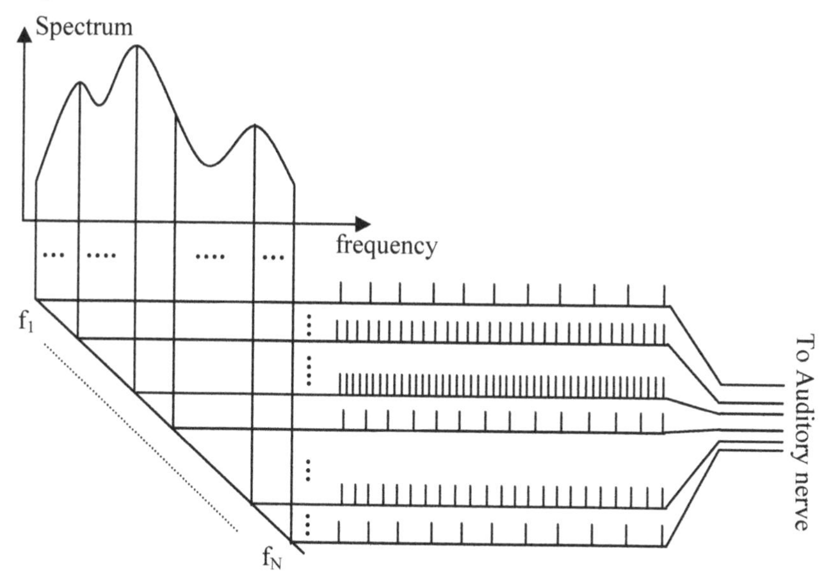


**Figure 2**. Captured spectrum converted into spike rates.

The human auditory system can capture understandable speech signal at variable intensities. This robustness is due to the Automatic Gain Control (AGC) mechanism, provided at different places in the auditory system, by the muscles supporting the ossicles chain and the outer hair cells in the organ of corti. In addition the spikes discharge also play role in signal normalisation. In Figure 2 each auditory nerve fibre is connected to one hair cell. When the basilar membrane is in its resting position, the auditory nerve fibres having a 'resting' discharge spike rate. When the hairs (stereocilia) of that hair cell 'bend' due to basilar membrane movement, an increased rate of spike discharge is transmitted via the associated auditory nerve fibre. The rate of spikes increases relatively to sound intensity, but saturate at normal conversational speech levels [15].

Beyond the cochlea, the exact form of signal processing employed by the human auditory system is currently uncertain. However, what is known is that the auditory cortex sections of the human brain are organised tonotopically [15] and that the phase locking of adjacent fibres is prominent within the auditory nerve [17]. Consequently, the known physiology of hearing leads us to the conclusion that the lower levels of the human auditory processing system can be approximated by an N component DFT spectrum vector. In the following section, a spike-based feature vector, derived from the DFT spectrum, is presented that is inspired by the physiology of the human audio processing system.

## 3 Delayed rank order coding

Rank order coding [18] is a common coding technique used to encode spike-based signals to be used in spiking neural networks [13]. One major disadvantage of using rank order coding is that it only takes into account the order of components of a feature vector and ignores the relative timing information among components. In this paper timing information is considered as well as the order of the components in a 'delayed' rank order coding feature vector. Taking the DFT spectrum, shown in Figure 2, as an example, a spike representing the frequency component with the largest value will be generated with zero delay time ($\Delta_3$) at a given onset point. A spike representing the second highest frequency component will be generated with a delay from this onset point ($\Delta_2$). This delay is equivalent to the difference between the intensities of the two frequency maxima. Here the delay is approximated by the difference between the spectrum components since it is proportional to the number of saturated fibres phase locked to the frequency maxima. Figure 3 shows the delayed rank order coding derived from the DFT spectrum.

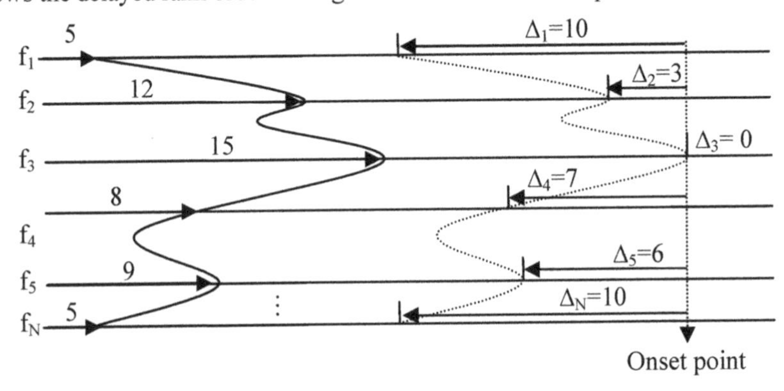


**Figure 3.** Delayed rank order coding extracted from DFT spectrum, $f_1$, $f_2$, …, $f_N$ are frequency positions along the basilar membrane, envelope on the left is the DFT spectrum values while the spikes on the right forms the delayed rank order coding feature vector.

To explain how the delay rank order coding feature vector is calculated from a DFT spectrum feature vector, as shown in Figure 3:

$$\text{DFT Spectrum vector} = [5, 12, 15, 8, 9, \ldots, 5] \tag{1}$$

The largest component magnitude value is represented by a zero delay, while the rest of the components are represented by their delay difference to the largest component value as follow:

$$\text{Delayed rank order coding} = [10, 3, 0, 7, 6, \ldots, 10] \tag{2}$$

One interesting aspect of the delayed rank order coding scheme is that it provides vector normalisation over the dynamic range of components values. In the standard DFT example the dynamic range is 15-5=10. The delayed rank order coding vector also has a dynamic range of 10, but this is normalised between 0 and $\Delta_{max}$ .This process compensates for any offset change in DFT spectrum (i.e. no volume normalisation pre-processing is required).

# 4 Proposed algorithm

Spiking neural networks for speaker recognition have been investigated in the literature using different structures [11-13, 19]. A Spiking Self Organising Map (SSOM) is suggested here as a speaker verification platform. The one-dimensional SSOM contains three spiking neurons, each working under an integrate-and-fire mechanism. The SSOM has an input of 64 inputs that represents the delayed rank order coding of the DFT spectrum of one speech frame, as explained in previous section. The choice of a 64 DFT spectrum component vector, rather than the 3600 component spike vector produced by the hair cells connected to the basilar membrane, is designed to approximate the frequency resolution down-scaling that is believed to occur as the signals move up through the various layers of the human auditory system [14]. Using a 64 DFT component input vector would also allow a direct comparison to be made between the results produced by the SSOM experiments presented here and those obtained in our previous SOM systems [20] and [21]. The structure of the SSOM (shown in Figure 4) is the same as that presented in [20] except that it uses the delayed rank order coding input vectors rather than the raw DFT spectrum vector used in [20].

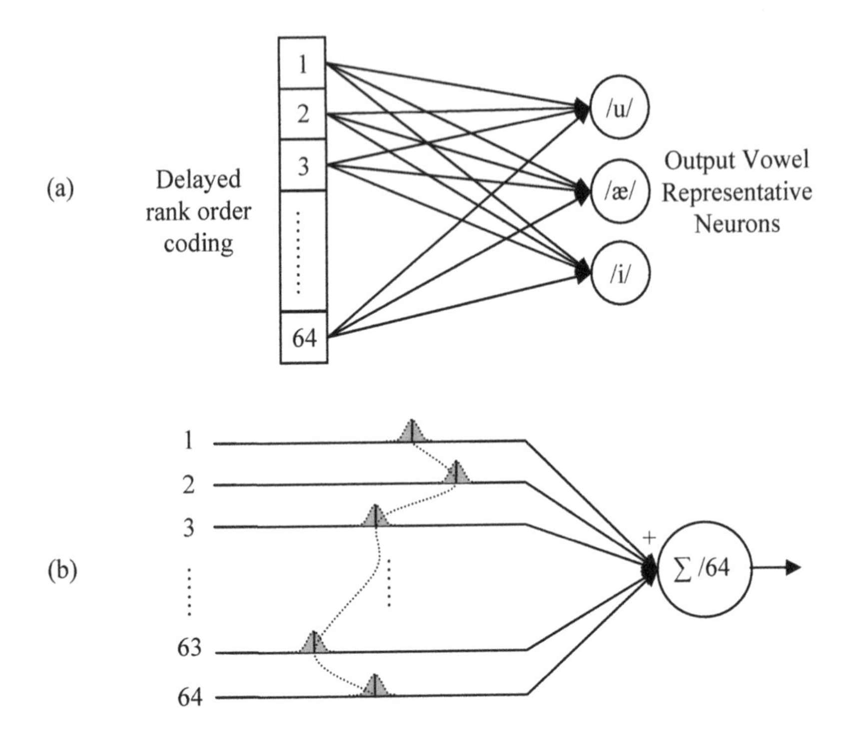


**Figure 4.** (a) SSOM structure. (b) Spiking neuron showing a fully synchronised input vector.

Each spiking neuron in Figure 4a is initially seeded with a target vector. This vector is the delayed rank order coding of a selected speech frame from each of the three vowels (/u/, /æ/ and /i/) as explained in section three. Generally, vowels are more stationary and periodic than other phonemes, making it easier to detect the formant frequencies [14]. In addition, these three specific vowels were chosen due to their intra-speaker and inter-speaker discrimination property [22]. The three vowels are contained in the words (two, five and eight) of the CSLU2002 database. The position of the target vector for each neuron is chosen after locating the vowel region within each word in the enrolment speech sample using the pre-processing linear correlation technique presented in [23]. When the spike timings of an input vector are fully synchronised with the spike timing of the target vector, each input synapse connected to the spiking neuron will respond with a maximum value of 1, resulting in an output of 1 as shown in Figure 4b. If a spike is off-synchronised with respect to its corresponding spike timing in the target vector, the response of the synapse will decrease according to a Gaussian distribution function (shaded area in Figure 4b); as in a biological auditory nerve response [17] and [24]. In other words, the more off-synchronised spikes the input vector is, the lower output response the spiking neuron produces. Based on this configuration, the output response of the spiking neuron to an input vector ranges between 1 (fully synchronised with target vector) and 0 (fully off-synchronised with target vector).

In the training phase, one SSOM is created for each enrolment sample. From that sample, the delayed rank order feature vector is calculated for each frame as explained above and three target vectors (one for each vowel) are selected and used to seed the three output neurons of the SSOM. All other enrolment sample frames are then presented repeatedly to the SSOM in order to optimise the target delay vectors. During this training of the SSOM, the winner spiking neuron is updated only when the response at its output exceeds a specific threshold of (0.7), this threshold being optimised empirically to ensure correct clustering and include only pure vowel information. The threshold criterion also prevents the SSOM from clustering silence and non-vowel information. The update formula of the spiking neuron is similar to the standard SOM formula with weights replaced by delays as follow:

$$\Delta_{new} = \Delta_{old} + \alpha\left(\Delta_{old} - \Delta_{input}\right) \tag{3}$$

Where $\Delta_{old}$ is the old delay value of the one synapse, $\Delta_{new}$ is the new modified delay value, $\Delta_{input}$ is the delayed rank order component of the input vector corresponding to the same synapse and $\alpha$ is the learning rate. The SSOM training parameters are similar to the modified SOM in [20] with 100 epochs and a learning rate of 0.1 which decreases linearly to zero over time. At the end of the training phase, each spiking neuron output of the trained SSOM represents a typical vowel model for the claimed speaker.

In the testing phase, the SSOM is used to verify a test sample. For each speech frame input, the spike timing of the input vector, of the test sample, are computed with respect to the spike timing of the target vector. By summing the synchronised spikes, a spike is generated at the output of the neuron only if the normalised summation exceeds an empirically optimised threshold value of (0.5). Each spiking neuron in the SSOM is more active when its related vowel information appears at the input. This is expected to be maximised when the test sample belongs to the claimed speaker. A lower activity output would be produced when the test sample belongs to an impostor.

To calculate a score for each vowel, the number of spikes generated at the output is then normalised over the duration of the vowel region within each word containing that vowel as follow:

$$S_i = \frac{number\ of\ spikes\ generated\ at\ neuron\ i}{number\ of\ frames\ within\ ith\ vowel\ region} \tag{4}$$

and:

$$S = (S_1 + S_2 + S_3)/3 \tag{5}$$

Where $S_i$ is the score of the *ith* vowel and $S$ is the final verification score averaged over the three individual output neuron scores.

## 5 System evaluation

The proposed algorithm was evaluated using speech samples of 50 speakers (27 females and 23 males), those speakers were arbitrarily selected from the CSLU2002 speaker verification database which consists of 91 speakers. The speech samples in this database were recorded over digital telephone lines with a sampling frequency of 8 kHz to produce 8-bit u-law files, which are then encoded into 8 kHz 16-bit wave format file. More information on the CSLU2002 database can be obtained on the website "http://www.cslu.ogi.edu". Two recording sessions samples are used for evaluation purposes, each session containing four samples for each speaker. Each speech sample contains the words (two, five & eight).

Using Equations 4 & 5 each sample in session one can be used as an enrolment sample to create one SSOM. The network is then tested against session two samples for both claimed speaker and impostors (the remaining 49 speakers in the dataset). By applying speaker dependent variable thresholds to the different scores values, the False Reject Rate (FRR) and the False Accept Rate (FAR) can be calculated for each speaker. The verification performance is obtained by means of Minimum Average Error Rate (MAER) = $_{\min}|(FRR+FAR)/2|$ as follow:

$$Performance(\%) = 100 - MAER \qquad (6)$$

Figure 5 shows the results for the proposed delayed rank order based SSOM together with results from the DFT spectrum based SOM presented in [20] and [21].

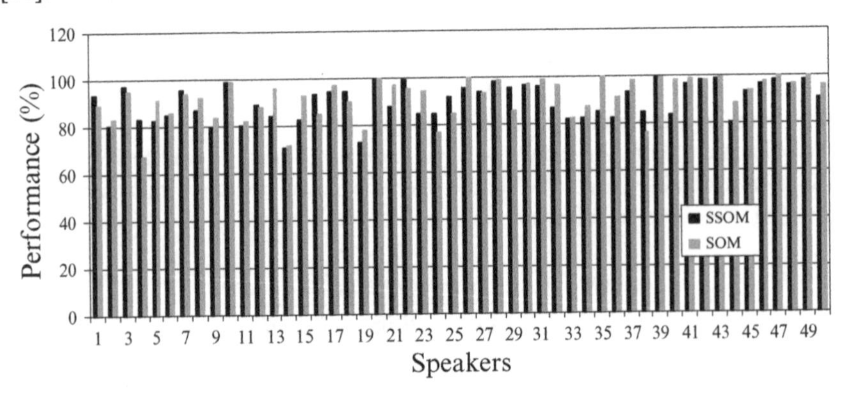


**Figure 5.** Performance of 50 speakers of CSLU2002 database.

It is clear from Figure 5 that both algorithms have similar performance. Although the DFT spectrum-based SOM outperforms on average, the delayed rank order coding SSOM improved speaker 4 significantly. The average performance of the 50 speakers for both algorithms is shown in Table 1.

**Table 1.** Average speaker verification performance.

| Method | Feature vector type | Performance (%) |
|---|---|---|
| SSOM | Delayed rank order coding | 90.1 |
| SOM | DFT spectrum | 91.7 |

The performance of the proposed algorithm is comparable to the DFT spectrum-based non-spiking SOM algorithm. No comparison has been made with the other algorithms presented in [20] and [21] since these use positive and negative samples during the training process. The proposed SSOM is trained using positive samples only.

## 6 Conclusion

This paper proposes a spiking neural network for speaker verification. The SSOM contains three spiking neurons representing the three vowel adopted in the experiment. Inspired by the physiology of the hearing in the human auditory system, the system uses delayed rank order coding as the feature vector. During the training phase the network update only the winner neuron and only if it is active beyond a certain level.

The proposed algorithm was evaluated using speech samples of 50 speakers from the CSLU2002 speaker verification database over two recording sessions. The algorithm shows an average speaker verification performance of 90.1%, while a non-spiking based SOM algorithm shows 91.7% in direct comparison using the same speech dataset. Due to the short duration of speech data used in the experiment at training and testing stages (~ 3 sec), the experimental environment can be classified as a very limited data condition scenario. Therefore, the proposed biologically inspired algorithm, even with a slightly lower verification performance, is still preferable to traditional speaker recognition systems, which's performance significantly decreases under such environment [25].

The results presented in this paper provide an encouraging baseline for further exploration of biologically plausible speech processing systems. Future work includes investigating the effects of reducing the number of input vector components as well including non-linear input signal processing derived from a more detailed model of the basilar membrane and the non-linear temporal encoding of spikes [14]. Additional work will also involve studying the dynamic role of outer hair cells in sound normalisation and including temporal information from the speech signal within the input feature vector.

## References

1. Reynolds, D.A. and R.C. Rose, *Robust text-independent speaker identification using Gaussian mixture speaker models.* Speech and Audio Processing, IEEE Transactions on, 1995. **3**(1): p. 72-83.
2. Campbell, W.M., et al., *Support vector machines for speaker and language recognition.* Computer Speech & Language, 2006. **20**(2-3): p. 210-229.
3. Seddik, H., A. Rahmouni, and M. Sayadi. *Text independent speaker recognition using the Mel frequency cepstral coefficients and a neural network classifier.* in *Control, Communications and Signal Processing, 2004. First International Symposium on.* 2004.
4. Oglesby, J. and J.S. Mason. *Radial basis function networks for speaker recognition.* in *Acoustics, Speech, and Signal Processing, 1991. ICASSP-91., 1991 International Conference on.* 1991.
5. Farrell, K.R., R.J. Mammone, and K.T. Assaleh, *Speaker recognition using neural networks and conventional classifiers.* Speech and Audio Processing, IEEE Transactions on, 1994. **2**(1): p. 194-205.
6. Kishore, S.P. and B. Yegnanarayana. *Speaker verification: minimizing the channel effects using autoassociative neural network models.* in *Acoustics, Speech, and Signal Processing, 2000. ICASSP '00. Proceedings. 2000 IEEE International Conference on.* 2000.
7. Mueen, F., et al. *Speaker recognition using artificial neural networks.* in *Students Conference, ISCON '02. Proceedings. IEEE.* 2002.
8. Kusumoputro, B., et al. *Speaker identification in noisy environment using bispectrum analysis and probabilistic neural network.* in *Computational Intelligence and Multimedia Applications, 2001. ICCIMA 2001. Proceedings. Fourth International Conference on.* 2001.
9. Monte, E., et al. *Text independent speaker identification on noisy environments by means of self organizing maps.* in *Spoken Language, 1996. ICSLP 96. Proceedings., Fourth International Conference on.* 1996.
10. George, S., et al. *Speaker recognition using dynamic synapse based neural networks with wavelet preprocessing.* in *Neural Networks, 2001. Proceedings. IJCNN '01. International Joint Conference on.* 2001.
11. Bing, L., W.M. Yamada, and T.W. Berger. *Nonlinear Dynamic Neural Network for Text-Independent Speaker Identification using Information Theoretic Learning Technology.* in *Engineering in Medicine and Biology Society, 2006. EMBS '06. 28th Annual International Conference of the IEEE.* 2006.
12. Timoszczuk, A.P. and E.F. Cabral, *Speaker recognition using pulse coupled neural networks*, in *2007 Ieee International Joint Conference on Neural Networks, Vols 1-6.* 2007, IEEE: New York. p. 1965-1969.
13. Wysoski, S.G., L. Benuskova, and N. Kasabov, *Text-independent speaker authentication with spiking neural networks*, in *Artificial Neural Networks - ICANN 2007, Pt 2, Proceedings*, J. MarquesDeSa, et al., Editors. 2007. p. 758-767.
14. Møller, A.R., *Hearing: anatomy, physiology, and disorders of the auditory system.* 2006: Academic Press.
15. Young, E.D., *Neural representation of spectral and temporal information in speech.* Philosophical Transactions of the Royal Society B: Biological Sciences, 2008. **363**(1493): p. 923-945.

16. Rabiner, L.R. and R.W. Schafer, *Theory and Applications of Digital Speech Processing.* 2010: Pearson.
17. Greenberg, S., et al., *Physiological Representations of Speech: Speech Processing in the Auditory System.* 2004, Springer New York. p. 163-230.
18. Thorpe, S. and J. Gautrais, *Rank order coding.* Computational Neuroscience: Trends in Research, ed. J.M. Bower. 1998, New York: Plenum Press Div Plenum Publishing Corp. 113-118.
19. George, S., et al. *Using dynamic synapse based neural networks with wavelet preprocessing for speech applications.* in *Neural Networks, 2003. Proceedings of the International Joint Conference on.* 2003.
20. Tashan, T., T. Allen, and L. Nolle. *Vowel based speaker verification using self organising map.* in *The Eleventh IASTED International Conference on Artificial Intelligence and Applications (AIA 2011).* 2011. Innsbruck, Austria: ACTA Press.
21. Tashan, T. and T. Allen, *Two stage speaker verification using Self Organising Map and Multilayer Perceptron Neural Network,* in *Research and Development in Intelligent Systems XXVIII,* M. Bramer, M. Petridis, and L. Nolle, Editors. 2011, Springer London. p. 109-122.
22. Rabiner, L.R. and R.W. Schafer, *Digital processing of speech signals.* Prentice-Hall signal processing series. 1978, Englewood Cliffs, N.J.: Prentice-Hall.
23. Tashan, T., T. Allen, and L. Nolle, *Speaker verification using heterogeneous neural network architecture with linear correlation speech activity detection.* In Press: Expert Systems, 2012.
24. Panchev, C. and S. Wermter, *Spike-timing-dependent synaptic plasticity: from single spikes to spike trains.* Neurocomputing, 2004. **58-60**(0): p. 365-371.
25. Jayanna, H.S. and S.R.M. Prasanna, *An experimental comparison of modelling techniques for speaker recognition under limited data condition.* Sadhana-Academy Proceedings in Engineering Sciences, 2009. **34**(5): p. 717-728.

# DATA MINING

# Parallel Random Prism: A Computationally Efficient Ensemble Learner for Classification

Frederic Stahl, David May and Max Bramer

**Abstract** Generally classifiers tend to overfit if there is noise in the training data or there are missing values. Ensemble learning methods are often used to improve a classifier's classification accuracy. Most ensemble learning approaches aim to improve the classification accuracy of decision trees. However, alternative classifiers to decision trees exist. The recently developed Random Prism ensemble learner for classification aims to improve an alternative classification rule induction approach, the Prism family of algorithms, which addresses some of the limitations of decision trees. However, Random Prism suffers like any ensemble learner from a high computational overhead due to replication of the data and the induction of multiple base classifiers. Hence even modest sized datasets may impose a computational challenge to ensemble learners such as Random Prism. Parallelism is often used to scale up algorithms to deal with large datasets. This paper investigates parallelisation for Random Prism, implements a prototype and evaluates it empirically using a Hadoop computing cluster.

## 1 Introduction

The basic idea of ensemble classifiers is to build a predictive model by integrating multiple classifiers, so called base classifiers. Ensemble classifiers are often used to

Frederic Stahl
Bournemouth University, School of Design, Engineering & Computing , Poole House, Talbot Campus, Poole, BH12 5BB e-mail: fstahl@bournemouth.ac.uk

David May
University of Portsmouth, School of Computing, Buckingham Building, Lion Terrace, PO1 3HE e-mail: David.May@myport.ac.uk

Max Bramer
University of Portsmouth, School of Computing, Buckingham Building, Lion Terrace, PO1 3HE e-mail: Max.Bramer@port.ac.uk

improve the predictive performance of the ensemble model compared with a single classifier [24]. Work on ensemble learning can be traced back at least to the late 1970s, for example the authors of [13] proposed using two or more classifiers on different partitions of the input space, such as different subsets of the feature space. However probably the most prominent ensemble classifier is the Random Forests (RF) classifier [9]. RF is influenced by Ho's Random Decision Forests (RDF) [18] classifier which aims to generalise better on the training data compared with traditional decision trees by inducing multiple trees on randomly selected subsets of the feature space. The performance of RDF has been evaluated empirically [18]. RF combines RDF's approach with Breiman's **b**ootstrap **agg**regat**ing** (Bagging) approach [8]. Bagging aims to increase a classifier's accuracy and stability. A stable classifier experiences a small change and an unstable classifier experiences a major change in the classification if there are small changes in the training data. Recently ensemble classification strategies have been developed for the scoring of credit applicants [19] and for the improvement of the prediction of protein structural classes [30]. Chan and Stolfo's Meta-Learning framework [11, 12] builds multiple heterogeneous classifiers. The classifiers are combined using a further learning process, a *meta-learning algorithm* that uses different combining strategies such as *voting*, *arbitration* and *combining*. Pocket Data Mining, an ensemble classifier for distributed data streams in mobile phone networks has recently been developed [25]. In Pocket Data Mining the base classifiers are trained on different devices in an ad hoc network of smart phones. Pocket Data Mining has been tested on homogeneous and heterogeneous setups of two different data stream classifiers, Hoeffding Trees [15] and incremental Naive Bayes, which are combined using weighted majority voting [27].

Most rule based classifiers can be categorised in the 'divide and conquer' (induction of decision trees) [23, 22] and the 'separate and conquer' approach [28]. 'Divide and conquer' produces classification rules in the intermediate form of a decision tree and 'separate and conquer' produces IF...THEN rules that do not necessarily fit into a decision tree. Due to their popularity most rule-based ensemble base classifiers are based on decision trees. Some ensemble classifiers consider heterogeneous setups of base classifiers such as Meta-Learning [11, 12], however in practice the base classifiers used are different members of the 'divide and conquer' approach. A recently developed ensemble classifier that is inspired by RF and based on the Prism family of algorithms [10, 5, 6] as base classifiers is Random Prism [26]. The Prism family of algorithms follows the 'separate and conquer' approach and produces modular classification rules that do not necessarily fit into a decision tree. Prism algorithms produce a comparable classification accuracy compared with decision trees and in some cases, such as if there is a lot of noise or missing values, even outperform decision trees.

Random Prism has been evaluated empirically in [26] and shows a better classification accuracy in most cases compared with its standalone base classifier. Furthermore unpublished empirical experiments of the authors show that Random Prism has also a higher tolerance to noise compared with its standalone base classifier. Yet some results presented in [26] show that Random Prism consumes substantially more CPU time compared with its standalone Prism base classifier on the same training data size. This is because like many ensemble classifiers such as RF, Random Prism builds multiple bags of the original training data and hence has to process a multiple of the training data compared with its stand alone base Prism classifier. However, as the runtime of ensemble learners is directly dependent to the training data size and the number of base classifiers, they could potentially be parallelised by training the individual classifiers on different processors. Google's MapReduce [14] framework and its free implementation named Hadoop [1] is a software framework for supporting distributed computing tasks on large datasets in a computer cluster. MapReduce's potential to scale up Random Prism is given through ensemble learning approaches that make use of MapReduce such as [21, 29, 3]. However most parallel ensemble approaches are based on decision trees. The work presented in this paper presents a computationally scalable parallel version of the Random Prism ensemble classifier that can be executed using a network of standard workstations utilising Google's MapReduce paradigm. The proposed *Parallel Random Prism* classifier is evaluated empirically using the free Hadoop [1] implementation of the MapReduce framework in a network of computer workstations. Parallelising Random Prism using Hadoop is particularly interesting as Hadoop makes use of commodity hardware and thus is also, from the hardware point of view, an inexpensive solution.

The remainder of this paper is organised as follows: Section 2 highlights the PrismTCS approach, a member of the Prism family of algorithms, which has been used as base classifier. Section 2 also highlights the Random Prism approach. The prototype of Parallel Random Prism is proposed in Section 3 and evaluated empirically in Section 4. Ongoing and possible future work is discussed in Section 5 and concluding remarks are presented in Section 6.

# 2 Random Prism

This section highlights first the basic rule induction approach: the Prism / PrismTCS classifier and compares the rulesets induced by Prism / PrismTCS with decision trees. The second part of this section introduces the Random Prism algorithm, and discusses its computational performance briefly.

## 2.1 The PrismTCS Approach

The intermediate representation of classification rules in the form of a decision tree is criticised in Cendrowska's original Prism paper [10]. Prism produces modular rules that do not necessarily have attributes in common such as the two rules below:

$$IF\ A = 1\ AND\ B = 1\ THEN\ class = x$$

*IF C = 1 AND D = 1 THEN class = x*

These modular rules cannot be represented in a decision tree without adding unnecessary rule terms. For this example it is assumed that that each of the four attributes represented in the rules above have three possible values *1*, *2* and *3*. Further it is assumed that all data instances matching any of the two rules above is labelled with class *x* and the remaining instances with class *y*. According to Cendrowska [10], forcing these rules into a tree structure would lead to the tree illustrated in Figure 1.

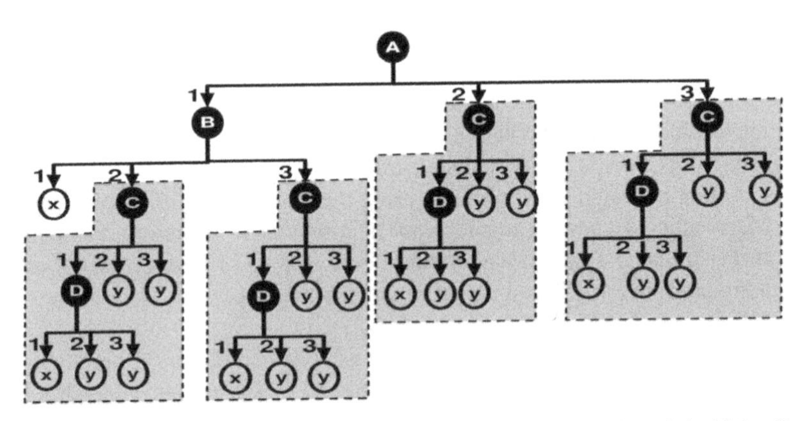

**Fig. 1** The replicated subtree problem based on Cendrowska's example in her original Prism Paper. The shaded subtrees highlight the replicated subtrees.

This will result into a large and needlessly complex tree with potentially unnecessary and possibly expensive tests for the user, which is also known as the replicated subtree problem [28].

All 'separate and conquer' algorithms follow the same top level loop. The algorithm induces a rule that explains a part of the training data. Then the algorithm separates the instances that are not covered by the rules induced so far and conquers them by inducing a further rule that again explains a part of the remaining training data. This is done until there are no training instances left [16].

Cendrowska's original Prism algorithm follows this approach. However it does not not scale well on large datasets. A version of Prism that attempts to scale up to larger datasets has been developed by one of the authors [6] and is also utilised for the Random Prism classifier.

There have been several variations of Prism algorithms such as PrismTC, PrismTCS and N-Prism which are implemented in the Inducer data mining workbench [7]. However PrismTCS (Target Class Smallest first) seems to have a better computational performance compared with the other Prism versions whilst maintaining the same level of predictive accuracy [26].

PrismTCS is outlined below using pseudocode [26]. $A_x$ denotes a possible attribute value pair and $D$ the training data. $Rule\_set\,rules = new\,Rule\_set()$ creates a new ruleset, *Rule rule = new Rule(i)* creates a new rule for class *i*, $rule.addTerm(A_x)$ adds attribute value pair $A_x$ as a new rule term to the rule, and $rules.add(rule)$ adds the newly induced rule to the ruleset.

```
        D' = D;
        Rule_set rules = new Rule_set();
Step 1: Find class i that has the fewest instances in the training
        set;
        Rule rule = new Rule(i);
Step 2: Calculate for each Ax p(class = i| Ax);
Step 3: Select the Ax with the maximum  p(class = i| Ax);
        rule.addTerm(Ax);
        Delete all instances in D' that do not cover rule;
Step 4: Repeat 2 to 3 for D' until D' only contains instances
        of classification i.
Step 5: rules.add(rule);
        Create a new D' that comprises all instances of D except
        those that are covered by all rules induced so far;
Step 6: IF (D' is not empty)
             repeat steps 1 to 6;
```

## 2.2 Random Prism Classifier

The basic Random Prism architecture is highlighted in Figure 2. The base classifiers are called R-PrismTCS and are based on PrismTCS. The prefix *R* denotes the random component in Random Prism which comprises Ho's [18] random feature subset selection and Breiman's bagging [8]. Both random components have been chosen in order to make Random Prism generalise better on the training data and be more robust if there is noise in the training data. In addition to the random components, J-pruning [6], a rule pre-pruning facility has been implemented. J-pruning aims to maximise the theoretical information content of a rule while it is being induced. J-pruning does that by triggering a premature stop of the induction of further rule terms if the rule's theoretical information content would decrease by adding further rule terms. In general, the reason for chosing PrismTCS as base classifier is that it is the computationally most efficient member of the Prism family of algorithms [26].

In Random Prism the bootstrap sample is taken by randomly selecting *n* instances with replacement from the training data *D*, if *n* is the total number of training instances. On average each R-PrismTCS classifier will be trained on 63.2 % of the original data instances [26]. The remaining data instances, on average 36.8 % of the original data instances, are used as validation data for this particular R-PrismTCS classifier.

The pseudocode below highlights the R-PrismTCS algorithm [26] adapted from the PrismTCS pseudocode in Section 2.1. *M* denotes the number of features in *D*:

```
        D' = random sample with replacement of size n from D;
        Rule_set rules = new Rule_set();
Step 1: Find class i that has the fewest instances in the training
        set;
        Rule rule = new Rule(i);
Step 2: generate a subset F of the feature space of size m where
        (M>=m>0);
```

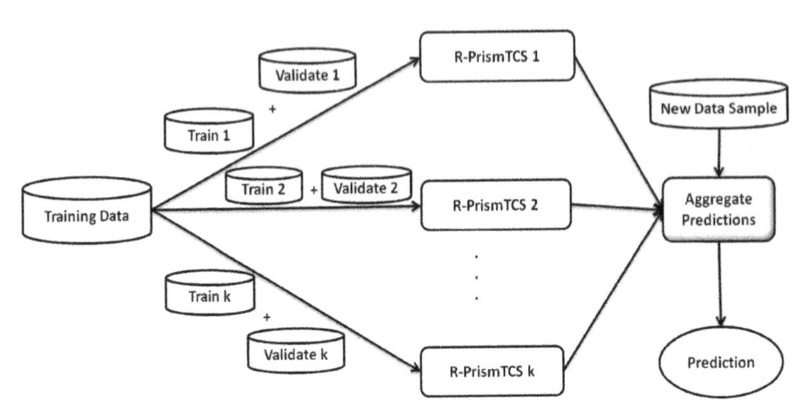


**Fig. 2** The Random Prism Architecture using the R-PrismTCS base classifiers and weighted majority voting for the final prediction.

```
Step 3: Calculate for each Ax in F  p(class = i| Ax);
Step 4: Select the Ax with the maximum  p(class = i| Ax);
        rule.addTerm(Ax);
        Delete all instances in D' that do not cover rule;
Step 5: Repeat 2 to 4 for D' until D' only contains instances
        of classification i.
Step 6: rules.add(rule);
        Create a new D' that comprises all instances of D except
        those that are covered by all rules induced so far;
Step 7: IF (D' is not empty)
              repeat steps 1 to 7;
```

For the induction of each rule term a different randomly selected subset of the feature space without replacement is drawn. The number of features drawn is a random number between 1 and *M*. The pseudocode outlined below highlights the Random Prism training approach [26], where *k* is the number of base classifiers and *i* is the *ith* R-PrismTCS classifier:

```
double weights[] = new double[k];
Classifiers classifiers = new Classifier[k];
for(int i = 0; i < k; i++)
   Build R-RrismTCS classifier r;
   TestData T = instances of D that have not been used to induce r;
   Apply r to T;
   int correct = number of by r correctly classified instances in T;
   weights[i] = correct/(number of instances in T);
```

The pseudocode highlighted above shows that for each *R-PrismTCS* classifier a weight is also calculated. As mentioned above, for the induction of each classifier only about 63.2% of the total number of training instances are used. The remaining instances, about 36.8% of the total number of instances are used to calculate the classifier's accuracy which we call its weight. As mentioned earlier in this section, Random Prism uses weighted majority voting, where each vote for a classification for a test instance corresponds to the underlying *R-PrismTCS* classifier's weight. This is different from RF and RDF which simply use majority voting. Random Prism also uses a user defined threshold $N$ which is the minimum weight a *R-PrismTCS* classifier has to provide in order to take part in the voting.

## 3 Parallelisation of the Random Prism Classifier

The runtime of Random Prism has been measured and compared with the runtime of PrismTCS as shown in Table 1 and as published in [26]. Intuitively one would expect that the runtime of Random Prism using 100 PrismTCS base classifiers is 100 times slower that PrismTCS. However, the runtimes shown in Table 1 clearly show that this is not the case, which can be explained by the fact that Random Prism base classifiers do not use the entire feature space to generate rules, which in turn limits the search space and thus the runtime. Nevertheless Random Prism is still multiple times slower than PrismTCS, hence parallelisation has been considered in [26] and is now described in this section.

**Table 1** Runtime of Random Prism on 100 base classifiers compared with a single PrismTCS classifier in milliseconds.

| Dataset | Runtime PrismTCS | Runtime Random Prism |
|---|---|---|
| monk1 | 16 | 703 |
| monk3 | 15 | 640 |
| vote | 16 | 672 |
| genetics | 219 | 26563 |
| contact lenses | 16 | 235 |
| breast cancer | 32 | 1531 |
| soybean | 78 | 5078 |
| australian credit | 31 | 1515 |
| diabetes | 16 | 1953 |
| crx | 31 | 2734 |
| segmentation | 234 | 15735 |
| ecoli | 16 | 734 |
| balance scale | 15 | 1109 |
| car evaluation | 16 | 3750 |
| contraceptive method choice | 32 | 3563 |

For the parallelisation of Random Prism we used Apache Hadoop, which is an application distribution framework designed to be executed in a computer cluster consisting of a large number of standard workstations. Hadoop uses a technique called *MapReduce*. MapReduce splits an application into smaller parts called *Mappers*. Each Mapper can be processed by any of the workstations in the nodes in the cluster. The results produced by the Mappers are then aggregated and processed by one or more *Reducer* nodes in the cluster. Hadoop also provides its own file system, the Hadoop Distributed File System (HDFS). HDFS distributes the data over the cluster and stores the data redundantly on multiple cluster nodes. This speeds up data access. Failed nodes are automatically recovered by the framework providing high reliability to the application.

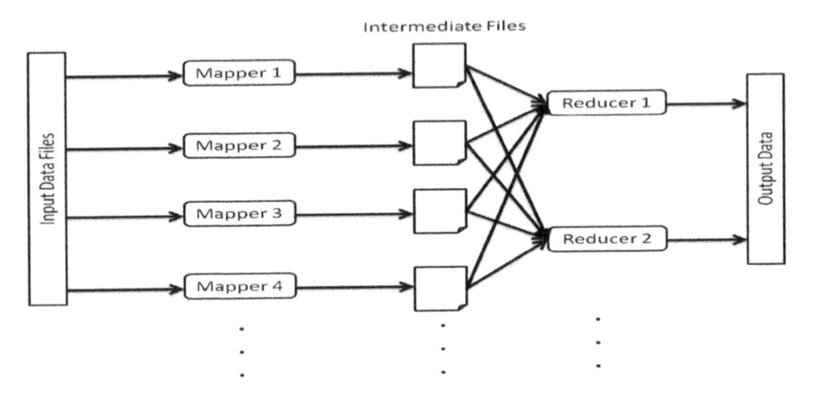


**Fig. 3** A typical setup of a Hadoop computing cluster with several Mappers and Reducers. A physical computer in the cluster can host more than one Mapper and Reducer.

Figure 3 highlights a typical setup of a Hadoop computing cluster. Each node in a Hadoop cluster can host several Mappers and Reducers. In Hadoop large amounts of data are analysed by distributing smaller portions of the data to the Mapper machines plus a function to process these data portions. Then the results of the Mappers (intermediate files) are aggregated by passing them to the Reducer. The Reducer uses a user defined function to aggregate them. Hadoop balances the workload as evenly as possible by distributing it as evenly as possible to the Mappers. The user is only required to implement the function for the Mapper and the Reducer.

We utilised Hadoop for the parallelisation of Random Prism as depicted in Figure 4. In Random Prism the Mapper builds a different bagged version of the training data and uses the remaining data as validation data. Furthermore each Mapper gets a R-PrismTCS implementation as a function to process the training and the validation data. The following steps describe the parallelisation of Random Prism using Hadoop:

```
Step 1: Distribute the training data over the computer cluster using the
        Hadoop Distributed File System (HDFS);
Step 2: Start x Mapper jobs, where x is the number of base PrismTCS
        classifiers desired. Each Mapper job comprises, in the following
        order:
          - Build a training and validation set using Bagging;
          - Generate a rulset by training the PrismTCS classifier on
            the training set;
          - Calculated the PrismTCS classifiers weight using the
            validation set;
          - Return the ruleset and the weight.
Step 3: Start the Reducer with a list of rulesets and weights produced
        each each Mapper (PrismTCS classifier);
Step 4: The Reducer retuns the final classifier which is a set of PrismTCS
        rulesets which perform weighted majority voting for each test
        instance.
```

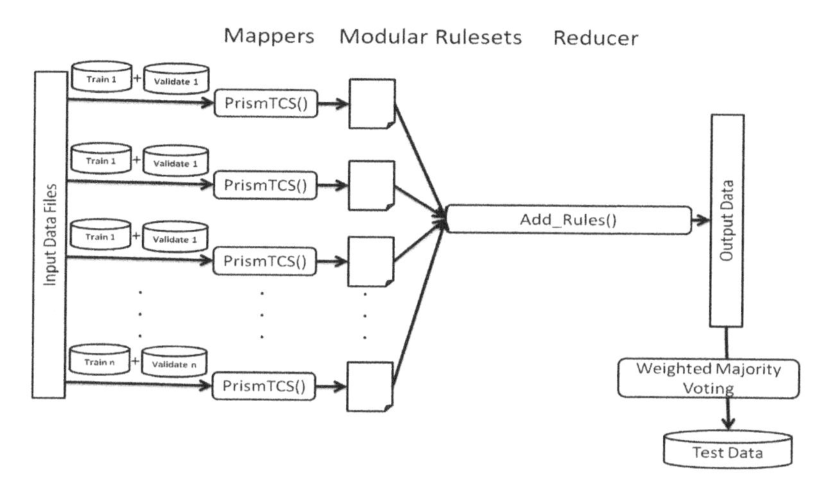


**Fig. 4** Parallel Random Prism Architecture

# 4 Evaluation of Parallel Random Prism Classification

The only difference between the parallel and the serical versions of Random Prism is that in the parallel version the R-PrismTCS classifiers are executed concurrently on multiple processors and in the serial version they are executed sequentially. Hence both algorithms have the same classification performance, but Parallel Random Prism is expected to be faster. For an evaluation of the classification performance of Random Prism the reader is referred to [26]. In this paper we evaluate Parallel Random Prism empirically with respect to its computational performance, using the four datasets outlined in Table 2. The datasets are referred to as tests in this paper. The data for tests 1 to 3 is synthetic biological data taken from the infobiotics data repository [2], and the data for test 4 is taken from the UCI repository [4]. The reason for using the infobiotics repository is that it provides larger datasets compared with the UCI repository. The computer cluster we used to evaluate Parallel Random Prism comprised 10 workstations, each providing a CPU of 2.8 GHz speed and a memory of 1 GB. The operating system installed on all machines is XUbuntu. In this paper the term node or cluster node refers to a workstation and each node hosts two Mappers. As for in the qualitative evaluation in [26], 100 base classifiers were used.

**Table 2** Evaluation datasets.

| Test | Test Dataset | Number of Data Instances | Number of Attributes | Number of Classes |
|---|---|---|---|---|
| 1 | biometric data 1 | 50000 | 5 | 5 |
| 2 | biometric data 2 | 15000 | 19 | 5 |
| 3 | biometric data 3 | 30000 | 3 | 2 |
| 4 | genetics | 2551 | 59 | 2 |

Parallel Random Prism has been evaluated in terms of its scalability to different sizes of the training data in 'size-up' experiments described in Section 4.1, and in term of its scalability to different numbers of computing nodes in 'Speed-up' experiments described in Section 4.2.

## *4.1 Size-up Behaviour of Parallel Random Prism*

The size-up experiments of the Parallel Random Prism system examine the performance of the system on a configuration with a fixed number of nodes/processors and an increasing data workload. In general we hope to achieve a runtime which is a linear function to the training data size. The two datasets with most instances, datasets test 1 and test 3 were chosen for the size-up experiments.

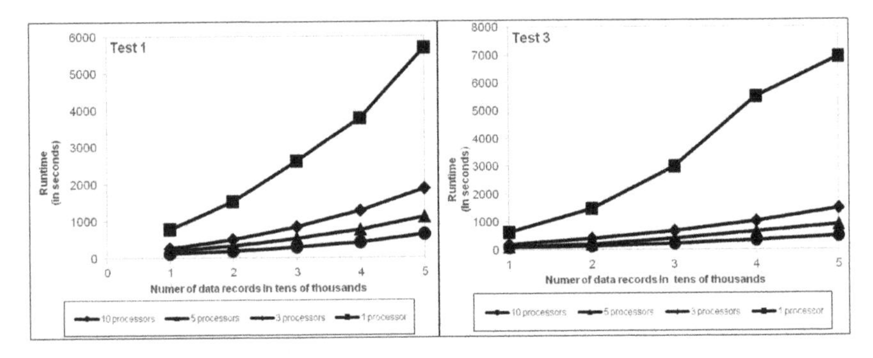


**Fig. 5** Size up behaviour of Parallel Random Prism on two test datasets.

For the size-up experiments we took a sample of 10000 data records from the dataset. In order to increase the number of data records we appended the data to itself in the vertical direction. The reason for appending the data sample to itself is that this will not change the concept hidden in the data. If we simply sampled a larger sample from the original data, then the concept may influence the runtime. Thus appending the sample to itself allows us to examine Parallel Random Prism's performance with respect to the data size more precisely. The calculation of the weight might be influenced by my appending the data to itself as a multiplied data instance may appear in both the test set and the training set. However this is not relevant to the computational performance examined here.

The linear regression equations below are corresponding to the size-up experiments illustrated in Figure 5 and support a linear size up behaviour of Parallel Random Prism on the two chosen datasets, where $x$ is the number of data records/instances in tens of thousands and $y$ the relative runtime in seconds:

$$1\ \ processor(test1):\quad y = 1206.8x - 748.8 \quad (R^2 = 0.972)$$
$$3\ \ processors(test1):\quad y = 396x - 242.6 \quad (R^2 = 0.972)$$
$$5\ \ processors(test1):\quad y = 226.2x - 99 \quad (R^2 = 0.968)$$

10 $processors(test1): \quad y = 128x - 52.8 \quad (R^2 = 0.959)$
1 $processor(test3): \quad y = 1662.2x - 1487 \quad (R^2 = 0.974)$
3 $processors(test3): \quad y = 315x - 193 \quad (R^2 = 0.977)$
5 $processors(test3): \quad y = 197.3x - 143.3 \quad (R^2 = 0.973)$
10 $processors(test3): \quad y = 103.7x - 61.5 \quad (R^2 = 0.965)$
In general we can observe a nice size-up of the system close to being linear.

## 4.2 Speed-up Behaviour of Parallel Random Prism

A standard metric to measure the efficiency of a parallel algorithm is the speed up factor [17, 20]. The speed up factor measures how much using a parallel version of an algorithm on $p$ processors is faster than using only one processor.

$$S_p = \frac{R_1}{R_p}$$

In this formula $S_p$ represents the speed up factor. $R_1$ is the runtime of the algorithm on a single machine and $R_p$ is the runtime on $p$ machines. The dataset size stays constant in all processor configurations. In the ideal case the speed up factor would be the same as the number of processors utilised. However, usually the speed up factor will be below the ideal value due to overheads, for example due to the data communication overhead.

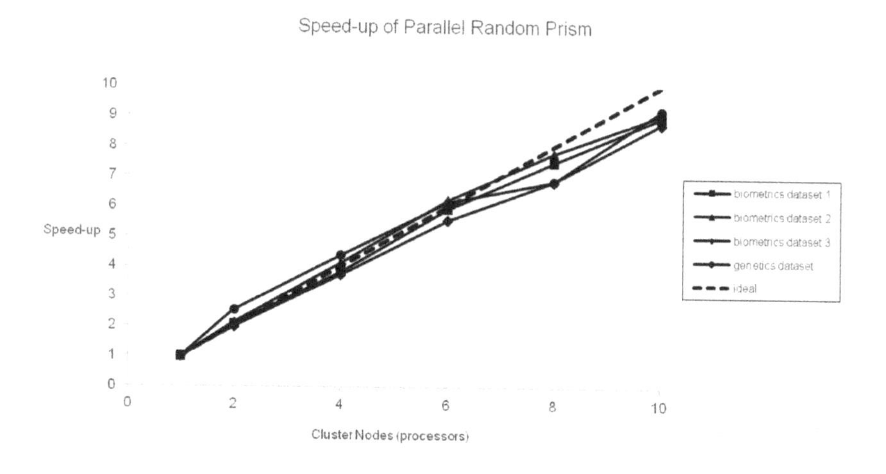


**Fig. 6** The Speed-up behaviour of Parallel Random Prism.

Figure 6 shows the speedup factors recorded for Parallel Random Prism on all four test datasets for different numbers of workstations (processors) used in the cluster, ranging from one workstation up to ten, the maximum number of workstations we had available for the experiments. The ideal speed-up is plotted as a dashed line. What can be seen in Figure 6 is that the speedup for all cases is close to the ideal case. However, there is a slightly increasing discrepancy the more workstations are utilised. However this discrepancy can be explained by the increased communication overhead caused by adding more nodes that retrieve bagged samples from the training data distributed in the network. There will be an upper limit of the speedup factors beyond which adding more workstations will decrease the speedup rather than increasing it. However considering the low discrepancy after adding 10 workstations suggests that we are far from reaching the maximum number of workstations that would still be beneficial. Two outliers can be identified for the speed-up factors for the genetics dataset. In fact the speedup is better than the ideal case. Such a 'superlinear' speedup can be explained by the fact that the operating system tends to move frequent operations or frequently used data elements into the cache memory which is faster than the normal system memory. The genetics dataset has many fewer data instances compared with the remaining datasets. Hence a plausible explanation for this superlinear speedup is that distributing the bagged samples between several workstations will result in a larger portion of the total data samples being held in the cache memory. This benefit may outweigh the communication overhead caused by using only two and four workstations. However adding more workstations will increase the communication overhead to a level that outweighs the benefit of using the cache memory.

In general we can observe a nice speedup behaviour suggesting that many more workstations could be used to scale up Parallel Random Prism further.

## 5 Ongoing and Future Work

The presented evaluation of the scalability of the proposed system was conducted with respect to the number of data instances. However, in some cases the data size might be determined by a large number of attributes, such as in gene expression data. Hence a further set of experiments that examine Parallel Random Prism's behaviour with respect to a changing number of attributes is planned.

A further development of the voting strategy is currently being considered. Depending on the sample drawn from the training data a base classifier may predict different classes with a different accuracy and hence a voting system that uses different weights for different class predictions is currently being developed. In addition to that a further version of Random Prism and thus Parallel Random Prism that makes use of different versions of Prism as base classifier will be developed in the future.

# 6 Conclusions

This work presents a parallel version of the Random Prism approach. Random Prism's classification accuracy and robustness to noise, has been evaluated in previous work and experiments and shown to improve PrismTCS's (Random Prism's base classifier's) performance. This work addresses Random Prism's scalability to larger datasets. Random Prism does not perform well on large datasets, just like any ensemble classifier that uses bagging. Hence a data parallel approach that distributes the training data over a computer cluster of commodity workstations is investigated. The induction of multiple PrismTCS classifiers on the distributed training data is performed concurrently by employing Google's Map/Reduce paradigm and Hadoop's Distributed File System.

The parallel version of Random Prism has been evaluated in terms of its size-up and speed-up behaviour on several datasets. The size-up behaviour showed a desired almost linear behaviour with respect to an increasing number of training instances. Also the speed-up behaviour showed an almost ideal performance with respect to an increasing number of workstations. Further experiments will be conducted in the future examining Parallel Random Prism's size-up and speed-up behaviour with respect to the number of attributes. In general the proposed parallel version of Random Prism showed excellent scalability with respect to large data volumes and the number of workstations utilised in the computer cluster.

The future work will comprise a more sophisticated majority voting approach taking the base classifier's accuracy for individual classes into account and also the integration of different 'Prism' base classifiers will be investigated.

**Acknowledgements** The research leading to these results has received funding from the European Commission within the Marie Curie Industry and Academia Partnerships & Pathways (IAPP) programme under grant agreement n° 251617.

# References

1. Hadoop, http://hadoop.apache.org/mapreduce/ 2011.
2. Jaume Bacardit and Natalio Krasnogor. The infobiotics PSP benchmarks repository. Technical report, 2008.
3. Justin D. Basilico, M. Arthur Munson, Tamara G. Kolda, Kevin R. Dixon, and W. Philip Kegelmeyer. Comet: A recipe for learning and using large ensembles on massive data. *CoRR*, abs/1103.2068, 2011.
4. C L Blake and C J Merz. UCI repository of machine learning databases. Technical report, University of California, Irvine, Department of Information and Computer Sciences, 1998.
5. M A Bramer. Automatic induction of classification rules from examples using N-Prism. In *Research and Development in Intelligent Systems XVI*, pages 99–121, Cambridge, 2000. Springer-Verlag.
6. M A Bramer. An information-theoretic approach to the pre-pruning of classification rules. In B Neumann M Musen and R Studer, editors, *Intelligent Information Processing*, pages 201–212. Kluwer, 2002.

7. M A Bramer. Inducer: a public domain workbench for data mining. *International Journal of Systems Science*, 36(14):909–919, 2005.
8. Leo Breiman. Bagging predictors. *Machine Learning*, 24(2):123–140, 1996.
9. Leo Breiman. Random forests. *Machine Learning*, 45(1):5–32, 2001.
10. J. Cendrowska. PRISM: an algorithm for inducing modular rules. *International Journal of Man-Machine Studies*, 27(4):349–370, 1987.
11. Philip Chan and Salvatore J Stolfo. Experiments on multistrategy learning by meta learning. In *Proc. Second Intl. Conference on Information and Knowledge Management*, pages 314–323, 1993.
12. Philip Chan and Salvatore J Stolfo. Meta-Learning for multi strategy and parallel learning. In *Proceedings. Second International Workshop on Multistrategy Learning*, pages 150–165, 1993.
13. B.V. Dasarathy and B.V. Sheela. A composite classifier system design: Concepts and methodology. *Proceedings of the IEEE*, 67(5):708–713, 1979.
14. Jeffrey Dean and Sanjay Ghemawat. Mapreduce: simplified data processing on large clusters. *Commun. ACM*, 51:107–113, January 2008.
15. Pedro Domingos and Geoff Hulten. Mining high-speed data streams. In *Proceedings of the sixth ACM SIGKDD international conference on Knowledge discovery and data mining*, KDD '00, pages 71–80, New York, NY, USA, 2000. ACM.
16. J Fuernkranz. Integrative windowing. *Journal of Artificial Intelligence Resarch*, 8:129–164, 1998.
17. John L Hennessy and David A Patterson. *Computer Architecture A Quantitative Approach*. Morgan Kaufmann, USA, third edition, 2003.
18. Tin Kam Ho. Random decision forests. *Document Analysis and Recognition, International Conference on*, 1:278, 1995.
19. Nan-Chen Hsieh and Lun-Ping Hung. A data driven ensemble classifier for credit scoring analysis. *Expert Systems with Applications*, 37(1):534 – 545, 2010.
20. Kai Hwang and Fay A Briggs. *Computer Architecture and Parallel Processing*. McGraw-Hill Book Co., international edition, 1987.
21. Biswanath Panda, Joshua S. Herbach, Sugato Basu, and Roberto J. Bayardo. Planet: massively parallel learning of tree ensembles with mapreduce. *Proc. VLDB Endow.*, 2:1426–1437, August 2009.
22. Ross J Quinlan. Induction of decision trees. *Machine Learning*, 1(1):81–106, 1986.
23. Ross J Quinlan. *C4.5: programs for machine learning*. Morgan Kaufmann, 1993.
24. Lior Rokach. Ensemble-based classifiers. *Artificial Intelligence Review*, 33:1–39, 2010.
25. F. Stahl, M.M. Gaber, M. Bramer, and P.S. Yu. Pocket data mining: Towards collaborative data mining in mobile computing environments. In *22nd IEEE International Conference on Tools with Artificial Intelligence (ICTAI)*, volume 2, pages 323 –330, October 2010.
26. Frederic Stahl and Max Bramer. Random Prism: An alternative to random forests. In *Thirty-first SGAI International Conference on Artificial Intelligence*, pages 5–18, Cambridge, England, 2011.
27. Frederic Stahl, Mohamed Gaber, Paul Aldridge, David May, Han Liu, Max Bramer, and Philip Yu. Homogeneous and heterogeneous distributed classification for pocket data mining. In *Transactions on Large-Scale Data- and Knowledge-Centered Systems V*, volume 7100 of *Lecture Notes in Computer Science*, pages 183–205. Springer Berlin / Heidelberg, 2012.
28. Ian H Witten and Frank Eibe. *Data Mining: Practical Machine Learning Tools and Techniques with Java Implementations*. Morgan Kaufmann, second edition, 2005.
29. Gongqing Wu, Haiguang Li, Xuegang Hu, Yuanjun Bi, Jing Zhang, and Xindong Wu. Mrec4.5: C4.5 ensemble classification with mapreduce. In *ChinaGrid Annual Conference, 2009. ChinaGrid '09. Fourth*, pages 249 –255, 2009.
30. Jiang Wu, Meng-Long Li, Le-Zheng Yu, and Chao Wang. An ensemble classifier of support vector machines used to predict protein structural classes by fusing auto covariance and pseudo-amino acid composition. *The Protein Journal*, 29:62–67, 2010.

# Questionnaire Free Text Summarisation Using Hierarchical Classification

Matias Garcia-Constantino, Frans Coenen, P-J Noble, and Alan Radford

**Abstract** This paper presents an investigation into the summarisation of the free text element of questionnaire data using hierarchical text classification. The process makes the assumption that text summarisation can be achieved using a classification approach whereby several class labels can be associated with documents which then constitute the summarisation. A hierarchical classification approach is suggested which offers the advantage that different levels of classification can be used and the summarisation customised according to which branch of the tree the current document is located. The approach is evaluated using free text from questionnaires used in the SAVSNET (Small Animal Veterinary Surveillance Network) project. The results demonstrate the viability of using hierarchical classification to generate free text summaries.

## 1 Introduction

The proliferation and constant generation of questionnaire data has resulted in substantial amounts of data for which traditional analysis techniques are becoming harder and harder to apply in an efficient manner. This is particularly the case with respect to the free text element that is typically included in questionnaire surveys. A technique used to understand and extract meaning from such free text is text summarisation. The motivations for text summarisation vary according to the field of study and the nature of the application. However, it is very clear that in all cases what is being pursued is the extraction of the main ideas of the original text, and the

Matias Garcia-Constantino · Frans Coenen
Department of Computer Science, The University of Liverpool, Liverpool, L69 3BX, UK, e-mail: mattgc,coenen@liverpool.ac.uk

P-J Noble · Alan Radford
School of Veterinary Science, University of Liverpool, Leahurst, Neston, CH64 7TE, UK, e-mail: rtnorle,alanrad@liverpool.ac.uk

consequent presentation of these ideas to some audience in a coherent and reduced form. Many text summarisation techniques have been reported in the literature; these have been categorised in many ways according to the field of study or to other factors inherent to the text. Jones *et al.* [16] proposed a categorisation dependent on: the input that is received, the purpose of the summarisation and the output desired. An alternative categorisation is to divide the techniques according to whether they adopt either a statistical or a linguistic approach.

Text classification, in its simplest form, is concerned with the assignation of one or more predefined categories to text documents according to their content [8]. The survey presented by Sebastiani [27] indicates that using machine learning techniques for automated text classification has more advantages than approaches that rely on domain experts to manually generate a classifier. As in the case of more established tabular data mining techniques, text classification techniques can be categorised according to whether they attach a single label (class) to each document or multiple labels. The approach proposed in this paper to generate text summaries is based on the concept of hierarchical text classification, which is a form of text classification that involves the use of class labels arranged in a tree structure. Hierarchical text classification is thus a form of multi-label classification. As Sun and Lim state [29], hierarchical classification allows a large classification problem to be addressed using a "divide-and-conquer" approach. It has been widely investigated and used as an alternative to standard text classification methods, also known as *flat classification* methods, in which class labels are considered independently from one another.

Although they are intended for different forms of application, text summarisation and text classification share a common purpose, namely to derive meaning from free text (either by producing a summary or by assigning labels). The reason why text summarisation can be conceived of as a form of text classification is that the classes assigned to text documents can be viewed as an indication (summarisation) of the main ideas, of the original free text, in a coherent and reduced form. Coherent because class names that are typically used to label text documents tend to represent a synthesis of the topic with which the document is concerned. It is acknowledged that a summary of this form is not as complete or as extensive as what many observers might consider to be a summary; but, if we assign multiple labels to each document then this comes nearer to what might be traditionally viewed as a summary. However, for anything but the simplest form of summarisation, the number of required classes will be substantial, to the extent that the use of flat classification techniques will no longer be viable, even if a number of such techniques are used in sequence. A hierarchical form of classification is therefore suggested. By arranging the potential class labels into a hierarchy multiple class labels can still be attached to documents in a more effective way than if flat classifiers were used. The effect is to permit an increase in the number of classes that can be used in the classification. Our proposed hierarchical approach uses single-label classifiers at each level in the hierarchy, although different classifiers may exist at the same level but in different branches in the hierarchy.

The advantages of using the proposed hierarchical text classification for text summarisation are as follows: (i) humans are used to the concept of defining things in a hierarchical manner, thus summaries will be produced in an intuitive manner, (ii) hierarchies are a good way of encapsulating knowledge, in the sense that each node that represents a class in the hierarchy has a specific meaning or significance associated with it with respect to the summarisation task, (iii) classification/summarisation can be achieved efficiently without having to consider all class labels for each unseen record, and (iv) it results in a more effective form of classification/summarisation because it supports the incorporation of specialised classifiers, at specific nodes in the hierarchy.

The rest of this paper is organised as follows. Related work is briefly reviewed in Section 2, and a formal definition of the proposed free text summarisation mechanism is presented in Section 3. Section 4 gives an overview of the SAVSNET (Small Animal Veterinary Surveillance Network) project questionnaire data used for evaluation purposes with respect to the work described in this paper. Section 5 describes the operation of the proposed approach. A comprehensive evaluation of the proposed approach, using the SAVSNET questionnaire data, is presented in Section 6. Finally, a summary of the main findings and some conclusions are presented in Section 7.

## 2 Related work

The text summarisation techniques proposed in the literature take into account the field of study and factors inherent to the text to be summarised. Afantenos *et al.* [1] provided what is referred to as a "fine grained" categorisation of text summarisation factors founded on the work of Jones *et al.* [16], and formulated the summarisation task in terms of input, purpose and output factors. The input factors considered were: (i) the number of documents used (single-document or multi-document), (ii) the data format in which the documents are presented (text or multimedia) and (iii) the language or languages in which the text was written (monolingual, multilingual and cross-lingual). The purpose factors were sub-divided according to: (i) the nature of the text (indicative or informative), (ii) how specific the summary must be for the intended audience (generic or user oriented) and (iii) how specific the summary must be in terms of the domain or field of study (general purpose or domain specific). The most significant output factors (amongst others) were sub-divided according to whether the summary needed to be: (i) complete, (ii) accurate and (iii) coherent. The "traditional" phases of text summarisation that most researchers follow are identified in [2], namely: (i) analysis of the input text, (ii) transformation of the input text into a form to which the adopted text summarisation technique can be applied and (iii) synthesis of the output from phase two to produce the desired summaries.

To the best knowledge of the authors, the generation of text summaries using text classification techniques has not been widely investigated. Celikyilmaz and

Hakkani-Tür [3] presented an "automated semi-supervised extractive summarisation" approach which used latent concept classification to identify hidden concepts in documents and to add them to the produced summaries. Previous work by the authors directed at text summarisation can be found in [10] and [11]. In [10] a summarisation classification technique, called Classifier Generation Using Secondary Data (CGUSD), was presented, it was directed at producing text summarisation classifiers where there was insufficient data to support the generation of classifiers from primary data. The technique was founded on the idea of generating a classifier for the purpose of text summarisation by using an alternative source of free text data and then applying it to the primary data. In [11], a semi-automated approach to building text summarisation classifiers, called SARSET (Semi-Automated Rule Summarisation Extraction Tool) was described, however this required substantial user intervention.

In [26], the integration of text summarisation and text classification is more synergic. Saravanan *et al.* proposed an approach to compose a summariser and a classifier integrated within a framework for cleaning and preprocessing data. They make the point that composition is invertible, meaning that summarisation can be applied first to increase the performance of the classifier or the other way around. As Saravanan *et al.* indicate, the use of classification improves the generation of summaries with respect to domain-specific documents. In [15], text classification is used to classify and select the best extracted sentences from text documents in order to generate summaries. In [14], a system is proposed that identifies the important topics in large document sets and generates a summary comprised of extracts related to identified topics.

Unlike the approach presented in this paper, the aforementioned approaches all use flat classification techniques to achieve free text summarisation. Hierarchical text classification makes use of the hierarchical relationships within an overall class structure to "boost" the effectiveness of text classification. The idea of using hierarchies for text classification can be effectively extended and customized for specific problems that involve the hierarchical representation of document sets. Typically, a hierarchy of a corpus of text documents is represented either as a decision tree or as a Directed Acyclic Graph (DAG). There are three main models in hierarchical classification: (i) big-bang, (ii) top-down and (iii) bottom-up. The big-bang model uses a single classifier to assign one or more class labels from the hierarchy to each document. The top-down and bottom-up models are based on the construction and application of classifiers in each level of the hierarchy where each classifier acts as a flat classifier within that level/branch. In the case of the top-down model, the taxonomy is traversed from the higher to the lower levels. In the bottom-up approach the taxonomy is traversed in the reverse manner to that adopted in the top-down model. The model adopted with respect to the work described in this paper is the top-down model.

There is a considerable amount of research that has been carried out concerning the top-down model using different classification techniques, such as: Support Vector Machine (SVM) [6, 19, 29], classification trees [21], path semantic vectors [9], Hierarchical Mixture Models [30], TF-IDF classification [4], k-nearest neighbour techniques [7], a variation of the Maximum Margin Markov Network framework [24], the Passive-Aggressive (PA) algorithm with latent concepts [22], neural networks [25], word clustering combined with Naive Bayes and SVM [5], multiple Bayesian classifiers [18] and boosting methods (BoosTexter and Centroid-Boosting) combined with Rocchio and SVM classifiers [12]. A comprehensive survey on hierarchical classification is presented in [28].

As was mentioned in the previous section, there are many advantages of using hierarchical text classification for text summarisation, advantages that indicate that the approach proposed in this paper is a viable alternative over existing text summarisation techniques. The main advantages over other text summarisation techniques are: (i) a more intuitive way of understanding the contents of a document because humans are used to the concept of defining things in a hierarchical way and (ii) the ability to handle large document sets due to the hierarchical approach's inherent divide-and-conquer strategy. It can be argued that a summary generated using this approach will be very similar to that generated using a multi-class flat classification or to systems that automatically assign tags to suggest topics [17]. However, our approach differs from these other techniques in that the resulting classes generated using the hierarchical text classification process are not isolated concepts. On the contrary they are related to each other due to their hierarchical nature, giving the domain expert a more coherent and clear insight of what a given document is about.

## 3 Proposed approach

The input to the proposed text summarisation hierarchical classifier generator is a "training set" of $n$ free text documents, $D = \{d_1, d_2, \ldots, d_n\}$, where each document $d_i$ has a sequence of $m$ summarisation class labels, $S = \{s_1, s_2, \ldots, s_m\}$, such that there is a one-to-one correspondence between each summarisation label $s_i$ and some class $g_j$. Thus the summarisation labels are drawn from a set of $n$ classes $G = \{g_1, g_2, \ldots, g_n\}$ where each class $g_j$ in $G$ has a set of $k$ summarisation labels associated with it $g_i = \{c_{j_1}, c_{j_2}, \ldots c_{j_k}\}$. The desired class hierarchy $H$ then comprises a set of nodes arranged into $p$ levels, $L = \{l_1, l_2, \ldots, l_p\}$, such as the one shown in Figure 1. Except at the leaf nodes each node in the hierarchy has a classifier associated with it.

Since we are using a top-down model, the classifiers can be arranged according to two approaches in terms of the scope and dependency between levels: (i) cascading and (ii) non-cascading. In the cascading case, the output of the classifier at the parent nodes influences the classification conducted at the child nodes at the next level of the hierarchy (we say that the classification process "cascades" downwards). Thus a classifier is generated for each child node (except the leaf nodes) depending on the resulting classification from the parent node. The classification process continues in this manner until there are no more nodes to be developed. In the case of the non-cascading model each classifier is generated independently from that of the parent node.

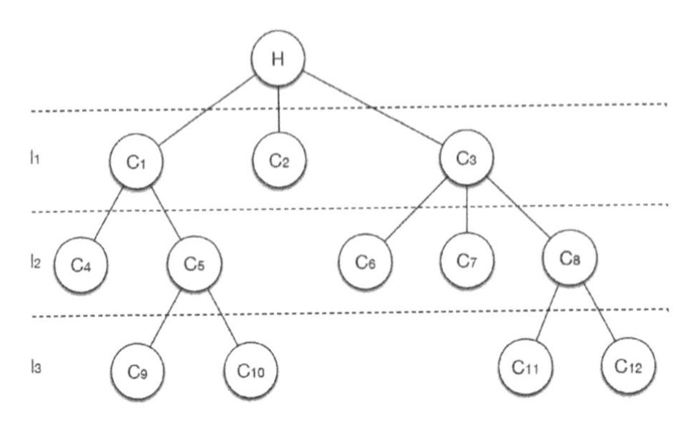


**Fig. 1** Hierarchy of classes.

In addition, two types of hierarchies are identified regarding the parent-child node relationship: single and multi-parent. The top-down strategy can be applied in both cases because, given a piece of text to be summarised, only one best child node (class) is selected per level. Examples of single and multi-parent hierarchies are shown in Figures 2 and 3.

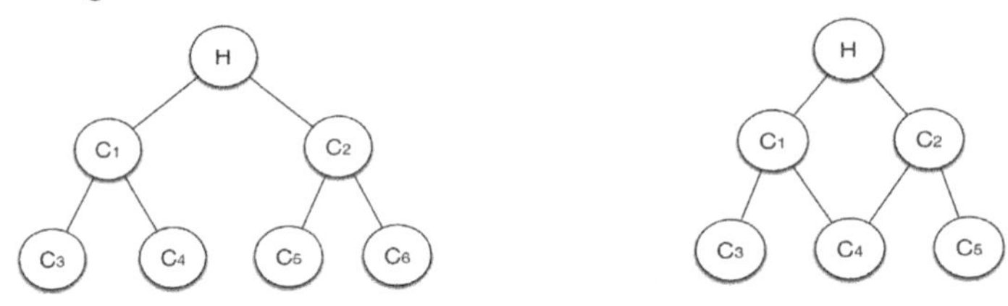


**Fig. 2** Single-parent hierarchy

**Fig. 3** Multi-parent hierarchy

The classifier generation process is closely linked to the structure of the hierarchy. When generating a cascading hierarchy we start by generating a classifier founded on the entire training set. For its immediate child branches we only used that part of the training set associated with the class represented by each child node. For non-cascading we use the entire training set for all nodes (except the leaf nodes), but with different labels associated with the training records according to the level we are at. The classifier generation process is described in more detail in Section 5.

Once the generation process is complete, the text summarisation classifier is ready for application to new data. New documents will be classified by traversing the hierarchical classifier in a similar manner to that used in the context of a decision tree. That is, at each level the process will be directed towards a particular branch in the hierarchy according to the current classification. The length of the produced summary will depend on the number of levels traversed within the hierarchy.

## 4 Motivational example

The collection of questionnaires used to evaluate our approach was generated as part of the SAVSNET project [23], which is currently in progress within the Small Animal Teaching Hospital at the University of Liverpool. The objective of SAVS-NET is to provide information on the frequency of occurrence of small animal diseases (mainly in dogs and cats). The project is partly supported by Vet Solutions, a software company whose software is used by some 20% of the veterinary practices located across the UK. Some 30 veterinary practices, all of whom use Vet Solutions' software, have "signed up" to the SAVSNET initiative.

The SAVSNET veterinary questionnaires comprise a tabular (tick box) and a free text section. Each questionnaire describes a consultation and is completed by the vet conducting the consultation. In the tabular section of the questionnaire, specific questions regarding certain veterinary conditions are asked (e.g. presence of the condition, severity, occurrence, duration), these questions define the hierarchy of classes. An example (also used for evaluation purposes in Section 5) is presented in Figures 4, 5, 6 and 7[1]. As shown in Figure 4, the first level in the hierarchy (in this example) distinguishes between GI (gastrointestinal) symptoms, namely: "diarrhoea" (D), "vomiting" (V) and "vomiting & diarrhoea" (V&D). The second level (Figure 5) distinguishes between the severity of the GI symptom presented: "haemorrhagic" (H), "non haemorrhagic" (NH) and "unknown severity" (US). The nodes at the next level (Figure 6) consider whether the identified symptom is: "first time" (1st), "nth time" (Nth) or "unknown occurrence" (UO). Finally (Figure 7), the nodes of the fourth level relate to the duration of the symptom: "less than one day" (<1), "between two and four days" (2-4), "between five and seven days" (5-7), "more than eight days" (8+) and "unknown duration" (UD). In this example, the child nodes for each of the GI symptoms are similar, making them hierarchically symmetric. However, our proposed method will work equally well on asymmetric hierarchies.

The tabular section of the questionnaires also includes attributes that are associated with general details concerning the consultation (e.g. date, consultation ID, practice ID), while others are concerned with the "patient" (e.g. species, breed, sex) and its owner (e.g. postcode). The classification/summarisation of the tabular element of the SAVSNET questionnaires is not the topic of interest with respect to this paper; this paper is concerned with the summarisation of the free text element of the questionnaires. The free text section of the questionnaires usually comprises notes made by vets, which typically describe the symptoms presented, the possible diagnosis and the treatment to be prescribed. It is the free text section that we are interested in summarising, although in some cases the free text element of the questionnaires is left blank.

[1] It should be noted that levels in the hierarchy are presented in separate figures for convenience only, they are in fact connected and should not be viewed as being independent.

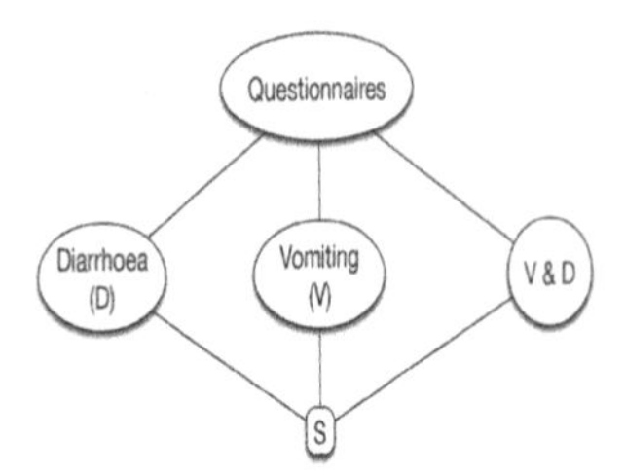


Fig. 4 Level 1, GI symptoms.

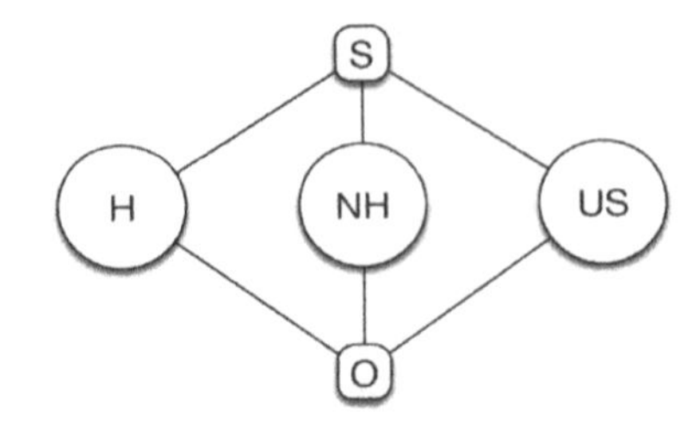


Fig. 5 Level 2, "Severity".

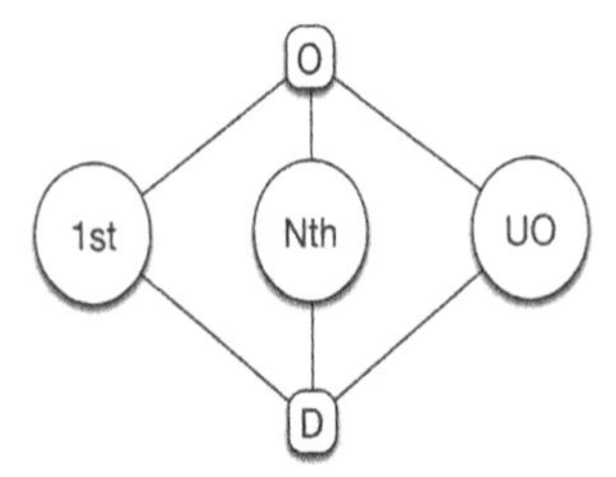


Fig. 6 Level 3, "Occurrence".

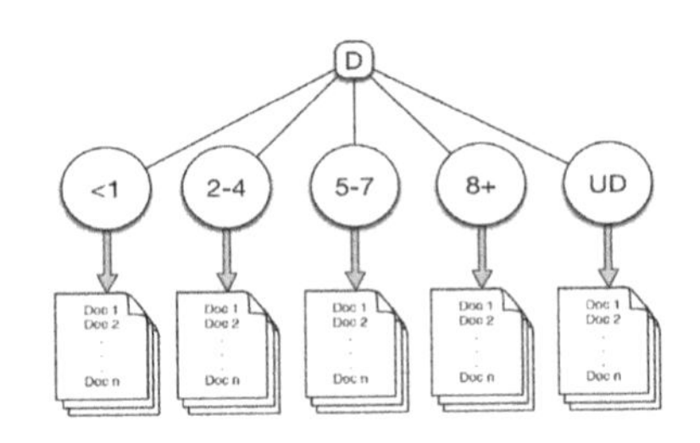


Fig. 7 Level 4, "Duration".

## 5 Classifier Generation and Application

In this section the two hierarchical text classifier generation approaches considered (cascading and non-cascading) are described in more detail. Recall that in the non-cascading approach the classification process is carried out independently in each node and, as its name implies, independently of the levels and the parent-child node relationship, in other words, flat classifiers are generated for each node; in the cascading approach the output of the classification of the parent nodes affects the classification of child nodes at the next level of the hierarchy. In both cases a 6 step classifier generation process is specified, as follows:

1. **Preprocessing of documents:** text is converted to lower case; numbers, symbols and stop words (common words that are not significant for the text classification/summarisation process) are removed, stemming is applied using an implementation of the Porter Stemming algorithm [31] and feature selection is performed using an implementation of the chi-square method [32].
2. **Classification of documents:** a classifier is generated for each node in the current level of the hierarchy. The classifier generation method chosen with respect to the evaluation included in this paper is the Support Vector Machine (SVM) method, via an implementation of Platt's Sequential Minimal Optimization (SMO) algorithm [20]. The nature of the classification depends on the approach taken:

   **(i) Cascading approach:** The output of the classification of the nodes in the current level will affect nodes at the next level down in the hierarchy.

**(ii) Non-cascading approach:** The classifier generation is carried out independently in each node. The classification results do not affect the classifier generation in nodes at the next level down in the hierarchy.

3. **Evaluation of the classification:** The generated classifier is evaluated and the results recorded. Based on the resulting evaluation metrics:

   **(i) Cascading approach:** Correctly and incorrectly classified instances are considered for the classification process in the next level down in the hierarchy.

   **(ii) Non-cascading approach:** The classification results from the previous level are not taken into account for the classifier generation conducted at the lower level down in the hierarchy.

4. **Verification of the existence of nodes in the next level down in the hierarchy:** Exit (hierarchical classifier is complete), if there are no more nodes to be developed at the next level down in the hierarchy. Otherwise continue with step 5.
5. **Classifier generation at the next level down in the hierarchy:** repeat process from step 2 for each node at the next level down in the hierarchy.

   **(i) Cascading approach:** Correctly and incorrectly classified instances for nodes in the previous level are taken into account for nodes in the current level.

   **(ii) Non-cascading approach:** Classifier generation at nodes in the current level will be performed regardless of the classification results produced at the previous level.

Once complete (and found to be effective when applied to an appropriately defined test set) the generated classifier may be applied to unseen data. A summary for each document is then produced using the resultant class labels generated at each level of the hierarchy. An example of such a summarisation, with respect to the text application considered in the following section, might be: {*Presented diarrhoea, not haemorrhagic, presented for the first time and the duration of the symptoms was between two and four days.*} Note that such a summarisation differs from traditional text summarisation techniques in that the words or phrases that comprise the resulting summary are not necessarily present in the original text. However the resulting summary is a concise, coherent and informative overview of the content of a document.

## 6 Evaluation

The evaluation of the proposed hierarchical technique for generating summaries from free text was carried out using a subset of the SAVSNET questionnaire corpus which we called SAVSNET-917; for the evaluation we also concentrated on summarising symptoms. This dataset is comprised of 917 records and several classes arranged over four different levels that were defined by specific questions, included in the questionnaire collection process, regarding certain veterinary conditions. The hierarchical arrangement of class labels is as shown in Figures 4, 5, 6 and 7 (described previously). Despite having a relatively small number of records in the SAVSNET-917 dataset, the four levels of the hierarchy were adequately taken into account for the experiments. The distribution of the documents per class for the four levels of the hierarchy is shown in Tables 1, 2, 3 and 4. From these tables it can be seen that the distribution of the documents per class is significantly unbalanced.

**Table 1** Number of records per class in the first level of SAVSNET-917 hierarchy.

| Class | Num. |
|---|---|
| *Diarrhoea* | 536 |
| *Vomiting* | 248 |
| *Vom&Dia* | 133 |
| Total | 917 |

**Table 2** Number of records per class in the second level of SAVSNET-917 hierarchy.

| Class | Num. |
|---|---|
| *Haemorrhagic* | 177 |
| *NotHaemorrhagic* | 604 |
| *UnknownSeverity* | 136 |
| Total | 917 |

**Table 3** Number of records per class in the third level of SAVSNET-917 hierarchy.

| Class | Num. |
|---|---|
| *FirstTime* | 573 |
| *NthTime* | 290 |
| *UnknownOccurrence* | 54 |
| Total | 917 |

**Table 4** Number of records per class in the fourth level of SAVSNET-917 hierarchy.

| Class | Num. |
|---|---|
| *LessThanOneday* | 273 |
| *BetweenTwoAndFourDays* | 411 |
| *BetweenFiveAndSevenDays* | 82 |
| *MoreThanEightDays* | 139 |
| *UnknownDuration* | 12 |
| Total | 917 |

The evaluation was conducted using Ten-fold Cross Validation (TCV). The evaluation metrics used were overall accuracy (Acc) expressed as a percentage, Area Under the receiver operating Curve (AUC) [13], sensitivity (Sn) and specificity (Sp). In relation to a confusion matrix, sensitivity measures the proportion of actual positives which are correctly identified, and specificity measures the proportion of negatives which are correctly identified. The AUC measure was used because it takes into consideration the "class priors" (the potential imbalanced nature of the input datasets).

Comparison with other types of summarisation tools, such as a NLP summariser, was not undertaken because of the nature of the different summaries produced; it did not make sense to compare a summary produced in the form of (say) a collection of keywords with a summary produced using our proposed hierarchical classification approach. We could have compared the operation of our approach with the result produced by the application of a sequence of "flat" classifiers, but this would simply have mimicked the operation of our non-cascading approach; thus such comparisons are not reported here. What we can say is that the validity of our summaries has been confirmed by domain experts working on the SAVSNET project.

Table 5 shows the results for the first level, which were the same regardless of the approach used because there was no parent level that had an influence on the classification process. The results for the other levels are shown in Tables 6, 7 and

8. The cascading and the non-cascading approaches are indicated in the tables using the abbreviations casc and ¬casc respectively.

**Table 5** Classification results for level 1.

| | Level 1 | | | |
|---|---|---|---|---|
| Approach | Acc (%) | AUC | Sn | Sp |
| Both approaches | 71.32 | 0.743 | 0.713 | 0.734 |

**Table 6** Classification results for level 2.

| Level 2 | Diarrhoea | | | | Vomiting | | | | Vomiting and diarrhoea | | | |
|---|---|---|---|---|---|---|---|---|---|---|---|---|
| | Acc (%) | AUC | Sn | Sp | Acc (%) | AUC | Sn | Sp | Acc (%) | AUC | Sn | Sp |
| casc | 60.81 | 0.60 | 0.61 | 0.54 | 75.57 | 0.50 | 0.76 | 0.23 | 59.26 | 0.62 | 0.59 | 0.65 |
| ¬casc | 73.69 | 0.64 | 0.74 | 0.55 | 89.11 | 0.49 | 0.89 | 0.10 | 98.50 | 0.50 | 0.98 | 0.02 |

**Table 7** Classification results for level 3.

| Level 3 | Haemorrhagic | | | | Not Haemorrhagic | | | | Unknown Severity | | | |
|---|---|---|---|---|---|---|---|---|---|---|---|---|
| | Acc (%) | AUC | Sn | Sp | Acc (%) | AUC | Sn | Sp | Acc (%) | AUC | Sn | Sp |
| casc | 62.73 | 0.59 | 0.63 | 0.54 | 62.58 | 0.61 | 0.63 | 0.58 | 67.19 | 0.58 | 0.67 | 0.49 |
| ¬casc | 62.71 | 0.60 | 0.63 | 0.58 | 65.23 | 0.60 | 0.65 | 0.54 | 58.82 | 0.60 | 0.59 | 0.62 |

**Table 8** Classification results for level 4.

| Level 4 | First Time | | | | Nth Time | | | | Unknown Occurrence | | | |
|---|---|---|---|---|---|---|---|---|---|---|---|---|
| | Acc (%) | AUC | Sn | Sp | Acc (%) | AUC | Sn | Sp | Acc (%) | AUC | Sn | Sp |
| casc | 50.08 | 0.60 | 0.50 | 0.66 | 44.28 | 0.63 | 0.44 | 0.72 | – | – | – | – |
| ¬casc | 49.21 | 0.59 | 0.49 | 0.66 | 45.86 | 0.65 | 0.46 | 0.75 | 51.85 | 0.44 | 0.52 | 0.35 |

In both approaches no incorrectly classified instances were removed, so the overall number of documents in each level is the same during the experiments. In the case of the cascading approach the instances that were correctly classified and were considered for the classifier generation in the next level down in the hierarchy improved the quality of the resulting summaries; providing for completeness, accuracy and coherency. However, a drawback of this approach is that incorrectly classified documents from a parent node will affect the resulting classification at child nodes and therefore the quality of the summaries produced. In the case of the non-cascading approach the generation of a classifier in each node is in isolation and only the number, quality and distribution of the instances per class will affect the quality of the classifier.

As can be seen from the results shown in Tables 5, 6, 7, and 8, while the accuracy increased from the first to the second level of the hierarchy, it decreased in the third and fourth levels. It can be conjectured that the unbalanced distribution of training texts per class effected the performance for both approaches and in the case of the cascading method the propagation of errors from parent to child was also found to have a considerable impact in the performance of the hierarchical text classification. If incorrectly classified records had been removed the accuracy and AUC values would have been increased from the higher to the lower nodes in the hierarchy due to not having wrongly classified records to consider. However it was conjectured that the removal of wrongly classified records might result in overfitting. Incorrectly classified documents were therefore not removed.

## 7 Conclusions and Future Work

This paper has presented an approach to the generation of text summaries using a hierarchical text classification approach. The main advantages of the proposed approach, over other text summarisation techniques, are: (i) a more intuitive way of understanding the contents of a document, because humans are used to the concept of defining things in a hierarchical way; and (ii) the ability to handle large document sets due to the hierarchical approach's inherent support for the divide-and-conquer strategy.

The approach was tested using the free text element of a subset of the SAVSNET dataset (SAVSNET-917). The reported experiments were carried out considering a four level hierarchy. The hierarchical classification was performed using two approaches: cascading and non-cascading. In the former approach the performance of a classifier at a parent node influenced the performance at its child nodes, in the latter case each node was considered independent of each other.

For evaluation purposes a Support Vector Machine (SVM) classifier implementation using Platt's Sequential Minimal Optimization (SMO) algorithm was adopted, but other types of classifier generator could equally well be used. The reported evaluation was conducted using TCV. We used a number of evaluation metrics so as to present a wide oversight of the performance of the proposed approaches: accuracy, AUC, sensitivity and specificity were used. The technique was evaluated in terms of the performance of the text classification because the summaries are generated with the labels found in the nodes of each level of the hierarchy. In other words, the generated summaries depend on how well the hierarchical text classification process performs. Besides the evaluation results of the proposed approach, domain experts reviewed the completeness, content accuracy (how accurate is the summary with respect of the original text) and coherency of the summaries generated. Although the domain experts reported that the summaries were complete and coherent, the accuracy of their content is expected to improve with a better performance of the hierarchical classification strategies. Results showed that both the cascading and the non-cascading hierarchical classification approaches performed relatively well when a considerable number of records were held at a given node. However, there is still work to be done with respect to the situation where we have few records at a node.

For future work, we also intend to consider extending the proposed technique to address multi-label classification at each hierarchy level in order to produce more comprehensive summaries than using just one label per hierarchy level. It may also be of interest to include the tabular component of the questionnaires in the hierarchical classification process so as to extend and improve the technique. Future work will also consider the application of the proposed technique to several other data sets (including benchmark data sets); a subsequent comparison of the obtained results will be carried out in order to evidence the generality of the technique. Extensive experiments comparing the proposed technique to other text summarisation techniques are also planned, although this will require derivation of appropriate comparison metrics.

## References


1. Afantenos, S. and Karkaletsis, V. and Stamatopoulos, P. (2005). Summarization from medical documents: a survey. Artificial Intelligence in Medicine Vol. 33, pp157-177.
2. Alonso, L. and Castellón, I. and Climent, S. and Fuentes, M. and Padró, L. and Rodríguez, H. (2004). Approaches to text summarization: Questions and answers. Inteligencia Artificial Vol. 8, pp22.
3. Celikyilmaz, A. and Hakkani-Tür, D. (2011). Concept-based classification for multi-document summarization. IEEE International Conference on Acoustics, Speech and Signal Processing (ICASSP), 2011, pp5540-5543.
4. Chuang, W. and Tiyyagura, A. and Yang, J. and Giuffrida, G. (2000). A fast algorithm for hierarchical text classification. Data Warehousing and Knowledge Discovery, pp409-418.
5. Dhillon, I.S. and Mallela, S. and Kumar, R. (2002). Enhanced word clustering for hierarchical text classification. Proceedings of the eighth ACM SIGKDD international conference on Knowledge discovery and data mining, pp191-200.
6. Dumais, S. and Chen, H. (2000). Hierarchical classification of web content. Proceedings of the 23rd annual international ACM SIGIR conference on Research and development in information retrieval, pp256-263.
7. Duwairi, R. and Al-Zubaidi, R. (2011). A Hierarchical K-NN Classifier for Textual Data. The International Arab Journal of Information Technology. Vol. 8, pp251-259.
8. Fragoudis, D. and Meretakis, D. and Likothanassis, S. (2005). Best terms: an efficient feature-selection algorithm for text categorization. Knowledge and Information Systems. Vol. 8, pp16-33.
9. Gao, F. and Fu, W. and Zhong, Y. and Zhao, D. (2004). Large-Scale Hierarchical Text Classification Based on Path Semantic Vector and Prior Information. CIS'09. International Conference on Computational Intelligence and Security. Vol. 1, pp54-58.
10. Garcia-Constantino, M. F. and Coenen, F. and Noble, P. and Radford, A. and Setzkorn, C. and Tierney, A. (2011). An Investigation Concerning the Generation of Text Summarisation Classifiers using Secondary Data. Seventh International Conference on Machine Learning and Data Mining. Springer, pp387-398.
11. Garcia-Constantino, M. F. and Coenen, F. and Noble, P. and Radford, A. and Setzkorn, C. (2012). A Semi-Automated Approach to Building Text Summarisation Classifiers. To be presented at the Eight International Conference on Machine Learning and Data Mining. Springer.
12. Granitzer, M. (2003). Hierarchical text classification using methods from machine learning. Master's Thesis, Graz University of Technology.
13. Hand, D.J. and Till, R.J. (2001). A Simple Generalisation of the Area Under the ROC Curve for Multiple Class Classification Problems. Machine Learning, 45, pp171-186.
14. Hardy, H. and Shimizu, N. and Strzalkowski, T. and Ting, L. and Zhang, X. and Wise, G.B. (2002). Cross-document summarization by concept classification. Proceedings of the 25th Annual International ACM SIGIR Conference on Research and Development in Information Retrieval, pp121-128.

15. Jaoua, M. and Hamadou, A. (2003). Automatic text summarization of scientific articles based on classification of extracts population. Computational Linguistics and Intelligent Text Processing, pp363-377.
16. Jones, K.S. and others. (1999). Automatic summarizing: factors and directions. Advances in automatic text summarization, pp1-12.
17. Katakis, I. and Tsoumakas, G. and Vlahavas, I. (2008). Multilabel text classification for automated tag suggestion. Proceedings of the ECML/PKDD 2008. Workshop in Discovery Challenge, pp75-83. Antwerp, Belgium.
18. Koller, D. and Sahami, M. (1997). Hierarchically Classifying Documents Using Very Few Words. Proceedings of the Fourteenth International Conference on Machine Learning, pp170-178.
19. Kumilachew, A. (2011). Hierarchical Amharic News Text Classification: Using Support Vector Machine Approach. VDM Verlag Dr. Müller.
20. Platt, J.C. (1999). Using analytic QP and sparseness to speed training of support vector machines. Advances in neural information processing systems, pp557-563.
21. Pulijala, A. and Gauch, S. (2004). Hierarchical text classification. International Conference on Cybernetics and Information Technologies, Systems and Applications: CITSA, pp21-25.
22. Qiu, X. and Huang, X. and Liu, Z. and Zhou, J. (2011). Hierarchical Text Classification with Latent Concepts. Proceedings of the 49th Annual Meeting of the Association for Computational Linguistics. Vol. 2, pp598-602.
23. Radford, A. and Tierney, Á. and Coyne, K.P. and Gaskell, R.M. and Noble, P.J. and Dawson, S. and Setzkorn, C. and Jones, P.H. and Buchan, I.E. and Newton, J.R. and Bryan, J.G.E. (2010). Developing a network for small animal disease surveillance. Veterinary Record. Vol. 167, pp472-474.
24. Rousu, J. and Saunders, C. and Szedmak, S. and Shawe-Taylor, J. (2005). Learning Hierarchical Multi-Category Text Classification Models. Proceedings of the 22nd International Conference on Machine Learning, pp744-751.
25. Ruiz, M.E. and Srinivasan, P. (2002). Hierarchical text categorization using neural networks. Information Retrieval. Vol. 5, pp87-118.
26. Saravanan, M. and Raj, P.C.R. and Raman, S. (2003). Summarization and categorization of text data in high-level data cleaning for information retrieval. Applied Artificial Intelligence, Vol. 17, pp461-474.
27. Sebastiani, F. (2002). Machine learning in automated text categorization. ACM computing surveys (CSUR). Vol. 34, pp1-47.
28. Silla, C.N. and Freitas, A.A. (2011). A survey of hierarchical classification across different application domains. Data Mining and Knowledge Discovery Vol. 22, pp31-72.
29. Sun, A. and Lim, E.P. (2001). Hierarchical text classification and evaluation. ICDM 2001, Proceedings IEEE International Conference on Data Mining. IEEE, pp521-528.
30. Toutanova, K. and Chen, F. and Popat, K. and Hofmann, T. (2001). Text classification in a hierarchical mixture model for small training sets. Proceedings of the tenth international conference on Information and knowledge management, pp105-113.
31. Willett, P. (2006). The Porter stemming algorithm: then and now. Program: electronic library and information systems Vol. 40, pp219-223.
32. Zheng, Z. and Wu, X. and Srihari, R. (2004). Feature selection for text categorization on imbalanced data. ACM SIGKDD Explorations Newsletter Vol. 6, pp80-89.

# Mining Interesting Correlated Contrast Sets

Mondelle Simeon, Robert J. Hilderman, and Howard J. Hamilton

**Abstract** Contrast set mining has been developed as a data mining task which aims at discerning differences across groups. These groups can be patients, organizations, molecules, and even time-lines. A valid correlated contrast set is a conjunction of attribute-value pairs that are highly correlated with each other and differ significantly in their distribution across groups. Although the search for valid correlated contrast sets produces a comparatively smaller set of results than the search for valid contrast sets, these results must still be further filtered in order to be examined by a domain expert and have decisions enacted from them. In this paper, we apply the minimum support ratio threshold which measures the ratio of maximum to minimum support across groups. We propose a contrast set mining technique which utilizes the minimum support ratio threshold to discover maximal valid correlated contrast sets. We also demonstrate how four probability-based objective measures developed for association rules can be used to rank contrast sets. Our experiments on real datasets demonstrate the efficiency and effectiveness of our approach.

## 1 Introduction

Discovering the differences between groups is a fundamental problem in many disciplines such as social sciences, medicine, psychiatry, and computer science. In computer science, most of the research in group differences is conducted in data mining. One data mining task whose goal is to find differences between groups is contrast set mining, which has been developed to detect all differences between contrasting groups from observational multivariate data [2]. Contrast set mining has been used to identify factors that distinguish between thrombolic and embolic stroke patients [8], identify the characteristics of news sites [19] and to contrast aircraft ac-

Mondelle Simeon · Robert J. Hilderman · Howard J. Hamilton
Department of Computer Science, University of Regina, Regina, Saskatchewan, Canada S4S 0A2
e-mail: {simeon2m, hilder, hamilton}@cs.uregina.ca

cident data with aircraft incident data in major safety databases then identify factors that are significantly associated with the accidents [11].

Existing contrast set mining techniques can produce a prohibitively large set of differences across groups with varying levels of interestingness [6] [15]. For example, if we have a reproductive health study of five demographic and socio-economic variables, each with two values, our search for differences in contraceptive methods used by married couples would cause us to examine all 242 possible combinations of attribute values. On larger datasets, the search space becomes inordinately large, and the search becomes prohibitively expensive producing a large number of results which ultimately must be analyzed and evaluated by a domain expert. In this paper we propose the use of mutual information and all confidence to select only the attributes and attribute-values that are highly correlated and the minimum support ratio threshold, which measures the ratio of maximum to minimum support across groups, to discover "more interesting" contrast sets from these highly correlated attributes and attribute-values.

## 2 Related Work

The STUCCO (Search and Testing for Understandable Consistent Contrasts) algorithm [2] is the original technique for mining contrast sets. The objective of STUCCO is to find statistically significant contrast sets from grouped categorical data. STUCCO forms the basis for a method proposed to discover negative contrast sets [18] that can include negation of terms in the contrast set. The main difference is their use of Holm's sequential rejective method [7] for the independence test.

The CIGAR (Contrasting Grouped Association Rules) algorithm [6] is a contrast set mining technique that specifically identifies which pairs of groups are significantly different and whether the attributes in a contrast set are correlated. CIGAR utilizes the same general approach as STUCCO, however it focuses on controlling Type II errors through increasing the significance level for the significance tests and by not correcting for multiple comparisons. Like STUCCO, CIGAR is only applicable to discrete-valued data.

Contrast set mining has also been applied to continuous data. Early work focussed on the formal notion of a time series contrast set and an efficient algorithm was proposed to discover contrast sets in time series and multimedia data [10]. Another approach utilized a modified equal-width binning interval, where the approximate width of the intervals is provided as a parameter to the model [13]. The methodology used is similar to STUCCO except that the discretization step is added before enumerating the search space.

The COSINE (Contrast Set Exploration using Diffsets) [14], GENCCS (Gener ate Correlated Contrast Sets) [15], and GIVE (Generate Interesting Valid contrast sEts) [16] algorithms are contrast set mining techniques that incorporate a vertical data format, a back tracking search algorithm, and simple discretization in mining maximal contrast sets and maximal correlated contrast sets from both discrete and continuous-valued attributes.

## 3 Problem Definition

Let $\mathcal{A} = \{a_1, a_2, \ldots, a_n\}$ be a set of $n$ distinct attributes. Let $\mathcal{Q}$ and $\mathcal{C}$ denote the set of *quantitative* attributes and the set of *categorical* attributes, respectively. Let $\mathcal{V}(a_k)$ be the domain of values for $a_k$. An *attribute-interval pair*, denoted as $a_k : [v_{kl}, v_{kr}]$, is an attribute $a_k$ associated with an interval $[v_{kl}, v_{kr}]$, where $a_k \in \mathcal{A}$ and $v_{kl}, v_{kr} \in \mathcal{V}(a_k)$. Further, if $a_k \in \mathcal{C}$ then $v_{kl} = v_{kr}$. Similarly, if $a_k \in \mathcal{Q}$, then $v_{kl} \leq v_{kr}$. Let $\mathcal{T} = \{x_1, x_2, \ldots, x_p\}$ where $x_k \in \mathcal{V}(a_k)$, $1 \leq k \leq p$, be a *transaction*. Let $\mathcal{D}$ be a set of transactions, called the *database*. Let $\{i_1, i_2, \ldots, i_m\}$ be a set of $m$ distinct values from the set $\{1, 2, \ldots, n\}$, $m < n$. Let $\mathcal{F} = \{a_{i_1}, a_{i_2}, \ldots, a_{i_m}\}$, $a_{i_k} \in \mathcal{A}$, be a set of $m$ distinct class attributes. Let $G = \{a_1 : [v_{1l}, v_{1r}], \ldots, a_m : [v_{ml}, v_{mr}]\}$, $a_k \in \mathcal{F}, 1 \leq k \leq m$, $a_i \neq a_j$, $\forall i, j$, be a set of $m$ distinct class attribute-interval pairs, called a *group*. Let $X = \{a_1 : [v_{1l}, v_{1r}], \ldots, a_q : [v_{ql}, v_{qr}]\}$, $a_i, a_j \in \mathcal{A} - \mathcal{F}$, $1 \leq k \leq q \leq n - m$, $a_i \neq a_j$, $\forall i, j$, be a set of distinct attribute-interval pairs, called a *contrast set*. A contrast set, $X$, is called $k$-specific, if $|X| = k$. The *support* of a contrast set, $X$, *in a database,* $\mathcal{D}$, denoted as $support(X)$, is the percentage of transactions in $\mathcal{D}$ containing $X$. The support of a contrast set, $X$, in a group, $G$, denoted as $support(X, G)$, is the percentage of transactions in $\mathcal{D}$ containing $X \cup G$.

A contrast set $X$ associated with $b$ mutually exclusive groups. $G_1, G_2, \ldots, G_b$ is called a *valid contrast set* if, and only if, the following four criteria are satisfied:

$$\exists i, j \; support(X, G_i) \neq support(X, G_j), \tag{1}$$

$$\exists i, j \; \max_{ij} |support(X, G_i) - support(X, G_j)| \geq \varepsilon, \tag{2}$$

$$support(X) \geq \sigma, \tag{3}$$

and

$$\exists i \; \max_{i} \left\{ 1 - \frac{support(X, G_i)}{support(Y, G_i)} \right\} \geq \kappa, \tag{4}$$

where $\varepsilon$ is the *minimum support difference threshold*, $\sigma$ is the *minimum frequency threshold*, $\kappa$ is the *minimum subset support ratio threshold*, and $X \subset Y$ such that $|X| = |Y| + 1$. Criteria 1 ensures that the contrast set represents a true difference between the groups. Contrast sets that meet this criteria are called *significant*. Criteria 2 ensures the effect size. Contrast sets that meet this criteria are called *large*. Criteria 3 ensures that the number of occurrences of the contrast set exceeds some number of transactions. Contrast sets that meet this criteria are called *frequent*. Criteria 4 ensures that the support of the contrast set in each group is different from that of its superset. Contrast sets that meet this criteria are called *specific*.

A valid contrast set is called a *maximal contrast set* if it is not a subset of any valid contrast set.

A valid contrast set is called a $\lambda$-*contrast set* (i.e., lambda contrast set) if, and only if,

$$\lambda(X) \geq \omega, \tag{5}$$

where $\omega$ is a user-defined *minimum support ratio threshold*, and the ratio of maximum to minimum support across the groups is sufficiently large. Formally, this ratio is defined as

$$\lambda(X) = \begin{cases} \infty, & \text{if } \min_{i=1}^{n}\{support(X,G_i)\} = 0, \\ \dfrac{\max_{i=1}^{n}\{support(X,G_i)\}}{\min_{j=1}^{n}\{support(X,G_j)\}}, & \text{otherwise.} \end{cases}$$

A large value for $\lambda(X)$ indicates that $X$ occurs in significantly more transactions in one group $G_i$ than in some other group $G_j$. Criteria 5 ensures that the ratio of maximum and minimum support across all groups is sufficiently large.

A $\lambda$-contrast set is called an $\infty$-*contrast set* (i.e., infinity contrast set) if, and only if,

$$\lambda(X) = \infty. \tag{6}$$

A value of $\infty$ indicates that $X$ is present in at least one group $G_i$, and absent from at least one other group $G_j$.

A valid contrast set, $X = \{a_1 = [v_{1l}, v_{1r}], a_2 = [v_{2l}, v_{2r}], \ldots, a_n = [v_{nl}, v_{nr}]\}$, is called a correlated contrast set if, and only if, the following two conditions are satisfied:

$$\forall a_i, a_j \in X, M_I(a_i; a_j) \geq \psi \tag{7}$$

$$\forall a_i, a_j \in X, \forall v_i \in \mathcal{V}(a_i), \forall v_j \in \mathcal{V}(a_j), A_C(a_i : [v_{il}, v_{ir}], a_j : [v_{jl}, v_{jr}]) \geq \xi \tag{8}$$

where $\psi$ is a minimum mutual information threshold, and $\xi$ is a minimum all-confidence threshold. Criterion 7 ensures that each attribute in a correlated contrast set tells a great amount of information about every other attribute. Criterion 8 ensures that each attribute-interval pair is highly correlated with every other attribute-interval pair.

Given two 1-specific contrast sets $X = P : [v_{il}, v_{ir}]$ and $Y = Q : [v_{jl}, v_{jr}]$, we can represent our knowledge of $X$ and $Y$ in the contingency table shown in Table 1. Thus, the mutual information [4] of $X$ and $Y$, is defined as:

$$M_I(X;Y) = log_2\left(\frac{n(X,Y) * N}{(n(X,Y) + n(X,\neg Y) \times (n(X,Y) + n(\neg X,Y)}\right),$$

and the mutual information of $P$ and $Q$ is defined as:

Table 1: Contingency table representing the relationship between $X$ and $Y$

| | $Y$ | $\neg Y$ | |
|---|---|---|---|
| $X$ | $n(X,Y)$ | $n(X,\neg Y)$ | $n(X)$ |
| $\neg X$ | $n(\neg X,Y)$ | $n(\neg X,\neg Y)$ | $n(\neg X)$ |
| | $n(Y)$ | $n(\neg Y)$ | $N$ |

$$M_I(P;Q) = \sum_{i=1}^{n}\sum_{j=1}^{m}\frac{n(X,Y)}{N}\times M_I(X;Y)$$

where $n = |\mathcal{V}(P)|$ and $m = |\mathcal{V}(Q)|$. We can define the all-confidence [4] of $X$ and $Y$ as follows:

$$A_C(X,Y) = \frac{n(X,Y)}{\max\{(n(X,Y)+n(X,\neg Y)),(n(X,Y)+n(\neg X,Y))\}}.$$

Given a database $\mathcal{D}$, a minimum support difference threshold $\varepsilon$, a minimum frequency threshold $\sigma$, a minimum subset support ratio threshold $\kappa$, and a minimum support ratio threshold $\omega$, a minimum mutual information threshold, $\psi$, and a minimum all-confidence threshold, $\xi$, our goal is to find all maximal correlated $\lambda$-contrast sets (i.e., all maximal correlated valid contrast sets that satisfy Equations 5 and 6).

## 4 Ranking Methods

A contrast set mining process typically returns many valid contrast sets. Consequently, measures are needed to rank the relative interestingness of the valid contrast sets prior to presenting them to the end-user. In our previous work we demonstrated the suitability of five measures for ranking contrast sets: distribution difference, unusualness, coverage, lift, and interestingness factor for ranking contrast sets [15] [16].

Probability-based objective measures have been studied extensively as a means for evaluating the generality and reliability of association rules [5] [9] [17]. Contrast set mining is closely related to association rule mining and thus, we propose that these probability-based objective measures can be adapted for the task of ranking contrast sets. In this section, we propose five measures and demonstrate their use in ranking contrast sets.

Here we define the variables used in the ranking methods described in this section. A contrast set, $X$, is represented by a set of association rules, $X \rightarrow G_1, X \rightarrow G_2, \ldots, X \rightarrow G_n$, where $G_1, G_2, \ldots, G_n$ are unique groups. Let $n(X,G_i)$ be the number of instances of $X$ in $G_i$. Let $n(X,\neg G_i)$ be the number of instances of $X$ in groups other than $G_i$ (that is, the number of times $X$ occurs in $G_1,\ldots,G_{i-1}$, $G_{i+1}$, $G_n$). Let $(\neg X,G_i)$ be the number of instances of contrast sets other than $X$ in $G_i$. Let

Table 2: Contingency table for $X \rightarrow G_i$

| | $G_i$ | $\neg G_i$ | $\Sigma$ *Row* |
|---|---|---|---|
| $X$ | $n(X,G_i)$ | $N(X,\neg G_i)$ | $n(X)$ |
| $\neg X$ | $n(\neg X,G_i)$ | $n(\neg X,\neg G_i)$ | $n(\neg X)$ |
| $\Sigma$ *Column* | $n(G_i)$ | $n(\neg G_i)$ | $N$ |

$n(\neg X,\neg G_i)$ be the number of instances of contrast sets other than $X$ in groups other than $G_i$. Let $N$ be the total number of instances.

The values $n(X,G_i), n(X,\neg G_i), n(\neg X,G_i)$, and $n(\neg X,\neg G_i)$ actually correspond to the observed frequencies at the intersection of the rows and columns in a $2 \times 2$ contingency table for the association rule $X \rightarrow G_i$, such as the one shown in Table 2. Rows represent the occurrence of the contrast set and the columns represent occurrence of the groups.

## 4.1 Leverage

The *leverage* (LEV) of an association rule $X \rightarrow G_i$ is the difference between the actual proportion of instances in which both $X$ and $G_i$ occur and the proportion that would be expected if $X$ and $G_i$ were statistically independent of each other [12] and is given by

$$\mathrm{LEV}(X \rightarrow G_i) = p(G_i|X) - P(X)P(G_i) = \frac{n(X,G_i)}{n(X)} - \frac{n(X)}{N} \times \frac{n(G_i)}{N}.$$

The values for LEV occur on the interval [-1, 1]. Values near one indicate strong independence between $X$ and $G_i$.

The leverage for a contrast set, $X$, is determined by the group $G_i$, for which the leverage is largest. Thus, the leverage of $X$ is given by

$$\mathrm{LEV}(X) = \max_i \mathrm{LEV}(X \rightarrow G_i).$$

## 4.2 Change of support

The *change of support* (CS) of of an association rule $X \rightarrow G_i$ measures the correlation between $X$ and $G_i$ [17] and is given by

$$\mathrm{CS}(X \rightarrow G_i) = p(G_i|X) - p(G_i) = \frac{n(X,G_i)}{n(X)} - \frac{n(G_i)}{N}.$$

The values for change of support occur in the interval [-0.5, 1]. Values near 1 indicate strong correlation between $X$ and $G_i$.

The CS for a contrast set, $X$, is determined by the group, $G_i$, for which change of support is largest, and is given by

$$\mathrm{CS}(X) = \max_i \mathrm{CS}(X \to G_i).$$

## 4.3 Yule's Q coefficient

The *Yule's Q coefficient* (YQ) of an association rule $X \to G_i$ measures the degree to which $X$ and $G_i$ are associated with each other [20] and is given by

$$\begin{aligned}\mathrm{YQ}(X \to G_i) &= \frac{p(X,G_i)p(\neg X \neg G_i) - p(X,\neg G_i)p(\neg X,G_i)}{p(X,G_i)p(\neg X \neg G_i) + p(X,\neg G_i)p(\neg X,G_i)} \\ &= \frac{n(X,G_i)n(\neg X \neg G_i) - n(X,\neg G_i)n(\neg X,G_i)}{n(X,G_i)n(\neg X \neg G_i) + n(X,\neg G_i)p(\neg X,G_i)}.\end{aligned}$$

The values for Yule's Q coefficient occur on the interval [-1,1]. Values near 1 indicate strong association between $X$ and $G_i$. Values near -1 indicate complete dissociation between $X$ and $G_i$. 0 indicates complete independence between $X$ and $G_i$.

The Yule's Q coefficient for a contrast set, $X$, is determined by the group, $G_i$, for which Yule's Q coefficient is largest. Thus, the Yule's Q coefficient of $X$ is given by

$$\mathrm{YQ}(X) = \max_i \mathrm{YQ}(X \to G_i).$$

## 4.4 Conviction

The *conviction* (CONV) of an association rule $X \to G_i$ measures true implication between $X$ and $G_i$ [3] and is given by

$$\mathrm{CONV}(X \to G_i) = \frac{p(X)p(\neg G_i)}{p(X,\neg G_i)} = \frac{n(X) \times n(\neg G_i)}{N \times n(X,\neg G_i)}$$

The values for conviction occur in the interval $[1, \infty]$. A value equal to 1 indicates $X$ and $G_i$ are completely unrelated. Values further from 1 indicate that $X$ and $G_i$ are more strongly related.

The conviction of a contrast set for all groups is the sum of the individual conviction values for the contrast set in each group. Thus the conviction of a contrast set $X$ is given by

$$\mathrm{CONV}(X) = \sum_i \mathrm{CONV}(X \to G_i).$$

# 5 Mining Interesting Valid Correlated Contrast Sets

Given a database $\mathcal{D}$, a minimum support difference threshold $\varepsilon$, a minimum frequency threshold $\sigma$, a minimum subset support ratio threshold $\kappa$, and a minimum support ratio threshold $\omega$, a minimum mutual information threshold, $\psi$, a minimum all-confidence threshold, $\xi$, and a ranking measure, $m$, the GENICCS (Generate Interesting Correlated Contrast Sets) algorithm, shown in Algorithm 1, traverses the search space to find valid contrast sets that satisfy Criteria 1, 2, 3, 4, 5, 7, and 8.

**Algorithm 1 GENICCS**$(\mathcal{D},\varepsilon,\sigma,\kappa,\omega,\psi,\xi,m)$

**Input:** Dataset $\mathcal{D}$, and parameters $\varepsilon,\sigma,\kappa,\omega,\psi,\xi$, and $m$
**Output:** Set of ranked correlated valid contrast sets $W$

1: $L_0$ =CREATE_CONTRAST_SETS($\mathcal{D}$)
2: $L_1$ =COMBINE($L_0,W,\varepsilon,\sigma,0,\omega,\psi,\xi,1$)
3: TRAVERSE($L_1,W,\varepsilon,\sigma,\kappa,\omega,\psi,\xi,1$)
4: RANK($W,m$)
5: **return** $W$

First, GENICCS creates all 1-specific contrast sets from the domain values of each attribute in the dataset, then stores them in $L_0$ (line 1). Quantitative attributes are discretized using the discretization algorithm previously described in [14]. Second, GENICCS calls the COMBINE subroutine to determine which of the 1-specific contrast sets in $L_0$ are valid, storing them in $L_1$ (line 2). Third, GENICCS calls the TRAVERSE subroutine to find correlated valid $k$-specific contrast sets using a modified generate-and-test approach, storing them in $W$ (line 3). Fourth, GENICCS calls the RANK subroutine to order the correlated valid contrast sets in $W$ according to a ranking measure $m$ (line 4). Finally, GENICCS returns the set of ranked correlated valid contrast sets in $W$ (line 5).

## 5.1 COMBINE

From a parent node at the current level in the search space, the COMBINE algorithm, shown in Algorithm 2, generates the set of child nodes at the next level of the search space whose elements satisfy Criteria 1, 2, 3, 4, 5, 7 and 8 .

COMBINE begins by combining prefix set Parent.P with each element $c$ of the combine set Parent.C, in turn, to create new contrast set, Child.P (line 2). COMBINE then determines whether Child.P satisfies Criteria 1, 2, 3, and 4 (line 3). COMBINE checks whether Child.P satisfies Criteria 6 (line 4). If so, it is potentially maximal, and is added to $W$ if it has no superset already in $W$ (lines 5 and 6). Otherwise, if Child.P satisfies Equation 5 (line 8), COMBINE adds Child to Children (line 9). After all combine sets in Parent.C have been considered, COMBINE updates the combine sets of each Child in Children so that they contain only valid contrast sets

**Algorithm 2 COMBINE**$(\text{Parent}, W, \varepsilon, \sigma, \kappa, \omega, \psi, \xi, l)$

**Input:** Node from which child nodes are generated, Parent
**Output:** Children, nodes at the next level of the search space and $W$.

```
 1: for each c ∈ Parent.C do
 2:    Child.P = Parent.P ∪ {c}
 3:    if IS_VALID(Child.P, ε, σ, κ, l) then
 4:       if λ(Child.P) == ∞ then
 5:          if ∄Y ∈ W : Child.P ⊆ Y then
 6:             W = W ∪ {Child.P}
 7:          end if
 8:       else if λ(Child.P) ≥ ω then
 9:          Children = Children ∪ {Child.P}
10:       end if
11:    end if
12: end for
13: UPDATE_COMBINE_SET(Parent.P, Children)
14: Sort Children
15: return Children
```

(line 13). At the first level of the search tree, COMBINE also ensures that these valid contrast sets when combined with Child.P satisfies Criteria 7 and 8. COMBINE then re-orders the contrast sets in descending order by the cardinality of the combine set, then in ascending order by support (line 14). Finally, COMBINE returns Children, the child node at the next level of the search space (line 15).

## *5.2 TRAVERSE*

The TRAVERSE algorithm, shown in Algorithm 3, traverses the search space to find contrast sets that satisfy Criteria 1, 2, 3, 4, 5, 7, and 8.

TRAVERSE returns to the previous invocation of the subroutine, pruning away the subtree of Parent.P if it is subsumed by a contrast set in $W_l$ (lines 2 and 3). Otherwise, TRAVERSE calls the COMBINE subroutine to generate the set of child nodes for the next level of the search space, $L_{l+1}$ (line 5). If $L_{l+1}$ is not empty (line 6), then TRAVERSE then calls itself to explore the next level of the search space with a subset of $W_l$, $W_{l+1}$, that contains only contrast sets from $W_l$ which can potentially subsume Parent.P (lines 7 and 8). After recursion completes, $W_l$, is updated with the elements from $W_{l+1}$ (line 9). Otherwise, if the prefix set Parent.P is not a subset of any contrast set in $W_l$ (line 10), TRAVERSE adds Parent.P to $W_l$ (line 11). Finally, TRAVERSE returns to the invocation of the subroutine from which it was called (line 14).

**Algorithm 3** **TRAVERSE**($L_l, W_l, \varepsilon, \sigma, \kappa, \omega, \psi, \xi, l$)

**Input:** Set of Parent nodes $L_l$
**Output:** $W_l$, updated with valid correlated contrast sets

1: **for** each Parent $\in L_l$ **do**
2: **if** $\exists Y \in W_l$ : Parent.P $\subseteq Y$ **then**
3: **return**
4: **end if**
5: $L_{l+1} = \text{COMBINE}(\text{Parent}, W_l, \varepsilon, \sigma, \kappa, \omega, \psi, \xi, l+1)$
6: **if** $|L_{l+1}| > 0$ **then**
7: $W_{l+1} = \{w \in W_l : \text{Parent.P} \in w\}$
8: $\text{TRAVERSE}(L_{l+1}, W_{l+1}, \varepsilon, \sigma, \kappa, \omega, \psi, \xi, l+1)$
9: $W_l = W_l \cup W_{l+1}$
10: **else if** $\nexists Y \in W_l$ : Parent.P $\subseteq Y$ **then**
11: $W_l = W_l \cup$ Parent.P
12: **end if**
13: **end for**
14: **return**

Table 3: A Sample Transaction Dataset

| *TID* | A | B | C | D | E | *Frequency* |
|---|---|---|---|---|---|---|
| $T_1$ | 1 | 1 | 1 | 20 | 0 | 10 |
| $T_2$ | 0 | 0 | 1 | 22 | 1 | 10 |
| $T_3$ | 1 | 0 | 1 | 25 | 1 | 10 |
| $T_4$ | 1 | 1 | 1 | 21 | 0 | 10 |

| *TID* | A | B | C | D | E | *Frequency* |
|---|---|---|---|---|---|---|
| $T_5$ | 0 | 1 | 1 | 23 | 1 | 10 |
| $T_6$ | 0 | 0 | 1 | 20 | 0 | 10 |
| $T_7$ | 0 | 1 | 1 | 26 | 1 | 10 |

| *TID* | A | B | C | D | E | *Frequency* |
|---|---|---|---|---|---|---|
| $T_8$ | 1 | 1 | 1 | 26 | 0 | 10 |
| $T_9$ | 1 | 1 | 1 | 21 | 0 | 10 |
| $T_{10}$ | 0 | 0 | 1 | 25 | 1 | 10 |

## *5.3 GENICCS Example*

GENCCS is initially called with parameters $\mathcal{D}$ = dataset in Table 3, $\mathcal{F}$ = attribute $E$, $\varepsilon = 0$, $\sigma = 0.1$, $\kappa = 0.0001$, $\omega = 0$, $\psi = 0.01$, $\xi = 0.25$, and $m$ = Conviction.
**GENICCS - Invocation:** GENICCS creates 8 1-specific contrast sets, $\{\texttt{A}:[0,0]$, $\texttt{A}:[1,1]$, $\texttt{B}:[0,0]$, $\texttt{B}:[1,1]$, $\texttt{C}:[0,0]$, $\texttt{C}:[1,1]$, $\texttt{D}:[20,22.9)\}$, and $\{\texttt{D}:[22.9,26]\}$ and stores them in $L_0$. GENICCS calls COMBINE with $L_0$, $W$, 0, 0.1, 0.0001, 0, and 1. $\texttt{C}:[1,1]$ fails to satisfy Criteria 1, $\{\texttt{C}:[0,0]\}$ fails to satisfy Criteria 3, and the remaining contrast sets satisfy Criteria 5, thus $L_1 = \{\texttt{A}:[0,0]$, $\texttt{A}:[1,1]$, $\texttt{B}:[0,0]$, $\texttt{B}:[1,1]$, $\texttt{D}:[20,22.9)$, $\texttt{D}:[22.9,26]\}$. We find that only $\{\texttt{A}:[0,0], \texttt{B}:[0,0]\}$, $\{\texttt{A}:[0,0]$, $\texttt{B}:[1,1]\}$, and $\{\texttt{A}:[1,1]$, $\texttt{B}:[1,1]\}$ satisfy Criteria 7 and 8. Thus, $\texttt{B}:[0,0]$, $\texttt{B}:[1,1]$, $\texttt{D}:[20,22.9)$, and $\texttt{D}:[22.9,26]$ have empty combine sets, $\texttt{A}:[0,0]$ has a combine set of $\{\texttt{B}:[0,0]$, $\texttt{B}:[1,1]\}$ and $\texttt{A}:[1,1]$ has a combine set of $\{\texttt{B}:[1,1]\}$. The frequencies of the contrast sets are $\{50,50,40,60,50,50\}$ and cardinalities of the combine sets are $\{2,1,0,0,0,0\}$, respectively of each element in $L_1$. Thus, $L_1$ keeps its existing order. We call TRAVERSE with $L_1$, $W_1$, 0, 0.1, 0.0001, 0, 0.01, 0.25, Conviction, and 1 (line 4).
**TRAVERSE - Invocation 1 - Iteration 1:** $W_1$ is empty so we call COMBINE with $\{\texttt{A}:[0,0]\}$, $W_1$, 0, 0.1, 0.0001, 0, and 2. $\{\texttt{A}:[0,0]$, $\texttt{B}:[1,1]\}$ satisfies Criteria 6. Thus it is potentially maximal and since $W_1$ is empty, we add it to $W_1$. $\{\texttt{A}:[0,0]$,

Table 4: Dataset Description

| Data Set | # Transactions | # Attributes | # Groups | Cardinality of Longest Combine Set | |
|---|---|---|---|---|---|
| | | | | Before MI&AC | After MI&AC |
| Census | 32561 | 14 | 5 | 145 | 16 |
| Mushroom | 8124 | 22 | 2 | 198 | 35 |
| Spambase | 4601 | 58 | 2 | 560 | 75 |
| Waveform | 5000 | 41 | 3 | 390 | 80 |

$\mathtt{B}:[0,0]\}$ satisfies Criteria 5, so we add it to $L_2$. Thus $L_2 = \{\{\mathtt{A}:[0,0], \mathtt{B}:[0,0]\}\}$. We update the combine set of the element in $L_2$. Since the current element in $W_1$, $\{\mathtt{A}:[0,0], \mathtt{B}:[1,1]\}$ cannot potentially be a superset of $\{\mathtt{A}:[0,0], \mathtt{B}:[0,0]\}$, $W_2$ is empty. We then call TRAVERSE with $L_2$, $W_2$, 0, 0.1, 0.0001, 0, Conviction, and 2. **TRAVERSE - Invocation 2:** $W_2$ is empty so we call the COMBINE subroutine with $\{\mathtt{A}:[0,0], \mathtt{B}:[0,0]\}$, $W_2$, 0, 0.1, 0.0001, 0, and 3, but $L_3$ is empty because $\{\mathtt{A}:[0,0], \mathtt{B}:[0,0]\}$ has an empty combine set. Since $W_2$ is empty, we add $\{\mathtt{A}:[0,0], \mathtt{B}:[0,0]\}$ to $W_2$ and this iteration of TRAVERSE is complete. $L_2$ has only one element, so this also ends the second invocation of TRAVERSE. GENICCS returns to Invocation 1 - Iteration 1 of TRAVERSE. It updates $W_1$ with the element, $\{\mathtt{A}:[0,0], \mathtt{B}:[0,0]\}$, from $W_2$.

## 6 Experimental Results

In this section, we present the results of an experimental evaluation of our approach. Our experiments were conducted on an Intel dual core 2.40GHz processor with 4GB of memory, running Windows 7 64-bit. Discovery tasks were performed on four real datasets obtained from the UCI Machine Learning Repository [1]. Table 4 lists the name, the number of transactions, the number of attributes, the number of groups for each dataset, and the cardinality of the longest combine set before and after mutual information and all confidence were applied. These datasets were chosen because of their use with previous contrast set mining techniques and the ability to mine valid contrast sets with high specificity.

Our experiments evaluate whether the use of mutual information and all confidence in limiting the size of the search space, and the use of the minimum support ratio threshold (MSRT) during the search process produces a smaller number of "more interesting" contrast sets in less time, than using either method separately. Thus we compare GENICCS with GENCCS, and GIVE. We do not use STUCCO and CIGAR for comparison, as they do not incorporate either mutual information, all confidence, or the MSRT. Additionally, we have previously demonstrated the efficiency and effectiveness of GENCCS over STUCCO and CIGAR [15].

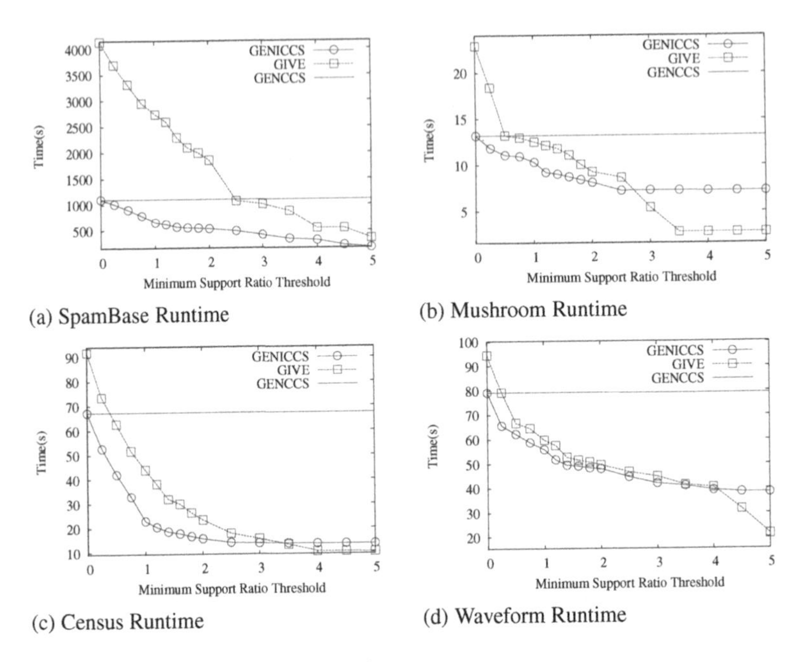


(a) SpamBase Runtime

(b) Mushroom Runtime

(c) Census Runtime

(d) Waveform Runtime

Fig. 1: Summary of runtime results

## 6.1 Performance of GENICCS

We first examine the efficiency of GENICCS by measuring the time taken to complete the contrast set mining task as the MSRT varies. For each of GENICCS, GENCCS, and, GIVE we set the significance level to 0.95, the minimum support difference threshold and minimum subset ratio threshold to 0, respectively, and averaged the results over 10 runs. Figure 1 shows the number of valid contrast sets discovered and the run time for each of the datasets as the MSRT is varied. For both GENICCS and GENCCS, we set $\psi$ and $\xi$ to be the mean mutual information, and mean all confidence values, respectively. For GENCCS, these mutual information and all confidence values were shown previously to be optimal [15].

Figure 1 shows that GENICCS has the same run time as GENCCS when the MSRT is 0, but decreases significantly as the MSRT is increased. Since the MSRT serves as a constraint, as we increase its value, fewer correlated contrast sets satisfy this constraint and GENICCS becomes more efficient than GENCCS. As the MSRT approaches 3, 3.5, and 4 for the Mushroom, Census, and Waveform datasets, respectively, the cost of using the mutual information and all confidence begins to outweigh the benefits of using the MSRT as GIVE begins to outperform GENICCS. GIVE never outperforms GENICCS on the spambase dataset.

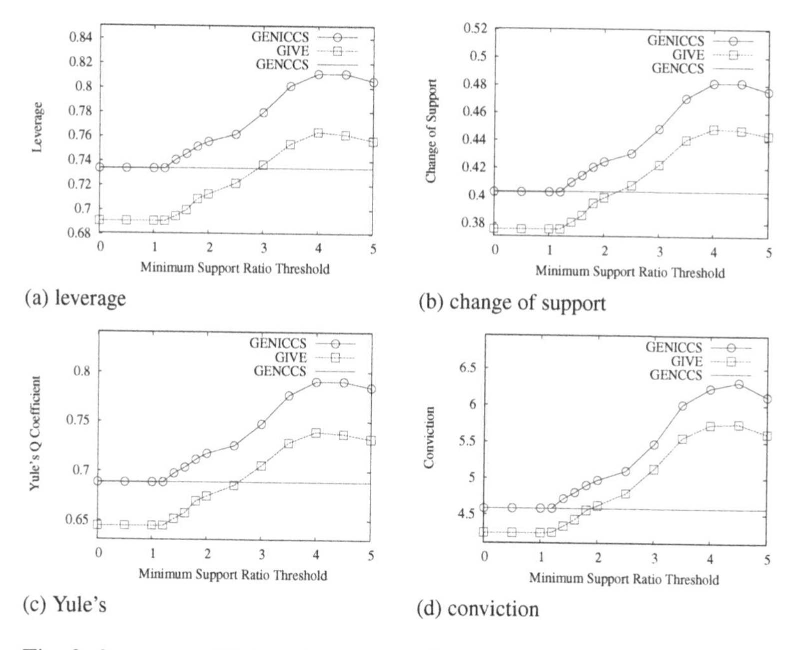


(a) leverage (b) change of support

(c) Yule's (d) conviction

Fig. 2: Summary of interestingness results

## *6.2 Interestingness of the Contrast Sets*

We examine the effectiveness of GENICCS by measuring the average leverage, change of support, yule's Q, and conviction for the valid contrast sets discovered as the MSRT varies, as shown in Figure 2. We focus on the Waveform dataset as it had not only the smallest run time difference between GENICCS and GIVE, 17%, but also the largest combine set, 80, after mutual information and all confidence were applied. The average leverage, change of support, yule's Q coefficient, and conviction of the valid contrast sets discovered by GIVE and GENCCS for each dataset is also provided for comparison.

Figure 2 shows that the maximal contrast sets discovered by GENICCS are more interesting, when measured by the average leverage, change of support, yule's Q coefficient, and conviction, than those discovered by either GIVE or GENCCS. The magnitude of the difference is significant even at higher MSRT values where GIVE outperforms GENICCS as shown in Figure 1d, which implies that even though GIVE is less expensive, GENICCS produces better quality contrast sets.

## 7 Conclusion

In this paper, we proposed a contrast set mining technique, GENICCS, for mining maximal valid correlated contrast sets that meet a minimum support ratio threshold.

We compared our approach with two previous contrast set mining approaches, GIVE and GENCCS, and found our approach to be comparable in terms of efficiency but more effective in generating interesting correlated contrast sets. We also introduced four probability-based interestingness measures and demonstrated how they can be used to rank correlated contrast sets. Future work will further incorporate space reduction techniques with additional interestingness measures.

## References


1. A. Asuncion and D.J. Newman. UCI machine learning repository, 2007.
2. Stephen D. Bay and Michael J. Pazzani. Detecting group differences: Mining contrast sets. *Data Min. Knowl. Discov.*, 5(3):213–246, 2001.
3. Sergey Brin, Rajeev Motwani, Jeffrey D. Ullman, and Shalom Tsur. Dynamic itemset counting and implication rules for market basket data. *SIGMOD Rec.*, 26(2):255–264, 1997.
4. Thomas M. Cover and Joy A. Thomas. *Elements of information theory*. Wiley-Interscience, New York, NY, USA, 2006.
5. Liqiang Geng and Howard J. Hamilton. Interestingness measures for data mining: A survey. *ACM Comput. Surv.*, 38(3):9, 2006.
6. R.J. Hilderman and T. Peckham. A statistically sound alternative approach to mining contrast sets. *Proceedings of the 4th Australasian Data Mining Conference (AusDM'05)*, pages 157–172, Dec. 2005.
7. S Holm. A simple sequentially rejective multiple test procedure. *Scandinavian Journal of Statistics*, 6:65–70, 1979.
8. Petra Kralj, Nada Lavrac, Dragan Gamberger, and Antonija Krstacic. Contrast set mining for distinguishing between similar diseases. In *AIME*, pages 109–118, 2007.
9. Nada Lavrac, Peter A. Flach, and Blaz Zupan. Rule evaluation measures: A unifying view. In *ILP*, pages 174–185, 1999.
10. Jessica Lin and Eamonn J. Keogh. Group sax: Extending the notion of contrast sets to time series and multimedia data. In *PKDD*, pages 284–296, 2006.
11. Zohreh Nazeri, Daniel Barbará, Kenneth A. De Jong, George Donohue, and Lance Sherry. Contrast-set mining of aircraft accidents and incidents. In *ICDM*, pages 313–322, 2008.
12. Gregory Piatetsky-Shapiro. Discovery, analysis, and presentation of strong rules. In *Knowledge Discovery in Databases*, pages 229–248. AAAI/MIT Press, 1991.
13. Mondelle Simeon and Robert J. Hilderman. Exploratory quantitative contrast set mining: A discretization approach. In *ICTAI (2)*, pages 124–131, 2007.
14. Mondelle Simeon and Robert J. Hilderman. COSINE: A Vertical Group Difference Approach to Contrast Set Mining. In *Canadian Conference on AI*, pages 359–371, 2011.
15. Mondelle Simeon and Robert J. Hilderman. GENCCS: A Correlated Group Difference Approach to Contrast Set Mining. In *MLDM*, pages 140–154, 2011.
16. Mondelle Simeon, Robert J. Hilderman, and Howard J. Hamilton. Mining interesting contrast sets. In *INTENSIVE 2012*, pages 14–21, 2012.
17. Pang-Ning Tan, Vipin Kumar, and Jaideep Srivastava. Selecting the right objective measure for association analysis. *Inf. Syst.*, 29(4):293–313, 2004.
18. Tzu-Tsung Wong and Kuo-Lung Tseng. Mining negative contrast sets from data with discrete attributes. *Expert Syst. Appl.*, 29(2):401–407, 2005.
19. Masaharu Yoshioka. Analyzing multiple news sites by contrasting articles. In *SITIS '08*, pages 45–51, Washington, DC, USA, 2008. IEEE Computer Society.
20. G. Udny Yule. On the association of attributes in statistics: With illustrations from the material of the childhood society, &c. *Philosophical Transactions of the Royal Society of London. Series A, Containing Papers of a Mathematical or Physical Character*, 194(252-261):257–319, 1900.

# DATA MINING AND MACHINE LEARNING

# eRules: A Modular Adaptive Classification Rule Learning Algorithm for Data Streams

Frederic Stahl, Mohamed Medhat Gaber and Manuel Martin Salvador

**Abstract** Advances in hardware and software in the past decade allow to capture, record and process fast data streams at a large scale. The research area of data stream mining has emerged as a consequence from these advances in order to cope with the real time analysis of potentially large and changing data streams. Examples of data streams include Google searches, credit card transactions, telemetric data and data of continuous chemical production processes. In some cases the data can be processed in batches by traditional data mining approaches. However, in some applications it is required to analyse the data in real time as soon as it is being captured. Such cases are for example if the data stream is infinite, fast changing, or simply too large in size to be stored. One of the most important data mining techniques on data streams is classification. This involves training the classifier on the data stream in real time and adapting it to concept drifts. Most data stream classifiers are based on decision trees. However, it is well known in the data mining community that there is no single optimal algorithm. An algorithm may work well on one or several datasets but badly on others. This paper introduces *eRules*, a new rule based adaptive classifier for data streams, based on an **e**volving set of **Rules**. *eRules* induces a set of rules that is constantly evaluated and adapted to changes in the data stream by adding new and removing old rules. It is different from the more popular decision tree based classifiers as it tends to leave data instances rather unclassified than forcing a classification that could be wrong. The ongoing development of eRules aims

---

Frederic Stahl
Bournemouth University, School of Design, Engineering and Computing, Poole House, Talbot Campus,BH12 5BB e-mail: fstahl@bournemouth.ac.uk

Mohamed Medhat Gaber
University of Portsmouth, School of Computing, Buckingham Building, Lion Terrace, PO1 3HE e-mail: Mohamed.Gaber@port.ac.uk

Manuel Martin Salvador
Bournemouth University, School of Design, Engineering and Computing, Poole House, Talbot Campus,BH12 5BB e-mail: msalvador@bournemouth.ac.uk

to improve its accuracy further through dynamic parameter setting which will also address the problem of changing feature domain values

## 1 Introduction

According to [13] and [2] streaming data is defined as high speed data instances that challenge our computational capabilities for traditional processing. It has greatly contributed to the recent problem of analysis of *Big Data*. Analysing these data streams can produce an important source of information for decision making in real time. In the past decade, many data stream mining techniques have been proposed, for a recent review and categorisation of data stream mining techniques the reader is referred to [12]. In this paper, we have used the sliding window [2, 9] to extend the rule-based technique *PRISM* [8] to function in the streaming environment. Motivated by the simplicity of *PRISM* and its explanatory power being a rule-based technique, we have developed and experimentally validated a novel data stream classification technique, termed *eRules*. The new technique is also able to adapt to concept drift, which is a well known problem in data stream classification. A concept drift occurs when the current data mining model is no longer valid, because of the change in the distribution of the streaming data. Our *eRules* classifier works on developing an initial batch model from a stored set of the data streams. This process is followed by an incremental approach for updating the model to concept drift. The model is reconstructed if the current model is unable to classify a high percentage of the incoming instances in the stream. This in fact may be due to a big concept drift. Although this may appear to be a drawback of our technique, it is one of its important strengths, as many other techniques fail to adapt to big drift in the concept when only an incremental update is used. The aim of this work is to contribute an alternative method for the data stream classification, as it is a well-known observation that established algorithms may work well on one or several datasets, but may fail to provide high performance on others. In fact eRules is being considered by the *INFER* project [1] to be included into its predictive methods toolbox. The INFER software is an evolving data mining system that is able to handle data streams. INFER uses a pool of possible data mining and pre-processing methods including data stream classifiers and evolves in order to find the optimal combination and parameter setting for the current prediction task [16]. eRules in its current state only allows static parameter settings and does not take the problem of changing domain values into account. However, ongoing developments address these issues.

The paper is organised as follows. Section 2 highlights related work and Section 3 gives the background on learning modular rules represented in this paper by the *Prism* algorithm. Our extension of the *Prism* algorithm enabling it to function in the streaming environment, developing what we have termed as *eRules* technique, is detailed in Section 4. We have experimentally validated our proposed technique in Section 5. In Section 6, we discuss our ongoing and future work related to the *eRules* technique. Finally, the paper is concluded by a summary in Section 7.

## 2 Related Work

Several approaches to adapt existing batch learning data mining techniques to data streams exist, such as reservoir sampling [15] and sliding window [2, 9]. The basic idea of reservoir sampling is to maintain a representative unbiased sample of the data stream without prior knowledge of the total number of data stream instances. Data mining techniques are then applied on the reservoir sample. The basic idea of sliding window is to only use recent data instances from the stream to build the data mining models. Among the many techniques proposed for data stream mining, a notable group of these algorithms are the *Hoeffding bound* based techniques [11]. The *Hoeffding bound* is a statistical upper bound on the probability that the sum of random variables deviates from its expected value. The basic *Hoeffding bound* has been extended and adopted in successfully developing a number of classification and clustering techniques that were termed collectively as *Very Fast Machine Learning* (VFML). Despite its notable success, it faces a real problem when constructing the classification model. If there are ties of attributes and the *Hoeffding bound* is not satisfied, the algorithm tends to increase the sample size. This may not be desirable in a highly dynamic environment. The more flexible approach in addressing this problem is using the sliding window.

## 3 Learning Modular Classification Rules

There are two main approaches to the representation of classification rules, the 'separate and conquer' approach, and the 'divide and conquer' approach, which is well known as the top down induction of decision trees. 'Divide and conquer' induces rules in the intermediate form of decision trees such as the C4.5 classifier [18], and'separate and conquer' induces IF...THEN rules directly from the training data. Cendrowska claims that decision tree induction grows needlessly large and complex trees and proposes the Prism algorithm, a 'separate and conquer' algorithm that can induce modular rules that do not necessarily fit into a decision tree [8] such as for example the two rules below:

$$IF\ a = 1\ and\ b = 2\ THEN\ class = x$$
$$IF\ a = 1\ and\ d = 3\ THEN\ class = x$$

We propose to base the rule induction of eRules on the Prism algorithm for several reasons. The classifier generated by Prism does not necessarily cover all possible data instances whereas decision tree based classifiers tend to force a classification. In certain applications, leaving a data instance unclassified rather than risking a false classification may be desired, for example, in medical applications or the control of production processes in chemical plants. Also Prism tends to achieve a better classification accuracy compared with decision tree based classifiers, if there is considerable noise in the data [5]. Algorithm 1 summarises the basic Prism approach for continuous data, assuming that there are $n$ $(> 1)$ possible classes [20]. $A$ denotes an attribute in the training dataset.

**Algorithm 1** Prism algorithm

```
1:  for i = 1 → n do
2:      W ← new Dataset
3:      delete all records that match the rules that have been derived so far for class i.
4:      for all A ∈ W do
5:          sort the data according to A
6:          for each possible split value v of attribute A do
7:              calculate the probability that the class is i
8:              for both subsets A < v and A >= v.
9:          end for
10:     end for
11:     Select the attribute that has the subset S with the overall highest probability.
12:     Build a rule term describing S.
13:     W ← S
14:     repeat lines 4 to 13
15:     until the dataset contains only records of class i.
16:                 ▷ The induced rule is then the conjunction of all the rule terms built at line= 12.
17:     repeat lines 2 to 16.
18:     until all records of class i have been removed.
19: end for
```

Even so Prism has been shown to be less vulnerable to overfitting compared with decision trees, it is not immune. Hence pruning methods for Prism have been developed such as J-pruning [6] in order to make Prism generalising better on the input data. Because of J-pruning's generalisation capabilities we use Prism with J-pruning for the induction of rules in eRules hence it is briefly described here. J-pruning is based on the J-measure which according to Smyth and Goodman [19] is the average information content of a rule *IF* $Y = y$ *THEN* $X = x$ and can be quantified as:

$$J(X;Y = y) = p(y) \cdot j(X;Y = y) \quad (1)$$

The first factor of the J-measure is $p(y)$, which is the probability that the antecedent of the rule will occur. The second factor is the cross entropy $j(X;Y{=}y)$, which measures the goodness-of-fit of a rule and is defined by:

$$j(X;Y = y) = p(x \mid y) \cdot log_2(\frac{p(x \mid y)}{p(x)}) + (1 - p(x \mid y)) \cdot log_2(\frac{(1 - p(x \mid y))}{(1 - p(x))}) \quad (2)$$

The basic interpretation of the J-value is that a rule with a high J-value also achieves a high predictive accuracy. If inducing further rule terms for the current rule results in a higher J-value then the rule term is appended as it is expected to increase the current rule's predictive accuracy; and if a rule term decreases the current rule's J-value then the rule term is discarded and the rule is considered as being in its final form. That is because appending the rule term is expected to decrease the current rule's predictive accuracy.

Having discussed the Prism algorithm for rule induction and the J-pruning for avoiding the model overfitting problem, the following section is devoted for a detailed discussion of our proposed technique *eRules* which is based on the concepts presented in this section.

# 4 Incremental Induction of Prism Classification Rules with eRules

Prism is a batch learning algorithm that requires having the entire training data available. However, often we do not have the full training data at hand. In fact the data is generated in real time and needs to be analysed in real time as well. So we can only look at a data instance once. It is one of the data stream mining techniques that is to be sublinear in the number of instances. This implies that the algorithm may only have one look at any data instance, or even some of the instances are discarded all together.

Data mining algorithms that incrementally generate classifiers have been developed, the most notable being Hoeffding Trees [10]. Inducing a Hoeffding Tree is a two stage process, first the basic tree is induced and second, once the tree is induced, it is updated periodically in order to adapt to possible concept drifts in the data. A concept drift is if the pattern in the data stream changes. However, incremental classifiers for data streams are based almost exclusively on decision trees, even though it has been shown that general classification rule induction algorithms such as Prism are more robust in many cases as mentioned in Section 3. Hence the development of an incremental version of Prism, *eRules*, may well produce better classification accuracy in certain cases.

Algorithm 2 describes the basic *eRules* approach. The basic algorithm consists of three processes. The first process learns a classifier in batch mode on a subset of the data stream, this is implemented in the very first stage when the algorithm is executed. This is done in order to learn the initial classifier. This process may also be repeated at any stage when the 'unclassification rate' is too high. The unclassification rate is discussed in 4.1. The second process adds new rules to the classifier in order to adapt to a concept drift, this is implemented in the ***if*** *inst is NOT covered by rules* statement in Algorithm 2. The third process validates whether the existing rules still achieve an acceptable classification accuracy and removes rules if necessary. This is implemented in the ***else if*** *inst wrongly classified* statement in

Algorithm 2. Removing rules aims to 'unlearn'/forget old concepts that are not valid anymore and adding rules aims to learn a newly appearing concept.

**Algorithm 2** eRules algorithm

```
1:  ruleset ← new RuleSet.
2:  buffer ← new DataBuffer.
3:  LEARNBATCHCLASSIFIER( ).
4:  while more instances available do
5:      inst ← take next data instance.
6:      if inst is NOT covered by rules then
7:          BUFFER.ADD(inst).
8:          if buffer fulfils Hoeffding Bound OR alternative metric then
9:              learn a new set of rules from buffer and add them to ruleset.
10:             BUFFER.CLEAR( ).
11:         end if
12:     else if inst wrongly classified then
13:         update rule information and delete rule if necessary.
14:     else if inst is correctly classified then
15:         update rule information.
16:     end if
17:     if unclassification rate too high then
18:         LEARNBATCHCLASSIFIER( ).
19:     end if
20: end while
21: procedure LEARNBATCHCLASSIFIER( )
22:     ruleset ← train classifier on n first incoming data instances.
23:     buffer ← new DataBuffer.
24: end procedure
```

The algorithm starts with the first process, collecting the first incoming data instances from the stream and induces an initial rule set in batch mode using the Prism algorithm. The classifier is reset and a new batch classifier is trained as soon as too many incoming test instances remain unclassified. How many data instances are used for the batch learning and how the unclassification rate is measured is discussed in Section 4.1. eRules uses each incoming labelled data instance in order to validate the existing rules and to update the classifier if necessary. If the incoming data instance is not covered by any of the rules, then the second process is invoked, i.e., the unclassified instance is added to a buffer containing the latest unclassified instances. If there are enough instances in this buffer then new rules are generated and added to the rule set, using the instances in the buffer. How one can measure if there are enough instances in the buffer is discussed in Section 4.2. In contrary, if the incoming data instance is covered by one of the rules then two cases are possible. The rule predicts the class label correctly and the concerning rule's accuracy is updated. In the second case the rule missclassifies the data instance, again the concerning rules classification accuracy is updated, and then removed if the accuracy is below a certain threshold. Section 4.3 discusses how such a threshold could be defined.

We have given an overview of our *eRules* technique in this section. The following subsections provide further details with respect to each of the three processes of *eRules*.

## 4.1 First Process: Learn Rules in Batch Mode

At the start of eRules' execution, or if the unclassification rate is too high, a set of rules is learned in batch mode using the Prism classifier. At the moment the eRules implementation allows the user to specify the number of data instances that are used to train the classifier in batch mode. By default the number of data instances is 50 as it seems to work well for all the experiments we conducted including those presented in this paper. However more dynamic methods could and will be used in future implementations, such as the Hoeffding bound, which is also used in the Hoeffding tree classifier [10] and other Hoeffding bound machine learning methods. The Hoeffding bound highlighted in equation (3) is used with different variations as an upper bound for the loss of accuracy dependent on the number of instances in each iteration of a machine learning algorithm. In equation (3) $\varepsilon$ is the value of the Hoeffding bound for an observed feature of range $R$ where $n$ is the number of independent observations.

$$\varepsilon = \sqrt{\frac{R^2 ln(\frac{1}{\delta})}{2n}} \tag{3}$$

The Hoeffding bound could be used in eRules to determine the optimal size of the batch of instances that minimises the error rate. This will be investigated by the authors for our future work in this area.

A second issue that remains to be discussed for the first process is how to determine the unclassification rate. A window of the most recent stream data instances is defined by the user, for which the current classifier's accuracy and its unclassification rate are defined. By default this window size is 50, however, it can be specified by the user. If there are for example 10 unclassified instances then the unclassification rate would be $\frac{10}{50}$. More dynamic methods to define the window size are currently being investigated.

## 4.2 Second Process: Adding New Rules

As discussed earlier in this section, the second process adds new rules to the rule set if there are 'enough' unclassified instances. If there are enough unclassified instances, then eRules will learn new rules only on the unclassified data instances using the Prism classifier. It will then add the new rules to the existing rule set. A window of unclassified instances with a certain size of the most recent unclassified data instances is defined in order to determine the current unclassification rate of the classifier. The oldest unclassified instance is always replaced by the newest unclassified instance. What remains to be discussed is the window size. The smaller the window size, the larger the risk that not the full concept describing the unclassified instances is represented in the window, and hence not learned. The larger the window size the higher the risk that older, no more valid concepts encoded in the unclassified instances, are learned. By default the algorithm uses window size 20. What needs to be determined as well is the optimal number of unclassified instances to induce new rules from, at the moment this is by default 30 unclassified instances. However, a possible improvement that is currently being investigated is the usage of the Hoeffding bound to determine the optimal number of unclassified instances to justify the induction of additional rules.

### 4.3 Third Process: Validation and Removal of Existing Rules

The removal of existing rules is determined by the classification accuracy of the individual rule but also on a minimum number of classification attempts the rule needs to fulfil in order to be discarded. Again both criteria can be predefined by the user but are in the current implementation by default 0.8 as the minimum accuracy and 5 as the minimum number of classifications attempted. The reason for not only using the minimum classification accuracy is the fact that a rule would be already discarded the first time it attempts to classify a test instance and actually assigns the wrong classification. For example, let us assume that the minimum number of classification attempts is 1, and a certain rule predicts a wrong class for the first instance it attempts to classify. In this case the rule would already be removed as its classification accuracy is 0. However if could happen that the rule only predicts the first test instance incorrectly and the following 4 instances correctly then the rule's accuracy would be 0.8 and also removed unless the minimum number of classification attempts is at least 5.

## 5 Evaluation

For the evaluation of eRules, three different stream generators have been used, the SEA, LED and Waveform generator. The three data streams used for experimentation can be typically found in the literature of concept drift detection. We provide the details of the three generators in the following.

1. **SEA Concepts Generator**: this artificial dataset, presented in [23], is formed by four data blocks with different concepts. Each instance has 3 numerical attributes and 1 binary class. Only the two first attributes are used for classification, and the third one is irrelevant. The classification function is $f_1 + f_2 =< \theta$, where $f_1$ and $f_2$ are the two first attributes and $\theta$ is a threshold that changes for each of the

four blocks in this order: 9,8,7,9.5. SEA data stream is the consecutive join of $SEA_9$, $SEA_8$, $SEA_7$ and $SEA_{9.5}$, where each block has 12500 instances and there is a sudden drift between blocks. It has been used in [4] [17] [14].

2. **LED Generator**: this data stream [7] has 24 binary attributes, but 17 of them are irrelevant for the classification. The goal is to predict the digit shown in a 7 segment LED display where each attribute has 10% of probability of being reversed. Concept drift can be included indicating a number *d* of attributes that change. It has been used in [24] [4].
3. **Waveform Generator**: this data stream was also presented in [7], and the goal is to distinguish between 3 different classes of waves. Each wave is generated by a combination of 2 or 3 basic waves. There are two variants of this problem: wave21 with 21 numerical attributes with noise, and wave40 with 19 extra irrelevant attributes. Concept drifts can be included by swapping a number *d* of attributes. It has been used in [24] [4] [14].

Taking these definitions as a starting point, and with the help of MOA software (Massive Online Analysis) [3], we have generated several datasets for the experimentation. A gradual concept drift has been included between data instances 450 and 550. The following parameters have been used for all experiments highlighted in this section: 50 instances to learn the initial classifier, an accuracy below 0.8 and a minimum of 5 classification attempts for each rule to be removed, a minimum of 30 unclassified instances have been accumulated before they are used to induce new rules and a window size of the last 20 data instances from which the unclassification rate and accuracy of the classifier are calculated.

Two versions of eRules have been evaluated, both use the parameters outlined above. The first version just tries to adapt to concept drifts by removing and adding new rules to the classifier. The second version re-induces the entire classifier, if the unclassification rate is too high (40% in the last 20 instances observed).

Figure 1 shows the unclassification rate, total accuracy and the accuracy on all instances for which eRules attempted a classification on the SEA data stream. It can be seen in both cases; eRules with and without re-induction, that there is a high unclassification rate and hence the total accuracy is relatively low. However, if we take only the classified instances into account it can be seen that the classification accuracy is relatively high. Also the concept drift is almost imperceptible. The unclassification rate for the version with re-induction is lower which in turn improves the total accuracy. However, the total classification accuracy for both versions of eRules is almost the same, only eRules with re-induction classifies more instances.

Figure 2 shows the unclassification rate, total accuracy and the accuracy on all instances for which eRules attempted a classification on the LED data stream. eRules for the LED data stream exhibits a similar behaviour as for the SEA data stream. It can be seen in both cases; eRules with and without re-induction, that there is a high unclassification rate and hence the total accuracy is relatively low. However, if we take only the classified instances into account it can be seen that the classification accuracy is relatively high. Also the concept drift is almost imperceptible. The unclassification rate for the version with re-induction is lower which in turn improves

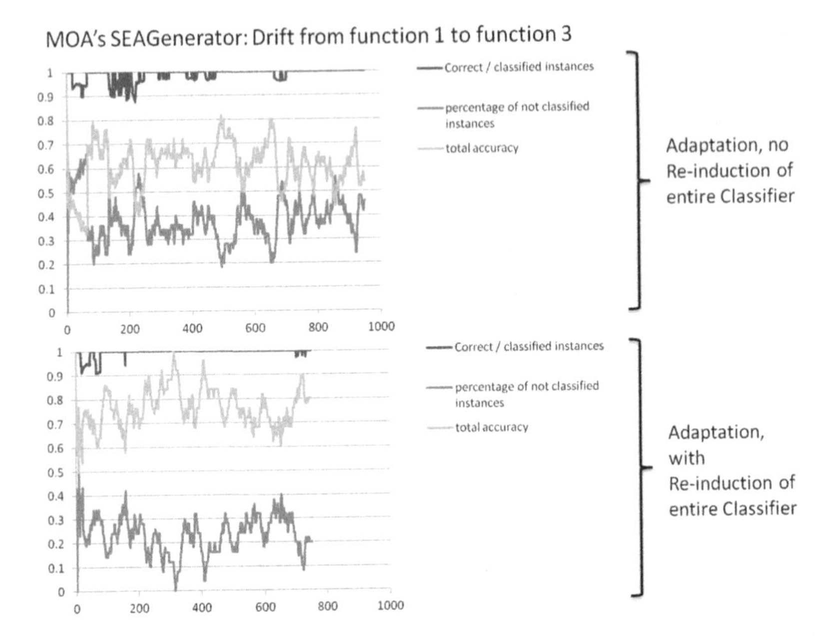


**Fig. 1** Evaluation with SEAGenerator: The horizontal axes show the number of iterations (the time stamp of the data instance received) and the vertical axes show the accuracy and percentage of unclassified instances. The graph on the top shows how the algorithm behaves without re-induction of the entire classifier and the graph on the bottom shows how the algorithm behaves with re-induction of the entire classifier. The concept drift takes place between instances 450 and 550.

the total accuracy. However, the total classification accuracy for both versions of eRules is almost the same, only eRules with re-induction classifies more instances.

Figure 3 shows the unclassification rate, the total accuracy and accuracy on all instances for which eRules attempted a classification on the Waveform data stream. In this particular case, eRules does not cope well with the concept drift. For the version of eRules without re-induction the total classification accuracy drops considerably and the unclassification rate increases. However, if we take only the instances in consideration that have been assigned a class label, then the drop in classification accuracy is relatively low. This high classification accuracy whilst having a high unclassification rate can be explained by the fact that due to the concept drift bad performing rules are removed and the newly induced rules are not describing the unclassified instances entirely. The reason for this could be that simply 30 unclassified instances are not enough to generate adequate rules. This will be resolved in the future by using metrics such as the Hoeffding bound in order to determine an adequate number of instances to induce new rules from. The version of eRules with re-induction performs better in this case as a completely new classifier is induced as soon as the unclassification rate reaches 40%. For the re-induction of the entire classifier the most recent 50 instances are used, which are 20 instances more compared with the induction of new rules covering unclassified cases. This again indicates that

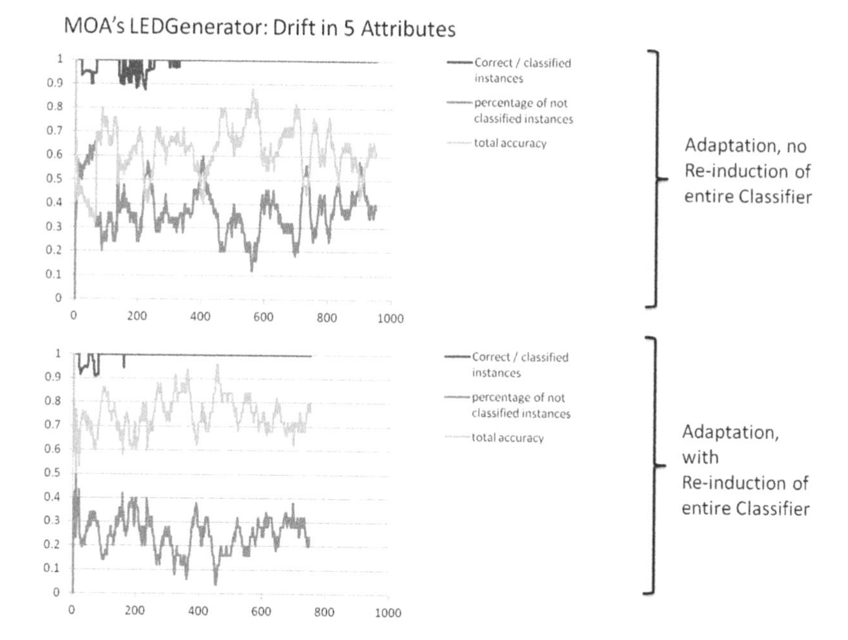


**Fig. 2** Evaluation with LEDGenerator: The horizontal axes show the number of iterations (the time stamp of the data instance received) and the vertical axes show the accuracy and percentage of unclassified instances. The graph on the top shows how the algorithm behaves without re-induction of the entire classifier and the graph on the bottom shows how the algorithm behaves with re-induction of the entire classifier. The concept drift takes place between instances 450 and 550.

in this case more than 30 unclassified instances are needed for the induction of new rules.

A comparison of the results obtained using eRules with previous approaches is not possible as our technique has unique features by having instances not classified as opposed to other methods that have only correct and incorrect classifications. However, in general it can be observed that eRules exhibits an excellent accuracy on the cases where classifications have been attempted. Sometimes eRules may leave a large number of instances unclassified, because they are not covered by the existing rules. eRules aims to reduce the unclassification rate by re-inducing the entire classifier once the unclassification rate becomes too high.

## 6 Ongoing and Future Work

All evaluation experiments outlined in Section 5 have been performed with a fixed parameter setting that seemed to work well in most cases and on data with stable feature domains. However, as it has been observed in Section 5 for the results displayed in Figure 3, there are cases in which the parameter setting could be improved.

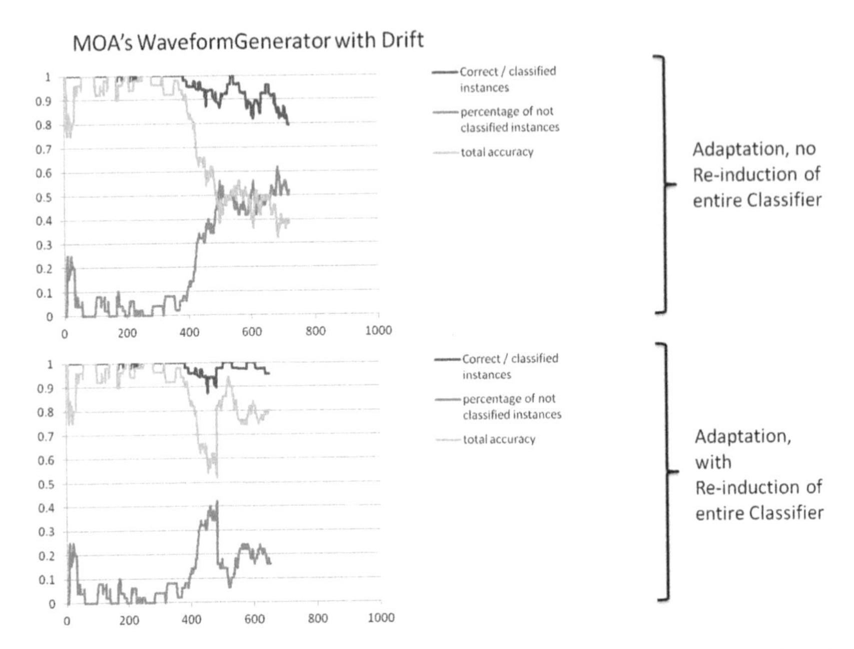


**Fig. 3** Evaluation with WaveformGenerator: The horizontal axes show the number of iterations (the time stamp of the data instance received) and the vertical axes show the accuracy and percentage of unclassified instances. The graph on the top shows how the algorithm behaves without re-induction of the entire classifier and the graph on the bottom shows how the algorithm behaves with re-induction of the entire classifier. The concept drift takes place between instances 450 and 550.

The ongoing work will comprise an investigation of metrics that could be used to dynamically adjust some of the parameters, for example the Hoeffding bound could be used to decrease the error by determining the optimal number of unclassified instances to be used to induce additional rules. Also the optimal size of the batch that is used to induce the initial classifier and to re-induce the entire classifier could be determined using the Hoeffding bound. The re-induction of the entire classifier also resets the classifier to changed feature domains. Thus it is expected that also the robustness of the classifier to changing feature domains is improved by the dynamic adjustment of the above mentioned parameters.

Further research will look into alternative implementations of the underlying modular classifier, with the aim to develop a rule induction method that delivers less unclassified cases. A possible candidate for that could be the PrismTCS classifier [6] which uses a default rule for unclassified instances.

Another possibility that will be investigated in future research is the usage of J-pruning. At the moment J-pruning is used for all rule induction processes in eRules, inducing the initial classifier in batch mode, re-induction of the entire classifier in batch mode and the induction of additional rules for the already existing classifier. However, J-pruning may not necessarily improve eRules's induction of additional classification rules. This is because the unclassified instances accumulated for inducing additional rules are polarised towards the concept that causes certain instances to be unclassified. Also a further pruning facility for Prism algorithms, Jmax-pruning [22] will be investigated for its usage in the eRules system.

We also consider improving eRules's computational scalability by making use of parallel processing techniques. The underlying Prism batch classifier has been parallelised successfully in previous research [21], and hence indicates eRules potential to be parallelised as well. A parallelisation, and hence speed up of the rule adaption could enable eRules to be applicable on high speed data streams.

## 7 Conclusions

A novel data stream classifier, eRules, has been presented based on modular classification rule induction. Compared with decision tree based classifiers this approach may leave test instances unclassified rather than giving them a wrong classification. This feature of this approach is highly desired in critical applications. eRules is based on three basic processes. The first process learns a classifier in batch mode, this is done at the beginning of eRules's execution in order to induce an initial classifier. However, batch learning is also performed to re-induce the classifier if it leaves too many instances unclassified. The second and third processes are used to adapt the classifier on the fly to a possible concept drift. The second process removes bad performing rules from the classifier and the third process adds rules to the classifier to cover instances that are not classifiable by the current rule set.

A first version of the classifier has been evaluated on three standard data stream generators, and it has been observed that eRules in general delivers a high classification accuracy on the instances it attempted to classify. However, eRules also tends to leave many test instances unclassified, hence a version of eRules that re-induces the entire classifier, if the unclassification rate is too high, has been implemented. eRules in general uses a fixed parameter setting, however, ongoing work on the classifier considers using the Hoeffding bound in order to dynamically adjust some of the parameters, while the algorithm is being executed.

**Acknowledgements** The research leading to these results has received funding from the European Commission within the Marie Curie Industry and Academia Partnerships & Pathways (IAPP) programme under grant agreement n° 251617.

## References

1. Computational intelligence platform for evolving and robust predictive systems, http://infer.eu/ 2012.
2. Brian Babcock, Shivnath Babu, Mayur Datar, Rajeev Motwani, and Jennifer Widom. Models and issues in data stream systems. In *In PODS*, pages 1–16, 2002.

3. Albert Bifet, Geoff Holmes, Richard Kirkby, and Bernhard Pfahringer. Moa: Massive online analysis. *J. Mach. Learn. Res.*, 99:1601–1604, August 2010.
4. Albert Bifet, Geoff Holmes, Bernhard Pfahringer, Richard Kirkby, and Ricard Gavaldà. New ensemble methods for evolving data streams. In *Proceedings of the 15th ACM SIGKDD international conference on Knowledge discovery and data mining*, KDD '09, pages 139–148, New York, NY, USA, 2009. ACM.
5. M A Bramer. Automatic induction of classification rules from examples using N-Prism. In *Research and Development in Intelligent Systems XVI*, pages 99–121, Cambridge, 2000. Springer-Verlag.
6. M A Bramer. An information-theoretic approach to the pre-pruning of classification rules. In B Neumann M Musen and R Studer, editors, *Intelligent Information Processing*, pages 201–212. Kluwer, 2002.
7. Leo Breiman, Jerome Friedman, Charles J. Stone, and R. A. Olshen. *Classification and Regression Trees*. Chapman & Hall/CRC, 1 edition, January 1984.
8. J. Cendrowska. PRISM: an algorithm for inducing modular rules. *International Journal of Man-Machine Studies*, 27(4):349–370, 1987.
9. Mayur Datar, Aristides Gionis, Piotr Indyk, and Rajeev Motwani. Maintaining stream statistics over sliding windows. In *ACM-SIAM Symposium on Discrete Algorithms (SODA 2002)*, 2002.
10. Pedro Domingos and Geoff Hulten. Mining high-speed data streams. In *Proceedings of the sixth ACM SIGKDD international conference on Knowledge discovery and data mining*, KDD '00, pages 71–80, New York, NY, USA, 2000. ACM.
11. Pedro Domingos and Geoff Hulten. A general framework for mining massive data stream. *Journal of Computational and Graphical Statistics*, 12:2003, 2003.
12. Mohamed Medhat Gaber. Advances in data stream mining. *Wiley Interdisciplinary Reviews: Data Mining and Knowledge Discovery*, 2(1):79–85, 2012.
13. Mohamed Medhat Gaber, Arkady Zaslavsky, and Shonali Krishnaswamy. Mining data streams: a review. *SIGMOD Rec.*, 34(2):18–26, 2005.
14. João Gama, Raquel Sebastião, and Pedro Pereira Rodrigues. Issues in evaluation of stream learning algorithms. In *Proceedings of the 15th ACM SIGKDD international conference on Knowledge discovery and data mining*, KDD '09, pages 329–338, New York, NY, USA, 2009. ACM.
15. Jiawei Han and Micheline Kamber. *Data Mining: Concepts and Techniques*. Morgan Kaufmann, 2001.
16. Petr Kadlec and Bogdan Gabrys. Architecture for development of adaptive on-line prediction models. *Memetic Computing*, 1:241–269, 2009.
17. J. Zico Kolter and Marcus A. Maloof. Dynamic weighted majority: An ensemble method for drifting concepts. *J. Mach. Learn. Res.*, 8:2755–2790, December 2007.
18. Ross J Quinlan. Induction of decision trees. *Machine Learning*, 1(1):81–106, 1986.
19. P. Smyth and R M Goodman. An information theoretic approach to rule induction from databases. 4(4):301–316, 1992.
20. F. Stahl and M. Bramer. Towards a computationally efficient approach to modular classification rule induction. *Research and Development in Intelligent Systems XXIV*, pages 357–362, 2008.
21. F. Stahl and M. Bramer. Computationally efficient induction of classification rules with the pmcri and j-pmcri frameworks. *Knowledge-Based Systems*, 2012.
22. F. Stahl and M. Bramer. Jmax-pruning: A facility for the information theoretic pruning of modular classification rules. *Knowledge-Based Systems*, 29(0):12 – 19, 2012.
23. W. Nick Street and YongSeog Kim. A streaming ensemble algorithm (sea) for large-scale classification. In *Proceedings of the seventh ACM SIGKDD international conference on Knowledge discovery and data mining*, KDD '01, pages 377–382, New York, NY, USA, 2001. ACM.
24. Periasamy Vivekanandan and Raju Nedunchezhian. Mining data streams with concept drifts using genetic algorithm. *Artif. Intell. Rev.*, 36(3):163–178, October 2011.

# A Geometric Moving Average Martingale method for detecting changes in data streams

**X.Z. Kong, Y.X. Bi and D.H. Glass** [1]

**Abstract** In this paper, we propose a Geometric Moving Average Martingale (GMAM) method for detecting changes in data streams. There are two components underpinning the GMAM method. The first is the exponential weighting of observations which has the capability of reducing false changes. The second is the use of the GMAM value for hypothesis testing. When a new data point is observed, the hypothesis testing decides whether any change has occurred on it based on the GMAM value. Once a change is detected, then all variables of the GMAM algorithm are re-initialized in order to find other changes. The experiments show that the GMAM method is effective in detecting concept changes in two synthetic time-varying data streams and a real world dataset 'Respiration dataset'.

## 1 Introduction

Volumes of data that are being increasingly generated by workflows, telecommunication and monitoring processes are often referred to as data streams. Such data can be characterized by: i) the order of tuples is not controlled; ii) the emission rates of data streams are not controlled; and iii) the available hardware resources are limited (i.e. RAM and CPU) [1]. In the past years many methods have been developed for managing and analyzing data streams, however the design of general, scalable and statistically abnormal change detection methods is a great challenge, which attracts significant attention within the research community [2]. The abnormal change detection technique can be used to solve a range of real world problems in areas such as finance, early warning of disasters, and network traffic monitoring. Specifically, detecting abnormal changes of stock prices over time could provide investors with some useful advice so as to reduce losses when stock prices fall abruptly [3]. In monitoring ground deformation and

---

1 University of Ulster at Jordanstown Newtownabbey, Northern Ireland, UK, BT37 0QB
kong-x@email.ulster.ac.uk;{y.bi,dh.glass}@ulster.ac.uk

landslides, detecting abrupt changes in water level and other measurements can be used to design intelligent systems. These intelligent systems would be capable of providing early warnings before landslides or other geological events so as to avoid massive loss [4]. Moreover in internet network monitoring, if the network traffic abruptly becomes larger and exceeds the given threshold, then the network could be attacked [5]. Additionally many methods have been developed to address natural hazards related problems, such as change detection of land disturbances [6-8], typhoon image analysis [9] and forest fire prediction [10].

A range of statistical work on abrupt change detection has been carried out [1]. Dries and Ruckert present an overview of change detection: change detection in the distribution can be considered as a statistical hypothesis test [11]. The Wald-Wolfowitz and Smirnov's test was generalized to multidimensional data sets in [12]. In [13], the authors analyze the electromagnetic intensity which was measured close to earthquakes and their results show that the changes of wave intensity are statistically significant. In [13], the authors use the Mann-Whitney *U* test to judge if populations of two different blocks have the same mean or instead of the more traditional Student *t*-Test [14]. This is because the Mann-Whitney *U* test is suitable for dealing with non-Gaussian distributions. In this paper we propose a Geometric Moving Average Martingale Test for deciding whether a change occurs in the data stream or not.

Shewhart control charts which need to choose a sample (or window) size are used to detect any significant deviation from a chosen products' quality [15, 16]. In [17-19], the authors use the likelihood techniques to detect changes in a parameter value such as the Gaussian mean. The methods of concept drift, such as ensemble learning [20-23], instance weighting [24], and a sliding window [25, 26], etc. can be used to detect changes. Most of these methods of detecting changes need sliding windows, but sizes of windows need to be selected on the basis of the accuracy of the methods. A typical problem with those methods is if a window size is larger than the changing rate, they delay the detection until the end of the window, whereas if a window is smaller than the changing rate, they may miss the detection or capture the change at a later time when gradual changes cumulate over time. It is difficult to decide an optimized window size. In order to solve the problem of "sliding window", we propose a Geometric Moving Average Martingale method, which does not need a sliding window and does not need to monitor the model accuracy.

The outline of this paper is as follows: in Section 2, we propose a Geometric Moving Average Martingale method for detecting changes in data streams. In Section 3, we introduce the GMAM Test; according to the GMAM Test we can decide whether any change has occurred, and we also provide the algorithm for the GMAM Test. In Section 4, we introduce the data description, experimental methodology, performance measures and experimental results in detail. In Section 5, we present conclusions and propose future work.

## 2 A Geometric Moving Average Martingale method

Control charts are specialized time series plots, which assist in determining whether a process is in control. Individuals' charts and "Shewhart" charts [15, 16] are widely-used forms of control charts; these are frequently referred to as "Shewhart" charts after the control charting pioneer Walter Shewhart who developed such techniques in 1924. These charts are suitable to detect relatively large shifts in the process. But Geometric (Exponentially Weighted) Moving Average (GMA/EWMA) control charts are more suitable for detecting smaller shifts. GMA/EWMA control charts may be applied for monitoring standard deviations in addition to the process mean and can be used to forecast values of a process mean. According to the GMA, the most recent observation gets the greatest weight and all previous observations' weights decrease in geometric progression from the most recent back to the first.

Martingale theory stems from the gambling industry and probability theory. It is the oldest model of financial asset pricing. So far the Martingale theory is widely used in investment optimization, decision making optimization, and survival analysis. In [27, 28], the research results show that Martingale theory can also be applied in statistical prediction of dynamic systems such as climate process statistical forecasts and modern epidemic statistical forecasts with a mathematical model for parameter estimation. In [29], the research results show the feasibility and effectiveness of the Martingale methodology in detecting changes in the data generating model for time-varying data streams. The Martingale theory [29] gets better precision and recall than the sequential probability ratio test (SPRT) for change detection described in [30]. But the Martingale theory method still can be improved. In order to improve the results for precision and recall, here we combine the Martingale theory which is used to measure the distance of a point to its cluster centre and the GMA algorithm to solve the problem of abnormal change detection in data streams. The benefit of the proposed GMAM method is that it gets better precision and recall than the Martingale method [29], but like the Martingale method it does not need a sliding window.

Based on the GMA method, the proposed decision function is as follows:

$$G_k = \sum_{i=0}^{k} \gamma_i M_{k-i} \tag{1}$$

where, $M_i$ is the Martingale value, $\{G_k \mid k = 1,2,3\cdots\}$ and $\gamma_i = \alpha(1-\alpha)^i$, $0 < \alpha \le 1$.The coefficient $\alpha$ acts as a forgetting factor. The decision function can be rewritten as follows:

$$G_k = (1-\alpha)G_{k-1} + \alpha M_k \text{ with } g_0 = 0 \tag{2}$$

We also defined the stopping rule:

$$G_k \ge h \tag{3}$$

where $h$ is a conveniently chosen threshold.

The Martingale $M_n$ is computed as follows [29]:
Given the training set $T = \{x_1, x_2, \cdots, x_n\}$ the strangeness which scores how much a data point is different from the other data points is defined as

$$s(T, x_i) = \|x_i - c_j\| \quad (4)$$

where $\|\cdot\|$ is a distance metric and $c_j$ is the cluster centre of cluster $S_j$, $x_i \in S_j$, in which the cluster $S = \{S_1, S_2, \cdots, S_k\}$ are obtained using a suitable clustering approach in this paper, K-means clustering has been used.

The strangeness measure is required for the randomized power Martingale which is defined in [31]:

$$M_n^{(\varepsilon)} = \prod_{i=1}^{n} (\varepsilon \hat{p}_i^{\varepsilon-1}) \quad (5)$$

where $\varepsilon \in [0,1]$, $\hat{p}_i$ is computed from the $\hat{p}$-value function as follows:

$$\hat{p}_i(\{x_1, x_2, \cdots x_n\}, \theta_i) = \frac{\#\{j \mid s_j > s_i\} + \theta_i \#\{j \mid s_j = s_i\}}{i} \quad (6)$$

where $\theta_i$ is randomly chosen from the interval of [0,1] at instance $i$ and, $s_j$ is the strangeness measure for $x_j$, $j = 1, 2, \cdots, i$. The initial Martingale value $M_0^{(\varepsilon)} = 1$.

## 3 A Geometric Moving Average Martingale Test

Consider the null hypothesis $\mathrm{H}_0$ : "no change in the data stream" against the alternative $\mathrm{H}_1$ : "a change occurs in the data stream." The GMAM Test continues to operate as long as

$$0 < G_k = \sum_{i=0}^{k} \gamma_i M_{k-i} < h \quad (7)$$

where $h$ is a positive number. One rejects the null hypothesis $\mathrm{H}_0$ when $G_k \geq h$. Suppose that $\{M_k : 0 \leq k < \infty\}$ is a nonnegative Martingale. The Doob's maximal inequality [32] states that for any $\lambda > 0$ and $n \in \square$ ,

$$\lambda P(\max_{0 \leq k < n}\{M_k \mid M_k \geq \lambda\}) \leq E(M_n) \quad (8)$$

hence, if $\mathrm{E(M_n)=E(M_0)=1}$, then

$$P(\max_{k \leq n} M_k \geq \lambda) \leq \frac{1}{\lambda} \quad (9)$$

for any $0 < k < \infty$ and $k \leq n$. We have $h = k \times \delta$ , thus we have the following property

$$P(\max_{k\le n} G_k \ge h) = P(\max_{k\le n}(\sum_{i=0}^{k}\gamma_i M_{k-i}) \ge h)$$

$$\le P(\max_{k\le n}(\sum_{i=0}^{k} M_{k-i}) \ge h) \le P(k\max_{k\le n}\max_{0<i<k} M_i) \ge h) \tag{10}$$

$$\le P(\max_{k\le n}\max_{0<i<k} M_i \ge \delta) \le \frac{1}{\delta}$$

Given a constant $c$, if the algorithm could not find any changes after computing $c$ data points, the algorithm will be re-initialized, so that the condition $k \le n \le c$ holds. We can add some data points which are before the re-initializing point to data points which are after the re-initializing point in order to find change point when there is a change at the re-initialized point. Because $h = k \times \delta \le c \times \delta$ and $c$ is a given constant, if $G_k$ has a high value, then $\delta$ also has a high value. From the Formula (10) we get $P(\max_{k\le n} G_k \ge h) \le \frac{1}{\delta}$, thus the probability that $G_k$ has a large value is very low. One rejects the null hypothesis when the GMAM value is greater than $h$. It is very likely the algorithm cannot detect a change when there is none. The proposed GMAM Test can decide whether a change occurs in the data stream according to hypothesis testing. Algorithm 1 presents the details of the GMAM algorithm. One may observe that $G_k = (1-\alpha)G_{k-1} + \alpha M_k$. Hence, no re-computation is needed $M_{k-i}, i = 0, \cdots k$, in Formula (1), when computing $G_k$ illustrated in Algorithm 1.

# 4 Experimental setting

## *4.1 Data description*

The experimental data sets used for this study are two synthetic time-varying data streams [29] and a real world dataset 'Respiration dataset'. The first data set is a simulated data stream using a rotating hyper-plane (RHP), a generated sequence of 100,000 data points, which consists of changes occurring at points $(1000 \times i) + 1$, for $i = 1, 2, \ldots, 99$. We randomly generate 1,000 data points with each component's value in the closed interval [0,1]. These data points are labeled positive and negative based on the following equation:

$$\sum_{i=1}^{m} w_i x_i \begin{cases} < c : negative \\ \ge c : positive \end{cases} \tag{11}$$

**Algorithm: The Proposed GMAM Test Algorithm**

**Input:** threshold $h$ ; forgetting factor $\alpha$ ; startnum which represents data point number of initial clusters.

**Output:** A set of change points.

1: Initialise: M[1]=1; i=1;flag=0;

2: Initialise: G(i)= $\alpha$ ;

3: **Repeat**

    If start a new change detection

        Initialize cluster

    Else

        Compute the strangeness of xi using Formula (4)

        Compute the $\hat{p}-$ value $\hat{p}_i$ using Formula (7)

        Compute the G(i) using Formula (1)

        If G(i)>h

            change detected

        Else continue

4: **Until** all the data points are processed

5: Output a set of the detected change points

Algorithm 1 Pseudo code for the proposed algorithm.

where $c \in \Re$ is an arbitrary fixed constant, $x_i$ is the $i$th component of a data point, and the fixed components $w_i, i = 1,\ldots,m,$ of a weight vector are randomly generated in the interval of [-1, 1]. Noise is added by randomly switching the class labels of 5% of the data points.

The second data set is a simulated data stream using the normally distributed clusters data generator (NDC). NDC generates a series of random centres for multivariate normal distributions. It randomly generates a fraction for this center, i.e. the fraction of points we will get from this center and randomly generates a separating plane. Based on the plane, NDC chooses classes for the centres, and then randomly generates the points from the distributions. The separability can be increased by increasing variances of distributions. NDC randomly generates data points from the normal distributions and assigns them labels based on the label of the centre to the corresponding distribution.

Linearly nonseparable binary-class data streams of 10500 data points consisting of changes occurring at points $(500 \times i) + 1$, for $i = 1, 2, \ldots, 20$. are simulated using the NDC in $\Re^{33}$ with randomly generated cluster means and variances. The values for each dimension are scaled to the range [-1, 1]. The generating process for the data stream is similar to that used for the rotating hyper plane data stream.

The Respiration dataset is a time series data showing a patient's respiration (measured by thorax extension), which is downloaded from, the UCR Time Series

Data Mining Archive. The data set consists of manually segmented data with labels of, 'awake' and 'sleep' [33]. We choose a discord length corresponding to 32 seconds because we want to span several breaths. An example of the original time-series as well as the annotation results are depicted in the top graph of Figure 6.

## *4.2 Experimental methodology and performance measures*

The experiments involve setting all the data points as training data, computing the GMAM values according to the order of data points to detect changes in the data stream, and then measuring the performance using Precision and Recall measures which are described below.

The performance is measured based on three performance metrics, namely recall, precision, and the mean delay time. Precision and recall are defined as follows:

$$\text{Precision} = \frac{\text{Number of Correct Detections}}{\text{Number of Detections}}$$

$$\text{Recall} = \frac{\text{Number of Correct Detections}}{\text{Number of True Changes}}$$

Precision is the probability of detection being correct, i.e., detecting a true change. Recall is the probability of a change being detected as a true change. The delay time for a detected change is the number of observed instances from the true change point to the detected change point. We also combine Precision and Recall together to define an integrated metric called the F-measure:

$$\text{F-measure} = \frac{2 \times \text{Precision} \times \text{Recall}}{\text{Precision+Recall}}$$

In fact the $F$-measure represents a harmonic mean between recall and precision. A high value of the $F$-measure ensures that the precision and recall are reasonably high.

## *4.3 Experimental results*

We analyze the effect of $\varepsilon \in [0.8,1)$ and the strangeness computed using the $K$-mean clustering on the GMAM performance for the NDC data set. Threshold $h$ is set to 5 and $\alpha$ is set to 0.1 for GMAM based on the strangeness measures defined in Formula 4 in this study. The experiment consists of 25 runs for each $\varepsilon$ value. Each run is a sequence of 10500 data points generated by the NDC data generator. And we also perform an experiment using the Martingale method for change detection as described in [29], $\lambda$ is also set to 5. We test some values $\varepsilon \in (0,0.8)$ with the GMAM method and Martingale method; we find the smaller $\varepsilon$ gets the worst precision and recall. So in this experiment we just select the $\varepsilon \in [0.8,1)$.

In this experiment, Precision-GMAM, Recall-GMAM and F-measure-GAMA represent the precision, recall and F-means of the proposed GMAM method, respectively. Precision-Mar, Recall-Mar and F-measure represent the precision, recall and F-means of the Martingale method which was proposed in [29]. Fig. 1 shows the experimental results of GMAM and the Martingale method. It can be seen that the GMAM method has better precision than the Martingale method in the interval of [0.8, 0.96] and their recalls are almost same.

As the proposed GMAM method uses the idea of exponential weighting of observations which can effectively reduce the false precision, it can get better precision and recall than the Martingale method. The mean delay time of the GMAM method is 33.16, and the mean delay time of Martingale method is 27.60.

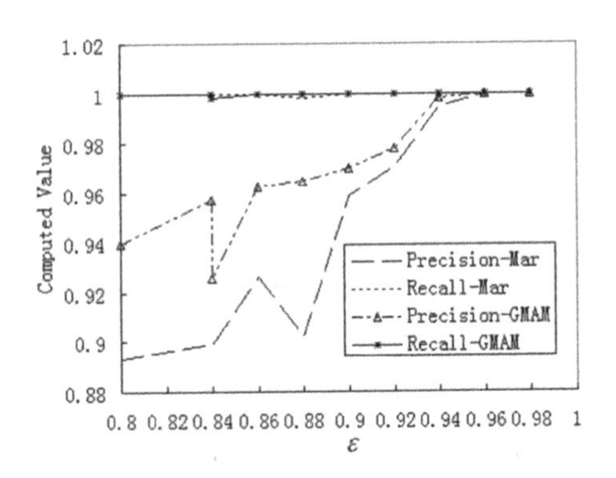


Figure 1 Experiment results of various $\varepsilon$ on the NDC data sequence.

Fig. 2 shows the recall and precision of the GMAM method and the Martingale method for parameter $h$ being between 1 and 10 on the data sequences simulated using the NDC data generator. Here $\varepsilon$ is set to 0.8 and $\alpha$ is set to 0.1 for GMAM based on cluster model strangeness measures. $\varepsilon$ also is set to 0.8 and $\lambda$ between 1 and 10 for the Martingale method. The experiment consists of 25 runs for each $h$ and $\lambda$ values.

The GMAM method using strangeness computed from the cluster model has better precision for the NDC data sequences in the x axis interval of [1, 6] and the recall in the x axis interval of [1, 2] in Figure 2. The mean delay time of the

GMAM method is 36.23, and the mean delay time of the Martingale method is 32.06.

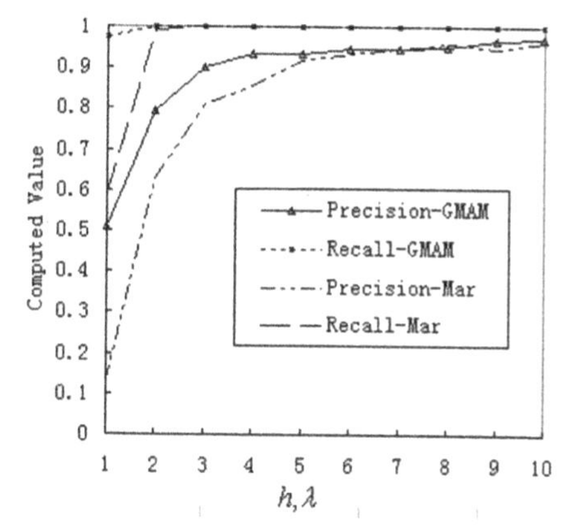


Figure 2 Experiment results of various $h$ and $\lambda$ on the NDC data sequence.

Fig. 3 shows the recall and precision of the GMAM method and the Martingale method for $h$ between 1 and 10 on the data sequences simulated using RHP with an arbitrary degree of changes and 5% noise. $\varepsilon$ is set to 0.8 and $\alpha$ is set to 0.1 for GMAM based on cluster model strangeness measures. $\varepsilon$ also is set to 0.8 and $\lambda$ between 1 and 10 for the Martingale method. The experiment consists of 10 runs for each $h$ and $\lambda$ values.

The GMAM method has better precision between 1 and 6 and recall between 1 and 4. The mean delay time of the GMAM method is 50.99, and the mean delay time of the Martingale method is 47.97.

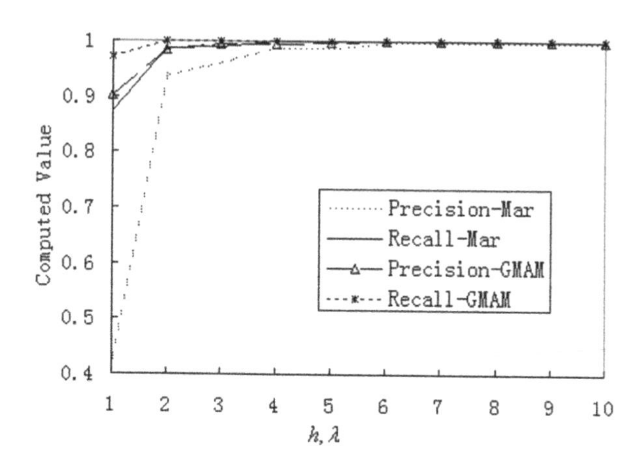


Figure 3 Experiment results of various $h$ and $\lambda$ on the RHP data sequence.

Fig. 4 shows the experimental results of GMAM, where Fig.4 (a) shows the precision, recall and F-means, and Fig.4 (b) shows the mean delay time. $h$ is set to 5 and $\varepsilon$ is set to 0.8 . GMAM has the better performance for NDC data sequence, and the smaller $\alpha$ , the better the performance. Fig.4 (b) shows that smaller $\alpha$ needs more delay time to identify the change. According to the Formula (1) and (3), we know the biggest weighted item is $M_k$ and the weight is $\alpha$ . For the same $h$ , the smaller $\alpha$ get the smaller main item value $\alpha M_k$, so it needs more data points to be used in Formula (1) in order to satisfy Formula (3). The mean delay time of the GMAM method is 34.27.

Fig. 5 shows the experimental result of data sequence from 33D NDC data sequence. GMAM with $h$=5, $\varepsilon$=0.96, $\alpha$=0.1, mean delay time is 36.60 with one false alarm near the point 4500.

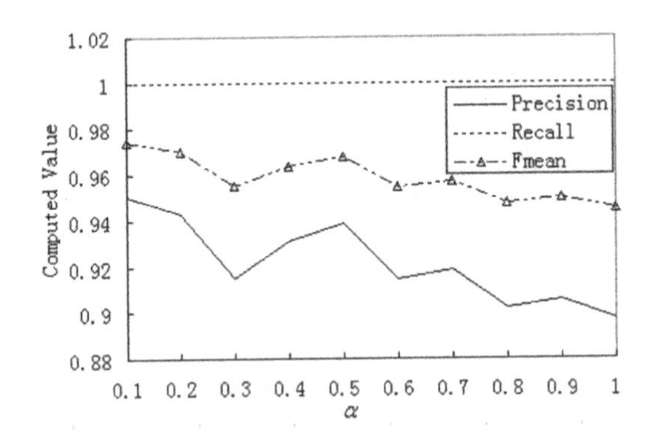


(a) Precision, recall, F-means

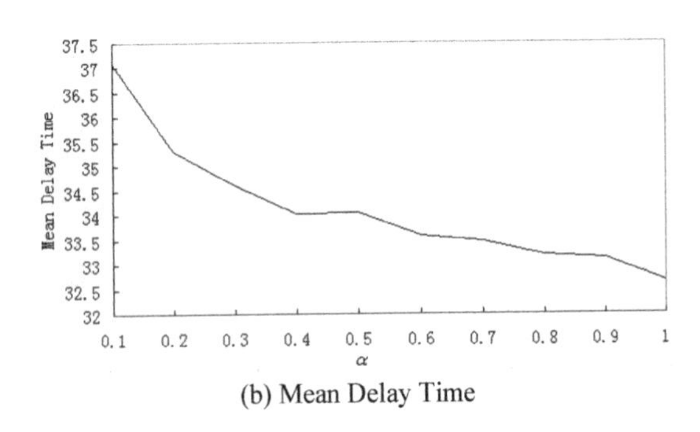


(b) Mean Delay Time

Figure 4 Experiment results of various $\alpha$ on the NDC data sequence.

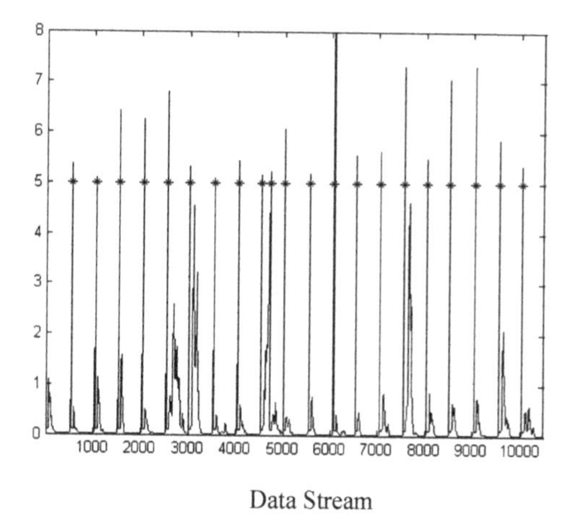


Figure 5 Data sequence from NDC data sequence. Y axis is GMAM value with $h$=5, $\varepsilon$=0.96, $\alpha$=0.1, mean delay time is 36.60 with one false alarms near the point 4500.

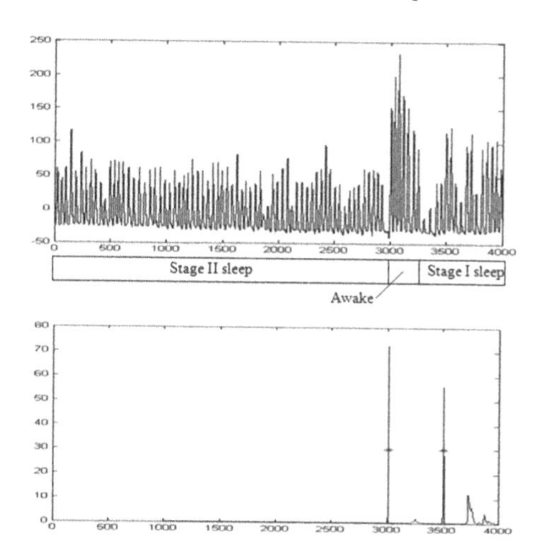


Figure 6 The raw time-series data of the respiration datasets (top) and the change-detection results by the GMAM method (bottom).

Figure 6 illustrates the original time-series (top) as well as the changes detected by the GMAM method (bottom). The change points detected are indicated with '*'. We set $h$=50, $\varepsilon$=0.8, $\alpha$=0.1. The task is to detect the time points at which the state of patients changes, in order to find the changes between 'sleep-awake' and 'awake-sleep', we need to set a threshold to determine if any change happens. Here we set the threshold as 30. The experiment consists of 1000 runs. For this real world dataset, many other methods have been applied, including Singular-Spectrum Analysis (SSA) [34], Change finder (CF) [35], Kullback-Leibler

Importance Estimation Procedure (KLIEP) [33] and One-class support vector machine (OSVM) [36].Table1 shows a comparison of the proposed method with the others.

Table 1. Experimental results.

| Change detection method | Precision/Accuracy rate (%) |
|---|---|
| CF | 23.7 |
| SSA | 46.5 |
| OSVM | 49.7 |
| KLIEP | 49.9 |
| GMAM | **51.2** |

## 5 Conclusions

In this paper, we propose an efficient Geometric Moving Average Martingale method for detecting changes in data streams. The experimental results show that the Geometric Moving Average Martingale method is effective in detecting concept changes in time-varying data streams simulated using two synthetic data sets and a real world dataset 'Respiration dataset'. In the experiment, we also need to set some parameters, such as the parameter $k$ in clustering. That parameter is different from the sliding window size. Firstly, the parameter does not cause missing the detection or capture the change, that is, if a sliding window size is smaller than the changing rate, they may miss the detection or capture the change at a later time when gradual changes accumulate over time. Secondly, the proposed method GMAM does not need to monitor the model accuracy, but most of the sliding window methods need to do so. The results demonstrate also that parameter $\alpha$ in Formula 2 is an important parameter; it can improve the results for precision and recall, the experiments show that the smaller $\alpha$ gets the better performance but needs the more mean delay time to identify the change. When $\alpha$ is set to 1, the Formula 2 can be rewritten as $G_k = M_k$, it is the same with the Martingale method Formula 6 in [29], so we can set $\alpha$ the smaller value than 1 in order to get the better performance than the Martingale method, but needs more delay time. The proposed Geometric Moving Average Martingale method uses the idea of exponential weighting of observations which can effectively reduce the number of false positives, so we can get better precision and recall than the Martingale method. The future work includes: 1) constructing more robust methods to reduce the mean delay time and 2) detecting changes in more real world data streams such as electromagnetic data streams.

## References

1. Bondu, M. Boullé: A supervised approach for change detection in data streams. , The 2011 International Joint Conference on Neural Networks (IJCNN), pp. 519 – 526 (2011).
2. Daniel Kifer, Shai Ben-David, Johannes Gehrke: Detecting Change in Data Streams. Proceedings of the 30th VLDB Conference,Toronto,Canada, pp. 180-191 (2004).
3. Leszek Czerwonka: Changes in share prices as a response to earnings forecasts regarding future real profits. Alexandru Ioan Cuza University of Iasi, Vol. 56, pp. 81-90 (2009).
4. Q. Siqing, W. Sijing: A homomorphic model for identifying abrupt abnormalities of landslide forerunners. Engineering Geology, Vol. 57, pp. 163–168 (2000).
5. Wei Xiong, NaixueXiong, Laurence T. Yang, etc.: Network Traffic Anomaly Detection based on Catastrophe Theory. IEEE Globecom 2010 Workshop on Advances in Communications and Networks, pp. 2070-2074 (2010).
6. Thomas Hilker , Michael A.Wulder , Nicholas C. Coops, etc. : A new data fusion model for high spatial- and temporal-resolution mapping of forest disturbance based on Landsat and MODIS. Remote Sensing of Environment, Vol. 113, pp. 1613–1627 (2009).
7. Ashraf M. Dewan , Yasushi Yamaguchi: Using remote sensing and GIS to detect and monitor land use and land cover change in Dhaka Metropolitan of Bangladesh during 1960–2005. Environ Monit Assess, Vol. 150, pp. 237-249 (2009).
8. Jin S. Deng, KeWang,Yang Hong,Jia G.Qi.: Spatio-temporal dynamics and evolution of land use change and landscape pattern in response to rapid urbanization. Landscape and Urban Planning, Vol. 92, pp. 187-198 (2009).
9. Asampbu Kitamoto: Spatio-Temporal Data Mining for Typhoon Image Collection.Journal of Intelligent Information Systems, Vol. 19(1), pp. 25-41 (2002).
10. Tao Cheng, Jiaqiu Wang: Integrated Spatio-temporal Data Mining for Forest Fire Prediction. Transactions in GIS. Vol. 12 (5), pp. 591-611 (2008).
11. A. Dries and U. Ruckert: Adaptive Concept Drift Detection. In SIAM Conference on Data Mining, pp. 233–244 (2009).
12. J.H. Friedman and L.C Rafsky: Multivariate generalizations of the Wald-Wolfowitz and Smirnov two-sample tests. Annals of Statistic, Vol. 4, pp. 697–717 (2006).
13. F. Nemec, O. Santolik, M. Parrot,and J. J. Berthelier: Spacecraft observations of electromagnetic perturbations connected with seismic activity. Geophysical Research Letters, Vol. 35(L05109), pp. 1-5 (2008).
14. Sheskin, D. J.: Handbook of Parametric and Nonparametric Statistical Procedures. 2nd ed. CRC Press, Boca Raton, Fla. pp. 513-727 (2000).
15. W.A. Shewhart: The Application of Statistics as an Aid in Maintaining Quality of a manufactured Product. Am.Statistician Assoc., Vol. 20, pp. 546-548 (1925).
16. W.A. Shewhart: Economic Control of Quality of Manufactured Product. Am. Soc. for Quality Control, (1931).
17. E.S. Page: On Problem in Which a Change in a Parameter Occurs at an Unknown Point. Biometrika, Vol. 44, pp. 248-252 (1957).
18. M.A. Girshik and H. Rubin: A Bayes Approach to a Quality Control Model, Annal of Math. Statistics, Vol. 23(1), pp. 114-125 (1952).
19. Ludmila I. Kuncheva: Change Detection in Streaming Multivariate Data Using Likelihood Detectors. IEEE Transactions on Knowledge and Data Engineering, Vol. 6(1), pp. 1-7 (2007).

20. F. Chu, Y. Wang, and C. Zaniolo: An Adaptive Learning Approach for Noisy Data Streams. Proc. Fourth IEEE Int'l Conf.Data Mining, pp. 351-354 (2004).
21. J.Z. Kolter and M.A. Maloof: Dynamic Weighted Majority: A New Ensemble Method for Tracking Concept Drift. Proc. Third IEEE Int'l Conf. Data Mining, pp. 123-130 (2003).
22. H. Wang, W. Fan, P.S. Yu, and J. Han: Mining Concept-Drifting Data Streams Using Ensemble Classifiers. Proc. ACM SIGKDD, pp. 226-235 (2003).
23. M. Scholz and R. Klinkenberg: Boosting Classifiers for Drifting Concepts.Intelligent Data Analysis, Vol. 11(1), pp. 3-28 (2007).
24. R. Klinkenberg: Learning Drifting Concepts: Examples Selection vs Example Weighting, Intelligent Data Analysis. special issue on incremental learning systems capable of dealing with concept drift, Vol. 8(3), pp. 281-300 (2004).
25. R. Klinkenberg and T. Joachims: Detecting Concept Drift with Support Vector Machines. Proc. 17th Int'l Conf. Machine Learning, P. Langley, ed., pp. 487-494 (2000).
26. G. Widmer and M. Kubat: Learning in the Presence of Concept Drift and Hidden Contexts.Machine Learning, Vol. 23(1), pp. 69-101 (1996).
27. Kong Fanlang: A Dynamic Method of System Forecast. Systems Engineering Theory and Practice, Vol. 19(3), pp. 58-62 (1999).
28. Kong Fanlang: A Dynamic Method of Air Temperature Forecast. Kybernetes, Vol. 33(2), pp. 282-287 (2004).
29. S. S. Ho, H. Wechsler: A Martingale Framework for Detecting Changes in Data Streams by Testing Exchangeability. IEEE transactions on pattern analysis and machine intelligence, Vol. 32(12), pp. 2113-2127 (2010).
30. S. Muthukrishnan, E. van den Berg, and Y. Wu: Sequential Change Detection on Data Streams, Proc. ICDM Workshop Data Stream Mining and Management, pp. 551-556 (2007)
31. V. Vovk, I. Nouretdinov, and A. Gammerman: Testing Exchangeability On-Line. Proc. 20th Int'l Conf. Machine Learning,T. pp. 768-775 (2003).
32. M. Steele: Stochastic Calculus and Financial Applications. SpringerVerlag, (2001).
33. E. Keogh, J. Lin, and A. Fu: HOT SAX: Efficiently finding the most unusual time series subsequences. In Proceedings of the 5th IEEE International Conference on Data Mining (ICDM'05), pp. 226-233 (2005).
34. V. Moskvina and A. A. Zhigljavsky: An algorithm based on singular spectrum analysis for change-point detection. Communication in Statistics: Simulation & Computation, Vol. 32(2), pp. 319-352 (2003).
35. Y. Takeuchi and K. Yamanishi: A unifying framework for detecting outliers and change points from non-stationary time series data. IEEE Transactions on Knowledge and Data Engineering, Vol. 18(4), pp. 482–489 (2006).
36. F. Desobry, M. Davy, and C. Doncarli: An online kernel change detection algorithm. IEEE Transactions on Signal Processing, Vol. 53(8), pp. 2961-2974 (2005).

# Using Chunks to Categorise Chess Positions

Peter C.R. Lane and Fernand Gobet

**Abstract** Expert computer performances in domains such as chess are achieved by techniques different from those which can be used by a human expert. The match between Gary Kasparov and Deep Blue shows that human expertise is able to balance an eight-magnitude difference in computational speed. Theories of human expertise, in particular the chunking and template theories, provide detailed computational models of human long-term memory, how it is acquired and retrieved. We extend an implementation of the template theory, CHREST, to support the learning and retrieval of categorisations of chess positions. Our extended model provides equivalent performance to a support-vector machine in categorising chess positions by opening, and reveals how learning for retrieval relates to learning for content.

## 1 Introduction

Building a machine which exhibits intelligent behaviour has been a long sought-after dream, particularly in the fields of artificial intelligence (AI) and cognitive science. Two main approaches may be identified. The first is to use whatever techniques are offered by computer science and AI, including brute force, to create artifacts that behave in an intelligent way. The second is to develop computational architectures that closely simulate human behaviour in a variety of domains. The difference between the two approaches separates the study of AI and cognitive science:

> AI can have two purposes. One is to use the power of computers to augment human thinking, just as we use motors to augment human or horse power. Robotics and expert systems are

Peter C.R. Lane
School of Computer Science, University of Hertfordshire, College Lane, Hatfield AL10 9AB, UK
e-mail: peter.lane@bcs.org.uk

Fernand Gobet
School of Social Sciences, Brunel University, Uxbridge UX3 8BH, UK
e-mail: fernand.gobet@brunel.ac.uk

major branches of that. The other is to use a computer's artificial intelligence to understand how humans think. In a humanoid way. If you test your programs not merely by what they can accomplish, but how they accomplish it, then you're really doing cognitive science; you're using AI to understand the human mind. (Herbert Simon, in an interview with Doug Stewart [37].)

In specific domains, AI can produce incredible performance. A long-standing goal of mathematicians and AI researchers was to develop a mechanical program that could play chess: the basic pattern for search algorithms was determined in the 1950s [36, 38], and recently algorithms such as Monte Carlo Tree Search [7] have dominated game-playing research. Humans, of course, are more selective in their search for good moves, and interactions of the search tree with other forms of knowledge were explored by Berliner [2] and Michie [31], amongst others. A more complete overview of this history can be found in Chapter 2 of [19].

In 1996 and 1997 IBM arranged two matches between its Deep Blue computer and the then World Chess Champion Gary Kasparov; Kasparov won the first match and Deep Blue won the second match – effectively a tie overall. These matches show the gulf between what we know of human intelligence and AI. The rapid indexing and sifting of an extensive pool of prior knowledge by the human counter-balanced an eight-magnitude difference in computational ability: where the human might calculate around 100 moves in 3 minutes, the computer would calculate of the order of 100 million moves per second. Chess has been described as the 'drosophilia of psychology' [6], and in these matches we see the need and opportunity for theories of intelligence to understand how humans efficiently perceive a stimulus, such as a chess board, and retrieve information suitable for problem solving.

We present a cognitive-science approach to solving this problem. Cognitive science has a long tradition of attempting to understand intelligence using an approach based around simulation (some of this history is discussed in [1, 16]). The earliest systematic work in this area is perhaps that of Newell and Simon [33]. The use of computational models for understanding cognition has grown over time, as their use offers many advantages. For example, their implementation as computer programs ensures a high degree of precision, and offers a sufficiency proof that the mechanisms proposed can carry out the tasks under study – something obviously desirable if practical artificial intelligence is the goal. The extent to which success is reached in simulating actual human behaviour can be assessed by using measures such as eye movements, reaction times, and error patterns, as well as the protocol analyses used by the earlier researchers, and, in more recent times, MRI scans of the working brain. Recent summaries have emphasised the popularity of this approach [29, 35].

As a concrete task, we consider the game of chess, and how chess players remember and retrieve information about specific chess positions. Chess is an ideal domain for studying complex learning because it provides a well-defined set of stimuli (the different board positions), there are numerous examples of games readily available for study, and, in particular, chess players are ranked on a numeric scale allowing gradations in levels of expertise to be identified in a quantitative manner. In competitive chess, players are limited in their thinking time, and success depends on an efficient retrieval of information from their prior study and experience to help understand the position in front of them. Chess positions present various problems to the player, enabling the researcher to probe the varying contributions of calculation and intuition in the problem-solving process.

We examine how a cognitive architecture based on the influential chunking and template theories [6, 20, 22] constructs and uses representations for visual stimuli using an incremental-learning technique. The form of the representation is governed both by physical demands of the domain, such as locality, and also by procedural requirements for an efficient indexing mechanism and to interact with visual attention. CHREST (Chunk Hierarchy and REtrieval STructures) [20, 23] has simulated data on human memory in a number of domains including expert behaviour in board games, problem solving in physics, first language acquisition, and implicit learning.

We extend CHREST to support the acquisition and retrieval of categories, and compare its performance with a statistical algorithm, the support-vector machine [9, 10]. In spite of the different algorithms employed, one using psychologically-plausible mechanisms and the other statistical optimisation techniques, the overall performance is very similar. We also analyse the trade off between learning an index for efficient retrieval of chunks against learning their content.

# 2 Chunking for Expert Performance

## *2.1 Chunking and Template Theories*

There is considerable evidence that much of the information we learn is stored as chunks – perceptual patterns that can be used as units and that also constitute units of meaning [20]. Perhaps the best evidence for this claim comes from the study of experts (individuals that vastly outperform others in a specific domain), and in particular chess experts. In their seminal study, Chase and Simon [6] showed a chess position for a few seconds, and then asked players to reconstruct it. The analysis of the latencies between the placements of the pieces, as well as of the patterns of relations between pairs of pieces, provided clear support for the idea of chunking. Additional support came from experiments where players had to copy a given position on another board [6], to sort positions [15], and to recall positions presented either randomly, by rows and columns, or by chunks [14]. The notion of chunk also in part explains why experts can remember a considerable amount of briefly-presented material, in spite of the limits of their short-term memory [6, 11, 32].

One limit of Chase and Simon's chunking theory is that it assumes that all knowledge is stored in relatively small units (maximum of 5-6 pieces in chess). In fact, there is substantial evidence that experts also encode information using larger representations. This evidence includes verbal protocols in recall and problem solving experiments [12], and experiments where the material to recall exceeds the amount of material that can be stored in short-term memory [22]. As the capacity of visual short-term memory is limited to about four chunks [18], the chunking theory predicts that no more than one board of about 24 pieces could be recalled. However, Gobet and Simon [22] showed that strong masters can recall up to five positions. To correct this as well as other weaknesses of the chunking theory, while keeping its strengths, Gobet and Simon developed the template theory. The key contribution of the theory, from the point of view of knowledge representation, is that it assumes that some chunks, which are met often when the agent deals with the environment, evolve into more complex data structures: templates. Templates encode both fixed information (the core) and variable information (the slots). Template theory, unlike most schema theories, proposes specific mechanisms by which templates are constructed incrementally when interacting with the environment [23]. While it takes a long time to learn a new chunk (8 seconds), information can be encoded rapidly in the slots of a template (250 milliseconds). Together with the notion of chunks, this explains experts' superiority in memory experiments with material taken from their domain of expertise, in particular when presentation times are brief.

## 2.2 Importance of specific perceptual knowledge

An important conclusion of research into expertise is that chunks encode specific and "concrete" information, as opposed to general and abstract information. While this is less the case with templates, which have slots to encode variable information, it is still the case that the scope of application of templates is limited by the information stored in their core, which is used to identify them in long-term memory. Several researchers have objected to this idea that chunks are specific e.g. [30], arguing that this is not efficient computationally – that is, abstract chunks would be more economical. However, the empirical evidence clearly supports the specificity hypothesis e.g. [3, 21, 34]. One example will suffice.

Bilalić [3] studied chess experts specialised in two different opening systems[1] with Black (the French and the Sicilian). When confronted with positions from the opening they did not play (e.g. "French" players with Sicilian positions), players saw their performance drop by one entire standard deviation in skill, compared to the case where they had to deal with the opening they were familiar with (e.g. "French" players with French positions). One standard deviation is a huge difference in skill. For example, when confronted with an opening she is not specialised in, the performance of a grandmaster would be similar to that of an international master.

[1] An opening system is a typical pattern for the first dozen or so moves of a chess game. Expert players must study opening systems in depth to compete successfully, and usually master five or six such systems with each colour [8].

## 3 CHREST

In this section, we summarise those aspects of CHREST which relate to information learning and retrieval. More detailed descriptions of CHREST are available in [12, 20]. Although we focus on chess, CHREST's mechanisms have been shown to generalise to other board games such as Awalé [17] and Go [4], as well as other domains such as language acquisition [13, 24]. The latest version of CHREST (version 4) is available at `http://chrest.info`.

### *3.1 Overview*

Fig. 1 shows the main components of the CHREST architecture. These include input/output, a long-term memory (LTM) and a short-term meory (STM). The STM holds the current working *representation* of the stimuli being perceived; this representation is a set of pointers into LTM. STM has a limited capacity – the typical size is 4 chunks, for visual information. Each chunk is a pointer to an item held in LTM, and this item can contain varying amounts of information. The STM is populated with chunks as the eye moves over the stimulus. As the eye perceives a set of pieces on a chess board, it sorts those pieces through an index, and places the retrieved chunk into STM. The information in STM is then available for further learning, problem solving, or other cognitive processing. An example of the representation and how it is retrieved can be found in the next section.

The index into LTM is represented as a discrimination network; Fig. 3 gives an example. Nodes within LTM are connected with *test links*, each test link containing one or more primitives from the target domain which must be matched for a sorted pattern to pass down the test link. Each node contains a pattern, its *image*; these node images form the chunks and templates stored within the network. This network is efficient, in the sense that it can rapidly index sufficient quantities of information to support expert performance [25]. CHREST supports multiple modalities (kinds

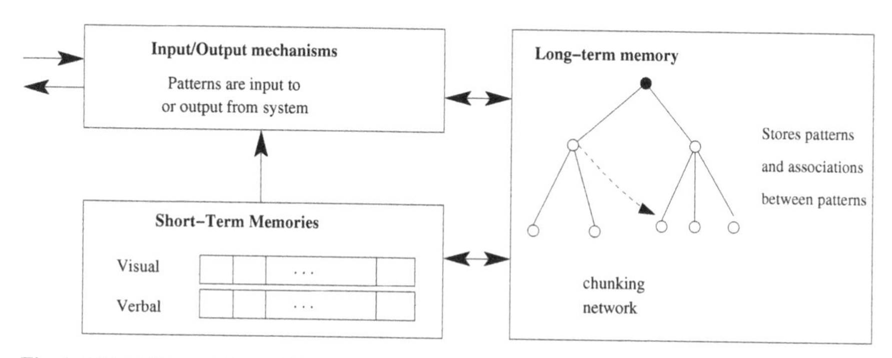


**Fig. 1** CHREST cognitive architecture

of input), and can form links between nodes of different types, such as visual and verbal information. The STM contains separate lists of chunks for visual and verbal information, and the root of the LTM first sorts patterns into the visual or verbal networks depending on their modality.

### 3.2 Learning and Representing Chunks

As with any modelling experiment, we begin with some assumptions about what a person knows. When modelling chess players, we assume that the player can recognise an individual chess piece and their location. The 'primitives' for the representation are known as 'item-on-square', and are represented as `[P 3 4]`, denoting the name of the piece, the row and column number.[2] Fig. 2 gives an example position. When the eye perceives the board, it sees any piece it is looking at, as well as any pieces in its periphery; the size of the periphery can be varied, but in our chess experiments the eye is assumed to perceive a $5 \times 5$ grid of the chess board. Thus, looking at square (3, 7) of the board in Fig. 2, the eye would retrieve the list of pieces < `[Q 1 6] [N 4 6] [P 4 5] [P 5 6] [P 1 7] [P 2 7]` >.

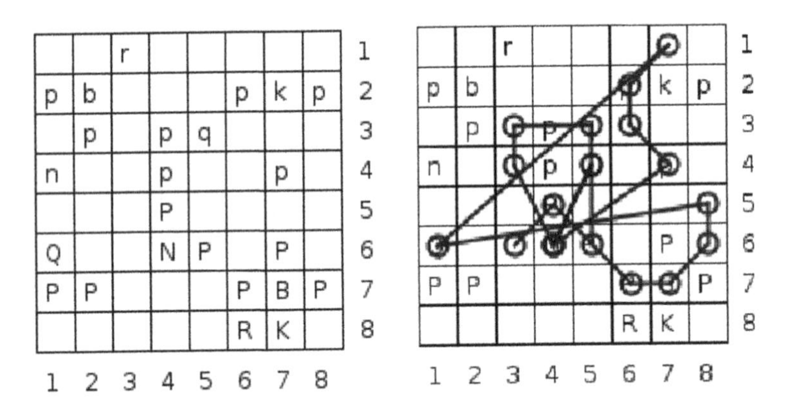


**Fig. 2** Example chess position (left) and sequence of eye fixations (right)

The list of pieces is known as a 'pattern'. The perceived pattern is sorted through the model's discrimination network by checking for the presence of pieces on the network's test links. When sorting stops, the node reached is compared with the perceived pattern to determine if further learning should occur. If the perceived pattern contains everything in the retrieved pattern and some more, then *familiarisation* occurs to add some more information to the stored chunk. If the perceived pattern contains some contradicting information to the retrieved pattern (such as a piece

[2] Although the item-on-square representation has its limitations, it is sufficient for modelling global properties of the chess board, see [27] for a discussion.

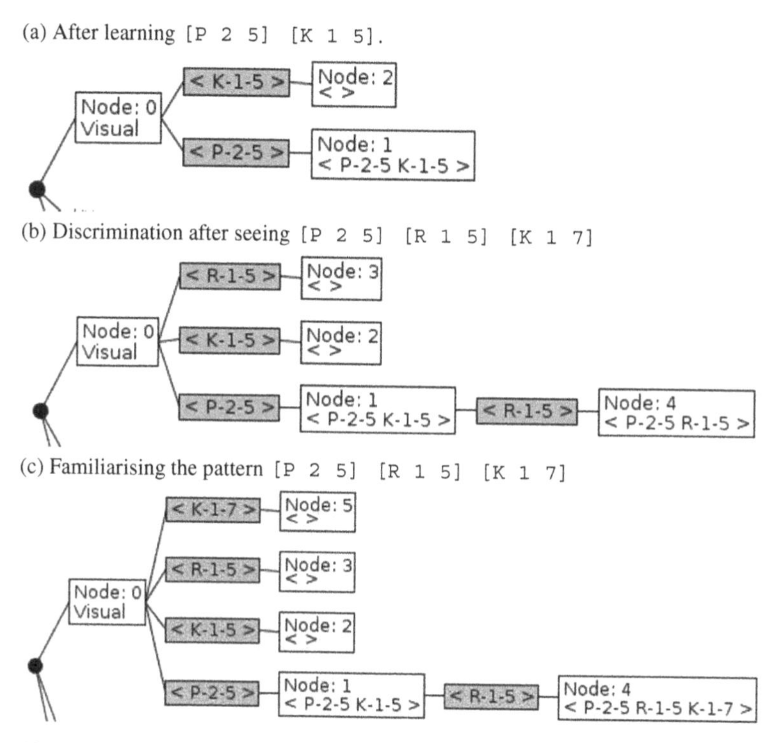


**Fig. 3** Illustration of familiarisation/discrimination learning process

on a different square), then *discrimination* occurs, to add a further test and node to the network. Thus, discrimination increases the number of distinct chunks that the model can identify, whereas familiarisation increases the amount of information that the model can retrieve about that chunk. Fig. 3 illustrates the two learning mechanisms.

## 3.3 Eye Fixations

CHREST combines the identification and learning of chunks with an active attention mechanism, for guiding its eye about the chess board. Essentially, the implicit goal in perception is to locate the largest chunks in LTM which represent the given stimulus. Once a chunk is identified and placed in STM, the location of a piece on one of the test links is used as the next fixation point; in this way, the LTM 'drives' the eye fixations to locations likely to contain information to recognise a chunk deeper in the LTM index, and hence containing more information.

As an illustration of how the eye fixations work, a model of chess positions was trained from a set of 8,644 positions, training continuing until the model's LTM contained 10,000 chunks, with each position scanned for 20 fixations. An indication

```
Fixations:
   (3, 6) First heuristic
   (4, 5) Random item heuristic
   (6, 7) Random item heuristic
   (7, 7) LTM heuristic
   (8, 6) Proposed movement heuristic
   (8, 5) Random place heuristic
   (1, 6) Random place heuristic
   (7, 1) LTM heuristic
   (6, 2) Random item heuristic
   (6, 3) Proposed movement heuristic
   (7, 4) Random item heuristic
   (4, 6) LTM heuristic
   (5, 4) Proposed movement heuristic
   (4, 6) LTM heuristic
   (5, 4) Proposed movement heuristic
   (4, 6) LTM heuristic
   (3, 4) Proposed movement heuristic
   (3, 3) LTM heuristic
   (5, 3) Random place heuristic
   (5, 6) Proposed movement heuristic
Chunks used:
   Node: 3371 < [P 4 5] [P 5 6] [P 6 7]
                [P 7 6] [P 8 7] >
   Node: 5780 < [P 1 7] [P 2 7] >
   Node: 435 < [P 4 5] [p 2 3] [p 4 4] [b 3 3] >
   Node: 788 < [P 5 6] [P 6 7] [P 7 6] [P 8 7]
               [K 7 8] [B 7 7] [b 7 5] [N 4 7] >
     Template:
        filled item slots:
         [R 6 8]
```

**Fig. 4** An illustration of the eye fixations made by the model

of the fixations is shown in Fig. 2 (right). Fig. 4 shows a trace of the model's eye fixations on the position in Fig. 2, where the coordinates on each line refer to the new location, and the description the reason for the move to that location. (Detailed analysis of the fit between the model and human behaviour is available [12].)

The eye fixations show the range of heuristics used in retrieving information: the eye initially is located at a random central square, and then proceeds using one of a set of heuristics. In this example, four heuristics are illustrated.

1. *Random item* fixates a previously unseen item within the field of view, but not at the centre.
2. *Random place* fixates a previously unseen square within the field of view, but not the centre.
3. *LTM* uses the largest chunk currently in STM to guide the next fixation to the location of information required to pass one of its test links.
4. *Proposed move* uses knowledge of how chess pieces move to guide the eye to objects of attack, even if located on the far side of the board.

At the end of its cycle of fixations, the model's STM contains pointers to nodes 3371, 5780, 435 and 788. The trace includes the images of these retrieved nodes; note that the last chunk listed was a template, and has one slot filled. The information in these four chunks and their links is the representation of the current stimulus.

### *3.4 Learning and retrieving categorisations*

We now extend CHREST's mechanisms to include the acquisition of interpretations of chess positions. The technique we use is based on that for learning links across modalities [28]. The idea is that information about the recognised chunks is used to determine the position's category. Being able to categorise positions is important for chess players, as this provides crucial information about the kind of plans to follow and moves to play. In particular, knowing the opening the position comes from is indicative of the strategic and tactical themes of the board position.

As each position is scanned, a count is maintained for each chunk determining how often that chunk is associated with the position's category: this count forms the *weighting* of the chunk to that category. For example, in the experiment described below the categories are the two openings, the French and Sicilian. Each chunk will have a count of how often it has been associated with each of these two openings. Smaller chunks (such as a pair of pawns) frequently occur in many categories, but other chunks are more indicative of particular openings (such as the line of central pawns characteristic of the French defence).

For categorising a position, the model records how many chunks it perceives which are indicative of a given category. If there are more chunks for one category, then that is the category for the position. If there is a tie, then the weightings for each chunk's association to a category are totalled, and the highest weighting wins.

## 4 Experiment in Categorising Chess Positions

We investigate how our model can learn to separate chess positions into their respective opening. Fig. 5 illustrates typical positions and their labels; note that positions are taken after the game has left its opening stage, and the models have no access to the moves of the game. A dataset of chess positions from two openings was created. The dataset contains 250 positions from the French defence and 250 positions from the Sicilian defence. We run a categorisation experiment on these data with a support-vector machine, to establish a base-line performance, and with CHREST.

As the test datasets may have an imbalanced distribution of classes, the geometric mean of accuracy is used as a performance measure: if the accuracy on the two individual classes is $a_1$ and $a_2$, then the geometric mean is $\sqrt{(a_1 \times a_2)}$. This measure has the advantage of treating the two classes equally, and severely penalising poor performance in any one [26].

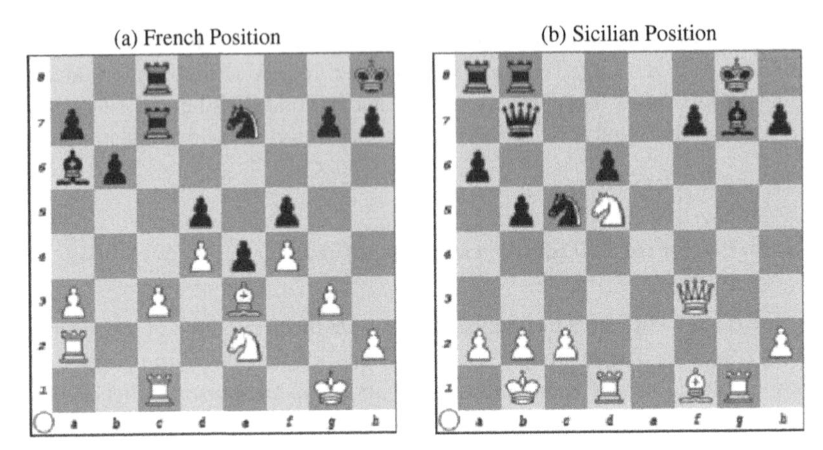


**Fig. 5** Example chess positions and their openings.

## 4.1 Support Vector Machine

For the support-vector machine experiment, the chess data were converted into features representing the 64-squares on the chess-board as a 64-element vector. Each element of the vector holds a number representing the piece at that location: 0 for empty, 1 for white pawn, up to 12 for black king. These numbers were then scaled into the range $[0,1]$, and a 0/1 class label applied representing if the position was from the French or Sicilian opening.

A support-vector machine model was created using the libsvm library [5]. The data were split randomly into 70% training and 30% testing. The training data were further split 4:1 for cross validation. A grid search for a radial-basis function kernel was performed over costs $\{2^{-5}, 2^{-3}, 2^{-1}, 2^{0}, 2^{1}, 2^{3}, 2^{5}, 2^{8}, 2^{10}, 2^{13}, 2^{15}\}$ and gammas $\{2^{-15}, 2^{-12}, 2^{-8}, 2^{-5}, 2^{-3}, 2^{-1}, 2^{1}, 2^{3}, 2^{5}, 2^{7}, 2^{9}\}$.

The best model had 112 support vectors and obtained a geometric mean of performance of 0.80 on the held-out test set.

## 4.2 CHREST

The dataset was split, randomly, into 70% for training and 30% for testing. A model was trained using the training data. Each position was scanned for 20 fixations, with the model learning both its discrimination network and weighted links to categories.

The model was then tested on the held-out test data. Each position was scanned for 20 fixations and the weighted categorisations retrieved for all of the perceived chunks to retrieve a category for the scanned position, as described above. This train/test process was repeated 10 times for each number of training cycles. The number of training cycles was varied from 1 to 40.

Fig. 6 shows the average over the 10 runs of the geometric mean of performance on the held-out test set, with error bars indicating $\pm 1$ standard deviation. The performance reaches and maintains a level around 0.80, which is equivalent to the performance of the support-vector machine model.

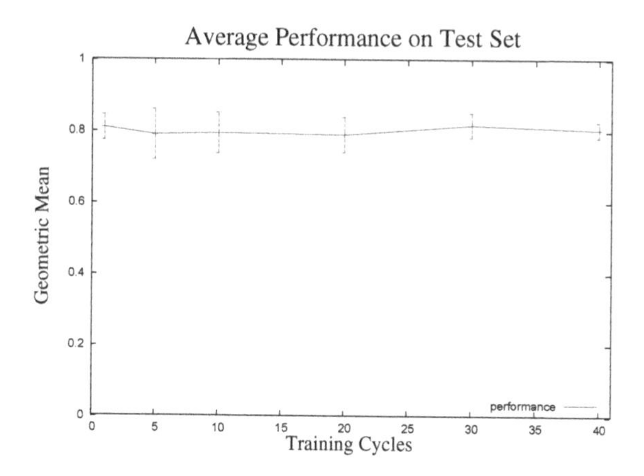


**Fig. 6** Average geometric-mean performance against training cycles.

## *4.3 Specialisation in long-term memory*

The long-term memory network grows both in size (the number of nodes) and in content (the size of the chunks stored in the nodes).

Fig. 7 shows a typical growth in the network size over time. The x-axis represents increasing number of training cycles, and the y-axis the average number of chunks within the network. The points are averaged from ten runs of the model, and error bars to $\pm 1$ s.d. are drawn. The graph shows a growth in network size over time, with increasing variation between models as the network gets larger. The rate of growth slows with increasing training cycles, as the model becomes familiar with the training data.

Fig. 8 shows a typical growth in the network content over time. The x-axis represents increasing number of training cycles, and the y-axis the average number of pieces. The solid line is the average size of chunks stored in all the nodes of the LTM, and the dashed line is the average depth of the nodes in the network. The depth approximates the amount of information required to *retrieve* a chunk, whereas the content gives the amount of information *stored* as a chunk. The points are averaged from ten runs of the model, and error bars to $\pm 1$ s.d. are drawn, but are relatively small (less than 0.5%).

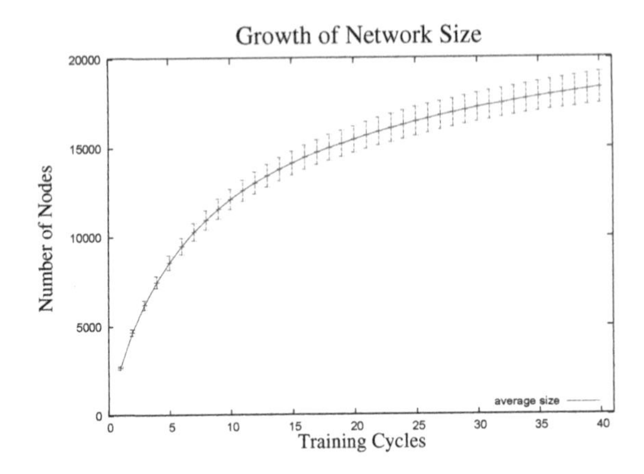


**Fig. 7** Typical increase in average network size.

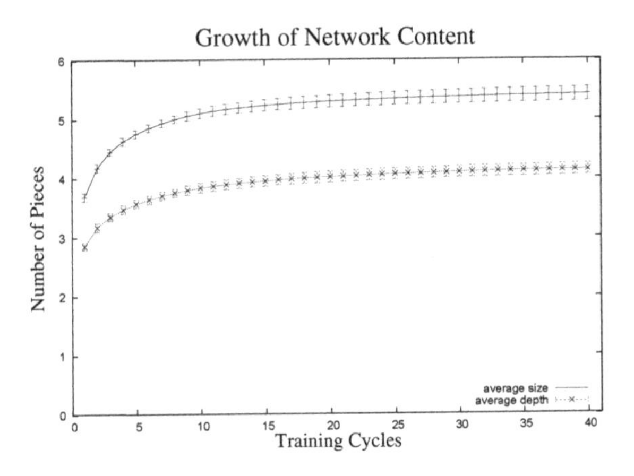


**Fig. 8** Typical increase in average network depth and average chunk size against training cycles.

The interesting aspect of this graph is the shape of the two lines. The line for the average depth increases sharply initially, as CHREST learns the general structure of the data. After about 5 passes through the data, this curve begins to flatten out, with an average depth of approximately 4. This indicates that CHREST has learnt the broad structure of the data, and correlates well with the change in rate of growth shown in Fig. 7.

The line for the average chunk size shows a similar increase over time, as the model acquires familiarity with the domain. The line for average chunk size is consistently higher than the average depth, showing that CHREST typically retrieves 30-50% more information than it requires to access that information.

## 5 Conclusion

We have discussed how human performance in the game of chess presents a challenge to theories of artificial intelligence: how do humans acquire and efficiently index a large store of knowledge for efficient problem solving? We have presented the CHREST model of human perception and learning, and extended it to handle a task of categorising chess positions based on the opening they come from. Experimental results demonstrate that the the process of learning chunks enables CHREST to categorise positions with a similar reliability to a statistical learning algorithm. Analysis of the network over time illustrates the trade off in learning how to *retrieve* chunks against learning the *content* of those chunks.

In future work, we will extend this model of categorisation to support more complex interpretations of chess positions, interpretations to support quality game playing with a minimum of look-ahead search.

## References

1. Anderson, J.R.: The Architecture of Cognition. Cambridge, MA: Harvard University Press (1983)
2. Berliner, H.J.: Search vs. knowledge: An analysis from the domain of games. Tech. Rep. No. CMU-CS-82-104. Pittsburgh: Carnegie Mellon University, Computer Science Department (1981)
3. Bilalić, M., McLeod, P., Gobet, F.: Specialization effect and its influence on memory and problem solving in expert chess players. Cognitive Science **33**, 1117–1143 (2009)
4. Bossomaier, T., Traish, J., Gobet, F., Lane, P.C.R.: Neuro-cognitive model of move location in the game of Go. In: Proceedings of the 2012 International Joint Conference on Neural Networks (2012)
5. Chang, C.C., Lin, C.J.: LIBSVM: a library for support vector machines (2001). Software available at `http://www.csie.ntu.edu.tw/~cjlin/libsvm`
6. Chase, W.G., Simon, H.A.: The mind's eye in chess. In: W.G. Chase (ed.) Visual Information Processing, pp. 215–281. New York: Academic Press (1973)
7. Chaslot, G., Bakkes, S., Szita, I., Sponck, P.: Monte-carlo tree search: A new framework for game AI. In: Proceedings of the Twentieth Belgian-Dutch Artificial Intelligence Conference, pp. 389–390 (2008)
8. Chassy, P., Gobet, F.: Measuring chess experts' single-use sequence knowledge using departure from theoretical openings: An archival study. PLos ONE **6** (2011)
9. Christianini, N., Shawe-Taylor, J.: An introduction to Support Vector Machines. The Press Syndicate of the University of Cambridge (2000)
10. Cortes, C., Vapnik, V.N.: Support-vector networks. Machine Learning **20**, 273–297 (1995)
11. de Groot, A.D.: Thought and Choice in Chess (First edition in 1946). The Hague: Mouton (1978)
12. de Groot, A.D., Gobet, F.: Perception and Memory in Chess: Heuristics of the Professional Eye. Assen: Van Gorcum (1996)
13. Freudenthal, D., Pine, J.M., Gobet, F.: Simulating the referential properties of Dutch, German and English root infinitives in MOSAIC. Language Learning and Development **15**, 1–29 (2009)
14. Frey, P.W., Adesman, P.: Recall memory for visually presented chess positions. Memory and Cognition **4**, 541–547 (1976)

15. Freyhoff, H., Gruber, H., Ziegler, A.: Expertise and hierarchical knowledge representation in chess. Psychological Research **54**, 32–37 (1992)
16. Gardner, H.: The Mind's New Science: A History of the Cognitive Revolution. New York: Basic Books (1987)
17. Gobet, F.: Using a cognitive architecture for addressing the question of cognitive universals in cross-cultural psychology: The example of awalé. Journal of Cross-Cultural Psychology **40**, 627–648 (2009)
18. Gobet, F., Clarkson, G.: Chunks in expert memory: Evidence for the magical number four... or is it two? Memory **12**, 732–47 (2004)
19. Gobet, F., de Voogt, A., Retschitzki, J.: Moves in Mind: The Psychology of Board Games. Hove and New York: Psychology Press (2004)
20. Gobet, F., Lane, P.C.R., Croker, S.J., Cheng, P.C.H., Jones, G., Oliver, I., Pine, J.M.: Chunking mechanisms in human learning. Trends in Cognitive Sciences **5**, 236–243 (2001)
21. Gobet, F., Simon, H.A.: Recall of random and distorted positions: Implications for the theory of expertise. Memory & Cognition **24**, 493–503 (1996)
22. Gobet, F., Simon, H.A.: Templates in chess memory: A mechanism for recalling several boards. Cognitive Psychology **31**, 1–40 (1996)
23. Gobet, F., Simon, H.A.: Five seconds or sixty? Presentation time in expert memory. Cognitive Science **24**, 651–82 (2000)
24. Jones, G.A., Gobet, F., Pine, J.M.: Linking working memory and long-term memory: A computational model of the learning of new words. Developmental Science **10**, 853–873 (2007)
25. Lane, P.C.R., Cheng, P.C.H., Gobet, F.: Learning perceptual schemas to avoid the utility problem. In: M. Bramer, A. Macintosh, F. Coenen (eds.) Research and Development in Intelligent Systems XVI: Proceedings of ES99, the Nineteenth SGES International Conference on Knowledge-Based Systems and Applied Artificial Intelligence, pp. 72–82. Cambridge, UK: Springer-Verlag (1999)
26. Lane, P.C.R., Clarke, D., Hender, P.: On developing robust models for favourability analysis: Model choice, feature sets and imbalanced data. Decision Support Systems **53**, 712–18 (2012)
27. Lane, P.C.R., Gobet, F.: Perception in chess and beyond: Commentary on Linhares and Freitas (2010). New Ideas in Psychology **29**, 156–61 (2011)
28. Lane, P.C.R., Sykes, A.K., Gobet, F.: Combining low-level perception with expectations in CHREST. In: F. Schmalhofer, R.M. Young, G. Katz (eds.) Proceedings of EuroCogsci, pp. 205–210. Mahwah, NJ: Lawrence Erlbaum Associates (2003)
29. Langley, P., Laird, J.E., Rogers, S.: Cognitive architectures: Research issues and challenges. Cognitive Systems Research **10**, 141–160 (2009)
30. Linhares, A., Freitas, A.E.T.A.: Questioning Chase and Simon's (1973) "Perception in chess": The "experience recognition" hypothesis. New Ideas in Psychology **28**, 64–78 (2010)
31. Michie, D.: King and rook against king: Historical background and a problem on the infinite board. In: M.R.B. Clarke (ed.) Advances in computer chess I, pp. 30–59. Edinburgh: University Press (1977)
32. Miller, G.A.: The magical number seven, plus or minus two: Some limits on our capacity for processing information. Psychological Review **63**, 81–97 (1956)
33. Newell, A., Simon, H.A.: The simulation of human thought. Mathematics Division, The RAND Corporation, P-1734 (1959)
34. Saariluoma, P.: Location coding in chess. The Quarterly Journal of Experimental Psychology **47A**, 607–630 (1994)
35. Samsonovich, A.: Toward a unified catalog of implemented cognitive architectures. In: Proceedings of the 2010 Conference on Biologically Inspired Cognitive Architectures, pp. 195–244. Amsterdam, The Netherlands: IOS Press (2010)
36. Shannon, C.E.: A chess-playing machine. Philosophical magazine **182**, 41–51 (1950)
37. Stewart, D.: Herbert Simon: Thinking machines. Interview conducted June 1994. Transcript available at http://www.astralgia.com/webportfolio/omnimoment/archives/interviews/simon.html (1997)
38. Turing, A.M.: Digital computers applied to games. In: B.V. Bowden (ed.) Faster than thought, pp. 286–310. London: Pitman (1953)

# PLANNING AND OPTIMISATION

# S-Theta*: low steering path-planning algorithm

Pablo Muñoz and María D. R-Moreno

**Abstract** The path-planning problem for autonomous mobile robots has been addressed by classical search techniques such as A* or, more recently, Theta*. However, research usually focuses on reducing the length of the path or the processing time. Applying these advances to autonomous robots may result in the obtained "short" routes being less suitable for the robot locomotion subsystem. That is, in some types of exploration robots, the heading changes can be very costly (i.e. consume a lot of battery) and therefore may be beneficial to slightly increase the length of the path and decrease the number of turns (and thus reduce the battery consumption). In this paper we present a path-planning algorithm called S-Theta* that *smoothes* the turns of the path. This algorithm significantly reduces the heading changes, in both, indoors and outdoors problems as results show, making the algorithm especially suitable for robots whose ability to turn is limited or the associated cost is high.

## 1 Introduction

Path-planning is focused on finding an obstacle-free path between an initial position and a goal, trying as far as possible that this path is optimal. The path-planning problem representation as a search tree over discretized environments with blocked or unblocked squares has been widely discussed. Algorithms such as A* [1] allow one to quickly find routes at the expense of an artificial restriction of heading changes of $\pi/4$. However, there have been many improvements such as its application to non-uniform costs maps in Field D* [2] or, more recently, Theta* [3] which aims

Pablo Muñoz
Departamento de Automática, Universidad de Alcalá, e-mail: pmunoz@aut.uah.es

María D. R-Moreno
Departamento de Automática, Universidad de Alcaá, e-mail: mdolores@aut.uah.es

to remove the restriction on heading changes that generates A* and gets better path lengths. The main difference between A* and Theta* is that the former only allows that the parent of a node is its predecessor, while in the last, the parent of a node can be any node. This property allows Theta* to find shorter paths with fewer turns compared to A*. However, this improvement implies a higher computational cost due to additional operations to be performed in the expansion nodes process as it will be explained in section 3. Other approximations want to reduce the processing time via heuristics [4] or improving the efficiency of the algorithms [5]. It is worth mentioning that these algorithms work on fully observable environments except Field D*, that can face partially observable environments applying a replanning scheme.

When designing the robot control system, we must take into consideration the morphology of the robot and the application environment. In this paper we focus on the path-planning problem for autonomous mobile robots in open environments with random obstacles as well as indoor areas with corridors and interconnected rooms.

In [6], an extensive comparison between Theta*, A*, Field D* and A*Post Smoothed (A*PS) [7] is performed. A*PS is a variant of A* that is based on smoothing the path obtained by A* at a later stage. As a result we can see that Theta* is the one that usually gets the shortest routes. However, depending on the features of the robot locomotion subsystem, it may be preferable solutions with longer distances but with lower number of heading changes.

Therefore, in this paper we propose the S-Theta* algorithm with an evaluation function that optimizes the heading changes of the route. The field of application is straight forward: robots whose rotation cost is greater than the movement in straight line (what is a vast majority). The paper is organized as follows. Next section shows the notation and the modelization of the terrain used. Then, section 3 describes the Theta* algorithm. The S-Theta* algorithm definition is given in section 4. The results obtained when comparing both algorithms will be shown in section 5. Finally, the conclusions will be outlined.

## 2 Grid Definition and Notation

The most common terrain discretization in path-planning is a regular grid with blocked and unblocked square cells [8]. For this kind of grids we can find two variants: (i) the center-node (fig. 1 left) in which the mobile element is in the center of the square; and (ii) corner-node (fig. 1 right), where nodes are the vertex of the square. For both cases, a valid path is that starting from the initial node reaches the goal node without crossing a blocked cell. In our experiments we have employed the corner-node approximation, but our algorithm also works with the center-node representation.

A node is represented as a lowercase letter, assuming $p$ a random node and, $s$ and $g$ the start and goal nodes respectively. Each node is defined by its coordinate pair $(x,y)$, being $x_p$ and $y_p$ for the $p$ node. A solution has the form $(p_1, p_2, ..., p_{n-1}, p_n)$

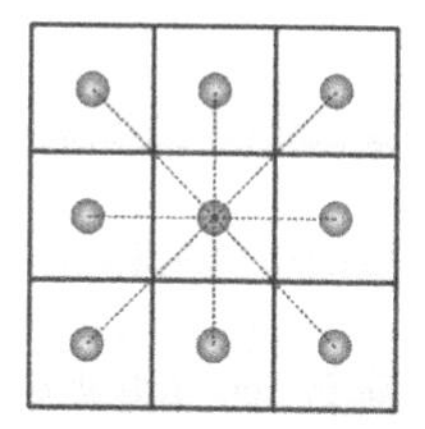
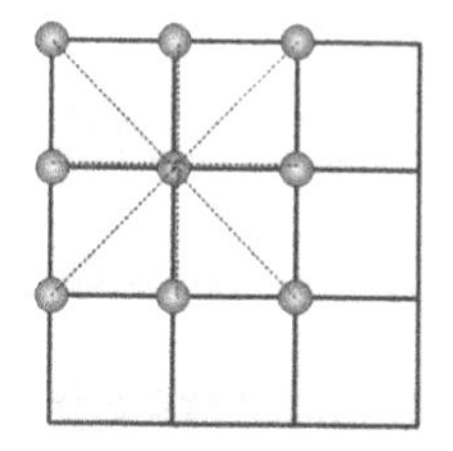

**Fig. 1** Possible node representations in grids. Left: center-node; right: corner-node.

with initial node $p_1 = s$ and goal $p_n = g$. As well, we have defined four functions related to nodes: (i) function $succ(p)$ that returns a subset with the visible neighbours of $p$; (ii) function $parent(p)$ that indicates who is the parent node of $p$; (iii) $dist(p,t)$ that represents the straight line distance between nodes $p$ and $t$ (calculated through the eq. 1); and (iv) function $angle(p,q,t)$ that gives as a result the angle (in degrees) formed by segments $\overline{pq}$ and $\overline{pt}$ (that is, the angle in opposition of the $\overline{qt}$ segment) of the triangle formed by $\triangle pqt$, in the interval $[0°, 180°)$.

$$\text{dist}(p,t) = \sqrt{(x_t - x_p)^2 + (y_t - y_p)^2} \quad (1)$$

## 3 Theta* Algorithm

Theta* [3, 6] is a variation of A* for any-angle path-planning on grids. It has been adapted to allow any-angle route, i.e. it is not restricted to artificial headings of $\pi/4$ as is the case of A* with 8 neighbors. Although there are works like [9] that subdivide cells easing this restriction, or others that use more neighbours, but they are ad-hoc solutions. There are two variants for Theta*: Angle-Propagation Theta* [3] and Basic Theta* [6]. We assume that talking about Theta* refers to the last one. The difference between these two versions is how the calculation of the line of sight between pairs of nodes is performed, being in Basic Theta* simpler but with a higher computational complexity in the worst case than the variant Angle-Propagation. Moreover, the latter sometimes indicates that two nodes do not have line of sight when in fact they do, slightly increasing the length of the resulting path.

Theta* shares most of the code of the A* algorithm, being the main block identical (alg. 1) and the only variation is in the `UpdateVertex` function (alg. 2). Thus, all the nodes will store the following values:

- $G(t)$: the cumulative cost to reach the node $t$ from the initial node.
- $H(t)$: the heuristic value, i.e, the estimated distance to the goal node. In the case of A* uses the Octile heuristic, while Theta* uses the Euclidean distance as shown in eq. 1.
- $F(t)$: the node evaluation function expressed by eq. 2.

$$\text{F}(t) = \text{G}(t) + \text{H}(t) \quad (2)$$

- *parent*$(t)$: the reference to the parent node. For A* we must have $parent(t) = p \Rightarrow t \in succ(p)$, however, Theta* eliminates this restriction and allows that the parent node of $t$ is any node $q$ accessible whenever there is a line of sight between $t$ and $q$.

Theta* will expand the most promising nodes in the order established in the open list, i.e. it first expands the nodes with lower values for $F(t)$ as shown in alg. 1. In the case that the expanded node is the goal, the algorithm will return the path by traversing the parents pointers backwards from the goal to the start node. If instead the open list is empty, it means that it is impossible to reach the goal node from the initial node and the algorithm will return a failure.

**Algorithm 1** Theta* algorithm

```
1  G(s) ← 0
2  parent(s) ← s
3  open ← ∅
4  open.insert(s, G(s), H(s))
5  closed ← ∅
6  while open ≠ ∅ do
7      p ← open.pop()
8      if p = g then
9          return path
10     end if
11     closed.insert(p)
12     for t ∈ succ(p) do
13         if t ∉ closed then
14             if t ∉ open then
15                 G(t) ← ∞
16                 parent(t) ← null
17             end if
18             UpdateVertex(p,t)
19         end if
20     end for
21 end while
22 return fail
```

When dealing with the successor nodes, the first thing that Theta* does is to check if among the successors of the current position $p$, being $t \in succ(p)$, and the parent of the current node $q = parent(p)$, there is a line of sight. The high computational cost associated to this checking process is due to the algorithm used (a variant for drawing lines [10] that only uses integer operations for verification) which linearly grows as a function of the distance between the nodes to evaluate. In case of no line of sight, the algorithm behaves as A* and the parent node of $t$ is $p$. However, if the node $t$ can be reached from $q$, the algorithm will check whether the node $t$ has already been reached from another node, and then, only update the node $t$ if the cost of reaching it from $q$ is less than the previous cost. In this case, the parent of $t$ will be $q$ and the node is inserted into the open list with the corresponding values of $G(t)$ and $H(t)$ as shown in alg. 2. Following the described expansion process, Theta* only has heading changes at the edges of the blocked cells.

**Algorithm 2** Update vertex function for Theta*

```
1  UpdateVertex(p, t)
2  if LineOfSight(parent(p),t) then
3      if G(parent(p)) + dist(parent(p),t) < G(t) then
4          G(t) ← G(parent(p)) + dist(parent(p),t)
5          parent(t) ← parent(p)
6          if t ∈ open then
7              open.remove(t)
8          end if
9          open.insert(t, G(t), H(t))
10     end if
11 else
12     if G(p) + dist(p,t) < G(t) then
13         G(t) ← G(p) + dist(p,t)
14         parent(t) ← p
15         if t ∈ open then
16             open.remove(t)
17         end if
18         open.insert(t, G(t), H(t))
19     end if
20 end if
```

## 4 S-Theta* Algorithm

The Smooth Theta* (S-Theta*) algorithm that we have developed from Theta*, aims to reduce the amount of heading changes that the robot should perform to reach the goal. To do this, we have based on a modified cost function $F(t)$, as shown in eq. 3.

$$\mathrm{F}(t) = \mathrm{G}(t) + \mathrm{H}(t) + \alpha(t) \tag{3}$$

The new term $\alpha(t)$ gives us a measure of the deviation from the optimal trajectory to achieve the goal as a function of the direction to follow, conditional to traversing a node $t$. Considering an environment without obstacles, the optimal path between two points is the straight line. Therefore, applying the triangle inequality, any node that does not belong to that line will involve both, a change in the direction and a longer distance. Therefore, this term causes that nodes far away from that line will not be expanded during the search. The definition of $\alpha(t)$ is given in def. 1 and is represented graphically in fig. 2.

**Definition 1.** $\alpha(t)$ represents the deviation in the trajectory to reach the goal node $g$ through the node $t$ in relation to the straight-line distance between the parent of its predecessor ($t \in succ(p)$ and $parent(p) = q$) and the goal node.

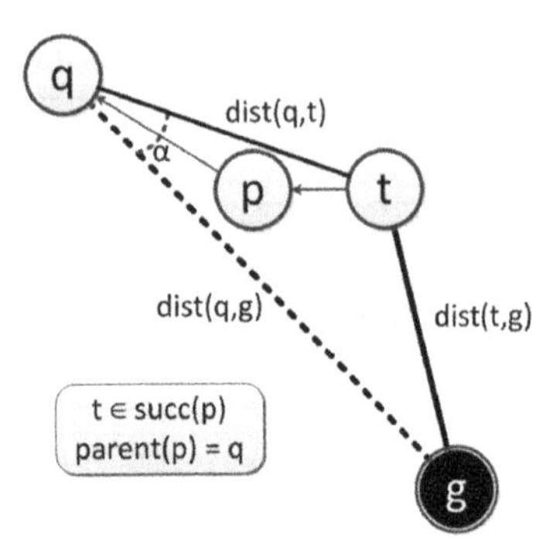


**Fig. 2** Graphical representation of $\alpha$. Actual position is $p$ with $parent(p) = q$. The successor considered is $t$.

To compute $\alpha(t)$ we have used eq. 4. A priori, it could seem interesting to use the triangle formed by the initial, goal and actual nodes. However, this line stays unchanged during the search, what implies that the algorithm would tend to follow the line connecting the initial node to the goal node. This would cause undesirable heading changes since one could rarely follow that route.

Then, we use the triangle $\triangle qtg$ for $\alpha(t)$ computation, where $q = parent(p)$ and $t \in succ(p)$. The result is that once the initial direction has changed, the algorithm tries to find the new shortest route between the successor to the current position, $t$, and the goal node. The shortest route will be, if there are no obstacles, the one with $\alpha(t) = 0$, i.e., the route in which the successor of the current node belongs to the line connecting the parent node of the current position and the goal node. Figure 3 shows how the value of $\alpha(t)$ evolves as the search progresses.

$$\alpha(t) = \arccos \frac{dist(q,t)^2 + dist(q,g)^2 - dist(t,g)^2}{2 \cdot dist(q,t) \cdot dist(q,g)} \quad \text{with } t \in succ(p) \text{ and } parent(p) = q \tag{4}$$

The alg. 3 shows the pseudocode of the function `UpdateVertex` for S-Theta*. For computation purposes, $\alpha(t)$ will be included as a cost in the evaluation function of the nodes, so the algorithm will also discriminate the nodes in the open list depending on the orientation of the search. Thus, a node in the open list may be replaced (which means that its parent will be changed) due to a lower value of $\alpha(t)$.

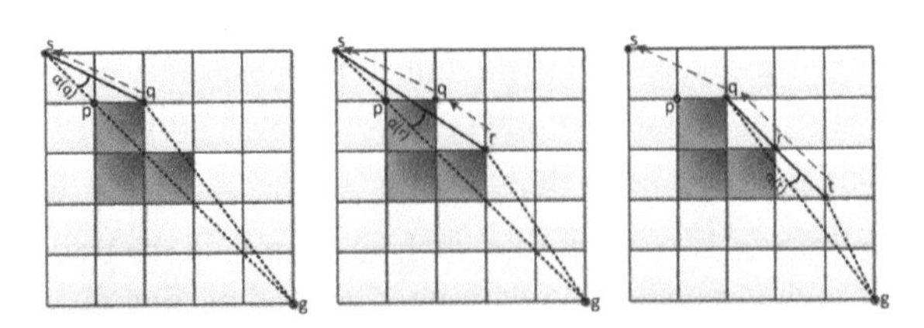

**Fig. 3** Representation of the evolution of $\alpha(t)$. Arrows are pointed to the parent of the node after expansion.

In contrast, Theta* updates a node depending on the distance to reach it, regardless of its orientation. As a result, the main difference with respect to Theta* is that S-Theta* can produce heading changes at any point, not only at the vertex of the obstacles as seen in fig. 4.

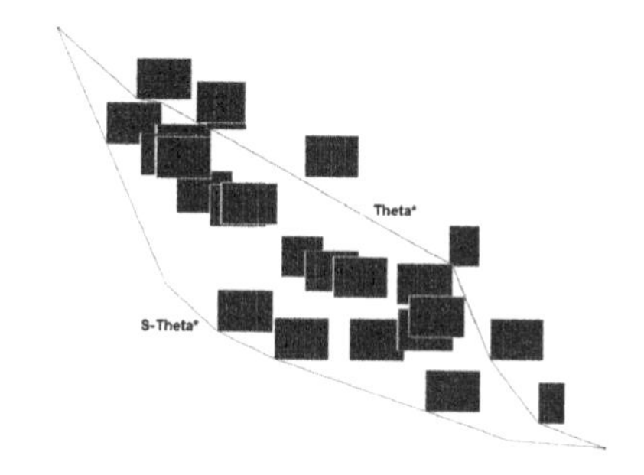


**Fig. 4** Solution paths for Theta* and S-Theta* in a random map. Theta* only has heading changes at vertices of blocked cells, while S-Theta* not.

**Algorithm 3** Update vertex function for S-Theta*

1 **UpdateVertex(p, t)**
2 $\alpha(t) \leftarrow angle(parent(p), t, g)$
3 **if** $\alpha = 0$ **or** $LineOfSight(parent(p), t)$ **then**
4 $\quad G_{aux} \leftarrow G(parent(p)) + dist(parent(p), t) + \alpha(t)$
5 $\quad$ **if** $G_{aux} < G(t)$ **then**
6 $\quad\quad G(t) \leftarrow G_{aux}$
7 $\quad\quad parent(t) \leftarrow parent(p)$
8 $\quad\quad$ **if** $t \in open$ **then**
9 $\quad\quad\quad open.remove(t)$
10 $\quad\quad$ **end if**
11 $\quad\quad open.insert(t, G(t), H(t))$
12 $\quad$ **end if**
13 **else**
14 $\quad G_{aux} \leftarrow G(p) + dist(parent(p), t) + \alpha(t)$
15 $\quad$ **if** $G_{aux} < G(t)$ **then**
16 $\quad\quad G(t) \leftarrow G_{aux}$
17 $\quad\quad parent(t) \leftarrow p$
18 $\quad\quad$ **if** $t \in open$ **then**
19 $\quad\quad\quad open.remove(t)$
20 $\quad\quad$ **end if**
21 $\quad\quad open.insert(t, G(t), H(t))$
22 $\quad$ **end if**
23 **end if**

$\alpha(t)$ affects the order of the open list, and thus, how the nodes are expanded. So we need to take into consideration the weight of this term in the evaluation function. If the relative weight in the evaluation function is to small, the algorithm works like the original Theta*, but if it is excessive, it possible that the deal between path-length and heading changes are not fine. $\alpha(t)$ takes values in the interval $[0°, 180°]$. Considering a map with 100x100 nodes, the cost of transversing from one corner to its opposite corner is $100\sqrt{2} \approx 141$, so the cost added by $\alpha(t)$ is usually less than $90°$, a little bit more than the half the cost for traversing among opposite corners of the map. This implies that the angle can grows near the 50% weight in the evaluation function, making that nodes with good $G+H$ and higher $\alpha(t)$ values go back into the open list due to nodes with worse $G+H$ values but with better $\alpha(t)$, and this is that we want to minimize. For this reason we consider that $\alpha(t)$ is well sized in relationship with the $G(t)+H(t)$ values. However, for smaller or bigger maps this shall not be valid. For example, for 50x50 nodes maps, the relative weight of $\alpha(t)$ is double than for a 100x100 nodes map and, for 500x500 nodes maps, is the fifth part. In the first case, the penalization implies an excessive cost to expand nodes that are a little bit far away from the line between $s$ and $g$, whereas in the second case $\alpha(t)$ has less effect in the search process. Then, the algorithm tends to behave like the original one, that is, both expand a similar number of nodes. In order to compensate this fact, we redefine the value of $\alpha(t)$ as shown in eq. 5, taking into consideration a map with NxN nodes. In the experiments this estimation works fine, so we consider that it is valid.

$$\alpha(t) = \alpha(t) \cdot \frac{\mathrm{N}}{100} \tag{5}$$

## 4.1 Implementation issues

We suggest two procedures to improve the implementation of the S-Theta* algorithm. The $\alpha(t)$ computation degrades the S-Theta* performance respect Theta* because of both, the cost of floating point operations and the increase of the cost of checking the lines of sight (to make less heading changes, the algorithm needs to check the line of sight for bigger map sections). However, the degradation is not significant in terms of CPU time as we will see in the experimental section. This is thanks to the optimization procedures that will be outlined here and that S-Theta* expands less nodes than the original Theta*.

Procedure 1 The distance between the node $t$ and the goal node ($dist(t,g)$) is static and we can save many operations if once it has been calculated is stored as a property of the node. Before computing the distance between a node and the goal node we will check if this operation has already been performed, so we can save this computation. We only need to initialize this data to a negative value in the instantiation of the nodes prior to the search.

Procedure 2 If $\alpha(t)$ is 0, it means that the node and the predecessor of its parent are in the same line, i.e. the nodes $parent(p)$, $p$ and $t$ are in the same line. Therefore, and given that $t \in succ(p)$ and being the set $succ(p)$ the neighbours reachable from $p$, it follows that there is a line of sight between $t$ and $parent(p)$. This saves the line of sight checking.

# 5 Experimental Results

In this section we show the results obtained by comparing the algorithms Theta* and S-Theta*. We also include as a reference the base of both algorithms, A*. The following subsections show the results obtained by running the algorithms on outdoors (random obstacles) maps and indoor maps with interconnected rooms and corridors. For both cases, we have taken into consideration the average values for the following parameters: (i) the length of the path, (ii) the total number of accumulated degrees by the heading changes, (iii) the number of expanded nodes during the search, (iv) the CPU time or search runtime, and, (v) the number of heading changes.

The algorithms are implemented in Java and all of them use the same methods and structures to manage the grid information. All non-integer operations use floating precision. The execution is done on a 2 GHz Intel Core i7 with 4 GB of RAM under Ubuntu 10.10 (64 bits). To measure the runtime we have employed `System.currentTimeMillis()`.

## 5.1 *Outdoors maps*

In the outdoors maps we have not exploited the Digital Elevation Model (DEM) to model the terrain. For classical algorithms such as A*, including the height of the points does not make any significant difference when working with the DEM, since we can know the height of all points since the movement is restricted to the vertices. However, in any-angle algorithms, we can traverse a cell at any point. Then, how to calculate or approximate that value is important. In this paper we have not considered the height, although extending the Theta algorithms it is not a difficult task but out of the scope of this paper.

Table 1 shows the results obtained from the generation of 10000 random maps of 500x500 nodes, gradually increasing the percentages of blocked cells to 5%, 10%, 20%, 30% and 40% (each obstacle group has 2000 problems). The way to generate the maps guarantee that there will be at least a valid path from any starting point to the goal. To do that, each time an obstacle is randomly introduced, we force that around the obstacle there are free cells and these free cells cannot overlap with another obstacle. Bold data represent the best values for each parameter measured. In all cases the initial position corresponds to the coordinates (0, 0) and the objective is to reach a node in the last column randomly chosen from the bottom fifth (499, 400-499). Figure 5 shows the last parameter considered to compare the algorithms (not shown in the table) that is the heading changes.

As we can see, the algorithm that obtains the shorter routes is Theta*, however, for the accumulated cost of the heading changes (total spin), the best values are obtained by S-Theta*. The path degradation that occurs in S-Theta* respect Theta* is 3.6% higher for the worse case (40% of blocked cells), while the improvement in the number of heading changes is around 29% in S-Theta*. For the same percentage of obstacles, Theta* performs, in average, 9.555 turns while the number of turns in

**Table 1** Experimental results for groups of 2000 random maps of 500x500 nodes

| *Blocked* | A* | Theta* | S-Theta* | A* | Theta* | S-Theta* |
|---|---|---|---|---|---|---|
| | **Path length** | | | **Total spin (degrees)** | | |
| **5 %** | 690.942 | **674.530** | 676.984 | 340.785 | **15.231** | 19.878 |
| **10 %** | 697.538 | **677.510** | 683.368 | 457.605 | 38.887 | **38.776** |
| **20 %** | 712.171 | **686.965** | 699.650 | 682.290 | 104.723 | **82.101** |
| **30 %** | 727.090 | **697.661** | 717.739 | 898.875 | 204.670 | **148.129** |
| **40 %** | 744.137 | **711.997** | 737.732 | 1151.685 | 350.847 | **249.910** |
| | **Expanded nodes** | | | **Runtime (msec)** | | |
| **5 %** | 18822.062 | 13193.082 | **9257.260** | 953.566 | 1074.314 | **829.230** |
| **10 %** | 23613.764 | 22844.558 | **16265.886** | **1175.865** | 1595.634 | 1271.088 |
| **20 %** | 31897.962 | 35410.890 | **25873.328** | **1439.859** | 2027.954 | 1803.116 |
| **30 %** | 39250.270 | 44620.867 | **33491.780** | **1516.181** | 2084.500 | 2111.310 |
| **40 %** | 43999.836 | 50526.354 | **41109.494** | **1390.862** | 1871.207 | 2375.552 |

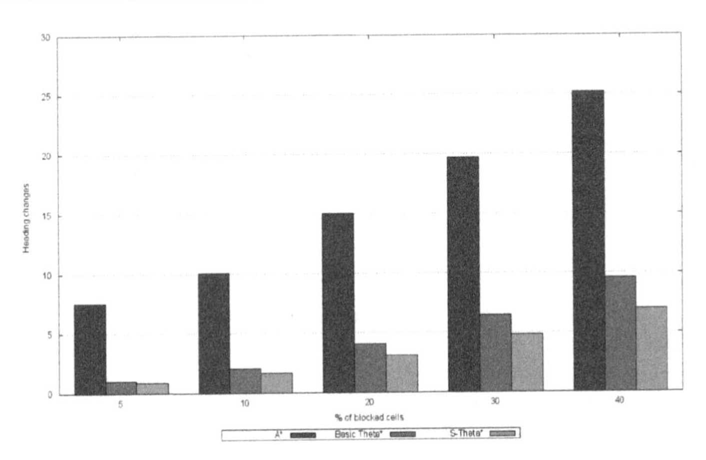


**Fig. 5** Average heading changes for groups of 2000 random maps of 500x500 nodes with different number of obstacles.

S-Theta* is reduced to 6.994 (26.8% less). Both algorithms always improve A* in all the parameters except the runtime. For the number of expanded nodes S-Theta* expands near 28% fewer nodes that Theta* except in the case of 40% of blocked cells where the improvement drops to 19%. In the case of time spent in searching the best results are for A*, except for the case of 5% of obstacles, in which S-Theta* gets the best result. However, S-Theta* shows being slightly faster than Theta* with less obstacles, only being surpassed by the original with 30% of obstacles (S-Theta* is only 0.013% slower) and with 40% of obstacles the runtime degradation is remarkable (21% slower). This is due to the increment in the number of expanded nodes: with 40% of blocked cells Theta* expands 5600 nodes more than his execution over maps with 30% of blocked cells, on the other hand, S-Theta* expands close to 7600 nodes more for the same conditions.

## *5.2 Indoor maps*

For indoor maps, we have run the algorithms over 1800 maps with different sizes, 150x150, 300x300 and 450x450 nodes (600 maps per size), always starting from the upper left corner (0, 0) and reaching the target in the opposite corner bottom. The indoor maps are generated from the random combination of 30x30 nodes square patterns that represent different configurations of corridors and rooms. These patterns are designed in a way that we can access to the next pattern through doors on each side, symmetrically placed. Table 2 shows the results for the four comparison criteria, where the best results are highlighted in bold. Figure 6 shows the heading changes using a plot representation.

**Table 2** Experimental results for groups of 600 indoor maps with different sizes

| *Nodes* | **A*** | **Theta*** | **S-Theta*** | **A*** | **Theta*** | **S-Theta*** |
|---|---|---|---|---|---|---|
| | **Path length** | | | **Total spin (degrees)** | | |
| **150x150** | 247.680 | **238.130** | 248.425 | 1151.025 | 650.224 | **584.231** |
| **300x300** | 497.740 | **480.626** | 513.876 | 2074.650 | 1313.495 | **1098.458** |
| **450x450** | 746.997 | **722.412** | 780.907 | 3025.950 | 1918.551 | **1590.245** |
| | **Expanded nodes** | | | **Runtime (msec)** | | |
| **150x150** | 8260.947 | 8943.435 | **4363.208** | 312.105 | 301.440 | **270.228** |
| **300x300** | 36379.942 | 41096.000 | **18203.616** | **1101.347** | 1281.430 | 1191.517 |
| **450x450** | 83993.192 | 96933.753 | **40562.192** | **3666.833** | 4530.158 | 5230.250 |

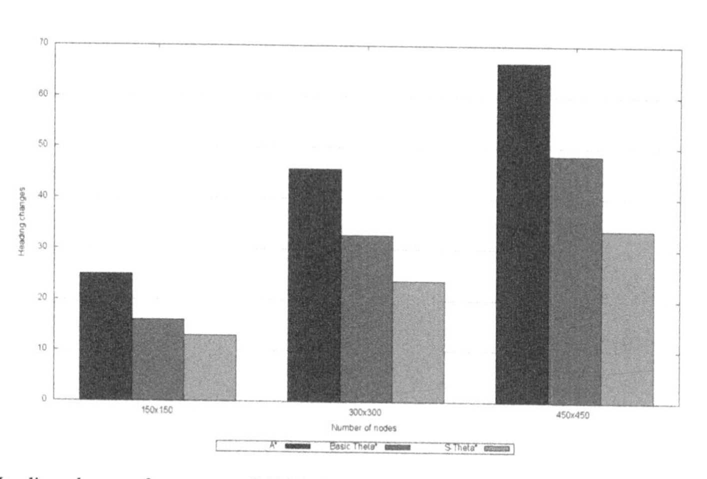


**Fig. 6** Heading changes for groups of 600 indoor maps with different sizes.

The data obtained for indoor maps are similar, in general terms, to those obtained in outdoor maps. In all cases the path length is shorter in Theta* than the rest of algorithms, although in this case A* obtains better results than S-Theta*. Furthermore, the path degradation in S-Theta* respect Theta* is 4% higher for the smallest size maps, rising up to 8% in the 450x450 nodes maps. However, for the accumulated cost of the heading changes, the difference is more significant than outdoors maps, getting S-Theta* the best results. For 150x150 nodes maps, S-Theta* uses 10% less turns, whereas in larger size maps this percentage is increased to 17%. In the heading changes there is a clear advantage of S-Theta* above the rest. In bigger maps we can appreciate the difference: Theta* performs on average 48.433 heading changes while this value in S-Theta* drops to 33.736 (about 30% lower). Also, it is even more drastic the reduction in the number of expanded nodes, Theta* expands more than double of nodes that S-Theta*, being S-Theta* who obtains the best results. Finally, the best execution time for small maps is obtained by S-Theta*, but is A* the winner for bigger maps. Comparing S-Theta* and Theta* for 300x300 nodes maps, S-Theta* is 7% faster than S-Theta*, while for 450x450 nodes maps, Theta is 13% faster than S-Theta*.

Finally, fig. 7 shows the sum of the length of the paths, and the total cumulative degree of heading changes for each algorithm. The data are normalized for comparison with Theta*, showing that for the A* algorithm the results are worse than for Theta*, while in the case of S-Theta* always keeps the values below, starting from 0.93 in 150x150 nodes maps and it gets stabilized around 0.89 from 300x300 nodes maps.

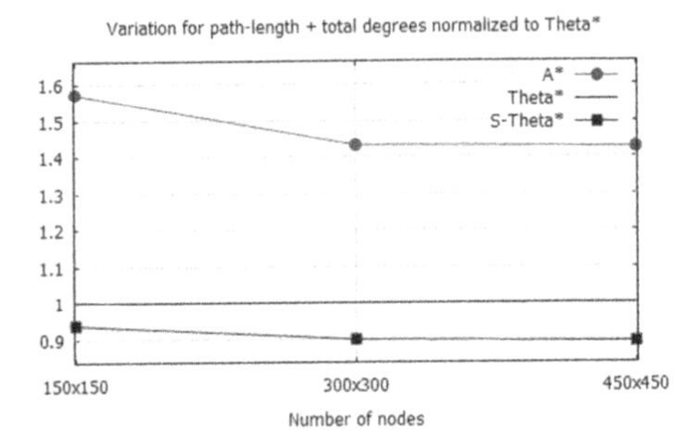


**Fig. 7** Normalized values for path length + total spin degrees for indoor maps (normalized to Theta*).

## 6 Conclusion

In this paper we have presented S-Theta*, an algorithm that is based on Theta*. The main motivation in this work was inspired by the idea that the cost of performing heading changes in real robots is high and it can be desirable to have longer paths with less turns (or at least with smoother turns) than shorter paths with abrupt direction changes. High degree heading changes in real robots are translated into high battery consumption and a lot of time in turning. Based on this idea, we have modified Theta* to compute the path deviation from the optimal path. For that, we have introduced a new parameter in the evaluation function called $\alpha(t)$, that effectively reduces the number of heading changes required by the search algorithm to achieve the objective, as well as the total cost associated with these turns.

As the experimental results show, the S-Theta* algorithm improves the original algorithm Theta* on the number of heading changes and its accumulate cost, in exchange for a slight degradation on the length of the path. This is true for both outdoors and indoor maps without considering the Digital Elevation Model (DEM) to model rough outdoor terrains. Taking into consideration that an optimal solution is the one that includes both the length of the path and the cost associated for turning, S-Theta* gets better results than Theta*, being small improvements for outdoor maps and more remarkable for indoor maps. In the case of outdoor maps, the best results are obtained with 40% of obstacles, in which S-Theta* gets 8% of improvement. In indoor with maps of 450x450 nodes, the improvement reaches the 10%.

In addition, since S-Theta* expands fewer nodes than the original Theta*, it will require less memory, which is always a plus in embedded systems, typically limited by memory and computation.

**Acknowledgements** Pablo Muñoz is supported by the European Space Agency (ESA) under the Networking and Partnering Initiative (NPI) *Cooperative systems for autonomous exploration missions.*

## References

1. I. Millington and J. Funge, *Artificial Intelligence for Games*, 2nd ed. Morgan Kaufmann Publishers, 2009.
2. D. Ferguson and A. Stentz, "Field D*: An interpolation-based path planner and replanner," in *Proceedings of the International Symposium on Robotics Research (ISRR)*, October 2005.
3. A. Nash, K. Daniel, S. Koenig, and A. Felner, "Theta*: Any-angle path planning on grids," in *In Proceedings of the AAAI Conference on Artificial Intelligence (AAAI)*, 2007, pp. 1177–1183.
4. P. M. noz and M. D. R-Moreno, "Improving efficiency in any-angle path-planning algorithms," in *6th IEEE International Conference on Intelligent Systems IS'12*, Sofia, Bulgaria, September 2012.
5. S. Choi, J. Y. Lee, and W. Yu, "Fast any-angle path planning on grid maps with non-collision pruning," in *IEEE International Conference on Robotics and Biomimetics*, Tianjin, China, December 2010, pp. 1051–1056.
6. K. Daniel, A. Nash, S. Koenig, and A. Felner, "Theta*: Any-angle path planning on grids," *Journal of Artificial Intelligence Research*, vol. 39, pp. 533–579, 2010.
7. A. Botea, M. Muller, and J. Schaeffer, "Near optimal hierarchical path-finding," *Journal of Game Development*, vol. 1, pp. 1–22, 2004.
8. P. Yap, "Grid-based path-finding," in *Advances in Artificial Intelligence*, ser. Lecture Notes in Computer Science, vol. 2338. Springer Berlin / Heidelberg, 2002, pp. 44–55.
9. G. Ayorkor, A. Stentz, and M. B. Dias, "Continuous-field path planning with constrained path-dependent state variables," in *ICRA 2008 Workshop on Path Planning on Costmaps*, May 2008.
10. J. Bresenham, "Algorithm for computer control of a digital plotter," *IBM Systems Journal*, vol. 4, pp. 25–30, 1965.

# Comprehensive Parent Selection-based Genetic Algorithm

**Hamid Ali**[1] **and Farrukh Aslam Khan**[2]

**Abstract** During the past few years, many variations of genetic algorithm (GA) have been proposed. These algorithms have been successfully used to solve problems in different disciplines such as engineering, business, science, and networking etc. Real world optimization problems are divided into two categories: (1) single objective, and (2) multi-objective. Genetic algorithms have key advantages over other optimization techniques to deal with multi-objective optimization problems. One of the most popular techniques of GA to obtain the Pareto-optimal set of solutions for multi-objective problems is the non-dominated sorting genetic algorithm-II (NSGA-II). In this paper, we propose a variant of NSGA-II that we call the comprehensive parent selection-based genetic algorithm (CPSGA). The proposed strategy uses the information of all the individuals to generate new offspring from the selected parents. This strategy ensures diversity to discourage premature convergence. CPSGA is tested using the standard ZDT benchmark problems and the performance metrics taken from the literature. Moreover, the results produced are compared with the original NSGA-II algorithm. The results show that the proposed approach is a viable alternative to solve multi-objective optimization problems.

## 1 Introduction

In our daily life, we want to optimize problems under some conditions, i.e., defined domain and objective functions. So, optimization has been a key area of research for many years. With the passage of time, real world problems have become increasingly complex; therefore, we always need a better optimization

[1] Hamid Ali, Department of Computer Science, National University of Computer and Emerging Sciences, A. K. Brohi Road H-11/4, Islamabad, Pakistan.
e-mail: hamid.ali@nu.edu.pk

[2] Farrukh Aslam Khan, Center of Excellence in Information Assurance, King Saud University Riyadh, Saudi Arabia.
e-mail: fakhan@ksu.edu.sa

algorithm [1]. To solve single and multi-objective problems many traditional optimization approaches are available such as steepest descent algorithm [2], linear programming [3], weighted sum method [4], etc. During the last few years, the use of evolutionary algorithms for optimization problems has significantly increased due to some weaknesses present in traditional optimization problems. There are a number of evolutionary algorithms available in the literature such as genetic algorithm (GA) [5], evolution strategies (ES) [6], differential evolution (DE) [7], ant colony optimization (ACO) [8], and particle swarm optimization (PSO) [9,10].

Genetic algorithms (GAs) [11] initialize a population of the candidate solutions. These candidate solutions are generated at random. The new population is generated using the genetic operators such as selection, variation, and replacement. The selection operator guaranties to select the best candidate solutions from the current population. These candidate solutions are used to generate the new population using variation operators (crossover and mutation) and at the end next the generation is produced using the replacement operator. Non-dominated sorting genetic algorithm-II (NSGA-II) [19] is a well-known variation of the genetic algorithm. However, in case of many optimization problems, it may get trapped into the local optimum. In order to avoid such a problem and improve the performance of the genetic algorithm, we develop a comprehensive parent selection-based genetic algorithm (CPSGA) by modifying the genetic operator method used in the standard NSGA-II algorithm. The proposed strategy ensures diversity to discourage the premature convergence. The performance of CPSGA is tested using the standard ZDT [15] test functions on two performance metrics taken from the literature. Also, the results produced are compared with the original NSGA-II algorithm and the results show that the proposed approach performs better and is well-suited to solve multi-objective optimization problems.

The rest of the paper is organized as follows: We briefly describe the related work in Section 2. In Section 3, the proposed CPSGA algorithm is described in detail. Section 4 presents the experimentation and evaluation of the proposed technique. Section 5 concludes the paper.

## 2 Related Works

To handle multi-objective optimization problems, a number of evolutionary algorithms have been proposed such a NSGA [12], MOGA [13], NPGA [14] etc. In these algorithms several operators are used for transforming a simple evolutionary algorithm (EA) to a multi-objective one. In all these algorithms, there were the following two common features present:

- Assigning the fitness based on non-dominated sorting.
- Maintaining the diversity within the same non-dominated front.

In [15] the authors illustrate that elitism improves the convergence speed of multi-objective EA. On the basis of elitism many algorithms are proposed in the literature such as PAES [16], SPEA [17], and Rudolph's elitist GA [18].

In [17] Zitzler and Thiele proposed an elitist multi-criterion EA with the concept of non-domination. In this paper an external population was used for genetic operator that was a collection of all non-dominated solutions discovered so far in the previous generations. Knowles and Corne [16] used a simple selection method. In this scheme the offspring and parent are compared; if parent dominates the offspring, the parent is accepted otherwise the offspring is selected as a parent in the next generation. Deb et al. [19] suggested a non-dominated sorting-based genetic algorithm. They suggested a fast non-dominated sorting approach and a selection method. In selection method the best individuals are selected from the parent and offspring on the basis of fitness and spread.

In this work, we present the proposed comprehensive parent selection-based GA approach, which uses a new parent selection method for generating each gene value of the new offspring from the selected parent.

# 3 The Proposed CPSGA Algorithm

In the literature, two state-of-the-art algorithms are available to find the Pareto optimal set of solutions for multi-objective problems: (1) non-dominated sorting genetic algorithm (NSGA-II) and (2) the improved strength Pareto evolutionary algorithm (SPEA2) [20]. The modifications in both algorithms take place in selection and replacement operators; the remaining algorithm is same as in case of standard single-objective GAs.

The premature convergence is still a main deficiency of the genetic algorithm NSGA-II. To handle the premature convergence, we modify NSGA-II to produce an offspring from the current population. In the NSGA-II algorithm only two parents are involved to create an offspring; this makes the genetic algorithm fast. However, if the selected parents are far from the global optimum, the offspring may easily get trapped into a local optimum. As we know that the fitness value of an individual is determined by the values of all genes. It may be possible that a best individual has low fitness value in some genes which produces a low quality offspring. To avoid this premature convergence, we have proposed a new strategy CPSGA to improve the original NSGA-II as explained in Algorithm 1. In CPSGA we select a new parent for crossover and mutation purposes to generate a new gene of an offspring besides the original NSGA-II in which a new parent is selected to generate a complete new offspring. So, in this case all the individuals are used to set the value of each gene of the new generated offspring. The diversity is preserved through this strategy to discourage the premature convergence.

**Algorithm 1:** CPSGA Algorithm

```
1. Initialize the population
2. Sort the initialized population
3. Start the evaluation process
4. for i=1 to number of generation do
       for j = 1 to P do  //population size
           for k = 1 to D do  //number of dimension
               If rand1 < 0.9 then  // perform crossover
                 • Select an individual randomly.
                 • Generate the kth gene of an offspring using the kth gene of
                   jth individual and the new selected parent by applying the
                   crossover.
                 • Make sure that the generated gene is within the decision
                   space.
               else // perform mutation
                 • Generate the kth gene of an offspring using the kth gene of
                   jth individual by applying the mutation.
                 • Make sure that the generated gene is within the decision
                   space.
               end if
           end for
           • Evaluate the fitness of generated offspring.
       end for
       • Combine the parent and offspring population.
       • Sort the combined population.
       • Select the best individuals on the basis of fitness.
   end for
5. Evaluate the performance metrics.
```

## 4 Experimentation and Evaluation

In this section, we illustrate the experimental setup and the results produced by our proposed method. We use five ZDT functions to test the quality of the proposed technique. We also compare the proposed technique with NSGA-II. The following goals are normally used to evaluate the performance of a multi-objective optimization algorithm:

- To find the distance between the Pareto front of the proposed algorithm and the global Pareto front (if we know earlier) of the given problem.
- To find average distance between individuals belonging to the same Pareto front.

## 4.1 Performance Measures

In literature several performance metrics have been recommended to evaluate the performance of a given algorithm [21]. To measure the performance of the proposed approach in the light of above goals, we use the following two performance metrics, Generational distance and Diversity metric.

### 4.1.1 Generational Distance (GD)

Veldhuizen and Lamont in [22] introduce the concept of generational distance to estimate the distance between the elements of generated non-dominated vector and the elements of the Pareto optimal set. The generational distance is calculated as follows:

$$GD = \frac{\sum_{i-1}^{n} d_i^2}{n} \tag{1}$$

Where $n$ is the number of elements in a Pareto front and $d$ is the minimum Euclidean distance between the elements of the Pareto front produced and the Pareto optimal front. The value of generational distance shows the average distance between The Pareto optimal set and Pareto front produced by any algorithm. The value shows that the solutions produced by the algorithm are in the Pareto optimal set and the value other than "zero" will show how "far" the produced solutions are from the Pareto optimal set.
The uniformity and smoothness within the produced solutions cannot be indicated by the generational distance. So to find the uniformity and smoothness in the produced solutions, we use another performance metric called the diversity metric.

### 4.1.2 Diversity Metric

The average distance between the produced solutions within a Pareto front is calculated using the diversity metric. We want to produce a Pareto front that uniformly covers the whole Pareto optimal set. The diversity metric is calculated as follows:

$$DM = \frac{d_f + d_l + \sum_{i=1}^{n} |d_i - \bar{d}|}{d_f + d_l + (N-1)\bar{d}} \tag{2}$$

Where $d_f$ and $d_l$ are the extreme points of the true Pareto optimal front, $\bar{d}$ is the average distance of the consecutive solutions and $N$ is the number of elements in a Pareto front. The value of DM shows that the generated solutions are widely and uniformly spread out along the true Pareto optimal front. The uniformity of any produced Pareto front can be calculated using the diversity metric.

## 4.2 Experimental setup

In the experimental setup, the parameters of NSGA-II and CPSGA are initialized as follows:

- The popolation size is set to 100;
- The maximum number of generations are set to 250;
- The tournament selection size is set to 2;
- The crossover probability $P_c$ is set to 0.9;
- The mutation probability is set to 10%.

where $n$ is the number of decision variables.

We perform the expirements using the ZDT test suite created by Zitzler et al., [15]. This test suite is one of the most widely used suite of benchmark multiobjective problems in the area of evolutionary algorithms. In this paper we used five real-valued problems as listed in Table 1. The detail of all the problems are given in Table 2. The ZDT test suite has two main advantages: (1) the true Pareto optimal fronts are well defined and (2) test results of other research papers are also available. We can easily compare different algorithms with the help of these results.

**Table 1.** Number of dimensions, variable range, and types of test problems.

| Problem | Dimensions | Variable bounds | Comments |
|---|---|---|---|
| ZDT1 | 30 | $[0,1]$ | Convex |
| ZDT2 | 30 | $[0,1]$ | Nonconvex |
| ZDT3 | 30 | $[0,1]$ | Convex, Disconnected |
| ZDT4 | 10 | $x_1 \in [0,1]$<br>$x_i \in [-5,5]$<br>$i = 2,...,n$ | Nonconvex |
| ZDT6 | 10 | $[0,1]$ | Nonconvex, Nonuniformly spaced |

**Table 2.** The definitions and global optimum of test problems used in this study.

| Problem | Objective functions | Optimal solutions |
|---|---|---|
| ZDT1 | $f_1(x) = x_1$<br>$f_2(x) = g(x)[1 - \sqrt{x_1 / g(x)}]$<br>$g(x) = 1 + 9(\sum_{i=2}^{n} x_i)/(n-1)$ | $x_1 \in [0,1]$<br>$x_i = 0$<br>$i = 2,\dots,n$ |
| ZDT2 | $f_1(x) = x_1$<br>$f_2(x) = g(x)[1 - (x_1 / g(x))^2]$<br>$g(x) = 1 + 9(\sum_{i=2}^{n} x_i)/(n-1)$ | $x_1 \in [0,1]$<br>$x_i = 0$<br>$i = 2,\dots,n$ |
| ZDT3 | $f_1(x) = x_1$<br>$f_2(x) = g(x)[1 - \sqrt{x_1 / g(x)} - \frac{x_1}{g(x)} \sin(10\pi x_1)]$<br>$g(x) = 1 + 9(\sum_{i=2}^{n} x_i)/(n-1)$ | $x_1 \in [0,1]$<br>$x_i = 0$<br>$i = 2,\dots,n$ |
| ZDT4 | $f_1(x) = x_1$<br>$f_2(x) = g(x)[1 - \sqrt{x_1 / g(x)}]$<br>$g(x) = 1 + 10(n-1) + \sum_{i=2}^{n}[x_i^2 - 10\cos(4\pi x_i)]$ | $x_1 \in [0,1]$<br>$x_i = 0$<br>$i = 2,\dots,n$ |
| ZDT6 | $f_1(x) = 1 - \exp(-4x_i)\sin^6(6\pi x_1)$<br>$f_2(x) = g(x)[1 - (f_1(x) / g(x))^2]$<br>$g(x) = 1 + 9[(\sum_{i=2}^{n} x_i)/(n-1)]^{0.25}$ | $x_1 \in [0,1]$<br>$x_i = 0$<br>$i = 2,\dots,n$ |

All objective functions are to be minimized.

### 4.2.1 Test function 1 (ZDT1)

We use the ZDT1 problem for the first experiment. The complete detail of ZDT1 is given in Tables 1 and 2. Figure 1 shows the graphical results generated by the proposed CPSGA and NSGA-II algorithms in case of first test function ZDT1. The continuous line in the figures represents the true Pareto front of the ZDT1 problem.

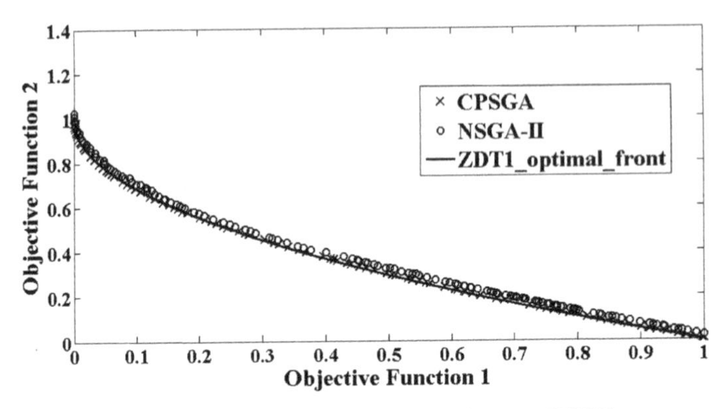


**Fig. 1.** Non-dominated solutions of CPSGA and NSGA-II in case of ZDT1.

Figure 1 clearly shows that CPSGA performs better than NSGA-II in case of the first test function ZDT1.

**Table 3.** Generational distance and diversity distance for ZDT1.

| Probability | Generational distance | Diversity distance |
|---|---|---|
| NSGA-II | 4.43e-002 ± 5.10e-002 | 8.01e-002 ± 6.42e-002 |
| CPSGA | 3.13e-003 ± 8.76e-004 | 4.02e-001 ± 1.01e-001 |

Table 3 shows the comparison of CPSGA and NSGA-II considering the generational distance and diversity distance metrics as described in Section 4.1. Table 3 shows that CPSGA outperforms NSGA-II with respect to generational distance while the NSGA-II produces better result than CPSGA in case of diversity metric. Since the generational distance has high priority over the diversity distance, we can conclude that CPSGA performs better than NSGA-II in this case.

#### 4.2.2 Test function 2 (ZDT2)

We use the ZDT2 problem for the second experiment. The complete detail of ZDT2 is given in Tables 1 and 2.

Figure 2 shows the graphical results generated by the proposed CPSGA and NSGA-II in case of the second test function ZDT2 chosen. The continuous line in the figure represents the true Pareto front of the ZDT2 problem. Figure 2 shows that CPSGA perform better than NSGA-II in case of the second test function ZDT2.

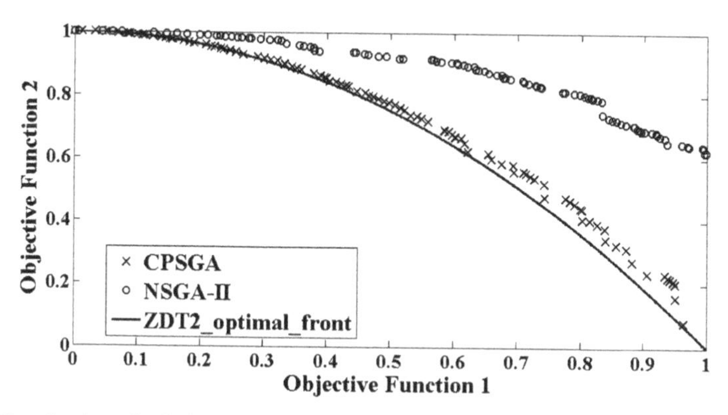


**Fig. 2.** Non-dominated solutions of CPSGA and NSGA-II in case of ZDT2.

Table 4 shows the comparison of CPSGA and NSGA-II considering the generational distance and diversity distance metrics as described in Section 4.1. Table 4 shows that CPSGA outperforms NSGA-II with respect to the generational distance as well as the diversity metric.

**Table 4.** Generational distance and diversity distance for ZDT2.

| Probability | Generational distance | Diversity distance |
|---|---|---|
| NSGA-II | 3.47e-002 ± 3.79e-002 | 8.62e-002 ± 9.62e-002 |
| CPSGA | 6.45e-003 ± 5.35e-003 | 2.85e-002 4.92e-002 |

### 4.2.3 Test function 3 (ZDT3)

We use the ZDT3 problem for the third experiment. The complete detail of the ZDT3 is given in the Tables 1 and 2.

Figure 3 shows the graphical results generated by the proposed CPSGA and NSGA-II in case of the third test function ZDT3 chosen. The continuous line in the figure represents the true Pareto front of the ZDT3 problem.

Figure 3 clearly shows that CPSGA performs better than NSGA-II in case of the third test function ZDT3.

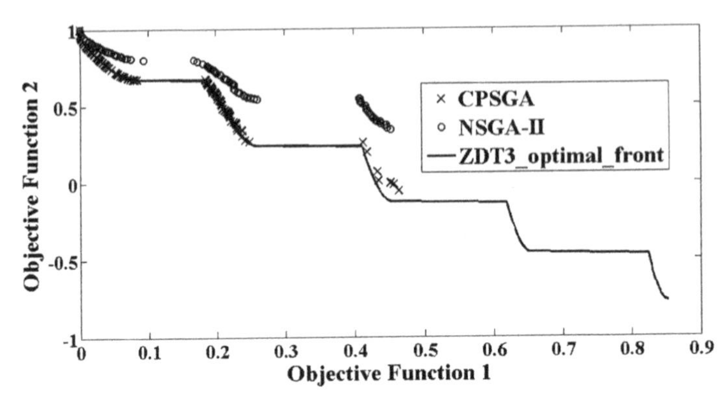


**Fig. 3.** Non-dominated solutions of CPSGA and NSGA-II in case of ZDT3.

**Table 5.** Generational distance and diversity distance for ZDT3.

| Probability | Generational distance | Diversity distance |
|---|---|---|
| NSGA-II | 5.73e-002 ± 4.93e-002 | 2.33e-001 ± 1.40e-001 |
| CPSGA | 8.53e-003 ± 4.33e-003 | 5.02e-001 ± 6.30e-002 |

Table 5 shows the comparison of CPSGA and NSGA-II considering the generational distance and diversity distance metrics as described in Section 4.1. Table 5 clearly shows that CPSGA outperforms NSGA-II with respect to the generational distance while with respect to the diversity metric it is slightly below NSGA-II, but with better standard deviation.

#### 4.2.4 Test function 4 (ZDT4)

We use the ZDT4 problem for our fourth experiment. The complete detail of the ZDT4 is given in Tables 1 and 2.

Figure 4 shows the graphical results generated by the proposed CPSGA and NSGA-II in case of the fourth test function ZDT4. The continuous line in the figure represents the true Pareto front of the ZDT4 problem.

Figure 4 shows that CPSGA performs better than NSGA-II in case of the fourth test function ZDT4.

Table 6 shows the comparison of CPSGA and NSGA-II considering the generational distance and diversity distance metrics as described in Section 4.1. Table 6 shows that CPSGA outperforms NSGA-II with respect to the generational distance as well as the diversity metric.

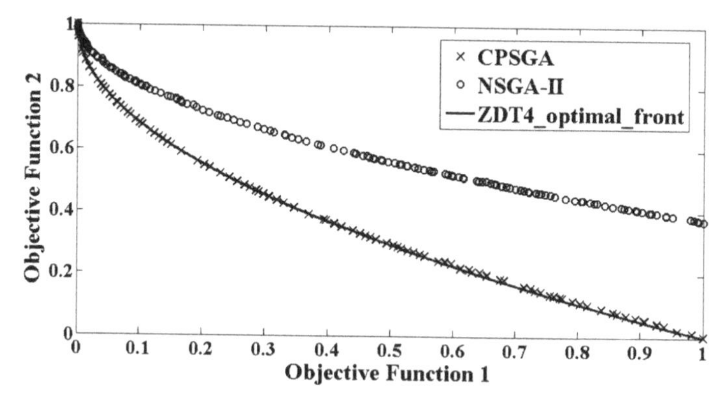


**Fig. 4.** Non-dominated solutions of CPSGA and NSGA-II in case of ZDT4.

**Table 6.** Generational distance and diversity distance for ZDT4.

| Probability | Generational distance | Diversity distance |
|---|---|---|
| NSGA-II | 2.61e-001 ± 3.63e-002 | 3.56e-001 ± 4.42e-002 |
| CPSGA | 1.65e-003 ± 1.09e-003 | 6.99e-002 ± 1.81e-001 |

### 4.2.5 Test function 5 (ZDT6)

We use the ZDT6 problem for the fifth experiment. The complete detail of the ZDT6 is given in Tables 1 and 2.

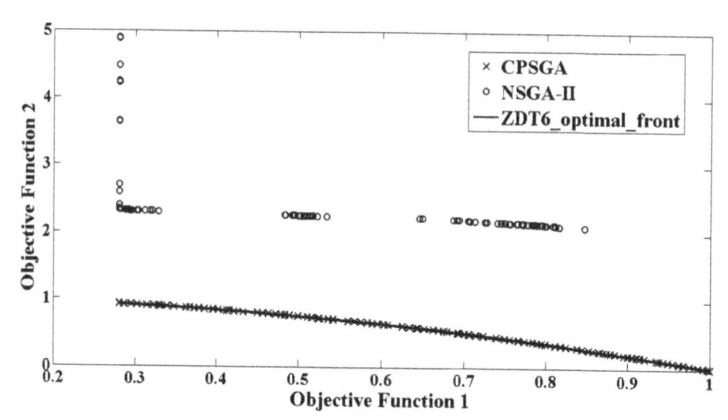


**Fig. 5.** Non-dominated solutions of CPSGA and NSGA-II in case of ZDT6.

Figure 5 shows the graphical results generated by the proposed CPSGA and NSGA-II in case of the fifth test function ZDT6 chosen. The continuous line in the figure represents the true Pareto front of the ZDT6 problem. Figure 5 clearly shows that CPSGA performs better than NSGA-II in case of the fifth test function ZDT6.

Table 7 shows the comparison of CPSGA and NSGA-II considering the generational distance and diversity distance metrics as described in Section 4.1. Table 7 shows that CPSGA outperforms NSGA-II with respect to the generational distance as well as the diversity metric.

**Table 7.** Generational distance and diversity distance for ZDT6.

| Probability | Generational distance | Diversity distance |
|---|---|---|
| NSGA-II | 3.10e-001 ± 3.43e-001 | 9.99e-001 ± 5.80e-001 |
| CPSGA | 2.24e-002 ± 2.38e-002 | 8.31e-001 ± 8.11e-001 |

## 5 Conclusion

In this paper we have presented a comprehensive parent selection-based genetic algorithm (CPSGA) in which an individual produces an offspring by selecting a new parent for each gene. The proposed strategy attempts to ensure diversity to discourage the premature convergence which is one of the major weaknesses of the genetic algorithm. The proposed approach is tested on five standard ZDT test functions using two performance metrics, i.e., generational distance and diversity metric. The results show that CPSGA performs better than NSGA-II in most of the cases and significantly improves the performance of the genetic algorithm. Additionally, CPSGA does not introduce any complex operations to the standard NSGA-II. The results clearly indicate that the proposed approach is a viable alternative to solve multi-objective optimization problems.

## References


1. Steuer, R. E.: Multiple Criteria Optimization: Theory, Computations, and Application. John Wiley and Sons, Inc., New York (1986)
2. Whitney, M. T., and Meany, R. K.: Two Algorithms Related to the Method of Steepest Descent. In: SIAM journal on Numerical Analysis. 4, no. 1, 109-118, March (1967)
3. Dantzig, G. B.: Linear Programming and Extensions. Princeton University Press, (1963)
4. Zadeh, L.: Optimality and Non-Scalar-Valued Performance Criteria. IEEE Transactions on Automatic Control, 8, no. 1, 59-60, January (1963)
5. Goldberg, D. E.: Genetic Algorithms in Search: Optimization and Machine Learning. Addison Wesley, Boston, MA: (1989)
6. Rechenberg, I.: Cybernetic Solution Path of an Experimental Problem. Royal Aircraft Establishment, Farnborough (1965)

7. Storn, R., and Price, K.: Differential Evolution- A Simple and Efficient Adaptive Scheme for Global Optimization over Continuous Spaces. Technical Report TR-95-012, International Computer Science Institute, Berkeley (1995)
8. Dorigo, M., Maniezzo, V., and Colorni.: The Ant System: Optimization by a Colony of Cooperating Agents. IEEE Transactions on Systems, Man, and Cybernetics Part B: Cybernetics, 26, no. 1, 29-41, February (1996).
9. Kennedy, J., and Eberhart, R. C.: Particle Swarm Optimization. in: Proceeding of IEEE International Conference on Neural Networks, pp. 1942-1948. Piscataway (1995)
10. Ali, H., Shahzad, W., and Khan, F. A.: Energy-efficient clustering in mobile ad-hoc networks using multi-objective particle swarm optimization. Applied Soft Computing, 12, no. 7, 1913-1928, July (2012)
11. Holland, J. H.: Adaptation in Natural and Artificial Systems. University of Michigan Press, Ann Arbor, MI(1975)
12. Srinivas, N., and Deb, K.: Multi-objective Optimization using Non-dominated Sorting in Genetic Algorithms. IEEE Transactions on Evolutionary Computation, 2, no. 3, 221-248, Fall (1994)
13. Fonseca, C. M., and Fleming, P. J.: Genetic algorithms for multi-objective optimization: Formulation, discussion and generalization. in: Proceedings of the Fifth International Conference on Genetic Algorithms, pp. 416-423. S. Forrest, Ed. San Mateo, CA: Morgan Kauffman (1993)
14. Horn, J., Nafploitis, N., and Goldberg, D. E.: A niched Pareto genetic algorithm for multi-objective optimization. in: Proceedings of the First IEEE Conference on Evolutionary Computation, pp. 82-87. Z. Michalewicz, (1994)
15. Zitzler, E., Deb, K., and Thiele, L.: Comparison of Multi-objective Evolutionary Algorithms: Empirical Results. IEEE Transactions on Evolutionary Computation, 8, no. 2, pp. 173-195, (2000)
16. Knowles, J., and Corne, D.: The Pareto archived evolution strategy: A new baseline algorithm for multi-objective optimization. in: Proceedings of the 1999 Congress on Evolutionary Computation, pp. 98-105. NJ:, Piscataway(1999)
17. Zitzler, E., and Thiele, L.: Multi-objective optimization using evolutionary algorithms-A comparative case study. in: Parallel Problem Solving From Nature, pp. 292-301. Berlin, Germany, (1998)
18. Rudolph, G.: Evolutionary search under partially ordered sets. Department of Computer Science/LS11, Univ. Dortmund, Dortmund, Tech. Report CI-67/99 Germany, (1999)
19. Deb, K., Pratap, A., Agarwal, S., and Meyarivan, T.: A Fast and Elitist Multi-objective Genetic Algorithm: NSGA-II. IEEE Transactions on Evolutionary Computation, 6, no. 2, pp. 182-197, April (2002)
20. Zitzler, E., Laumanns, M., and Thiele, L.: SPEA2: Improving the strength Pareto evolutionary. Swiss Federal Institute of Technology (ETH) Zurich, Zurich, TIK-Report 103, Switzerland, (2001)
21. Fonseca, C. M., and Fleming, P. J.: On the Performance Assessment and Comparison of Stochastic Multi-objective Optimizers. in: Proceedings of the 4th International Conference on Parallel Problem Solving from Nature IV, pp. 584-593. London, UK(1996)
22. Veldhuizen, D. A. V., and Lamont, G. B.: Multi-objective Evolutionary Algorithm Research: A History and Analysis. Technical Report, TR-98-03, Department of Electrical and Computer Engineering, Graduate School of Eng., Air Force Inst., Wright Patterson AFB, OH, (1998).

# Run-Time Analysis of Classical Path-Planning Algorithms

Pablo Muñoz, David F. Barrero and María D. R-Moreno

**Abstract** Run-time analysis is a type of empirical tool that studies the time consumed by running an algorithm. This type of analysis has been successfully used in some Artificial Intelligence (AI) fields, in paticular in Metaheuristics. This paper is an attempt to bring this tool to the path-planning community. In particular, we analyse the statistical properties of the run-time of the A*, Theta* and S-Theta* algorithms with a variety of problems of different degrees of complexity. Traditionally, the path-planning literature has compared run-times just comparing their mean values. This practice, which unfortunately is quite common in the literature, raises serious concerns from a methodological and statistical point of view. Simple mean comparison provides poorly supported conclusions, and, in general, it can be said that this practice should be avoided.
After our analysis, we conclude that the time required by these three algorithms follows a lognormal distribution. In low complexity problems, the lognormal distribution looses some accuracy to describe the algorithm run-times. The lognormality of the run-times opens the use of powerful parametric statistics to compare execution times, which could lead to stronger empirical methods.

## 1 Introduction

Path-planning is a well known problem in Artificial Intelligence (AI) with many practical applications. Several classical AI algorithms, such as A* [12], have been

Pablo Muñoz
Departamento de Automática, Universidad de Alcalá, Spain, e-mail: pmunoz@aut.uah.es

David F. Barrero
Departamento de Automática, Universidad de Alcalá, Spain e-mail: david@aut.uah.es

María D. R-Moreno
Departamento de Automática, Universidad de Alcalá, spain, e-mail: mdolores@aut.uah.es

applied to solve the path-planning problem as the notable literature about this topic shows. Nonetheless, despite the research interest that this issue has attracted, the statistical properties of the run-time of these algorithms has not been, to the authors' knowledge, studied before.

The study of the run-time from a statistical perspective is known as *run-time analysis*. The common practice in the literature is to report the run-time of the algorithm with means and, sometimes, some dispersion measure. However, this practice has several drawbacks, mainly due to the loose of valuable information that this reporting practice involves such as asymmetries in the run-time, or the shape of its distribution.

Run-time analysis overcomes this limitation by considering all the data with its focus on the statistical properties of the algorithm run-time. The main tool of run-time analysis is the *Run-Time Distribution* (RTD) which is the empirical statistical distribution of the run-time; in this way, the RTD is fully characterized. Other statistics such as the mean can be obtained from the RTD, and it opens exciting parametric statistical analysis tools. For instance, once the run-time distribution of an algorithm is identified, parametric hypothesis tests can be used to compare the run-time of two algorithms, providing a more rigorous comparison methodology.

This paper is an attempt to provide a first run-time analysis of some classical path-planning algorithms (A* [12] and Theta* [19]), and a new path-planning algorithm (S-Theta*) [18] that we have developed. We follow an empirical approach, running the algorithms in grid-maps of different difficulties and fitting the resulting run-time distribution with some statistical distributions. We show that, in the same line as the related run-time literature in other fields, the run-time of the three algorithms under study follow a lognormal distribution. We also observe a dependence between the run-time distribution and the problem difficulty. Our experiments show that very easy problems are not well characterized by a lognormal RTD.

The paper is structured as follows. First, we introduce the path-planning algorithms that are object of study. Then we describe the RTD analysis, including some related literature mainly from Metaheuristics. Next, in section 4 we perform the run-time analysis. The paper finishes with some conclusions.

## 2 Path planning

Path-planning or pathfinding [20] is a widely discussed problem in robotics and video games [11, 17]. In a simple manner, the path-planning problem aims to obtain a feasible route between two points. Feasible means that the route can not violate constraints such as traversing obstacles. This is a classical problem that can be addressed with several search algorithms such as classical graph-search [20], Evolutionary Algorithms [21] or multiagent systems [8], just to mention some of them.

In this work we have considered one classical algorithm, A* [12], and two variations. These algorithms use a grid of cells to discretize the terrain. And of course, they work over nodes. A* is a simple and fast search algorithm that can be used to solve many AI problems, path-planning among them. It combines an heuristic and a cost function to achieve optimal paths. However, it has an important limitation: it typically uses 8 neighbours nodes, so it restricts the path headings to multiples of $\pi/4$, causing that A* generates a sub-optimal path with zig-zag patterns. Other approximations use more adjacent nodes or use framed cells [1] to solve (or relax) this limitation, and thus requiring, in most cases, more computational effort. For each node, A* maintains three values: (i) the length of the shortest path from the start node to actual node; (ii) the heuristic value for the node, an estimation of the distance from the actual node to the goal; and (iii) the parent of the node. The heading changes limitation makes that the best heuristic for A* is the octile distance.

There are some variations of A* that try to generate smoother heading changes. Some proposals smooth the path using framed cells [1] while others post-process the result of A* [4]. Another group of algorithms called "any-angle" allow heading changes in any point of the environment. Most popular examples of this category are Field D* [9] and Theta* [19], that allows any node to be the parent of a node, not only its predecessor like A*. An alternative to Theta* is our algorithm called S-Theta*, which modifies the cost and heuristic functions to guide the search in order to obtain similar paths to Theta*, but with less heading changes. An exhaustive comparison over some path-planning algorithms can be found in [6].

Given its popularity and presence in the literature, we selected A*. It is a very well studied algorithm and it is quite popular in path-planning research, so it seemed reasonable to chose it. Additionally we have also included Theta* and S-Theta*, which are two algorithms designed specifically to path-planning in order to solve the heading changes problem. Before begining the RTD analysis of those algorithms, we first introduce in more detail the RTD analysis and related literature.

## 3 Introduction to run-time analysis

The basic tool used for run-time analysis is the RTD, term that was introduced by Hoos and Stützle [16]. Let us name $rt(i)$ the run-time of the ith successful run, and $n$ the number of runs executed in the experiment, then the RTD is defined as the empirical cumulative probability $\hat{P}(rt \leq t) = \#\{i|rt(i) \leq t\}/n$ [15]. It is simply the empirical probability of finishing the algorithm run before $t$. There is an extensive literature about RTDs, mainly in Metaheuristics, but there are also several studies in classical AI problems.

Hoos and Stützle applied RTD analysis to different algorithms and problems. They found that the RTD of WSAT algorithms applied to the 3SAT problem is exponential when the parameters setting is optimal, shifted exponential or Weibull when the parameters setting is not optimal [16]. Analogously, they studied in [13] the RTD of some other stochastic local search algorithms, such as GWSAT, GSAT with tabu-lists, TMCH and WMCH, to solve instances of SAT and CSPs, finding that, again, when parameters are optimal, the RTD follows an exponential, and otherwise RTD fits a Weibull distribution. Curiously, this result only holds for hard instances, in easy instances they did not find statistical significance. In a later work [14], they observed that the RTD of hard 3SAT instances solved with WalkSAT algorithm also follows an exponential distribution, and more interestingly, the higher the difficulty of the problem, the higher the fit is found. In a more recent work, Hoos and Stützle [15] compared various versions of GSAT and WalkSAT algorithms to solve some problems coded as 3SAT (random 3-SAT, graph coloring, block world and logistic planning), finding that these algorithms also have exponential RTDs for hard problems and high values of noise parameter. More importantly, they found that RTDs of easy problems are no exponential, despite their tails are still exponential.

Chiarandini and Stützle [5] studied the RTD of ILS, ACO, Random Restart Local Search and two variants of SA applied to the course timetabling problem, finding that the Weibull distribution approximates well the RTDs. They report, however, that in SA, the RTD in hard problems can be approximated, at least partially, using a shifted exponential distribution. On the contrary than Hoos, Stützle and Chiarandini, Frost *et al.* [10] studied the RTD using the same algorithm, backtracking, with different problem instances of the CSP. The RTD of the algorithm running on solvable instances [7] follows a Weibull distribution, while unsolvable instances generate lognormal run-times. However, only the lognormal distribution for solvable problems had statistical significance. In addition, Barrero *et al.* studied the RTDs in tree-based Genetic Programming [3], finding lognormal RTDs whose goodness of fit depends on the problem difficulty.

In summary, we conclude that despite the variety of algorithms and problems studied using RTDs, the are three omnipresent distributions: Exponential, Weibull and Lognormal. Curiously, these three distributions play a central role in Reliability Theory, suggesting a link. In the following section we study the presence of these three distributions in the RTD of A*, Theta* and S-Theta* for path-planning problems of varying difficulty.

## 4 RTD analysis of path-planning algorithms

In order to carry out the run-time analysis, we need to gather empirical data in a controlled way. Given the need of a uncertainty source, experiments may vary two factors, the random seed and the problem. The former is common in the Metaheuristics and random search literature, while the latter is used with deterministic algorithms. Even though this distinction does not use to be clear in the literature, we think that it is critical since it determines the comparability of the experiments and their interpretation.

## 4.1 Experimental design

In order to assess the run-time behavior of the path-planning algorithms we have performed an experimental approximation. These algorithms are deterministic, so, on the contrary than other run-time analysis performed in Metaheuristics, there is no variability due to the random seed. In this case, the variation comes from the problem, to be more specific, we used a random map generator that can control the percentage of obstacles in the map, and therefore, we can set the complexity of the problem. It provides a mechanism to study how the problem difficulty influences the run-time behavior of the algorithms[1].

We have generated random maps with different ratio of random obstacles, in particular we used $5\%, 10\%, 20\%, 30\%$ and $40\%$ of obstacles. For each ratio of obstacles we generated $2,000$ random maps of $500 \times 500$ nodes and solved the map with each of the three algorithms under study. This procedure yields $2,000 \times 5 = 10,000$ runs for each one of the three algorithms. The initial point in the map is always the corner at the top left of the map, and the goal node is placed at the right, locating between the bottom corner and randomly $20\%$ up nodes. The map generator was implemented with guarantees to keep at least one path from the start node to the goal node. To do this, when an obstacle is set, his periphery is protected to avoid overlapping obstacles. Thus, it is always possible surround an obstacle to avoid it.

The algorithms were implemented in Java. In order to make a fair comparison, the implementation of the three algorithms use the same methods and structures to manage the information grid[2]. To measure the runtime we have employed `System.currentTimeMillis()`. Reporting time in this way has several drawbacks [2], but in this case we think it is justified because the algorithms contains computations that are not well captured by machine-independent time measures, such us the number of expanded nodes. Of course, the price we have to pay is an increased difficulty to repeat and compare these results, but given that our interest is studying the statistical properties of the run-time rather than compare algorithms, we think that in this case it is an acceptable drawback.

## 4.2 Experimental results

We have firstly performed an exploratory analysis of the results. To this end we depicted several histograms of the run-times, as can be seen in Fig. 1. The histograms were grouped by algorithm and problem hardness and they show some interesting facts.

---

[1] In case of acceptance, all the code, datasets and scripts needed to repeat the experiments reported in this paper would be published on a web site under an open licence.

[2] The execution was done on a 2 GHz Intel Core i7 with 4 GB of RAM under Ubuntu 10.10 (64 bits).

We observe that the shape of the run-time histograms varies in function of the algorithm and problem hardness. S-Theta* has a longer tail and its shape is more asymmetrical in comparison to the rest of the algorithms, this fact is more clear in hard problems than in easy ones. The run-time of A* presents a smaller variance than the other two algorithms, and it increases with the problem hardness. The run-time required to solve hard problems has more variance than easy problems. In any case, the shape and range of values of Theta* and S-Theta* are similar, which seems reasonable given that they are variations of the same algorithm.

Looking for a statistical distribution able to fit data is more interesting. With the results reported in the literature we have tried to fit data to the three distributions that appear to play a more important role in RTD analysis, the Exponential, Weibull

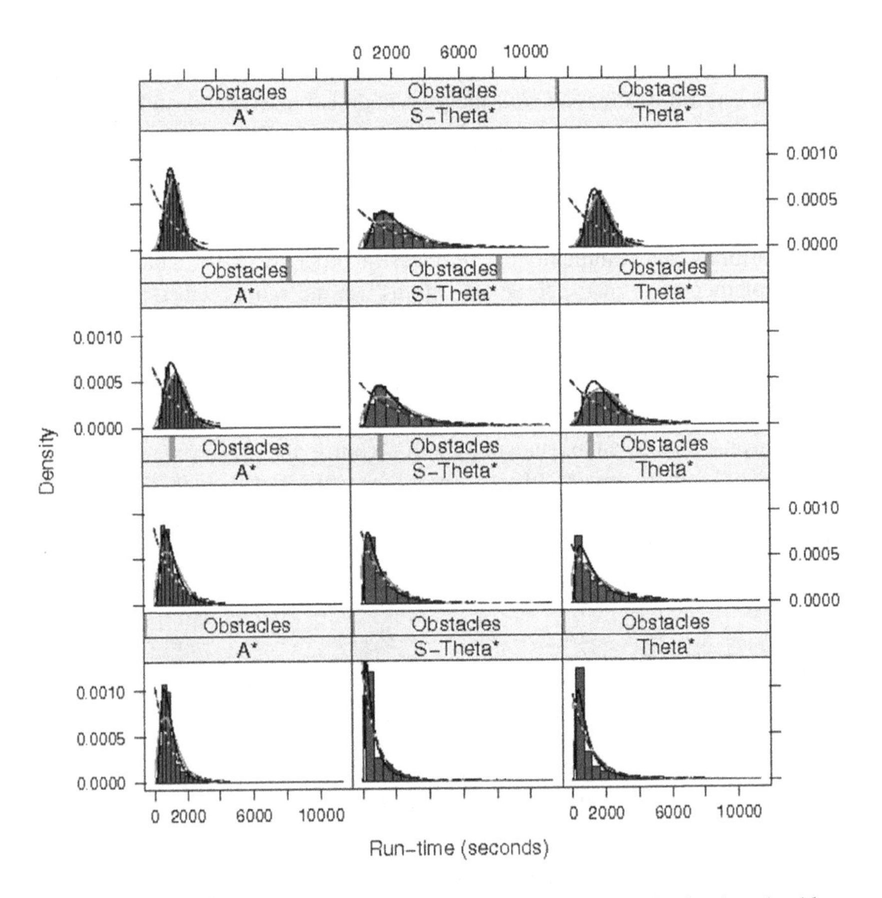


**Fig. 1** Histogram of the run-time, measured in seconds, of the three path-planning algorithms under study. Histograms have been grouped by algorithm and ratio of obstacles. The histograms have been adducted with three distributions that appear overlapped: Lognormal (black), exponential (dashed) and Weibull (gray).

and Lognormal distribution. So, the histograms shown in Fig. 1 are depicted with a set of overlapping distributions, which corresponds to the three statistical distributions previously mentioned. The parameters of these distributions were fitted by maximum likelihood. As the reader can observe in Fig. 1, the distribution that fits better our data in most cases is the lognormal. The exceptions to this observation are Theta* and S-Theta* solving maps with a very low number of obstacles; in that case the exponential distribution seems to describe the run-time behavior better than the lognormal. Even in this case, the A* run-time is well described by the lognormal, that is able to describe its peaks better than the exponential.

**Table 1** Estimated parameters of the lognormal distribution fitted by maximum likelihood. Each row corresponds to an experiment with 2,000 problems with the given rate of obstacles in the map.

| Algorithm | Obstacles | $\hat{\mu}$ | $\hat{\sigma}$ |
|---|---|---|---|
| A* | 5 | 6.690 | 0.567 |
| A* | 10 | 6.886 | 0.600 |
| A* | 20 | 7.131 | 0.542 |
| A* | 30 | 7.228 | 0.449 |
| A* | 40 | 7.176 | 0.361 |
| Theta* | 5 | 6.520 | 0.883 |
| Theta* | 10 | 6.989 | 0.887 |
| Theta* | 20 | 7.395 | 0.700 |
| Theta* | 30 | 7.515 | 0.534 |
| Theta* | 40 | 7.464 | 0.390 |
| S-Theta* | 5 | 6.202 | 0.975 |
| S-Theta* | 10 | 6.749 | 0.916 |
| S-Theta* | 20 | 7.220 | 0.744 |
| S-Theta* | 30 | 7.445 | 0.648 |
| S-Theta* | 40 | 7.601 | 0.589 |

Given the observations of the exploratory analysis, it seems reasonable to focus the study on the lognormal distribution and hypothetise that the run-time of the A*, Theta* and S-Theta* algorithms in path-planning problems follow a lognormal distribution. The parameters of the lognormal distributions that we obtained using maximum-likelihood are shown in Table 1. In order to evaluate the hypothesis about the RTDs lognormality, it is desirable to provide additional evidences. One of the most interesting properties of the lognormal distribution is its close relationship to the normal distribution, actually, lognormal data can be converted to normal data by simply taking logarithms. We can use this property to verify whether the RTD is lognormal or not in a more formal way.

To verify the lognormality of the RTD, we first plotted a QQ-plot of the logarithm of the run-time against a normal distribution, which is shown in Fig 2. If the logarithm of the run-time is normal, we can conclude that the run-time is lognormal. Fig. 2 confirms our initial suspicion about the lognormality of the RTD. In addition, the QQ-plots also show the relationship between the distribution and the problem hardness. Theta* and S-Theta* algorithms produce less lognormal RTDs in very easy problems. On the contrary, the influence of the problem hardness on A*, if it has any, is not so evident, the QQ-plot is almost a line for all the ratios of obstacles. In summary, A* RTDs seem lognormal for any number of obstacles, while the run-time of Theta* and S-Theta* algorithms seem to follow a lognormal distribution, but easy problems fit worse.

A more rigorous analysis of the lognormality of the run-time is desirable. For this reason we have performed a Shapiro-Wilk test of normality of the logarithm of the run-time, whose result is shown in Table 2. In order to avoid undesirable effects of the large number of samples, we have computed the test using 30 random runs. Results shown in the table confirms our previous observations in the histograms and the QQ-plot. The p-values in almost all the cases are quite high, which means that the hypothesis of lognormality is compatible, with a high probability, with our data.

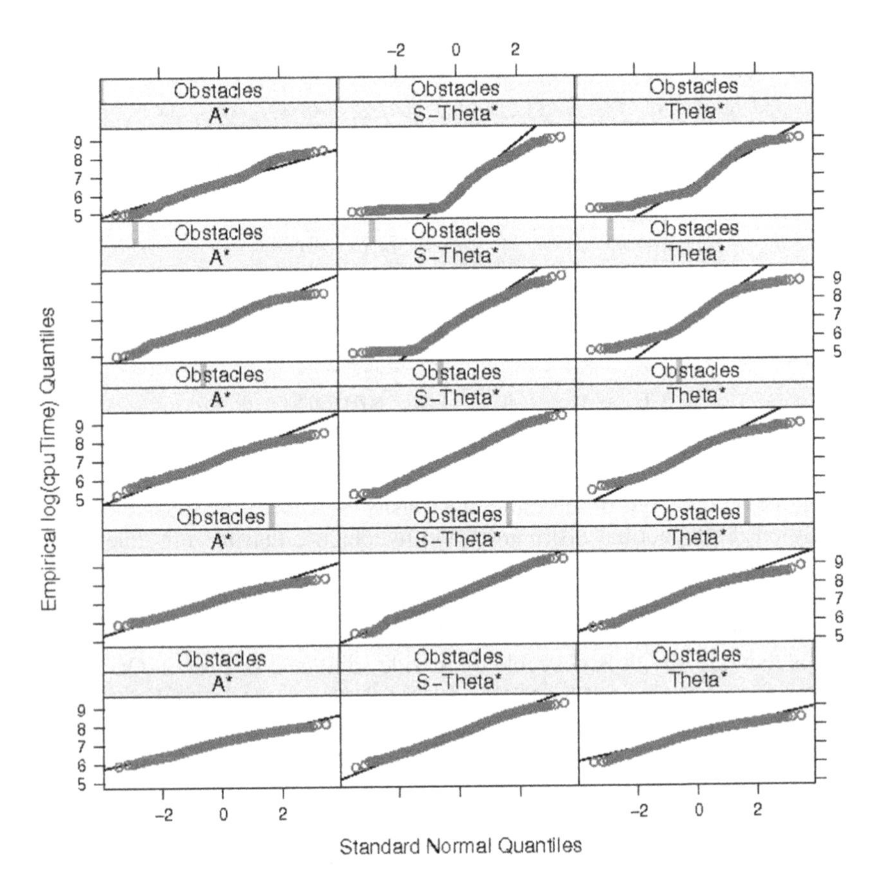


**Fig. 2** Quantile plot that assesses the normality of the logarithm of the run-time of the three algorithms under study: A*, Theta* and S-Theta*. Given the relationship between the lognormal and normal distributions, the logarithm of lognormal data must transform it into normal data.

**Table 2** Shapiro-Wilk test of normality of the logarithm of the run-time. With these p-values we cannot reject the hypothesis of normality of the logarithm of the run-time, which means that we cannot reject the lognormality of the run-time. Only Theta* and S-Theta* with a ratio of 5% of obstacles provide evidence to let us reject the normality of the RTD.

| Algorithm | Obstacles | W | p-value | Significance |
|---|---|---|---|---|
| A* | 5 | 0.9772 | 0.7473 | |
| A* | 10 | 0.9579 | 0.2743 | |
| A* | 20 | 0.9808 | 0.8475 | |
| A* | 30 | 0.9378 | 0.0792 | |
| A* | 40 | 0.9546 | 0.2237 | |
| Theta* | 5 | 0.8998 | 0.0083 | $\alpha = 0.001$ |
| Theta* | 10 | 0.9437 | 0.1142 | |
| Theta* | 20 | 0.9479 | 0.1487 | |
| Theta* | 30 | 0.9343 | 0.0641 | |
| Theta* | 40 | 0.9444 | 0.1194 | |
| S-Theta* | 5 | 0.8322 | 0.0003 | $\alpha = 0.001$ |
| S-Theta* | 10 | 0.9409 | 0.0965 | |
| S-Theta* | 20 | 0.9771 | 0.7428 | |
| S-Theta* | 30 | 0.9829 | 0.8968 | |
| S-Theta* | 40 | 0.9876 | 0.9725 | |

However, there are two exceptions, the p-value of Theta* and S-Theta* with a ratio of 5% of obstacles is quite small, 0.0083 and 0.0003 respectively, which drives us to reject the null hypothesis (i.e. the lognormality of the data) with $\alpha = 0.001$.

We were curious about the different RTDs reported by the literature and the reason of those differences. In order to obtain some clue for its answer, we performed a simple experiment. Instead of plotting run-time histograms grouped by algorithm and problem hardness, we tried to plot histograms joining all the runs belonging to the same algorithms, in this way, problems of different hardness were merged. As in the exploratory analysis, we plotted the three main distributions in RTD literature to check rapidly whether data fits or not that distribution. The result can be seen in Fig. 3. We have to consider that this figure shows in fact an overlapping of several distributions, it can be seen as the sum of each column in Fig. 3.

Fig. 3 is a quite interesting result. The lognormal distribution still seems to fit very well the A* run-time, however, the Theta* algorithms are not so clear. The left tails of their histogram seems to have disappeared, or at least it is sufficiently small to not appear clearly in the histogram. This fact introduces the exponential distribution in the discussion, it can be seen that this distribution is able to fit data quite well, nonetheless, the lognormal distribution still provides a excellent fit. So, it makes us conjecture that depending on how the run-time is visualized the statistical model that fit the RTD may change. In particular, it is well known that joining random variables of a certain distribution may produce a random variable of another distribution. This fact could explain, in part, the diversity of RTDs found in the literature, and it is an additional motivation to take care about how experimentation and data processing are done.

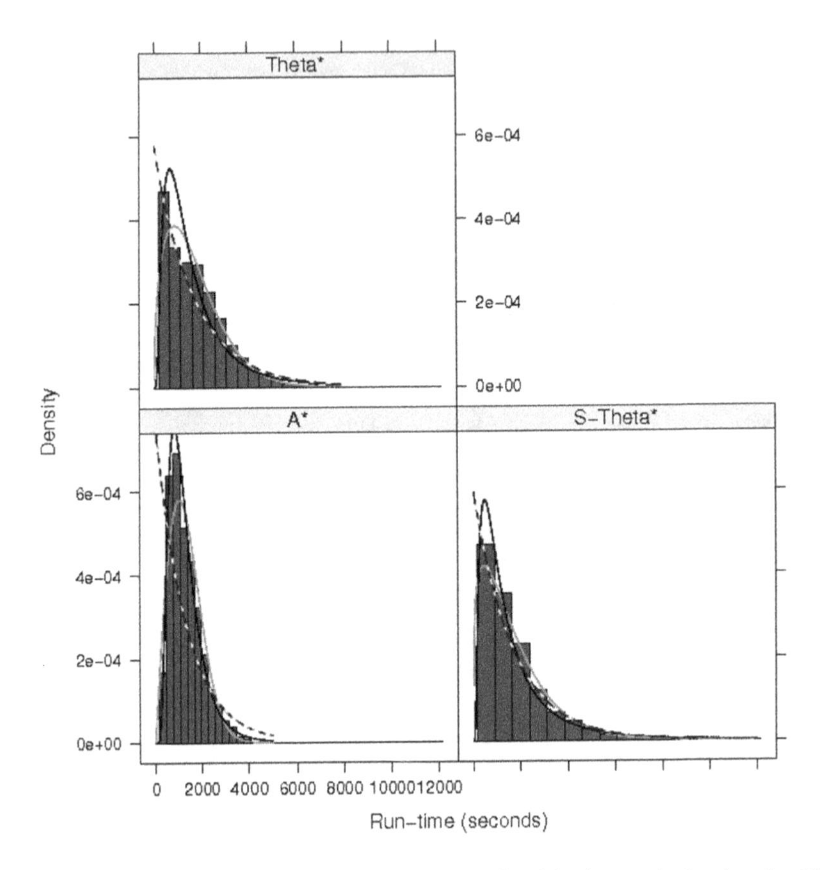


**Fig. 3** Histogram of the run-time, measured in seconds, of the three path-planning algorithms under study. Histograms have been grouped by algorithm, but not by ratio of obstacles, so each histogram contains runs of different problem hardness. The histograms have been adjusted with three distributions that appear overlapped: Lognormal (black), Weibull (gray) and exponential (dashed).

With this evidence, it seems reasonable to assume the lognormality of the RTD in problems with a ratio of obstacles higher than 5% of the three algorithms under study.

## 5 Conclusions

Along this paper we have performed a run-time analysis of A*, Theta* and S-Theta* to study their RTD statistical properties and the influence of the problem hardness. The RTD of those algorithms, when applied to a set of non-trivial path-planning problems of similar hardness, follows a lognormal distribution. The evidence we have found in this line is strong. However, we also observed that the goodness-of-fit is weaker for the Theta* algorithms when the problem is easy.

This observation leads to a better knowledge about A*, Theta* and S-Theta algorithms, but it also has a practical implications. The common procedure to compare run-times followed in the literature is a naïve comparison of means, which is a weak method from a statistical point of view. If the RTD of the algorithm is known, strong parametric methods could be used, and in particular lognormal RTDs open the use of well-known (and easy) statistics to enhance the experimetal methods used to study path-planning algorithms. So, in order to compare the run-time of two algorithms with a sound statistical basis we can -and should- use hypothesis testing (Student's t-test with the logarithm of the run-time) or ANOVA if several algorithms are involved.

In particular, the results shown in this paper open powerful parametric statistics to analyze run-times in path-planning problems. Even without the knowledge about the lognormal distribution of the run-time, it is still possible to use non-parametric statistics to compare run-times, however, parametric statistics are more powerful, and therefore finding sounded conclussions is easier.

From a more general perspective, our results are clearly aligned with previous results reported in the literature, mainly in Metaheuristics, but also in some classical AI problems. Very different algorithms applied to different problems have shown a similar run-time behaviour, which turns out an intriguing fact. So, a natural question that raises at this point is whether this behaviour is as general as it seems to be and, more importantly, why RTDs are so well described by only three distributions. These are questions that we think deserve some research.

**Acknowledgements** Pablo Muñoz is supported by the European Space Agency (ESA) under the Networking and Partnering Initiative (NPI) *Cooperative systems for autonomous exploration missions*.

## References

1. G. Ayorkor, A. Stentz, and M. B. Dias. Continuous-field path planning with constrained path-dependent state variables. In *ICRA 2008 Workshop on Path Planning on Costmaps*, May 2008.
2. R. Barr and B. Hickman. Reporting Computational Experiments with Parallel Algorithms: Issues, Measures, and Experts' Opinions. *ORSA Journal on Computing*, 5:2–2, 1993.
3. D. F. Barrero, B. Castaño, M. D. R-Moreno, and D. Camacho. Statistical Distribution of Generation-to-Success in GP: Application to Model Accumulated Success Probability. In *Proceedings of the 14th European Conference on Genetic Programming, (EuroGP 2011)*, volume 6621 of *LNCS*, pages 155–166, Turin, Italy, 27-29 Apr. 2011. Springer Verlag.
4. A. Botea, M. Muller, and J. Schaeffer. Near optimal hierarchical path-finding. *Journal of Game Development*, 1:1–22, 2004.
5. M. Chiarandini and T. Stützle. Experimental Evaluation of Course Timetabling Algorithms. Technical Report AIDA-02-05, Intellectics Group, Computer Science Department, Darmstadt University of Technology, Darmstadt, Germany, April 2002.
6. K. Daniel, A. Nash, S. Koenig, and A. Felner. Theta*: Any-angle path planning on grids. *Journal of Artificial Intelligence Research*, 39:533–579, 2010.

7. S. Epstein and X. Yun. From Unsolvable to Solvable: An Exploration of Simple Changes. In *Workshops at the Twenty-Fourth AAAI Conference on Artificial Intelligence*, 2010.
8. M. Erdmann and T. Lozano-Perez. On multiple moving objects. *Algorithmica*, 2:477–521, 1987.
9. D. Ferguson and A. Stentz. Field D*: An interpolation-based path planner and replanner. In *Proceedings of the International Symposium on Robotics Research (ISRR)*, October 2005.
10. D. Frost, I. Rish, and L. Vila. Summarizing CSP Hardness with Continuous Probability Distributions. In *Proceedings of the Fourteenth National Conference on Artificial Intelligence and Ninth Conference on Innovative Applications of Artificial Intelligence (AAAI'97/IAAI'97)*, pages 327–333. AAAI Press, 1997.
11. O. Hachour. Path planning of autonomous mobile robot. *International Journal of Systems, Applications, Engineering & Development*, 2(4):178–190, 2008.
12. P. Hart, N. Nilsson, and B. Raphael. A formal basis for the heuristic determination of minimum cost paths. *IEEE Transactions on Systems Science and Cybernetics.*, 4:100–107, 1968.
13. H. Hoos and T. Stützle. Characterizing the Run-Time Behavior of Stochastic Local Search. In *Proceedings AAAI99*, 1998.
14. H. Hoos and T. Stützle. Towards a Characterisation of the Behaviour of Stochastic Local Search Algorithms for SAT. *Artificial Intelligence*, 112(1-2):213–232, 1999.
15. H. Hoos and T. Stützle. Local Search Algorithms for SAT: An Empirical Evaluation. *Journal of Automated Reasoning*, 24(4):421–481, 2000.
16. H. H. Hoos and T. Stützle. Evaluating Las Vegas Algorithms – Pitfalls and Remedies. In *Proceedings of the Fourteenth Conference on Uncertainty in Artificial Intelligence (UAI-98)*, pages 238–245. Morgan Kaufmann Publishers, 1998.
17. I. Millington and J. Funge. *Artificial Intelligence for Games*. Morgan Kaufmann Publishers, 2 edition, 2009.
18. P. Muñoz and M. D. R-Moreno. S-Theta*: low steering path-planning algorithm. In *Thirty-second SGAI International Conference on Artificial Intelligence (AI-2012)*, Cambridge, UK, 2012.
19. A. Nash, K. Daniel, S. Koenig, and A. Felner. Theta*: Any-angle path planning on grids. In *In Proceedings of the AAAI Conference on Artificial Intelligence (AAAI)*, pages 1177–1183, 2007.
20. N. Nilsson. *Principles of Artificial Intelligence*. Tioga Publishing Company, Palo Alto, CA. ISBN 0-935382-01-1, 1980.
21. K. Sugihara and J. Smith. A genetic algorithm for 3-d path planning of a mobile robot. Technical report, Tech. Rep. No. 96-09-01. Software Engineering Research Laboratory,University of Hawaii at Manoa, 1996.

# KNOWLEDGE MANAGEMENT AND PREDICTION

# A Generic Platform for Ontological Query Answering

Bruno Paiva Lima da Silva and Jean-François Baget and Madalina Croitoru

**Abstract** The paper presents ALASKA, a multi-layered platform enabling to perform ontological conjunctive query answering (OCQA) over heterogeneously-stored knowledge bases in a generic, logic-based manner. While this problem knows today a renewed interest in knowledge-based systems with the semantic equivalence of different languages widely studied, from a practical view point this equivalence has been not made explicit. Moreover, the emergence of graph database provides competitive storage methods not yet addressed by existing literature.

## 1 Introduction

The ONTOLOGICAL CONJUNCTIVE QUERY ANSWERING problem (OCQA) (also known as ONTOLOGY-BASED DATA ACCESS (ODBA) [15]) knows today a renewed interest in knowledge-based systems allowing for expressive inferences. In its basic form the input consists a set of facts, an ontology and a conjunctive query. The aim is to find if there is an answer / all the answers of the query in the facts (eventually enriched by the ontology). Enriching the facts by the ontology could either be done (1) previous to query answering by fact saturation using the ontology (forward chaining) or (2) by rewriting the query cf. the ontology and finding a match of the rewritten query in the facts (backwards chaining).

While existing work focuses on logical properties of the representation languages and their equivalence [1, 9], the state of the art employs a less principled approach when implementing such frameworks. From a practical view point, this equivalence has not yet been put forward by the means of a platform allowing for a logical, uniform view to all such different kinds of paradigms. There is indeed existing work studying different cases for optimizing the reasoning efficiency [17, 5], however, none of them look at it as a principled approach where data structures are deeply

LIRMM, University Montpellier II, France, e-mail: bplsilva, croitoru@lirmm.fr
LIRMM, INRIA, France, e-mail: baget@lirmm.fr

investigated from a storage and querying retrieval viewpoint. The choice of the appropriate encoding is then left to the knowledge engineer and it proves to be a crafty task.

In this paper we propose a generic, logic-based architecture for OCQA allowing for transparent encoding in different data structures. The proposed platform, ALASKA (**A**bstract and **L**ogic-based **A**rchitecture for **S**torage systems and **K**nowledge bases **A**nalysis) allows for storage and query time comparison of relational databases, triple stores and graph databases.

## *1.1 Motivation and Use Case*

We will motivate and explain the contribution of ALASKA by the means of a real world case from the ANR funded project Qualinca. Qualinca is aiming at the validation and manipulation of bibliographic / multimedia knowledge bases. Amongst the partners, the ABES (French Higher Education Bibliographic Agency) and INA (French National Broadcasting Institute) provide large sets of RDF(S) data containing referencing errors. Examples of referencing errors include oeuvres (books, videos, articles etc.) mistakenly associated with the wrong author, or authors associated with a wrong / incomplete subset of their oeuvres. In order to solve such referential errors Qualinca proposed a logic based approach (as opposed to the large existing body of work mostly employing numerical measures) [11]. However, when it comes to retrieving the RDF(S) based data from ABES (or INA) and storing this data for OCQA, no existing tool in the literature could help the knowledge engineer decide what is the best system to use. Furthermore, the recent work in graph databases has surprisingly not been yet investigated in the context of OCQA (while graphs give encouraging results when used in central memory as shown in [9]).

The main contribution of ALASKA is to help make explicit different storage system choices for the knowledge engineer. More precisely, ALASKA can encode a knowledge base expressed in the positive existential subset of first order logic in different data structures (graphs, relational databases, 3Stores etc.). This allows for the transparent performance comparison of different storage systems.

The performance comparison is done according to the (1) storage time relative to the size of knowledge bases and (2) query time to a representative set of queries. On a generic level the storage time need is due to forward chaining rule mechanism when fast insertion of new atoms generated is needed. In Qualinca, we also needed to interact with the ABES server for retrieving their RDF(S) data. The server answer set is limited to a given size. When retrieving extra information we needed to know how fast we could insert incoming data into the storage system. Second, the query time is also important. The chosen queries used for this study are due to their expressiveness and structure. Please note that this does not affect the fact that we are in a semi-structured data scenario. Indeed, the nature of ABES bibliographical data with many information missing is fully semi-structured. The generic querying allowing for comparison is done across storage using a generic backtrack algorithm to implement the backtracking algorithm for subsumption. ALASKA also allows for direct querying using native data structure engines (SQL, SPARQL).

This paper will explain the ALASKA architecture and the data transformations needed for the generic logic-based view over heterogeneously stored data over the following systems:

→ **Relational databases:** Sqlite 3.3.6[1], MySQL 5.0.77[2]
→ **Graph databases:** Neo4J 1.5[3], DEX 4.3[4]
→ **Triples stores:** Jena TDB 0.8.10[5], Sesame 2.3.2[6]

Tests were also performed over Oracle 11g. A major drawback of using it for our tests is that the software requires the initial setting of the memory usage. Since the server used [7] is not as performant as needed, Oracle did not have a lot of resources available.

## 2 State of the Art

Let us consider a database $F$ that consists of a set of logical atoms, an ontology $\mathcal{O}$ written in some (first-order logic) language, and a conjunctive query $Q$. The OCQA problem stated in reference to the classical *forward chaining* scheme is the following: "Can we find an answer to $Q$ in a database $F'$ that is built from $F$ by adding atoms that can be logically deduced from $F$ and $\mathcal{O}$?"

Recent works in databases consider the language Datalog$^+$ [6] to encode the ontology. This is a generalization of Datalog that allows for existentially quantified variables in the hypothesis (that do not appear in the conclusion). Hence the forward chaining scheme (called here *chase*) can generate new variables while building $F'$. This is an important feature for ontology languages (called *value invention*), but is the cause of the undecidability of Datalog$^+$. An important field of research today, both in knowledge representation and in databases, is to identify decidable fragments of Datalog$^+$, that form the Datalog$^{\pm}$ family [7, 4]. In description logics, the need to answer conjunctive queries has led to the definition / study of less expressive languages (e.g. $\mathcal{EL}$ [2] and DL-Lite families [8]). Properties of these languages were used to define profiles of the Semantic Web OWL 2 [8] language. In the following we briefly present the logical language we are interested in this paper and the notation conventions we use.

---

[1] http://www.sqlite.org/
[2] http://www.mysql.com/
[3] http://www.neo4j.org/
[4] http://www.sparsity-technologies.com/dex
[5] http://jena.sourceforge.net/
[6] http://www.openrdf.org/
[7] 64-bit Quadcore AMD Opteron 8384 with 512 Kb of cache size and 12 Gb of RAM
[8] www.w3.org/TR/owl-overview

A vocabulary $W$ is composed of a set of predicates and a set of constants. In the vocabulary, constants are used to identify all the individuals one wants to represent, and predicates represent all the interactions between individuals.

**Definition 1 (Vocabulary).** Let $C$ be a set of constants, and $P = P_1 \cup P_2 ... \cup P_n$ a set of predicates of arity $i = 1,...,n$ ($n$ being the maximum arity for a predicate). We define $W = (P,C)$ as a vocabulary.

An atom on $W$ is of form $p(t_1,...,t_k)$, where $p$ is a predicate of arity $k$ in $W$ and the $t_i \in T$ are terms (i.e. constants in $W$ or variables). For a formula $\phi$, we note $terms(\phi)$ and $vars(\phi)$ respectively the terms and variables occurring in $\phi$. We use the classical notions of semantic consequence ($\models$), and equivalence ($\equiv$). A conjunct is a (possibly infinite) conjunction of atoms. A fact is the existential closure of a conjunct. We also see conjuncts and facts as sets of atoms.

The full fact w.r.t. a vocabulary $W$ contains all ground atoms that can be built on $W$ [9]. (thus any fact on $W$ is a semantic consequence of it). A $\forall$-rule is a formula $\forall X(H \rightarrow C)$ where $H$ and $C$ are conjuncts and $vars(C) \subseteq vars(H) \subseteq X$. A $\forall\exists$-rule $R = (H,C)$ is a closed formula of form $\forall x_1 ... \forall x_p (H \rightarrow (\exists z_1 ... \exists z_q C))$ where $H$ and $C$ are two finite non empty conjuncts respectively called the hypothesis and the conclusion of $R$. In examples, we omit quantifiers and use the form $H \rightarrow C$.

**Definition 2 (Facts, KB).** Let $F$ a set of facts, and $\mathcal{O}$ a set of rules, we define a knowledge base (KB) $K = (F,\mathcal{O})$.

The complexity of OCQA may vary according to the type of rules of $\mathcal{O}$. Also, according to the set of rules we choose, we retrieve a semantical equivalence from one problem onto others that are (or have already been) studied in the literature. Choosing $\mathcal{O}$ as an empty set makes the problem equivalent to ENTAILMENT in the RDF [13] language. Using a set of $\forall$-rules instead will get us into the RDFS and Datalog scope. Finally, if one chooses to define $\mathcal{O}$ as a set of $\forall\exists$-rules, the problem becomes similar to the ones we find in the Datalog$\pm$ [7] and Conceptual Graphs languages [9].

While the above semantic equivalences have been shown, storage structures implementing these languages are compared in an ad-hoc manner and only focus on relational databases and 3Stores [12]. Such methodology does not ensure for logical soundness and completeness because the knowledge is not viewed across platform in a logical manner. More importantly, graph databases have not yet been considered as a storage solution in the OCQA context. In the next section we present the ALASKA architecture: a generic, logic based platform for OCQA.

[9] In this paper we use the notion of fact for any subset of atoms, contrary to the Datalog notation. and justified by "historical" reasons [9].

## 3 Alaska Architecture

The ALASKA core (data structures and functions) is written independently of any language used by storage systems it will access. The advantage of using a subset of First Order Logic to maintain this genericity is to be able to access and retrieve data stored in any system by the means of a logically **sound** common data structure. **Local encodings** will be transformed and translated into any other **representation language** at any time. The operations that have to be implemented are the following: (1) retrieve all the terms of a fact, (2) retrieve all the atoms of a fact, (3) add a new atom to a fact, (4) verify if a given atom is already present in the facts.

The platform architecture is multi-layered. Figure 1 represents its class diagram, highlighting the different layers.

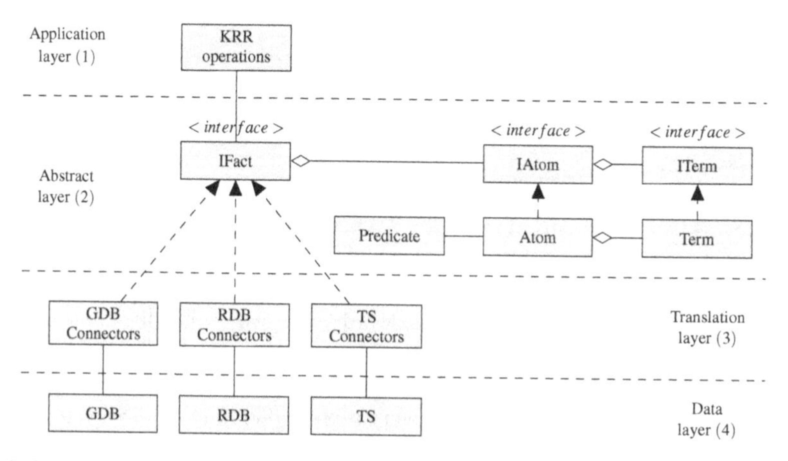


**Fig. 1** Class diagram representing the software architecture.

The first layer is (1) the **application** layer. Programs in this layer use data structures and call methods defined in the (2) **abstract** layer. Under the abstract layer, the (3) **translation** layer contains pieces of code in which logical expressions are translated into the languages of several storage systems. Those systems, when connected to the rest of the architecture, compose the (4) **data** layer. Performing higher level KRR operations within this architecture consists of writing programs and functions that use exclusively the formalism defined in the abstract layer. Once this is done, every program becomes compatible to any storage system connected to architecture.

Let us consider the workflow used by ALASKA in order to store new facts. The fact will first be parsed in the application layer (1) into the set of atoms corresponding to its logical formula, as defined in the abstract layer (2). Then, from this set, a connector located in layer (3) translates the fact into a specific representation language in (4). The set of atoms obtained from the fact will be translated into a graph database model and into the relational database model. Both models are detailed in the next Subsection. Please note that the forward chaining rule application process generates new atoms and thus this process respects the insertion workflow detailed below.

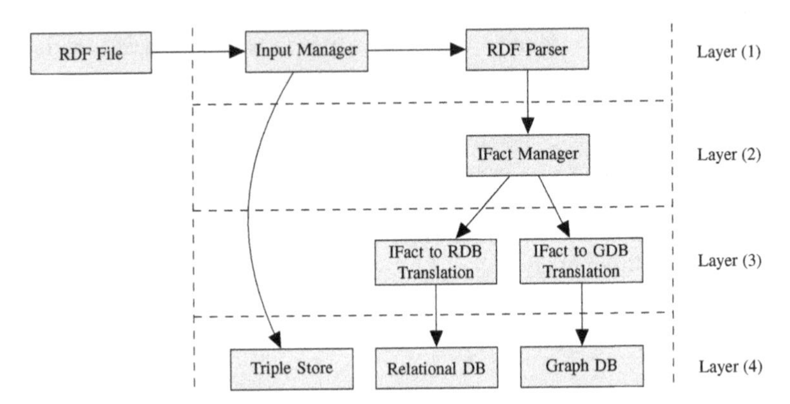


**Fig. 2** Testing protocol workflow for storing a knowledge base in RDF with ALASKA.

This workflow is visualised in Figure 2 where a RDF file is stored into different storage systems. In this workflow, we have chosen not to translate the knowledge base into IFact when storing the knowledge base into a triples store since it would require the exact opposite translation afterwards. This way, when the chosen system of destination is a triples store, the Input Manager does not operate a translation but sends the information from the RDF file directly to the triples store. Such direct storage is, of course, available for every representation in ALASKA (relational databases or graphs). When evaluating the storage systems, for equity reasons, such "shortcuts" have not been used.

Finally, the querying in our architecture takes place exactly in the same manner as the storage workflow. In Figure 3, on the left hand side we show the storing workflow above and on the right hand side the querying workflow (generic algorithm or native querying mechanism).

## *3.1 Transformations*

In this section we detail the transformations used in order to store the logical facts into the different encodings: relational databases (traditional or 3Stores) or graph databases.

Encoding facts for a relational database needs to be performed in two steps. First, the relational database schema has to be defined according to the knowledge base vocabulary. The information concerning the individuals in the knowledge base can only be inserted once this is done. According to the arities of the predicates given in

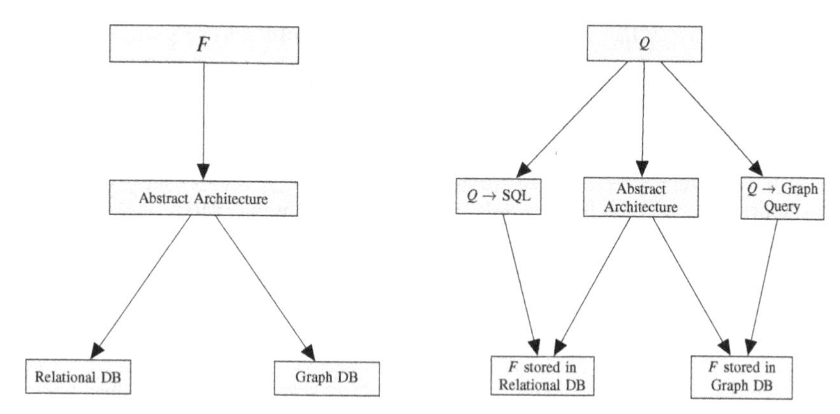


**Fig. 3** ALASKA storage and querying workflow.

the vocabulary, there are two distincts manners to build the schema of the relational database: in the first case, one relation is created for each predicate in the vocabulary. The second case can only be used when all the predicates in the vocabulary share the same arity (cf. RDF [13]). In this case, only one relation is defined. This encoding is similar to the ones used for Triples Stores [14].

It has to be remembered that there are no variable terms in a relational database. Indeed, variables are frozen into fresh constants before being inserted into a table. In order to maintain this information within the base, two distinct methods exist: in the first method, every term $t$ in the database is renamed and receives an identifier prefix: $c{:}t$ or $v{:}t$ according to its type. In the second case, no changes are made to the label of the terms, but an extra unary relation, $R_{vars}$ is created and added to the schema. Variable terms are then stored inside this table. The schema of the database remains unchanged when using the first method.

**Definition 3.** Let $R$ the set of relations of the relational database. $R = \bigcup_{i=1}^{n} R_i$ and $R_i = \{R_p(A_1,...,A_i) \mid \forall p \in P_i\}$. If variables are stored in an additional relation, then $R = \bigcup_{i=1}^{n} R_i \cup R_{vars}(A_1)$.

Once the database schema is defined, relation tuples are defined according to the atoms of the KB facts. $(x_1,...,x_j) \in R_{p_j}$ iff $p_j(x_1,...,x_j)$. In the particular case in which all predicates do share the same arity, the set of relations of the database $R$ = $\{R_{k+1}(A1,...,A_{k+1})\}$. In this case, relation tuples are defined with the following: $(x_1,...,x_j,p_j) \in R_{k+1}$ iff $p_j(x_1,...,x_j)$.

Once the fact in the knowledge base is stored in the relational database, query answering can be done through SQL requests. By the definitions above, applying a SQL statement $S_Q$ over the relational database corresponds to compute a labeled homomorphism from a conjunctive query $Q$ into the fact [1].

**Definition 4.** Let $G = (N,E,l)$ a hypergraph: $N$ is its set of vertices, $E$ its set of hyperedges and $l$ a labelling function.

When encoding a fact into a hypergraph, the nodes of the hypergraph correspond to the terms of the fact. Hence, $N = T$. The (hyper)edges that connect the nodes of the hypergraph correspond to the atoms in which are present the corresponding terms. $E = \bigcup_{p \in P} E_p$ with $E_p = \{(t_1, \ldots, t_j) \mid p(t_1, \ldots, t_j) \in A \text{ and } l(t_1, \ldots, t_j) = p\}$.

Let us illustrate the processes detailed above by the means of an example. The knowledge base we take for example contains the following facts:

1. $\exists a, b, c, d, e \; ( \; p(a,b) \wedge p(c,e) \wedge q(b,c) \wedge q(e,d) \wedge r(a,c) \wedge q(d,c) \; )$

| $R_p$ | |
|---|---|
| 1 | 2 |
| c:a | c:b |
| c:c | c:e |

| $R_q$ | |
|---|---|
| 1 | 2 |
| c:b | c:c |
| c:e | c:d |

| $R_r$ | |
|---|---|
| 1 | 2 |
| c:a | c:c |
| c:d | c:c |

| $R_3$ | | |
|---|---|---|
| 1 | 2 | 3 |
| c:a | c:b | p |
| c:c | c:e | p |
| c:b | c:c | q |
| c:e | c:d | q |
| c:a | c:c | r |
| c:d | c:c | r |

**Fig. 4** Encoding a fact in a relational database.

Figure 4 represents the encoding of the knowledge base facts in a relational database. As there are no variables in the example, no additional relation is added to the database schema. As all the predicates in the example share the same arity, both possibilities of schema for the relational database are represented on the figure. The three tables on the left list the content of the database when there is one table created per predicate. The table on the right shows the content of the database when one single table is created and an extra attribute is created to store the predicate of the atoms.

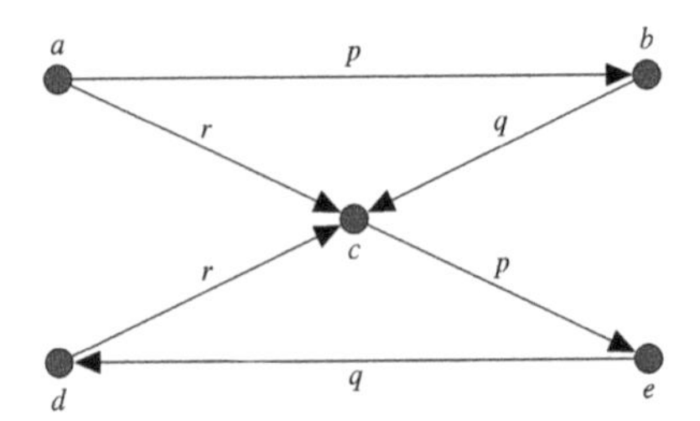


**Fig. 5** Graph containing the facts (1) from Section 3.1.

Figure 5 represents the generated hypergraph after having encoded the knowledge base fact.

### 3.2 *Storage Algorithms*

In order to store large knowledge bases and to prevent the issues previously described, we have implemented the generic storage algorithm below.

**Algorithm 1**: Input Manager store method

```
   Input: S a stream of atoms, f an IFact, bSize an integer
   Output: a boolean value
1  begin
2    buffer <-- an empty array of size bSize;
3    counter <-- 0;
4    foreach Atom a in S do
5      if counter = bSize then
6        f.addAtoms(buffer,null);
7        counter <-- 0;
8      buffer[counter] = a;
9      counter++;
10   f.addAtoms(buffer,counter); return true;
11 end
```

Algorithm 1 illustrates the manner the Input Manager handles the stream of atoms received as input, as well as the manner it creates a buffer and sends atoms for the storage system in groups and not one-by-one nor all-at-once.

As one may notice, the algorithm calls the *addAtoms* method of the fact $f$ also given as input. The genericity of ALASKA platform certifies that any storage system connected to the platform must implement this method (among others). A fact is differently encoded according to the storage system at target according to the transformations explained in Sections 3.1.

### 3.3 *Querying Algorithms*

In order to measure the querying efficiency of each of the different storage paradigms (relational databases or graphs) we have implemented a generic backtrack algorithm that makes use of the fact retrieval primitives of each system. Please note that numerous improvements are possible for this algorithms (cf. [9] or [10]) but this work is out of the scope of this paper. We are currently investigating the use of a constraint satisfaction solver (Choco) for the optimisation of the backtrack algorithm as such.

## 4 Evaluation

In this section we describe our strategy for evaluating ALASKA and its capability of comparing different storage systems. The results of this evaluation are presented once the data sets and respective queries for the tests are chosen.

**Algorithm 2**: Backtrack algorithm

```
Input: K a knowledge base
Output: a boolean value
1  begin
2      if mode = graph then
3          g ⟵ empty graph;
4      else
5          while Atom a in A do
6              n ⟵ empty array of nodes;
7              foreach Term t in a.terms do
8                  if exists node with label t then
9                      id ⟵ node.id;
10                 else
11                     create new node with id newId;
12                     id ⟵ newId;
13                 n.push(id);
14             create hyperedge with label a.predicate between all nodes in n;
15         return true;
16 end
```

As already stated in Section 1, our initial choice was to perform our tests using parts of data already available with ABES. Unfortunately such information were not available for this paper due to confidentiality reasons. By browsing the literature in the SPARQL benchmarking we found different datasets that were still pertinent to our problem. We thus used the knowledge base generator supplied by the SP2B project [16]. The generator enables the creation of knowledge bases with a certain parametrised quantity of triples maintaining a similar structure to the original DBLP knowledge base. Please refer to the paper [16] for a discussion on the relevance of this dataset. Using those generated knowledge bases then requires an initial translation from RDF into first order logic expressions (done offline, according to [3]). Please note that the initial RDF translation step has not been taken into account when reporting on storage times.

### *4.1 Queries*

Another very interesting feature of the SP2B project is the fact that it also provides a set of 10 SPARQL queries that covers the whole specifications of the SPARQL language. As we decided to use their generated knowledge within our storage tests, we have also planned to use their set of queries for our querying tests. Using such set of queries however was not possible due to the fact that half of their queries use some SPARQL keywords (OPTIONAL, FILTER) which makes that the query cannot be directly translated into a Datalog query, which is the query language we have chosen for our generic backtrack algorithm. We thus created our own set of queries for our knowledge bases structurally similar to the ones proposed in SP2B:

1. $type(X, Article)$
2. $creator(X, PaulErdoes) \wedge creator(X,Y)$
3. $type(X, Article) \wedge journal(X, 1940) \wedge creator(X,Y)$
4. $type(X, Article) \wedge creator(X, PaulErdoes)$

| Size of the stored knowledge bases | | | | | |
|---|---|---|---|---|---|
| System | 5M | 20M | 40M | 75M | 100M |
| DEX | 55 Mb | 214.2 Mb | 421.7 Mb | 785.1 | 1.0 Gb |
| Neo4J | 157.4 Mb | 629.5 Mb | 1.2 Gb | 2.3 Gb | 3.1 Gb |
| Sqlite | 767.4 Mb | 2.9 Gb | 6.0 Mb | 11.6 Gb | 15.5 Mb |
| Jena TDB | 1.1 Gb | 3.9 Gb | 7.5 Gb | - | - |
| RDF File | 533.2 Mb | 2.1 Gb | 4.2 Gb | 7.8 Gb | 10.4 Gb |

**Fig. 6** Table comparing knowledge bases sizes in different systems.

## *4.2 Results*

For our tests, RDF files of 1, 5, 20, 40, 75 and 100 million triples were generated then stored on disk with storage systems selected from the available systems in ALASKA. On such tests, we not only measure the time elapsed during the storing process, but also the final size of the stored knowledge base. As time results are important in order to see which is the most efficient storage system, disk usage results are equally important in the context of OCQA.

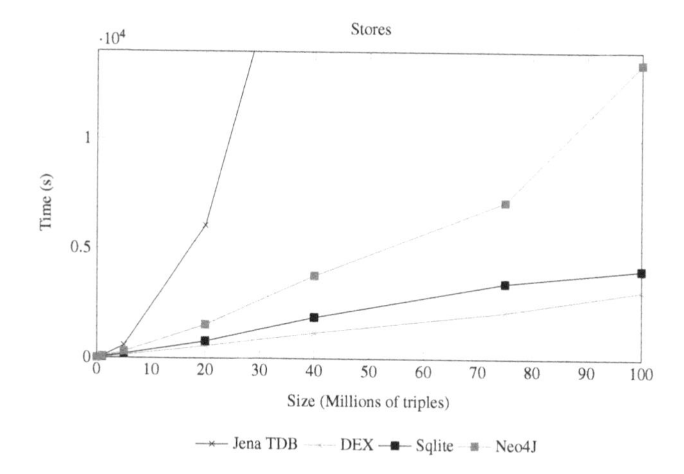


**Fig. 7** Storage performance using ALASKA for large knowledge bases

Figure 7 shows the time results of the tests performed on the storage systems selected due to their intermediate performance on smaller data sets. On such tests, we observe that DEX is the most efficient system from all the systems selected. Sqlite is second, followed by Neo4J. Jena TDB results are far from convincing. Not only its storing time is very high, but also the testing program could not go beyond the 40M triples knowledge base. We believe that such inefficiency may come from the parsing issues linked to the Jena framework. Our aim is not to optimize a storage system in particular but to be able to give a fast and good overview of available storage solutions.

Finally, Figure 8 shows the number of answers for each query in different sizes knowledge bases, while Figure 9 show the amount of time needed to perform those queries using the generic backtrack algorithm over the two best storage systems SQL Lite and DEX.

| ... | | | | |
|---|---|---|---|---|
| Size | Q1 | Q2 | Q3 | Q4 |
| 1 K | 129 answers | 104 answers | 30 answers | 50 answers |
| 20 K | 1685 answers | 458 answers | 30 answers | 100 answers |
| 40 K | 3195 answers | 567 answers | 30 answers | 100 answers |
| 80 K | 6084 answers | 678 answers | 30 answers | 100 answers |
| 100 K | 7441 answers | 721 answers | 30 answers | 100 answers |
| 200 K | 14016 answers | 838 answers | 30 answers | 100 answers |
| 400 K | 25887 answers | 967 answers | 30 answers | 100 answers |

**Fig. 8** Number of answers for each query

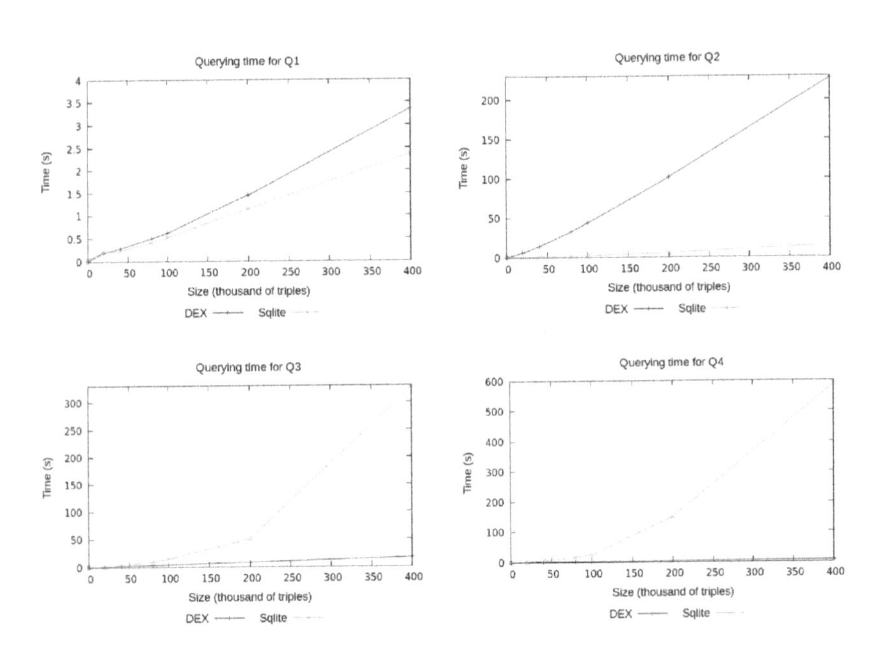


**Fig. 9** Results of querying tests after having performed the 4 selected queries on the tested systems

### 4.3 Implementation Aspects

Storing large knowledge bases using a straight-forward implementation of the testing protocol has highlighted different issues. We have distinguished three different issues that have appeared during the tests: (1) memory consumption at parsing level, (2) use of transactions, and (3) garbage collecting time.

**Memory consumption** at parsing level depends directly of the parsing method chosen. A few experiences have shown that some parsers/methods use more memory resources than others while accessing the information of a knowledge base and transforming it into logic. We have initially chosen the Jena framework parsing functions in order to parse RDF content, but we have identified that it loads almost the whole input file in memory at reading step. We have thus implemented an RDF parser which does not store the facts in main memory, but feeds them one at a time to the ALASKA input file.

The use of **transactions** also became necessary in order to store large knowledge bases properly.

**Garbage collecting** (GC) issues have also appeared as soon as preliminary tests were performed. Several times, storing not very large knowledge bases resulted in a GC overhead limit exception thrown by the Java Virtual Machine. The exception indicates that at least 98% of the running time of a program is consumed by garbage collecting.

In order to address both transaction and garbage collection issues, an atom buffer was created. The buffer is filled with freshly parsed atoms at parsing level. At the beginning, the buffer is full and then every parsed atom is pushed into the buffer before being stored. Once the buffer is full, parsing is interrupted and the atoms in the buffer are sent to the storage system for being stored. Once all atoms are stored, instead of cleaning the buffer by destroying all the objects, the first atom of the buffer is moved from the buffer into a stack of atoms to be recycled. Different stacks are created for each arity of predicates. In order to replace this atom, a new atom is only created if there is no atom to be recycled from the stack of the arity of the parsed atom. If there is an atom to be recycled, then it is then put back in the buffer, with its predicate and terms changed by attribute setters. The buffer is then filled once again, until it is full and the atoms in it are sent to storage system.

## 5 Discussion

A novel abstract platform (ALASKA) was introduced in order to provide an unified logic-based architecture for ontology-based data access. In its current state ALASKA can be considered as the bottom layer of a generic KR architecture. Today, ALASKA is the only software tool that provides for a logical abstraction of ontological languages and the possibility to encode factual knowledge (DB, ABox) in one's storage system of choice. The transparent encoding in different data structures made thus possible the comparison of storage and querying capabilities of relational databases and graph databases.

While the initial aim of devising and describing a platform able to compare systems for OCQA has been archived, we are currently taking this work further in two main directions.

First we are optimising the backtrack algorithm and comparing the optimised version with the native SQL / SPAQRL engines. We are currently using a constraints satisfaction solver in order to benefit from the different backtrack optimisations in the literature. Second, we will extend our various tests over different size knowledge bases and different expressivity queries in order to be able to give "best storage system recipes" to knowledge engineers. Obtaining such comparative results is a long and tedious process (due to different implementation adaptations ALASKA needs to take in account), but the final decision support system it could generate could be of great interest for application aspects of OCQA community.

## References

1. S. Abiteboul, R. Hull, and V. Vianu. *Foundations of Databases*. Addison-Wesley, 1995.
2. F. Baader, S. Brandt, and C. Lutz. Pushing the el envelope. In *Proc. of IJCAI 2005*, 2005.
3. J.-F. Baget, M. Croitoru, A. Gutierrez, M. Leclère, and M.-L. Mugnier. Translations between rdf(s) and conceptual graphs. In *ICCS*, pages 28–41, 2010.
4. J.-F. Baget, M.-L. Mugnier, S. Rudolph, and M. Thomazo. Walking the complexity lines for generalized guarded existential rules. In T. Walsh, editor, *Proceedings of the 22nd International Joint Conference on Artificial Intelligence*, pages 712–717. IJCAI / AAAI, 2011.
5. C. Basca and A. Bernstein. Avalanche: Putting the spirit of the web back into semantic web querying. In *ISWC Posters&Demos*, 2010.
6. A. Calì, G. Gottlob, and T. Lukasiewicz. A general datalog-based framework for tractable query answering over ontologies. In *Proceedings of the Twenty-Eigth ACM SIGMOD-SIGACT-SIGART Symposium on Principles of Database Systems*, pages 77–86. ACM, 2009.
7. A. Calì, G. Gottlob, T. Lukasiewicz, B. Marnette, and A. Pieris. Datalog+/-: A family of logical knowledge representation and query languages for new applications. In *LICS*, pages 228–242, 2010.
8. D. Calvanese, G. De Giacomo, D. Lembo, M. Lenzerini, and R. Rosati. Tractable reasoning and efficient query answering in description logics: The dl-lite family. *J. Autom. Reasoning*, 39(3):385–429, 2007.
9. M. Chein and M.-L. Mugnier. *Graph-based Knowledge Representation and Reasoning—Computational Foundations of Conceptual Graphs*. Advanced Information and Knowledge Processing. Springer, 2009.
10. M. Croitoru and E. Compatangelo. A tree decomposition algorithm for conceptual graph projection. In *Tenth International Conference on Principles of Knowledge Representation and Reasoning*, pages 271–276. AAAI Press, 2006.
11. M. Croitoru, L. Guizol, and M. Leclère. On link validity in bibliographic knowledge bases. In *Proc. of IPMU*, 2012.
12. B. Haslhofer, E. M. Roochi, B. Schandl, and S. Zander. Europeana RDF store report. Technical report, University of Vienna, Vienna, Mar. 2011.
13. P. Hayes, editor. *RDF Semantics*. W3C Recommendation. W3C, 2004. http://www.w3.org/TR/rdf-mt/.
14. A. Hertel, J. Broekstra, and H. Stuckenschmidt. Rdf storage and retrieval systems. *Handbook on Ontologies*, pages 489–508, 2009.
15. M. Lenzerini. Data integration: A theoretical perspective. In *Proc. of PODS 2002*, 2002.
16. M. Schmidt, T. Hornung, G. Lausen, and C. Pinkel. Sp2bench: A sparql performance benchmark. *CoRR*, abs/0806.4627, 2008.
17. M. Sensoy, G. de Mel, W. W. Vasconcelos, and T. J. Norman. Ontological logic programming. In *WIMS*, page 44, 2011.

# Multi-Agent Knowledge Allocation

Sebastian Rudolph[1] and Madalina Croitoru[2]

**Abstract** Classical query answering either assumes the existence of just one knowledge requester, or knowledge requests from distinct parties are treated independently. Yet, this assumption is inappropriate in practical applications where requesters are in direct competition for knowledge. We provide a formal model for such scenarios by proposing the Multi-Agent Knowledge Allocation (MAKA) setting which combines the fields of query answering in information systems and multi-agent resource allocation. We define a bidding language based on exclusivity-annotated conjunctive queries and succinctly translate the allocation problem into a graph structure which allows for employing network-flow-based constraint solving techniques for optimal allocation.

## 1 Introduction

Knowledge exchange between actors is a central aspect of multi-agent systems. Often, such exchange is demand-driven and the participating agents can be divided into *knowledge requesters* and *knowledge providers. Conjunctive query answering* constitutes the de-facto standard for interacting with structured information resources, thoroughly addressed in the databases area [8, 1], but later also adopted for ontological knowledge-based systems [7, 6]. The classical setting in query answering assumes just one knowledge requester; in the case of multiple requesters, queries posed by different parties are processed and answered independently from each other, thus making the multi-requester scenario an easy extension of the individual case: if several knowledge requesters ask for (possibly overlapping) information, this information would (potentially duplicately) be distributed according to the queries.

Sebastian Rudolph
KSRI/AIFB, Karlsruhe Institute of Technology, Germany e-mail: Sebastian.Rudolph@kit.edu

Madalina Croitoru
LIRMM, Montpellier, France e-mail: Madalina.Croitoru@lirmm.fr

While this practice is natural in some cases, the underlying independence assumption is questionable in practical applications where requesters are in direct competition for information. In some multi-agent settings where requester agents concurrently demand information from a provider agent (such as in military applications, news agencies, intelligence services, etc.), requester agents will not be willing to share "sensitive" information with other agents.

A structurally related problem is the *multi-agent resource allocation* (MARA) setting [9]. However, in such a setting (i) the agents ask for resources (not knowledge) and (ii) agents *a priori* know the pool of available resources. Work in this field is aimed at bidding language expressiveness (e.g., what preferences over resource subsets can be correctly represented by the language) or algorithmic aspects of the allocation problem [12, 3, 11]. Multiplicity of resources, or resources used exclusively or shared has also been recently investigated in a logic-based language [13].

This paper proposes the *Multi-Agent Knowledge Allocation* (MAKA) setting, where $n$ requester agents ask for knowledge (and not resources). They express their requests in the form of queries that are endowed with exclusivity constraints and valuations, which indicate the subjective value of potentially allocated query answers. Knowledge allocation poses interesting inherent problems not only from a bidding and query answering viewpoint, but also in terms of mechanism design.

While a first informal sketch of this idea has been presented in [10], the contributions of this paper can be summarized as follows: On the conceptual side, the aim of this paper is

- to formally introduce the novel problem setting of Multi-Agent Knowledge Allocation,
- to define the notion of exclusivity-aware queries to account for a setting where knowledge requesters compete for information and
- to identify directions for future work opened up by this novel setting: increased expressivity, dynamic allocations, fairness, multiple providers etc.

On the technical side, drawing from the fields of query answering in information systems and MARA, we

- define syntax and semantics of a bidding language featuring exclusivity-annotated queries as well as valuation functions and
- show a way to succinctly translate the multi-agent knowledge allocation problem into a network flow problem allowing to employ a wide range of constraint solving techniques to find the optimal allocation of answers to agents.

The MAKA framework goes beyond current information sharing mechanisms by proposing an alternative modeling based on economic principles. It is particularly well-suited for the new challenges of data sharing and exclusivity issues in the data-driven journalism domain, where open data markets (as represented by the Linked Open Data initiative[1]) and closed information markets may interface. In this

[1] http://www.w3.org/wiki/SweoIG/TaskForces/CommunityProjects/LinkingOpenData

context, economically justified decisions have to be made regarding whether (RDF-)structured data is either to be openly published online (e.g., as part of the Linked Open Data cloud) or to be shared on a peer-to-peer basis. Contracts containing exclusivity guarantees will have to be automatically concluded and endorsed.

The paper is structured as follows: In Section 2 we conceptually extend the classical setting of query answering by ways of handling exclusivity demands. Section 3 introduces and formally defines the Multi-Agent Knowledge Allocation problem. Consequently, Section 4 defines knowledge allocation networks and shows that the knowledge allocation problem can be reduced to a maximum flow problem in these networks. Section 5 concludes and discusses avenues for future work.

## 2 Querying with Exclusivity Constraints

We introduce a framework of exclusivity-aware querying as a basis for an adequate MAKA bidding formalism. For illustration purposes, we accompany the notions introduced herein by an example about celebrity news. We define the term of a *knowledge base* to formally describe a pool of information (or knowledge) that a *knowledge provider agent* holds and that is to be delivered to *knowledge requester agents* in response their bids expressed via queries.

**Definition 1 (Knowledge Base).** Let $C$ be a set of constants and $P = P_1 \cup P_2 \ldots \cup P_n$ a set of predicates of arity $i = 1,2,\ldots,n$, respectively. Given some $i \in \{1,\ldots,n\}$, $p \in P_i$ and $c_1,\ldots,c_i \in C$ we call $p(c_1,c_2\ldots,c_i)$ a *ground fact*. We denote by $\mathcal{GF}_{P,C}$ the set of all ground facts. A *knowledge base* $\mathcal{K}$ is defined as a set of ground facts: $\mathcal{K} \subseteq \mathcal{GF}_{P,C}$.

*Example 1.* Consider the unary predicates `actor`, `director`, `singer`, the binary predicates `marriage` and `act` as well as five constants `AJ` (Angelina Jolie), `BP` (Brad Pitt), `MMS` (Mr. and Ms. Smith), `JB` (Jessica Biel), `JT` (Justin Timberlake). Our example knowledge base contains the facts

```
actor(AJ) director(AJ) marriage(AJ,BP)
actor(BP) singer(JT)   act(AJ,MMS)
actor(JB)              act(BP,MMS)
```

While in our version, the notion of knowledge base is very much alike a classical database, we will extend this notion in future work to include more general logical statements.

To describe a knowledge requester's information need, we must provide for a sort of templates where *(query) variables* take the place of the pieces of information that a knowledge requester is interested in. Following the standard terminology of logical querying, we call these templates *atoms*.

**Definition 2 (Terms, Atoms).** Let $V$ be a countably infinite set of *(query) variables*. We define the set of *terms* by $T = V \cup C$. As usual, given $i \in \{1\ldots n\}$, $p \in P_i$ and $t_1,\ldots,t_i \in T$ we call $p(t_1,\ldots,t_i)$ an *atom*. We denote by $\mathcal{A}_{T,P}$ the set of all atoms constructed from $P$ and $T$. Note that ground facts are just special atoms.

*Example 2.* For instance, if we consider the set of variables $V = \{x, y, \ldots\}$ and the set of constants $C$ as above then $\texttt{actor}(x)$, $\texttt{act}(y,\texttt{MMS})$, $\texttt{marriage}(\texttt{AJ},\texttt{BP})$ are atoms over the previously defined sets $P$ and $C$.

Since in the MAKA scenario, requesters might be competing for certain pieces of knowledge, we have to provide them with the possibility of asking for an atom exclusively (exclusive) or not (shared). This additional information is captured by the notion of *exclusivity-annotated* atoms, ground facts and queries defined next.

**Definition 3 (Exclusivity-Annotated Atoms).** An *exclusivity-annotated atom* is an element from $\mathcal{A}^e_{P,T} := \mathcal{A}_{P,T} \times \{\mathsf{shared}, \mathsf{exclusive}\}$. Let $\{\mathsf{shared}, \mathsf{exclusive}\}$ be ordered as to $\mathsf{shared} \leq \mathsf{exclusive}$. In particular, an *exclusivity-annotated ground fact* is an element from $\mathcal{GF}^e_{P,C} := \mathcal{GF}_{P,C} \times \{\mathsf{shared}, \mathsf{exclusive}\}$.

*Example 3.* Following our above example, $\langle\texttt{actor}(x), \mathsf{shared}\rangle$, $\langle\texttt{act}(y,\texttt{MMS}), \mathsf{shared}\rangle$, $\langle\texttt{marriage}(\texttt{AJ},\texttt{BP}), \mathsf{shared}\rangle$, and $\langle\texttt{marriage}(\texttt{AJ},\texttt{BP}), \mathsf{exclusive}\rangle$ would be exclusivity-annotated atoms.

Exclusivity annotation is a novel concept going beyond the classical query answering framework. The purpose of the order defined between exclusive and shared will become clear in the subsequent development just like the actual semantics of exclusive and shared will only be made explicit when defining what an allocation is. For specifying structurally complex information needs, (exclusivity-annotated) atoms need to be logically combined, giving rise to the central notion of (exclusivity-annotated) *queries*.

**Definition 4 (Exclusivity-Annotated Queries).** An *exclusivity-annotated query* (EAQ) is an element of *boolexp*($\mathcal{A}^e_{P,T}$), i.e., a positive Boolean expression (an expression with Boolean operators $\wedge$ and $\vee$) over exclusivity-annotated atoms.[2]

*Example 4.* For example, a query $q_{\text{TheSun}}$ asking for exclusive marriages between actors and directors (where only the "marriage" itself is required as exclusive information, but the "actor" and "director" knowledge is sharable with other knowledge requester agents) would be written as:

$$\begin{aligned} q_{\text{TheSun}} = \langle\texttt{marriage}(x,y), \mathsf{exclusive}\rangle \wedge \\ ((\langle\texttt{actor}(x), \mathsf{shared}\rangle \wedge \langle\texttt{director}(y), \mathsf{shared}\rangle) \vee \\ (\langle\texttt{actor}(y), \mathsf{shared}\rangle \wedge \langle\texttt{director}(x), \mathsf{shared}\rangle)). \end{aligned}$$

Apart from the exclusivity annotation, this query formalism captures the core of common querying formalisms like SQL and SPARQL [14]. We omit filtering as well as left or right joins for the sake of brevity, however, these could be easily added.

As usual, when an EAQ $q$ is posed to a knowledge base, answers are encoded as bindings of the variables to elements from $C$ that make $q$ true in $\mathcal{K}$. For the further

[2] Note that this generalizes the classical notion of conjunctive queries where only $\wedge$ is allowed.

presentation, it is convenient to also identify parts $W$ of $\mathcal{K}$ that allow to derive that $\mu$ is an answer to $q$; such $W$ are called *witnesses.*[3]

**Definition 5 (Query Answers & Witnesses).** For an EAQ $q$, let $vars(q)$ denote the set of variables in the atoms of $q$ and $dup(\mathcal{K}) := \mathcal{K} \times \{\mathsf{shared}, \mathsf{exclusive}\}$ a knowledge base $\mathcal{K}$'s full enrichment with possible annotations. An *answer* to $q$ w.r.t. $\mathcal{K}$ is a mapping $\mu : vars(q) \cup C \rightarrow C$ with $\mu(c) = c$ for all $c \in C$ such that $eval(q, \mu, dup(\mathcal{K})) = true$. Thereby, $eval : boolexp(\mathcal{A}^e_{P,T}) \times C^T \times 2^{\mathcal{GF}^e_{P,C}} \rightarrow \{true, false\}$ is the evaluation function defined as follows: For an exclusivity-annotated atom $\langle p(t_1,\ldots,t_i), e\rangle$ and some set $A \subseteq \mathcal{GF}^e_{P,C}$, we let $eval(\langle p(t_1,\ldots,t_i), e\rangle, \mu, A) = true$ exactly if we have $\langle p(\mu(t_1),\ldots,\mu(t_i)), f\rangle \in A$ for an $e \preceq f$; further, the truth-value assignment is lifted to Boolean expressions in the usual way. Given an answer $\mu$ to a query $q$, a *witness* for $\mu$ is a set $W \subseteq dup(\mathcal{K})$ of exclusivity-annotated ground atoms for which $eval(q, \mu, W) = true$. Moreover, $W$ is called *minimal*, if $eval(q, \mu, W') = false$ holds for all $W' \subset W$. We let $\mathcal{W}^{\min}_{\mathcal{K},q,\mu}$ denote the set of all minimal witnesses for $\mu$ and $\mathcal{W}^{\min}_{\mathcal{K},q}$ the set of minimal witnesses for all answers to $q$.

Note that the handling of $\preceq$ implements the fact that if some piece of knowledge is requested as shared information (i.e. without demanding exclusivity), it is acceptable to nevertheless get it assigned exclusively.

*Example 5.* Let us consider the previous example query for marriages between actors, where the marriage information was asked for exclusively. There is only one answer $\mu$ to this query w.r.t. our previously introduced knowledge base: $\mu = \{x \mapsto \mathtt{AJ}, y \mapsto \mathtt{BP}\}$. There are four minimal witnesses for $\mu$ arising from different combinations of exclusivity annotations:

| atom | $W_1$ | $W_2$ | $W_3$ | $W_4$ |
|---|---|---|---|---|
| `marriage(AJ,BP)` | exclusive | exclusive | exclusive | exclusive |
| `director(AJ)` | shared | exclusive | shared | exclusive |
| `actor(BP)` | shared | shared | exclusive | exclusive |

This means that the ground atom `marriage(AJ,BP)` can only be exclusively allocated (as ⟨`marriage(AJ,BP)`, exclusive⟩) whereas the ground atoms `director(AJ)` and `actor(BP)` can be either "shareably" allocated with one or several other requesters (⟨`actor(BP)`, shared⟩) or exclusively allocated only to one requester agent (⟨`director(AJ)`, exclusive⟩) (cf. Definition 3).

# 3 The Knowledge Allocation Problem

*Multi Agent Knowledge Allocation* (MAKA) can be seen as an abstraction of a market-based approach to query answering. In a MAKA system, there is central node $a$, the *auctioneer* (or *knowledge provider*), and a set of $n$ nodes, $I = \{1, \ldots, n\}$,

[3] Since our querying formalism is monotone, we can be sure that $\mu$ is an answer to $q$ w.r.t. $W$ whenever it is an answer w.r.t. $\mathcal{K}$.

the *bidders* (or *knowledge requesters*), who express their information need (including exclusivity requirements) via queries to be evaluated against a knowledge base $\mathcal{K}$ held by the auctioneer. Depending on the allocation made by the auctioneer, the bidders are provided with minimal witnesses for answers to their queries.

Initially, the auctioneer asks each bidder to submit (in a specified common *bidding language*) their *bid*, called *knowledge request* in our framework.

**Definition 6 (Knowledge Request).** The *knowledge request* of bidder $i$, denoted by $Q_i$ is a set of pairs $\langle q, \varphi \rangle$ where $q$ is an EAQ and $\varphi : \mathbb{N} \to \mathbb{R}_+$ is a monotonic function. Thereby, $\varphi(k)$ expresses the *individual* interest (value) of bidder $i$ in obtaining $k$ distinct answers to $q$.

*Example 6.* Following the ongoing example in the paper, a knowledge request for an exclusively known marriage between a known actor and a known director, where each such marriage information is paid 30 units for would be the singleton set $\{\langle q_{\mathrm{TheSun}}, \varphi \rangle\}$ with $\varphi = k \mapsto 30 \cdot k$.

The valuation function $\varphi : \mathbb{N} \to \mathbb{R}_+$ can be defined in several ways and also captures the option of formulating discount models.

Consequently, given the knowledge request $Q_j$ of some bidder $j$, the individual prize $\mathfrak{v}_j(S)$ the agent is willing to pay on receiving a portion $S \subseteq dup(\mathcal{K})$ of facts from the knowledge base endowed with exclusivity guarantees is calculated by summing up the costs for the individual query matches arising from $\langle q, \varphi \rangle \in Q_j$, which, in turn, are determined by counting the answers for $q$ w.r.t. the partial knowledge base $S$ and applying the function $\varphi$ to that number, i.e.:

$$\mathfrak{v}_j(S) = \sum_{\langle q, \varphi \rangle \in Q_j} \varphi(|\{\mu \mid eval(q, \mu, S) = true\}|).$$

Based on bidders' valuations, the auctioneer will determine a *knowledge allocation*, specifying for each bidder her obtained knowledge bundle and satisfying the *exclusivity constraints* (ensuring that exclusivity annotations associated to atoms in the respective bundle are indeed complied with).

**Definition 7 (Knowledge Allocation).** Given a knowledge base $\mathcal{K}$ and a set $\{1, \ldots, n\}$ of bidders, a *knowledge allocation* $\mathbf{O}$ is defined as an n-tuple $(O_1, \ldots, O_n)$, with $O_i \subseteq dup(\mathcal{K})$ for all $i \in \{1, \ldots, n\}$ such that

- $\{\langle a, \mathsf{shared} \rangle, \langle a, \mathsf{exclusive} \rangle\} \nsubseteq O_1 \cup \ldots \cup O_n$ for all ground atoms $a \in \mathcal{K}$, and
- $O_i \cap O_j \cap (\mathcal{K} \times \{\mathsf{exclusive}\}) = \emptyset$ for all $i, j$ with $1 \leq i < j \leq n$.

The first condition guarantees that the same ground atom $a$ has not been given both exclusively and shared to the same, or two different agents. The second condition ensures that two atoms have not been exclusively allocated to two different agents.

*Example 7.* In our example, satisfaction of the conditions makes sure that once we have allocated $\langle \texttt{director(AJ)}, \textsf{exclusive} \rangle$ to some requester, then we cannot additionally allocate $\langle \texttt{director(AJ)}, \textsf{shared} \rangle$ (condition 1) nor can we allocate $\langle \texttt{director(AJ)}, \textsf{exclusive} \rangle$ to another requester (condition 2).

Given a knowledge allocation, one can compute its *global value* by summing up the individual prizes paid by the bidders for the share they receive. Obviously, the knowledge allocation problem aims at an *optimal allocation*, which maximizes this value.[4]

**Definition 8 (Global Value of a Knowledge Allocation).** Given an allocation $\mathbf{O}$, its *global value* $\mathfrak{v}(\mathbf{O})$, is defined by $\mathfrak{v}(\mathbf{O}) = \sum_{j=1,n} \mathfrak{v}_j(O_j)$. An allocation $\mathbf{O}$ is called *optimal* if $\mathfrak{v}(\mathbf{O}') \leq \mathfrak{v}(\mathbf{O})$ holds for all allocations $\mathbf{O}'$.

*Example 8.* Let us again consider the knowledge base $\mathcal{K}$ introduced in Section 2 and three agents 1, 2, 3 with knowledge requests $Q_1, Q_2, Q_3$ asking for information. Assume $Q_2 = \{\langle q_2, \varphi_2 \rangle\}$ where $q_2$ asks for marriages between an actor and a director but requests the marriage information exclusively, i.e.,

$$
\begin{aligned}
q_1 = & \langle \texttt{marriage}(x,y), \textsf{exclusive} \rangle \wedge \\
& ((\langle \texttt{actor}(x), \textsf{shared} \rangle \wedge \langle \texttt{director}(y), \textsf{shared} \rangle) \vee \\
& (\langle \texttt{actor}(y), \textsf{shared} \rangle \wedge \langle \texttt{director}(x), \textsf{shared} \rangle)).
\end{aligned}
$$

and $\varphi_2(k) = 120 \cdot k$. Further assume $Q_2 = \{\langle q_2, \varphi_2 \rangle\}$, where $q_2$ asks for marriages where there is at least one actor involved, i.e.,

$$
\begin{aligned}
q_2 = & \langle \texttt{marriage}(x,y), \textsf{shared} \rangle \wedge \\
& (\langle \texttt{actor}(x), \textsf{shared} \rangle \vee \langle \texttt{actor}(y), \textsf{shared} \rangle).
\end{aligned}
$$

and $\varphi_2(k) = 50 \cdot k$, i.e. Agent 2 is willing to pay 50 units for every answer. Finally let $Q_3 = \{\langle q_3, \varphi_3 \rangle\}$ with $q_3$ asking for marriages between people acting in the same movie (with the latter information requested exclusively, but not the former), i.e.,

$$
\begin{aligned}
q_3 = & \langle \texttt{marriage}(x,y), \textsf{shared} \rangle \wedge \\
& \langle \texttt{act}(x,z), \textsf{exclusive} \rangle \wedge \langle \texttt{act}(y,z), \textsf{exclusive} \rangle.
\end{aligned}
$$

and $\varphi_3(k) = 100 \cdot k$. Executing the bidders' knowledge requests against the knowledge base, the knowledge provider identifies one answer per posed request: $q_1$ yields the answer $\mu_1 = \{x \mapsto \texttt{AJ}, y \mapsto \texttt{BP}\}$, while $q_2$ yields the same answer $\mu_2 = \{x \mapsto \texttt{AJ}, y \mapsto \texttt{BP}\}$, and for $q_3$ we obtain the answer $\mu_3 = \{x \mapsto \texttt{AJ}, y \mapsto \texttt{BP}, z \mapsto \texttt{MMS}\}$. The corresponding minimal witnesses are displayed in the following table.

[4] Note that in general, the valuation function subject to maximization needs not to be based on monetary units but can be a utility function in the general sense, factoring in privacy concerns and other aspects.

| exclusivity-annotated atom | $\mathcal{W}^{\min}_{\mathcal{K},q_1,\mu_1}$ | | | | $\mathcal{W}^{\min}_{\mathcal{K},q_2,\mu_2}$ | | | | | | | | $\mathcal{W}^{\min}_{\mathcal{K},q_3,\mu_3}$ | |
|---|---|---|---|---|---|---|---|---|---|---|---|---|---|---|
| ⟨`marriage(AJ,BP)`,shared⟩ | | | | | × | × | × | × | | | | | × | |
| ⟨`marriage(AJ,BP)`,exclusive⟩ | × | × | × | × | | | | | × | × | × | × | | × |
| ⟨`director(AJ)`,shared⟩ | × | × | | | | | | | | | | | | |
| ⟨`director(AJ)`,exclusive⟩ | | | × | × | | | | | | | | | | |
| ⟨`actor(AJ)`,shared⟩ | | | | | × | | | | × | | | | | |
| ⟨`actor(AJ)`,exclusive⟩ | | | | | | × | | | | × | | | | |
| ⟨`actor(BP)`,shared⟩ | × | | × | | | | × | | | | × | | | |
| ⟨`actor(BP)`,exclusive⟩ | | × | | × | | | | × | | | | × | | |
| ⟨`act(AJ,MMS)`,exclusive⟩ | | | | | | | | | | | | | × | × |
| ⟨`act(BP,MMS)`,exclusive⟩ | | | | | | | | | | | | | × | × |

From the values above it is clear that, for an optimal allocation, `marriage(AJ,BP)` (the ground fact that poses shareability problems) must be allocated shareably to the agents 2 and 3 that did not ask for exclusivity and will bring a global valuation of 150, while allocating it exclusively to agent 1 would only bring 120. One optimal allocation, satisfying this consideration, would be $\mathbf{O} = (O_1, O_2, O_3)$ with $O_1 = \emptyset$, $O_2 =$ {⟨`marriage(AJ,BP)`,shared⟩,⟨`actor(AJ)`,shared⟩,⟨`actor(BP)`,shared⟩}, and $O_3 =$ {⟨`marriage(AJ,BP)`,shared⟩,⟨`act(AJ,MMS)`,exclusive⟩,⟨`act(BP,MMS)`,exclusive⟩}.

The auctioneer's task of finding an allocation with maximal global value for a given set of bidders' knowledge requests $\{Q_1, \ldots, Q_n\}$, is known as the *Winner Determination Problem* (WDP) in the Combinatorial Auctions' field. It is known to be NP-hard, equivalent to weighted-set-packing, yet tends to be solvable in many practical cases, but care is often required in formulating the problem to capture structure that is present in the domain [15]. Usually, the WDP is expressed as an integer linear programming problem (ILP); there are standard methods for solving this type of problems [9], which can also be employed for the MAKA scenario.

However, the specifics of the WDP instances generated by MAKA settings suggest an alternative formulation in terms of network flows which will be introduced in the next sections. While not being more efficient in terms of worst-case complexity (which is impossible due to the known NP-hardness), this approach enables a condensed representation of the problem by means of *structure sharing* and allows for capitalizing on solution methods stemming from the area of network flows [2].

## 4 Knowledge Allocation Networks

We propose a solution to the MAKA problem that is based on a network representation which naturally captures the notion of containment of an atom in a witness allocated to a given agent. The above containment relation is represented using paths in the network flow (an atom belongs to a witness if and only if there is a certain path from the atom vertex to a vertex representing the witness) and a mechanism to express which path must be considered in order to instantiate a given witness. This mechanism is based on a simple extension of network flows, which is described below. We first define the generic terminology of allocation networks before we specify how to create a knowledge allocation network representing a specific MAKA problem.

**Definition 9 (Allocation Network).** An *allocation network* is defined as a tuple $\mathcal{N} = (V_a, V_o, E, \text{START}, \text{END}, \mathsf{cap}, \mathsf{lb}, \mathsf{flow}, \mathsf{val})$ where:

1. $(V_a \cup V_o, E)$ is an acyclic digraph with edges $E$ and two types of nodes, called *allocatable* ($V_a$) and *other* nodes ($V_o$) and two distinguished nodes $\text{START}, \text{END} \in V_o$; Every node in $\mathcal{N}$ lies on a directed path from START to END. We use $V_{\text{int}} = (V_a \cup V_o) \setminus \{\text{START}, \text{END}\}$ to denote *internal nodes*.
2. $\mathsf{cap}, \mathsf{lb} : E \to \mathbb{N} \cup \{+\infty\}$ are functions defined on the set of directed edges; for a directed edge $\langle i, j\rangle \in E$ we call $\mathsf{cap}(\langle i, j\rangle)$, denoted $c_{ij}$, the *capacity* and we call $\mathsf{lb}(\langle i, j\rangle)$, denoted $l_{ij}$, the *lower bound* on $\langle i, j\rangle$ and additionally require $l_{ij} \leq c_{ij}$. All edges $\langle a, \text{END}\rangle$ have capacity $+\infty$ and lower bound 0.
3. $\mathsf{flow} : V_{\text{int}} \to \{\mathsf{conservation}, \mathsf{multiplier}\}$ is a labeling function associating each internal node $v$ with a *flow rule*.
4. $\mathsf{val} : V^- \times \mathbb{N} \to \mathbb{R}^+$, a *flow valuation function* with $V^- = \{v \mid \langle v, \text{END}\rangle \in E\}$.

As discussed before, we now assume that flow is pushed through the network starting from the arcs leaving the start node. The flow on each arc is a non-negative integer value. If the flow $f_{ij}$ on an arc $\langle i, j\rangle$ is positive, it must satisfy the constraints imposed by the specified *lower bound* $l$ and *upper bound* (also called *capacity*) $c$. Finally, the internal nodes of the network obey the flow rules associated to them.

**Definition 10 (Allocation Flow).** Let $\mathcal{N} = (V_a, V_o, E, \text{START}, \text{END}, \mathsf{cap}, \mathsf{lb}, \mathsf{flow}, \mathsf{val})$ be an allocation network. An *allocation flow* in $\mathcal{N}$ is a function $f : E \to \mathbb{N}$ satisfying the following properties, where we let $f_{ij}$ abbreviate $f(\langle i, j\rangle)$:

1. For each $\langle i, j\rangle \in E$ we have $f_{ij} = 0$ or $l_{ij} \leq f_{ij} \leq c_{ij}$
2. If $v \in V_{\text{int}}$ has $\mathsf{flow}(v) = \mathsf{conservation}$ then $\sum_{\langle i,v\rangle \in E} f_{iv} = \sum_{\langle v,i\rangle \in E} f_{vi}$.
3. If $v \in V_{\text{int}}$ has $\mathsf{flow}(v) = \mathsf{multiplier}$ and $f_{iv} > 0$ for some $i$, then $l_{vj} < f_{vj} < c_{vj}$ for all $j$ with $\langle v, j\rangle \in E$.
4. If $v \in V_{\text{int}}$ has $\mathsf{flow}(v) = \mathsf{multiplier}$ and $f_{iv} = 0$ for all $i$, then $f_{vj} = 0$ for all $j$ with $\langle v, j\rangle \in E$.

The set of all allocation flows in $\mathcal{N}$ is denoted by $\mathcal{F}^{\mathcal{N}}$.

After specifying the network and flows on it, we define how the network gives rise to a function associating values to subsets $S$ of the network's allocatable nodes. This *valuation* function is derived from the maximal flow possible when only nodes from $S$ are allowed to receive nonzero in-flow.

**Definition 11 (Valuation of an Allocation Network).** For an allocation network $\mathcal{N} = (V_a, V_o, E, \text{START}, \text{END}, \mathsf{cap}, \mathsf{lb}, \mathsf{flow}, \mathsf{val})$, the *value* of an allocation flow $f$ in $\mathcal{N}$, denoted by $\mathsf{val}(f)$, is defined by $\mathsf{val}(f) = \sum_{v \in V^-} \mathsf{val}(v, f_{v\text{END}})$. The *grant* associated to $f$, denoted by $G_f \subseteq V_a$ is defined by $G_f = \{a \in V_a \mid f_{va} > 0 \text{ for some } \langle v, a\rangle \in E\}$. The *valuation* associated to $\mathcal{N}$ is the function $\mathfrak{v}_{\mathcal{N}} : 2^{V_a} \to \mathbb{R}_+$, where for each $X \subseteq V_a$, $\mathfrak{v}_{\mathcal{N}}(X) = \max\{\mathsf{val}(f) \mid f \in \mathcal{F}^{\mathcal{N}}, G_f = X\}$.

In the following, we will describe how the framework of allocation networks can be applied to our problem of multi-agent knowledge allocation. To that end, we define two types of networks. *Single bidder allocation networks* capture the payments obtainable from separate bidders for a given portion of the knowledge base $\mathcal{K}$ according to their requests, whereas *knowledge allocation networks* integrates these separate representations into one in order to determine an optimal allocation.

**Definition 12 (Single Bidder Allocation Network).** Given a knowledge base $\mathcal{K}$ and some bidder's knowledge request $Q$, the associated *single bidder allocation network $\mathcal{N}_{\mathcal{K},Q}$* is the allocation network $(V_a, V_o, E, \text{START}, \text{END}, \mathsf{cap}, \mathsf{lb}, \mathsf{flow}, \mathsf{val})$ where

1. $V_a = \bigcup\{W \mid W \in \mathcal{W}^{\min}_{\mathcal{K},q}, \langle q,\varphi\rangle \in Q\}$, i.e. the allocatable nodes contain every exclusivity-annotated ground fact occurring in any minimal witness for any answer to any query posed by the bidder. We let $\mathsf{flow}(a) = \mathsf{multiplier}$ for every $a \in V_a$
2. $V_o = Q \cup Ans \cup Wit \cup \{\text{START}, \text{END}\}$ where:
   - For every $\langle q,\varphi\rangle \in Q$ we let $\mathsf{flow}(\langle q,\varphi\rangle) = \mathsf{conservation}$ and $\mathsf{val}(\langle q,\varphi\rangle, n) = \varphi(n)$.
   - *Ans* contains all pairs $\langle q,\mu\rangle$ with $\mu$ an answer to $q$ and $\langle q,\varphi\rangle \in Q$ for some $\varphi$. We let $\mathsf{flow}(\langle q,\mu\rangle) = \mathsf{multiplier}$.
   - *Wit* contains all witnesses for *Ans* i.e. $Wit = \bigcup_{\langle q,\mu\rangle \in Ans} \mathcal{W}^{\min}_{\mathcal{K},q,\mu}$, we let $\mathsf{flow}(W) = \mathsf{conservation}$ for every $W \in Wit$.
3. $E$ contains the following edges:
   - $\langle \text{START}, a\rangle$ for every $a \in V_a$, whereby $l_{\text{START},a} = 0$ and $c_{\text{START},a} = 1$
   - $\langle a, W\rangle$ for $a \in V_a$ and $W \in Wit$ whenever $a \in W$, moreover $l_{a,W} = 1$, $c_{a,W} = 1$.
   - $\langle W, \langle q,\mu\rangle\rangle$ for $W \in Wit$ and $\langle q,\mu\rangle \in Ans$ whenever $W \in \mathcal{W}^{\min}_{\mathcal{K},q,\mu}$. We let $l_{W,\langle q,\mu\rangle} = k$ and $c_{W,\langle q,\mu\rangle} = k$ where $k = |\{a \mid \langle a, W\rangle \in E\}|$.
   - $\langle\langle q,\mu\rangle, \langle q,\varphi\rangle\rangle$ for $\langle q,\mu\rangle \in Ans$ and $\langle q,\varphi\rangle \in Q$. We let $l_{\langle q,\mu\rangle,\langle q,\varphi\rangle} = 1$ as well as $c_{\langle q,\mu\rangle,\langle q,\varphi\rangle} = 1$.
   - $\langle\langle q,\varphi\rangle, \text{END}\rangle$ for all $\langle q,\varphi\rangle \in Q$. We let $l_{\langle q,\varphi\rangle,\text{END}} = 0, c_{\langle q,\varphi\rangle,\text{END}} = +\infty$.

The following theorem guarantees, that the established network definition indeed captures the bidder's interest.

**Theorem 1.** *Given a knowledge base $\mathcal{K}$, some bidder's knowledge request $Q_i$, and the respective single bidder allocation network $\mathcal{N}_{\mathcal{K},Q_i}$, the valuation $\mathfrak{v}_{\mathcal{N}_{\mathcal{K},Q_i}}$ associated to $\mathcal{N}_{\mathcal{K},Q_i}$ coincides with the bidder's valuation $\mathfrak{v}_i$.*

*Proof.* (Sketch) Consider a set $S$ of exclusivity-annotated ground facts. By definition, $\mathfrak{v}_{\mathcal{N}_{\mathcal{K},Q_i}}(S)$ equals the maximal $\mathsf{val}(f)$ among all flows $f$ for which exactly those allocatable nodes that correspond to $S$ receive non-zero in-flow. For such a flow $f$, the structure of the $\mathcal{N}_{\mathcal{K},Q_i}$ ensures that

- exactly those nodes from *Wit* receive full input flow which correspond to the witnesses contained in $S$,
- consequently, exactly those nodes from *Ans* receive non-zero input flow which correspond to answers for which $S$ contains some witness,
- thus, the flow that END receives from all $\langle q,\varphi\rangle \in Q$ equals the number of answers to $q$ that can be derived from $S$.

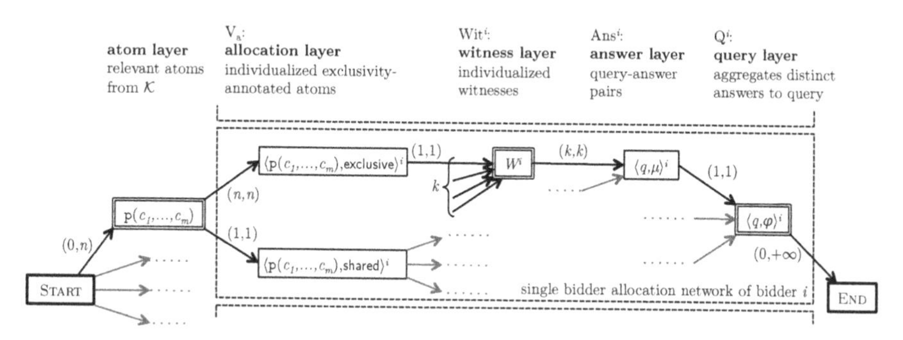


**Fig. 1** Structure of a Knowledge Allocation Network. Doubly boxed nodes indicate the conservation flow rule, all other internal nodes are multipliers. Edges $\langle i, j\rangle$ are associated with lower bounds and capacities by labels $(l_{ij}, c_{ij})$.

Given these correspondences, we can conclude:

$$\begin{aligned}\mathfrak{v}_{\mathcal{N}_{\mathcal{K},Q_i}}(S) &= \textstyle\sum_{\langle q,\varphi\rangle\in Q_i} \varphi(f_{\langle q,\varphi\rangle\text{END}}) \\ &= \textstyle\sum_{\langle q,\varphi\rangle\in Q_i} \varphi(|\{\mu \mid eval(q,\mu,S) = true\}|) \\ &= \mathfrak{v}_i(S)\end{aligned}$$

□

Toward an integrated representation of the MAKA problem, we conjoin the individual bidders' networks into one large network which also takes care of the auctioneer's actual allocation problem including the enforcement of the exclusivity constraints. Figure 1 illustrates the overall structure of the Knowledge Allocation Network formally defined next.

**Definition 13 (Knowledge Allocation Network).** Given a knowledge base $\mathcal{K}$ and knowledge requests $Q_1,\ldots,Q_n$ of bidders $1,\ldots,n$, the associated *knowledge allocation network* $\mathcal{N}_{\mathcal{K},Q_1,\ldots,Q_n}$ is the tuple $(V_a, V_o, E, \text{START}, \text{END}, \mathsf{cap}, \mathsf{lb}, \mathsf{flow}, \mathsf{val})$ which is constructed from the respective $n$ single bidder allocation networks $\mathcal{N}_{\mathcal{K},Q_i} = (V_a^i, V_o^i, E^i, \text{START}^i, \text{END}^i, \mathsf{cap}^i, \mathsf{lb}^i, \mathsf{flow}^i, \mathsf{val}^i)$ in the following way:

1. $V_a \cup V_o = Atoms \cup \{\text{START}, \text{END}\} \cup \bigcup_i \{v^i \mid v \in V_{\text{int}}^i\}$ i.e. except for the start and end nodes, the knowledge allocation network contains one copy of every single bidder network node, further we let $\mathsf{flow}(v^i) = \mathsf{flow}^i(v)$ and $V_a = \bigcup_i \{v^i \mid v \in V_a^i\}$, i.e., the allocatable nodes in the new network are exactly the allocatable nodes from all single bidder allocation networks. We let $\mathsf{val}(v^i) = \mathsf{val}^i(v)$ where applicable. As new nodes we have $Atoms = \{a \mid \{\langle a, \text{exclusive}\rangle, \langle a, \text{shared}\rangle\} \cap \bigcup_i V_a^i \neq \emptyset\}$, i.e., annotation-free atoms whose annotated versions are part of witnesses for answers to some of the bidder-posed queries. We let $\mathsf{flow}(a) = \text{conservation}$ for all $a \in Atoms$.
2. $E$ contains the following edges:

- $\langle v_1^i, v_2^i\rangle$ for all $\langle v_1, v_2\rangle \in E^i$, i.e. all edges between internal nodes of the single bidder networks stay the same, likewise their capacity and lower bound values,

- $\langle \text{START}, a \rangle$ for all $a \in Atoms$, we let $l_{\text{START}a} = 0$ and $c_{\text{START}a} = n$,
- $\langle a, \langle a, \mathsf{shared} \rangle^i \rangle$ for all $\langle a, \mathsf{shared} \rangle \in V_{\text{a}}^i$, we let $l_{a,\langle a,\mathsf{shared}\rangle^i} = 1$ as well as $c_{a,\langle a,\mathsf{shared}\rangle^i} = 1$,
- $\langle a, \langle a, \mathsf{exclusive} \rangle^i \rangle$ for all $\langle a, \mathsf{exclusive} \rangle \in V_{\text{a}}^i$, and we let $l_{a,\langle a,\mathsf{exclusive}\rangle^i} = n$ and $c_{a,\langle a,\mathsf{exclusive}\rangle^i} = n$,
- $\langle \langle q, \varphi \rangle^i, \text{END} \rangle$ for every $\langle \langle q, \varphi \rangle, \text{END} \rangle \in E^i$, as required we let $l_{\langle q,\varphi\rangle^i,\text{END}} = 0$ and $c_{\langle q,\varphi\rangle^i,\text{END}} = +\infty$.

Figure 2 displays the knowledge allocation network associated to the allocation problem described in Example 8. We arrive at the following theorem, which ensures that the MAKA problem can be reformulated as a maximal flow problem in our defined knowledge allocation network. Therefore, once the network has been constructed, any maximum flow will represent an optimal allocation.

**Theorem 2.** *Given a knowledge base $\mathcal{K}$, knowledge requests $Q_1, \ldots, Q_n$, and the associated knowledge allocation network $N_{\mathcal{K},Q_1,\ldots,Q_n}$, the valuation $\mathfrak{v}_{N_{\mathcal{K},Q_1,\ldots,Q_n}}$ associated to $N_{\mathcal{K},Q_1,\ldots,Q_n}$ is related with the global valuation $\mathfrak{v}$ of the knowledge allocation problem in the following way:*

1. *for every network flow $f \in \mathcal{F}^{N_{\mathcal{K},Q_1,\ldots,Q_n}}$, we have that $\mathbf{O}_f := (\{a \mid a^1 \in G_f\}, \ldots, \{a \mid a^n \in G_f\})$ is a knowledge allocation (in particular, the exclusivity constraints are satisfied),*
2. *conversely, for every knowledge allocation $\mathbf{O} = (O_1, \ldots, O_n)$, there is a network flow $f \in \mathcal{F}^{N_{\mathcal{K},Q_1,\ldots,Q_n}}$ with $G_f = \bigcup_{1 \leq i \leq n} \{a^i \mid a \in O_i\}$,*
3. *for any allocation $\mathbf{O} = (O_1, \ldots, O_n)$, we have*

$$\mathfrak{v}_{N_{\mathcal{K},Q_1,\ldots,Q_n}}\Big(\bigcup_{1 \leq i \leq n} \{a^i \mid a \in O_i\}\Big) = \mathfrak{v}(\mathbf{O}).$$

*Proof.* (Sketch) To show the first claim, one has to show that $(\{a \mid a^1 \in G_f\}, \ldots, \{a \mid a^n \in G_f\})$ satisfies the two conditions from Definition 7. That this is indeed the case follows from the fact that by construction, if an allocation layer node of the shape $\langle \mathsf{p}(c_1, \ldots, c_m), \mathsf{exclusive} \rangle^i$ receives nonzero in-flow it must consume all in-flow that the according atom layer node $\mathsf{p}(c_1, \ldots, c_m)$ receives which forces all other edges leaving $\mathsf{p}(c_1, \ldots, c_m)$ to have a flow of zero. To show the second claim, one assumes a knowledge allocation $(O_1, \ldots, O_n)$ and constructs a valid flow for $N_{\mathcal{K},Q_1,\ldots,Q_n}$ along the correspondences shown in the proof of Theorem 1. Finally, given the one-to-one correspondence between flows and allocations established by the two preceding claims, we use Theorem 1 to show that the two functions indeed coincide. □

## 5 Conclusion and Future Work

We have introduced the Multi-Agent Knowledge Allocation framework, drawing from the fields of query answering in information systems and combinatorial auctions. We defined an appropriate exclusivity-aware bidding language and provided a translation of the allocation problem into flow networks.

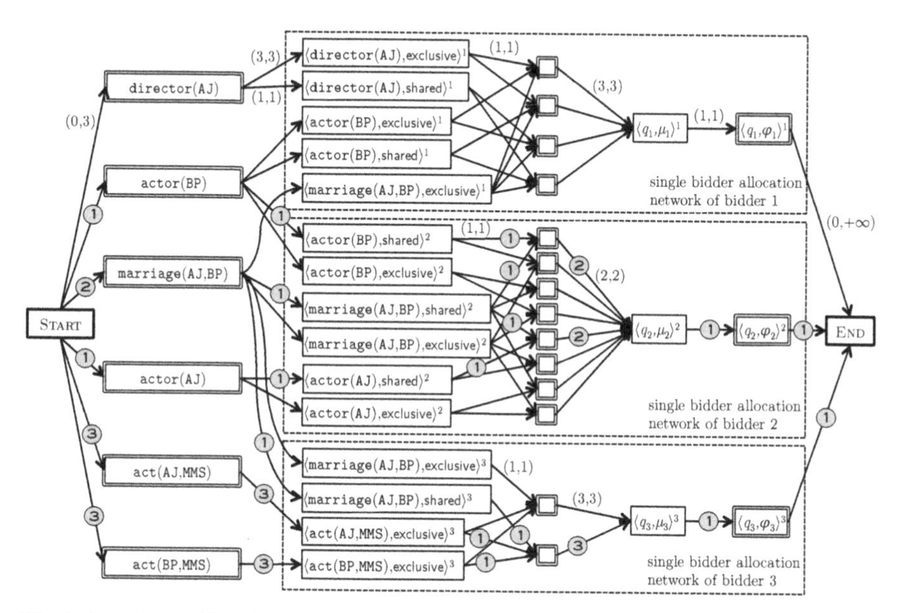


**Fig. 2** Knowledge Allocation Network for allocation problem from Example 8. Upper and lower bounds are left out when redundant. Circled numbers describe the flow associated to the optimal allocation described in Example 8.

An implementation of the existing framework – combining existing query-answering with constraint solving methods – is one immediate next step, as well as an experimental comparison of traditional WDP solving techniques with our preferred approach based on a formulation as max-flow problem. Conceptual future work on the subject will include the following topics:

**Extending knowledge base expressivity.** Instead of just ground facts, more advanced logical statements such as Datalog rules (used in deductive databases) or ontology languages (such as RDFS [5] or OWL [16]) could be supported. In that case, one has to distinguish between propositions which are explicitly present in the knowledge base and those that are just logically entailed by it. If the latter kind of propositions can be delivered to the requester, exclusivity constraints will have to be propagated to them.

**Extending the bidding language.** One might, e.g. allow for *non-distinguished* variables, i.e. variables whose concrete binding is not of interest (like variables in SQL that do not occur in the SELECT part). An appropriate extension of the framework would then allow not just for ground facts (like `marriage(AJ,BP)`) to be delivered to the requester but also for "anonymized" facts (like $\texttt{marriage(AJ},*)$ or, more formally $\exists x.\texttt{marriage(AJ},x)$), requiring an adaption of exclusivity handling. On the other hand, exclusivity annotations might be extended to allow for a more fine-grained specification: a requester might be willing to share some information with certain other requesters but not with all of them.

**Non-financial allocation criteria.** Note that the valuation function subject to maximization needs not to be based on monetary units but can be a utility function in the general sense, factoring in privacy concerns and other aspects.

**Fairness.** As mentioned before, an agent might receive the empty set given the fact that all the answers have been given to competitors who asked for the same information but paid more. This raises fairness [4] problems that should be investigated under this setting.

**Multiple Providers.** Finally, it might be useful to extend the setting to the case where multiple agents offer knowledge; in that case different auctioning and allocation mechanisms would have to be considered. This would also widen the focus towards distributed querying as well as knowledge-providing web-services and the corresponding matchmaking and orchestration problems.

We believe that the introduced MAKA approach might be a first step toward a cross-disciplinary new field combining several subdisciplines of AI (including Knowledge Representation, Multi-Agent Systems and Constraint Satisfaction Problem solving) with adjacent areas as diverse as economics, sociology, databases, optimization, and Web science.

## References

1. S. Abiteboul, R. Hull, and V. Vianu. *Foundations of Databases*. Addison-Wesley, 1995.
2. Ravindra K. Ahuja, Thomas L. Magnanti, and James B. Orlin. *Network Flows: Theory, Algorithms, and Applications*. Prentice Hall, 1993.
3. C. Boutilier and H. Hoos. Bidding languages for combinatorial auctions. In *Proc. IJCAI'01*, pages 1211–1217, 2001.
4. S. J. Brams. On envy-free cake division. *J. Comb. Theory, Ser. A*, 70(1):170–173, 1995.
5. Dan Brickley and Ramanathan V. Guha, editors. *RDF Vocabulary Description Language 1.0: RDF Schema*. W3C Recommendation, 2004.
6. A. Calì, G. Gottlob, and M. Kifer. Taming the infinite chase: Query answering under expressive relational constraints. In *Proc. KR'08*, pages 70–80, 2008.
7. D. Calvanese, G. De Giacomo, D. Lembo, M. Lenzerini, and R. Rosati. Tractable reasoning and efficient query answering in description logics: The *dl-lite* family. *J. Autom. Reasoning*, 39(3):385–429, 2007.
8. A. K. Chandra and P. M. Merlin. Optimal implementation of conjunctive queries in relational data bases. In *Proc. STOC'77*, pages 77–90, 1977.
9. P. Cramton, Y. Shoham, and R. Steinberg. *Combinatorial Auctions*. MIT Press, 2006.
10. M. Croitoru and S. Rudolph. Exclusivity-based allocation of knowledge. In *Proc. AAMAS 2012*, pages 1249–1250, 2012.
11. A. Giovannucci, J. Rodriguez-Aguilar, J. Cerquides, and U. Endriss. Winner determination for mixed multi-unit combinatorial auctions via Petri nets. In *Proc. AAMAS'07*, 2007.
12. N. Nisan. Bidding and allocations in combinatorial auctions. In *Proc. EC-2000*, 2000.
13. D. Porello and U. Endriss. Modelling combinatorial auctions in linear logic. In *Proc. KR'10*, 2010.
14. Eric Prud'hommeaux and Andy Seaborne, editors. *SPARQL Query Language for RDF*. W3C Recommendation, 2008.
15. M. Rothkopf, A. Pekec, and R. Harstad. Computationally manageable combinational auctions. *Management Science*, 44:1131–1147, 1998.
16. W3C OWL Working Group. *OWL 2 Web Ontology Language: Document Overview*. W3C Recommendation, 2009.

# A Hybrid Model for Business Process Event Prediction

Mai Le, Bogdan Gabrys and Detlef Nauck

**Abstract** Process event prediction is the prediction of various properties of the remaining path of a process sequence or workflow. The prediction is based on the data extracted from a combination of historical (closed) and/or live (open) workflows (jobs or process instances). In real-world applications, the problem is compounded by the fact that the number of unique workflows (process prototypes) can be enormous, their occurrences can be limited, and a real process may deviate from the designed process when executed in real environment and under realistic constraints. It is necessary for an efficient predictor to be able to cope with the diverse characteristics of the data. We also have to ensure that useful process data is collected to build the appropriate predictive model. In this paper we propose an extension of Markov models for predicting the next step in a process instance. We have shown, via a set of experiments, that our model offers better results when compared to methods based on random guess, Markov models and Hidden Markov models. The data for our experiments comes from a real live process in a major telecommunication company.

## 1 Introduction

Large service providers like telecommunication companies run complex customer service processes in order to provide communication services to their customers. The flawless execution of these processes is essential since customer service is

Mai Le
Bournemouth University, School of Design, Engineering & Computing, Bournemouth, UK. Currently on placement at BT Research & Technology, Ipswich, UK e-mail: mai.phuong@bt.com

Bogdan Gabrys
Bournemouth University, School of Design, Engineering & Computing, Bournemouth, UK e-mail: bgabrys@bournemouth.ac.uk

Detlef Nauck
BT, Research and Technology, Ipswich, UK e-mail: detlef.nauck@bt.com

an important differentiator for these companies. In order to intervene at the right time, pre-empt problems and maintain customer service, operators need to predict if processes will complete successfully or run into exceptions. Very often processes are poorly documented because they have evolved over time or the legacy IT systems that were used to implement them may have changed. This adds the additional challenge to identify the actual process model. Process mining [12] can reconstruct process models from workflow data to some extent by searching for a model that explains the observed workflow. This process model contains the list of all unique process prototypes (workflow sequences). Data from this mining process can help to address the problem of pro-active process monitoring and process event prediction.

In order to predict process events, we look to approaches known from data mining [2]. Due to the nature of the data, approaches that can deal with temporal sequence data are most relevant. One example from this class of approaches are Markov models (MMs) which have been used to study stochastic processes. In [8], for example, it has been shown, that Markov models are well suited to the study of web-users' browsing behaviour. Event sequences are used to train a Markov model which encodes the transition probabilities between subsequent events. Event sequences can represent customer behaviour, customer transactions, workflow in business processes etc. The prediction of the continuation of a given sequence is based on the transition probabilities encoded in the Markov model.

The order of a Markov model represents the number of past events that are taken into account for predicting the subsequent event. Intuitively, higher order Markov models are more accurate than lower order Markov models because they use more information present in the data. This has been shown to be true in the prediction of web browsing behaviour. Lower order models may not be able to discriminate the difference between two subsequences, especially when the data is very diverse, i.e. if there are a large number of unique sequences. In [9] it has been shown that higher order Markov models can be more accurate than lower order models in the context of customer service data.

However, higher order models can suffer from weak coverage in particular when the data is diverse and not clustered. For sequences which are not covered by the model, a default prediction is required. Default predictions typically reduce the accuracy of the model. It is obvious that with increasing order of the model, the computational complexity also grows. In the case of plain Markov models, there are already approaches to deal with the trade-off between coverage and accuracy. These approaches have been introduced in the studies of [4], [5] and [8]. The general idea is to merge the transition states of different order Markov models, which is followed by pruning 'redundant' states. A particular method is the selective Markov model, an extension of $K^{th}$ order Markov models [4].

In this paper, we present a hybrid approach to address the lack of coverage in higher order Markov models. This approach is a combination of pure Markov models and sequence alignment technique [13]. The sequence alignment technique is applied when the default prediction is required. This is the case when the given sequence cannot be found in the transition matrix. The matching procedure is applied in order to extract the given sequence's most similar sequences (patterns) from the transition matrix. Based on the prediction for the obtained sequences, the prediction for the given sequence is found.

We are testing our approach on workflow data from a telecommunication business process, but it is likely to be applicable in a variety of other sequential prediction problems – especially if Markov models have already been used and shown to be relevant.

The rest of this paper is organised as follows. Section 2 introduces the predictive models which are used in this study. It is followed by Section 3 which presents the experiments' results and evaluation. Finally, in Section 4 our conclusions and future work direction are discussed.

## 2 Predictive Models

This section presents Markov models and their extension that can help enhancing the current approaches to process management in a telecommunication business context by predicting future events based on the current and past data. Here, a business process instance $(S_j)$ is a composition of discrete events (or tasks) in a time-ordered sequence, $S_j = \left\{s_1^j, s_2^j \dots s_{n_j}^j\right\}$, $s_j$ takes values from a finite set of event types $E = \{e_1, \dots, e_L\}$. Apart from its starting time $t_i^j$ and duration of $T_i^j$, each of these events has its own attributes. For simplicity, we assume that a process does not contain any overlapping events, that means there are no parallel structures.

The goal of our predictive models is to predict the next event $s_{i+1}^{N+1}$ following event $s_i^{N+1}$ in a given process instance $S_{N+1} = \{s_1^{N+1}, s_2^{N+1}, \dots, s_{i-1}^{N+1}\}$ based on the data from closed process instances $S = \{S_1, S_2 \dots S_N\}$. A simple predictive model can be illustrated in Figure 1.

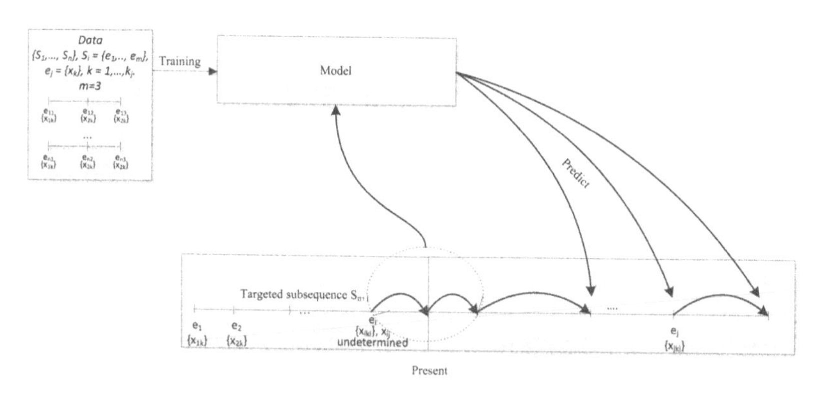


**Fig. 1** A simple predictive model.

In the following subsections, we will discuss some examples of process prediction models, namely Markov models (MMs), hybrid Markov models (MSAs) and Hidden Markov models (HMMs).

## 2.1 Markov Models

Markov models are a kind of stochastic model. They are based on the mathematical theory of Markov chains. Markov models [7] have been used in studying stochastic processes in term of modelling and predicting customer behaviour. The idea of the model is to use $k > 0$ steps to predict the following $(k+1)^{\text{th}}$ step. Given a sequence of random variables $\{X_n\}$, a first order Markov model uses the current step $x_{i-1}$ to predict the next step $x_i$ and to generalise, a $k^{th}$ order Markov model uses the last $k$ steps to predict the next one:

$$p(x_i|x_{i-1},\ldots,x_{i-n}) = p(x_i|x_{i-1},\ldots,x_{i-k}). \tag{1}$$

To construct a Markov model a matrix $M$, $M = (m_{ij})$ is created containing the probabilities for moving from the previous $k$ steps to the next. The columns of the matrix correspond to the unique tasks, $i = (1,\ldots,I)$, and the rows of the matrix correspond to the unique patterns of length $k$ which are present in the training data set that is used to construct the model. These unique patterns are called states $j = (1,\ldots,J)$. Elements $m_{ij}$ of matrix $M$ are the probabilities $p_{ij}$:

$$p_{ij} = \frac{n_{ij}}{n_i}. \tag{2}$$

where $n_i$ is the number of times pattern $i$ occurs in the data set, and $n_{ij}$ is the number of times pattern $i$ is followed by step $j$ in the data set.

Higher order Markov models can be expected to be more accurate. However, we have to take into account their weak coverage property. For example, consider using a Markov model to predict web-page access and consider a sequence of web pages $\{P_1,P_2,P_1,P_3,P_2,P_1\}$ that is used to create the model. In a second order Markov model, there would be no sample $\{P_2,P_3\}$. Consequently, the prediction for this pattern would have to be the default prediction of the model, i.e. the most frequent step ($P_1$ in this example). One approach to overcome the problem is to merge the transition states of different order Markov models, after pruning the 'redundant' states. A particular method is the selective Markov model, an extension of $K^{th}$ order Markov models [4].

Markov models are often not dynamic and once the model is trained, the obtained matrix is fixed and used for predictions. It is important that the training set data is large enough to be representative for the encountered patterns. In order to accommodate changes in data the matrix can be rebuilt from scratch after a certain period of time. It is straightforward to create a dynamic Markov model that adapts to new data by storing the counts $n_{ij}$ and $n_i$ and updating $M$ with additional rows and/or columns if new patterns and/or steps are encountered in new data. We can also apply a discounting factor to the current counts to give more weight to the pattern frequencies in new data.

## 2.2 *Hidden Markov Models*

In the field of sequential data, hidden Markov models (HMMs) are a very powerful tool. Similar to Markov models, HMMs are a type of stochastic model. They are also based on the mathematical theory of Markov processes which further developed into the theory of HMMs by Baum in the 1960s [1]. HMMs received much attention due to their application in speech recognition. Moreover, there is a wide range of applications ranging from telecommunication, recognition (gesture, speech, etc.) to finance, biology, etc.

Many approaches in temporal data series only deal with observed variables. We can consider how does the performance of the model change after adding an unobservable or hidden variable which we assume to have an influence on the observed variables. HMMs assume that there is a latent variable that influences the values of the observed variable. The benefits of this approach are shown, for example, in [3].

As an example for an application domain consider modelling customer behaviour based on customer event sequences (e.g. historic purchases, contacts with customer service etc). A latent variable that influences customer behaviour could be the level of customer satisfaction.

Let $O = \{o_1, \dots, o_N\}$ be the sample of observed variables, and $Q$ a sequence of hidden states $\{q_1, \dots, q_N\}$. The observed variable can be continuous and follow a (usually Gaussian) distribution, or it can be discrete, and take values from a finite set $V = \{v_1, \dots, v_K\}$. The hidden states are discrete, and their value space is $\Omega$, $\Omega = \{s_1, \dots, s_M\}$. The expression of the joint probability of observed sequence is as follows:

$$p(o_1, o_2, \dots, o_n | q_1, \dots, q_n) = p(q_1) \prod_{i=2}^{n} p(q_i | q_{i-1}) \prod_{i=1}^{n} p(o_i | q_i). \tag{3}$$

Due to the complexity of the computation, lower order hidden Markov models are more popular. The most popular one is the first order hidden Markov model which assumes that the current state depends only on the preceding state and is independent from the earlier states.

$$p(q_n | q_{n-1}, o_{n-1}, \dots, q_1, o_1) = p(q_n | q_{n-1}), \tag{4}$$

Intuitively, higher order HMMs can be expected to be more accurate [11]. However, the lack of coverage problem needs to be considered, as mentioned in the secion on Markov models. It is obvious that with increasing order of the model we are facing more complex computation. A HMM is defined by a set of three parameters $\lambda = (A, B, \pi)$:

$$A = (a_{ij}) = (p(q_i|q_j)), \tag{5}$$

where $A$ is the transition matrix and an element $a_{ij}$ is the probability to move from state $j$ to state $i$. It is necessary to point out that only homogeneous HMMs are considered here, implying the time independence of the transitions matrix. The non-homogeneous type is discussed in Netzer [6]. Furthermore,

$$B = (b_{ij}) = (p(o_i|q_j)), \tag{6}$$

where $B$ is the matrix containing probabilities of having $o_t = v_i$, denoted as $o_i$, if the hidden state is $q_j$ at time $t$. Finally,

$$\pi_j = p(q_1 = s_j), \tag{7}$$

is the probability that $s_j$ is the initial state at time $t = 1$.

Creating an HMM from data is more complex than building a Markov model. In order to use HMMs, three problems have to be solved: estimation, evaluation and decoding. The estimation problem is about finding the optimal set of parameters $\lambda = (A, B, \pi)$ which maximize $p(O|\lambda')$ given a sequence $O$ of observed states and some inital model $\lambda'$. The Baum-Welch algorithm is used for this. The evaluation problem means finding $p(O|\lambda)$ given the sequence of observations $O$ and the model $\lambda$ using the forward (backward) algorithm. To solve the decoding problem the Viterbi algorithm is applied for finding the most realisable, suitable sequence of hidden states $q_1, \ldots, q_k$ for the associated sequence $O$ and model $\lambda$.

Researchers have been working on improvements of the performance of HMMs. Similar to Markov models, there are approaches to deal with the trade-off between coverage and accuracy. However, merging the transition states of higher-order hidden Markov models is much more complicated than for Markov models.

### *2.3 Hybrid Markov Models*

The work presented in this paper investigates the idea of combining higher order Markov models with a sequence alignment technique [13] in order to maintain the high accuracy of the higher order Markov models and, at the same time, compensate for their lack of coverage. When a higher order Markov model is employed to predict process events and the given sequence has no match in the transition matrix $M$, we would be forced to produce a default prediction which would be detrimental to the model accuracy. To this end, we present our approach, MSA for Markov Sequence Alignment, which is based on the assumption that similar sequences are likely to produce the same outcome. Basically, the steps that MSA took to identify the next step for a previously unseen sequence $S_{new}$ has no match in the transition matrix as follows:

- Compare $S_{new}$ with all sequences (patterns, states) in the transition matrix, and extract $l$ states which are most similar.
- Generate prediction based on these $l$ selected states. For each state $i$ in the transition matrix $M = (m_{ij})$ there is a corresponding predictive row vector, $(m_{i1}, \ldots, m_{iJ})$. The predicted next step of state $i$ is the task $j$ with $m_{ij} = \max(m_{i1}, \ldots, m_{iJ})$. Thus, the predictions for the $l$ selected states contain $l^*$ unique events, $l^* <= l$, $E^* = \{e_1, \ldots, e_{l^*}\}$. Then, the final prediction for $S_{new}$ is the most frequent event in $E^*$. If, however, there is no single most frequent event in $E^*$ then the one with highest probability is selected. Alternatively, we go back to the $l$ most similar states and select the sequence that is most similar to $S_{new}$ and use its predicted step as the prediction for $S_{new}$. If there is no single most similar sequence, we select the one with the highest frequency from this set.

In order to compare $S_{new}$ with the sequences in the transition matrix, we need to be able to define and measure their similarities or distances. Here, the sequence alignment technique is applied [13]. In particular, we adopt the edit (deletion and insertion) weighted distance function for its flexibility in determining the degree of similarity. Given two sequences, a weight is assigned to each pair of events at the same position in the two sequences. The weight is 0 if the two events are identical, 1 or $\delta$ (a specified value) if they are different. In case both events of the pair must be deleted in order to keep the longest possible similar subsequences, the corresponding weight is 1. If only one of the two events must be deleted the weight is $\delta$. The sum of weights of the comparison is then used to determine the similarity of sequences. The lower the weight is, the more similar two sequences are.

This algorithm starts by constructing a matrix where the elements represent similarity between any pair of events from the two sequences to be matched. This matrix is then used to find the most suitable editing of the given sequences where deletions and insertions are penalised. The optimal editing is chosen based on the total score computed. For example, given two sequences $ABCDE$ and $EBCAD$, the matrix looks as follows:

$$\begin{pmatrix} & A & B & C & D & E \\ E & 0 & 0 & 0 & 0 & 1 \\ B & 0 & 1 & 0 & 0 & 0 \\ C & 0 & 0 & 1 & 0 & 0 \\ A & 1 & 0 & 0 & 0 & 0 \\ D & 0 & 0 & 0 & 1 & 0 \end{pmatrix}$$

Next, the first events $A$ and $E$ from both sequences are deleted which results in a weight of 1. The second event in both sequences is $B$, and has weight 0. The fourth event $A$ is deleted from the sequence $EBCAD$ in order to match two $D$s in fourth and fifth positions in the two sequences. We add $\delta$ to the sum of weights. Finally, we delete the last event $E$ from the sequence $ABCDE$, adding another $\delta$. The total weight for matching the two sequences is $w = 1 + 0 + 0 + \delta + \delta = 1 + 2\delta$.

The following example will illustrate our method for the case of the transition matrix of a third order Markov model:

$$\begin{pmatrix} & A & B & C \\ ABC & 0.1 & 0.2 & 0.7 \\ CBC & 0.1 & 0.5 & 0.4 \\ BAA & 0.2 & 0.5 & 0.3 \end{pmatrix}$$

and a given sequence $BCC$. This sequence has not occurred before and is not stored in the matrix. The aim is to find the most similar sequences from the matrix and use their predictions to generate the prediction for the given sequence $BCC$. We first match $BCC$ to $ABC$:

$$\begin{pmatrix} & A & B & C \\ B & 0 & 1 & 0 \\ C & 0 & 0 & 1 \\ C & 0 & 0 & 1 \end{pmatrix}$$

the weight of this comparison is $w = \delta + 0 + \delta = 2\delta$. Similarly, the resulting weights of matching $CBC$ and $BAA$ against $BCC$ are $w = 2\delta$ and $w = 2$, respectively. Let $\delta = 0.4$ and $l = 2$, the two chosen sequences in this case are then $ABC$ and $CBC$ and their predictive vectors are $(0.1, 0.2, 0.7)$ and $(0.1, 0.5, 0.4)$, respectively. Based on these vectors, the predictive next steps are $\{B, C\}$. The weights and frequency of occurrence of these two events are equal hence, the next step of the given sequence $BAC$ is predicted to be $C$ due to higher transition probability, $m_{13} = 0.7$ which is greater than $m_{22} = 0.5$.

Algorithms 1 and 2 illustrate the procedure of matching two sequences by deletion and insertion of symbols with penalties.

## 3 Evaluation

Having defined MSA, we now present a discussion of our empirical evaluation aimed at assessing the efficiency as well as exploring some potential future research direction for employing our Markov extension approach in business process event forecasting. Our model has been implemented in Matlab and run against two data sets of process instances from the records of a major telecom company. The first dataset (DS1) consists of telecommunication line fault repair records for a nine month period. The second (DS2) covers a one month period and represents a process for fixing broadband faults.

### 3.1 Process analysis and data preprocessing

In these datasets, the population of process instances turned out to be very diverse and not straightforward to work with. In particular, $DS1$ is very diverse and contains 28963 entries, 2763 unique process instances, and 285 unique tasks. The process instance length varies from 1 to 78. On the other hand, $DS2$ represents a significantly

**Algorithm 1** Function scan(*row, col, length*): identifies the optimal editing process to match two sequences via deletion/insertion of symbols

```
1: commonIndexes1(length) = row
2: commonIndexes2(length) = col
3: if (length > maxCommonLength) then                ▷ store the longest path
4:     maxCommonLength = length
5:     for i = 1 : maxCommonLength do
6:         maxCommonIndexes1(i) = commonIndexes1(i)
7:         maxCommonIndexes2(i) = commonIndexes2(i)
8:     end for
9:     maxCommonScore =
10:    calculateScore(n, maxCommonIndexes1, maxCommonIndexes2, maxCommonLength)
11: else
12:    if (length == maxCommonLength) then
13:        score = calculateScore(n, commonIndexes1, commonIndexes2, length)
14:        if (score < maxCommonScore) then   ▷ if the paths are equal, store the smallest score
15:            for i = 1 : maxCommonLength do
16:                maxCommonIndexes1(i) = commonIndexes1(i)
17:                maxCommonIndexes2(i) = commonIndexes2(i)
18:            end for
19:            maxCommonScore = score
20:        end if
21:    end if
22: end if
```

**Algorithm 2** Funciton calculateScore(n, maxCommonIndexes1, maxCommonIndexes2, maxCommonLength): computes the score of matching two sequences

```
1: totalPoints = 0
2: totalDeltas = 0
3: maxCommonIndexes1(maxCommonLength + 1) = n + 1
4: maxCommonIndexes2(maxCommonLength + 1) = n + 1
5: last1 = 0
6: last2 = 0
7: for i = 1 : maxCommonLength + 1 do
8:     dist1 = maxCommonIndexes1(i) − last1 − 1
9:     dist2 = maxCommonIndexes2(i) − last2 − 1
10:    if (dist1 > dist2) then
11:        noPoints = dist2
12:        noDeltas = dist1 − dist2
13:    else
14:        noPoints = dist1
15:        noDeltas = dist2 − dist1
16:    end if
17:    totalPoints = totalPoints + noPoints
18:    totalDeltas = totalDeltas + noDeltas
19:    last1 = maxCommonIndexes1(i)
20:    last2 = maxCommonIndexes2(i)
21: end for
22: return totalPoints + totalDeltas * deltaValue
```

smaller scale set with only 5194 entries, 794 process instances, and 10 unique tasks. The process instance length varies from 2 to 32.

Figure 2 shows a visualisation of 10 percent of set $DS1$. The diagram has been created by BT's process mining tool Aperture [10] which uses workflow data to create a process diagram. For this process, the diagram illustrates the complexity of the workflow data. It is basically impossible to visually analyse or understand the process from this figure. This is a typical scenario for processes that have evolved over time and are poorly documented.

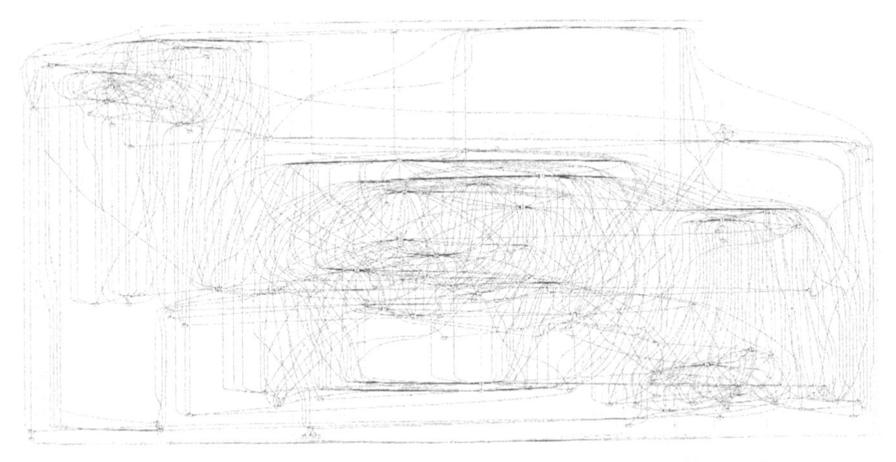

**Fig. 2** Process model obtained by using Aperture for DS1 visualising a highly complex process.

The data we used for our experiment was stored in flat files and each row (record) contains an unique event instance with its corresponding process name, date and other required attributes. When reading data into memory, each record also stored the three immediately preceding events as well as the next three future events for ease of access in the implementation of our algorithm.

## 3.2 Results

To evaluate MSA, we benchmarked our model with three other approaches:

- *RM - Random Model*: in order to find the next task following the current one, we randomly select from the set of potential next tasks. For example, if from the historical data, we know that tasks following A belong to the set $\{C,D,E\}$, we random select a value from that set as the predicted next step.

- *All $K^{th}$ Markov Models*: a number of different order Markov models are generated from first order to $K^{th}$ order. Given a sequence, we start with the highest order Markov model. If the given sequence cannot be found in the transition matrix, we create a new shorter sequence by removing the first event in the sequence. We then continue the procedure with the next lower order Markov model until we either find a match in a transition matrix or after trying the first order MM a default prediction is required.
- *HMM - Hidden Markov Models*: we tested several first order HMMs with different lengths for the input sequences and different number of hidden states and picked the best one.

For each comparison, we used 90% entries of the data sets as training data and the remaining 10% were used as test data. Basically, for each sequence in the test data, we checked if our model can correctly predict the next step of a particular task (i.e. we measured the number of successful predictions). The results were evaluated using 10-fold cross-validation. The results are displayed as a percentage. After the $1^{st}$ order to $7^{th}$ order MSAs were built, we investigated different values for $l$, the number of sequences (patterns) in the transition matrix which are most similar to the given sequence. $l$ was varied from 1 to 7, and the results show that $l = 5$ is optimal for DS1 and $l = 3$ is optimal for DS2.

We now look at the specific results.

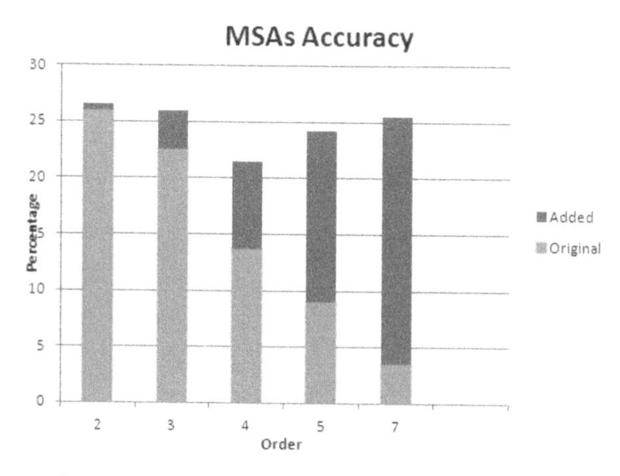


**Fig. 3** Percentage of correct predictions before and after applying the default prediction module in Markov models using data set DS1.

Figures 3 and Figure 4 illustrate MSAs accuracy against the two data sets, DS1 and DS2 respectively. The results show that by incorporating the default prediction improvement module, MSA outperforms other comparable models, especially when the order of the Markov model increases. This is because the relevance of the comparison between sequences is directly proportionate to their lengths. As can

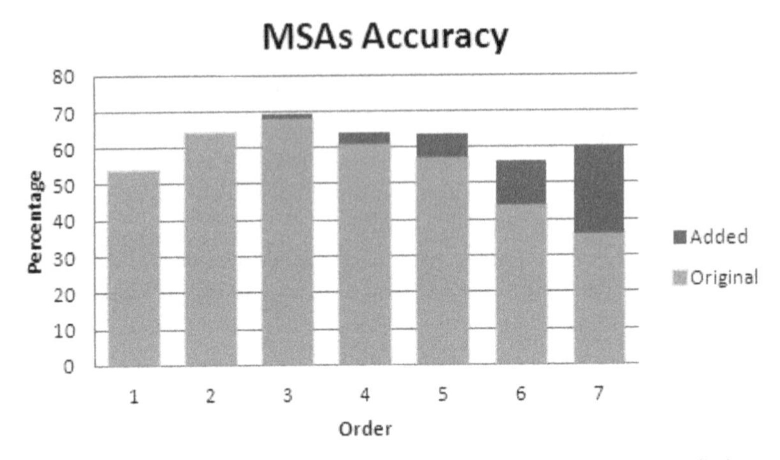


**Fig. 4** Percentage of correct predictions before and after applying the default prediction module in Markov models using data set DS2.

be seen, the second order MSA performs best among a range of different orders of MSA. For the case of DS1 it is about 27% correct. The third order MSA works best in the case of DS2, it provides correct predictions in about 70%. Nonetheless, the highest performance order depends on the data set. For the data set which provides a good performance with a relatively high Markov order (i.e. $5^{th}$ and above), the role of our default prediction improvement module becomes highly significant.

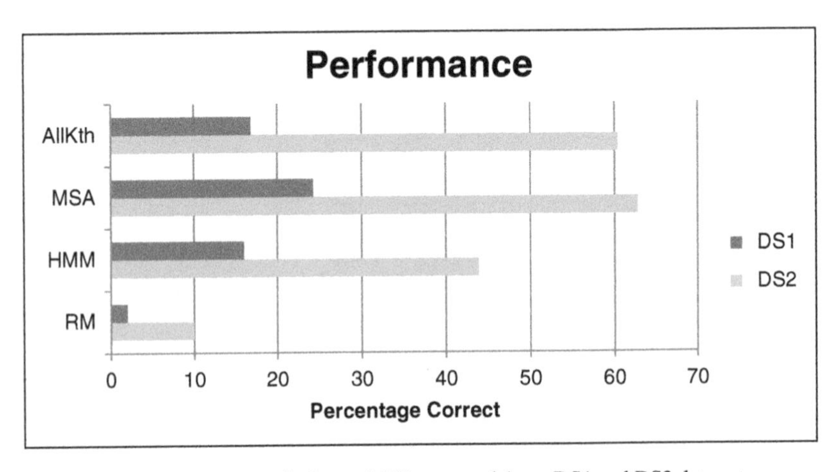


**Fig. 5** Percentage of correct predictions of different models on DS1 and DS2 data sets.

Figure 5 shows the performances of all models against the two datasets. As can be seen, RM performs worst with only around 10% success for DS2. When applied to the bigger dataset DS1, the result goes down to nearly 2% (14 times worse than that of MSAs' average performance). This can be explained by the fact that with DS2, each task can have an average of 7 potential next tasks whereas this figure is nearly 30 for DS1. Thus, the probability of picking the right next task reduces as the set size increases. The results highlight the difficulty of handling complex data sets. When the data is not too diverse (i.e. DS2), MSA (fifth order) obtains the highest number of correct predictions with a result of 63%. Compared to other benchmarks, MSA results are better than a fifth-order Markov model which achieves 57%, an All $K^{th}$ ($K = 5$) Markov model, which achieves 60%, and an HMM which achieves 43%. Please note that we did not pick the most accurate MSA and All $K^{th}$ Markov model to compare to the HMM and the RM. The reason is that All $K^{th}$ Markov models and MSAs only have an advantage with higher orders. In order to see how well they cope with default predictions, we chose a $5^{th}$ order MSA and an All $5^{th}$ Markov model.

When the data is more fragmented as in the case of DS1, the effectiveness of our model is reduced. However, even though the performance on DS1 is not as good as the performance on DS2, our model still performs an important role considering the requirement of selecting a single correct task from 30 potential candidates on average. Even in this case we still manage to outperform the HMM result by 20% and also outperform the All $5^{th}$ Markov model result by 3%.

## 4 Conclusion

In this paper, we have introduced several models for predicting the next step in business processes in a telecommunication services context where no formal description of the processes exists or the real process significantly deviates from the designed path prototype when being applied in real environments and under realistic constraints. Specifically, we first applied Markov and Hidden Markov models to generate probability matrices of moving from one event to others; and then using them to provide predictions. Next, we have proposed a novel predictive model (MSA) which is an extension of Markov models employing sequence alignment. In order to analyse its effectiveness, we have empirically evaluated MSA against a two datasets from a major telecommunication company with varying degrees of diversification. The results have clearly demonstrated that MSAs outperforms both Markov models and HMMs by at least 10%. In addition, we have also shown that higher order Markov models outperform a first order Markov model. The default prediction module we introduced significantly improves the corresponding original Markov model result when the order of model is high (starting from $5^{th}$ order MM). Nonetheless, although the improvement provided by the new default prediction module is quite significant, higher order MSAs do not outperform second order MSA in the case of the data set DS1 and third order MSA in the case of the data set DS2. These experiments motivate us to use sequence alignment technique as a similarity measure to group sequences which have similar characteristic and then tackle them separately. Our future research will look into: (1) using sequence alignment, sliding window and K-nearest neighbour to predict the outcome of the process and (2) $K$-means clustering to cluster data into $K$ groups and then treat each group with suitable approaches.

## References


1. LE Baum, T Petrie, G Soules, and N Weiss, 'A maximization technique occurring in the statistical analysis of probabilistic functions of markov chains', *The Annals of Mathematical Statistics*, **41**(1), 164–171, (1970).
2. MJA Berry and GS Linoff, *Data mining techniques: for marketing, sales, and customer relationship management*, Wiley New York, 2004.
3. LA Cox Jr and DA Popken, 'A hybrid system-identification method for forecasting telecommunications product demands', *International Journal of Forecasting*, **18**(4), 647–671, (2002).
4. M Deshpande and G Karypis, 'Selective markov models for predicting web page accesses', *ACM Transactions on Internet Technology (TOIT)*, **4**(2), 163–184, (2004).
5. M Eirinaki, M Vazirgiannis, and D Kapogiannis, 'Web path recommendations based on page ranking and markov models', in *Proceedings of the 7th annual ACM international workshop on Web information and data management*, WIDM '05, pp. 2–9, (2005).
6. O Netzer, J Lattin, and VS Srinivasan, 'A hidden markov model of customer relationship dynamics', (2007).
7. A Papoulis, *Probability, Random Variables and Stochastic Processes*, 1991.
8. J Pitkow and P Pirolli, 'Mining longest repeating subsequences to predict world wide web surfing', in *The 2nd USENIX Symposium on Internet Technologies & Systems*, pp. 139–150, (1999).
9. D Ruta and B Majeed, 'Business process forecasting in telecommunication industry', 1–5, (2010).
10. P Taylor, M Leida, and BA Majeed, 'Case study in process mining in a multinational enterprise', in *Lecture Notes in Business Information Processing*. Springer, (2012).
11. S Thede and M Happer, 'A second-order hidden markov model for part-of-speech tagging', 175–182, (1999).
12. W van der Aalst, *Process mining: Discovery, conformance and enhancement of business processes*, Springer-Verlag, Berlin, 2011.
13. M Waterman, 'Estimating statistical significance of sequence alignments', *Philosophical transactions-Royal society of London series B biological sciences*, **344**, 383–390, (1994).

# SHORT PAPERS

# A Comparison of Machine Learning Techniques for Recommending Search Experiences in Social Search

Zurina Saaya, Markus Schaal, Maurice Coyle, Peter Briggs and Barry Smyth

**Abstract** In this paper we focus on one particular implementation of social search, namely *HeyStaks*, which combines ideas from web search, content curation, and social networking to make recommendations to users, at search time, based on topics that matter to them. The central concept in HeyStaks is the *search stak*. Users can create and share staks as a way to curate their search experiences. A key problem for HeyStaks is the need for users to pre-select their active stak at search time, to provide a context for their current search experience so that HeyStaks can index and store what they find. The focus of this paper is to look at how machine learning techniques can be used to recommend a suitable active stak to the user at search time automatically.

## 1 Introduction

Recently researchers have recognised the potential of web search as a fertile ground for a more social and collaborative form of information retrieval. For example, researchers in the area of *collaborative information retrieval* have sought to make web search more collaborative and more social in a variety of ways; see for example [2, 6]. The basic insight driving this research is that search is a fundamentally collaborative affair – people frequently search for similar things in similar ways [9] – and by embracing collaboration mainstream search can be improved. This is one important objective with so-called *social search* approaches [2] in an effort to bridge the gap between query-based search and sharing within social networks.

In our research we are interested in the intersection between web search, social sharing/collaboration, and content curation to explore ways to support com-

Zurina Saaya, Markus Schaal, Maurice Coyle, Peter Briggs, Barry Smyth
CLARITY: Centre for Sensor Web Technologies, University College Dublin, Ireland e-mail: {Zurina.Saaya,Markus.Schaal,Maurice.Coyle,Peter.Briggs,Barry.Smyth}@ucd.ie

munity collaboration on individual topics of interest through existing mainstream search engines. To this end the HeyStaks social search service has been developed. Briefly, HeyStaks allows users to create and share topical collections of content called *search staks* or just *staks*. These staks are filled by the searching activities of members; queries, tags, and pages are added to staks as users search on topics of interest. And these staks serve as a source of new result recommendations at search time as *relevant* pages that have been found by other *reputable* stak members for similar queries are recommended to new searchers. A unique feature of HeyStaks is that it is fully integrated into mainstream search engines (e.g. Google, Bing, Yahoo), via a browser plugin, so that searchers can continue to search as normal whilst enjoying the benefits of a more collaborative search experience.

One of the main points of friction in HeyStaks is the need for users to pre-select an active stak at search time. However, although every effort has been made to make stak selection as easy and natural task as possible, inevitably many users forget to select a stak as they search. In our work we address this problem by comparing various approaches to automate stak selection. In particular, the main contribution of this paper is to extend the work of [8] which described a term-based approach to stak recommendation, by proposing an alternative machine learning approach.

## 2 Framing the Stak Recommendation Task

To frame the stak recommendation task we can consider a typical user $U$ to be a member of a number of different staks, $staks(U) = S_1, ..., S_n$, and the task is to recommend one or more of these staks as potential active staks; in what follows we will often focus on the recommendation of a single best stak without loss of generality. What context information is available to guide this recommendation task? Obviously the user has submitted a search query, $q_T$, which provides a strong recommendation signal, but this is just one element of the current search context. The search query is submitted directly to an underlying search engine – it is intercepted by HeyStaks in parallel – which responds with a set of results (including titles, snippets, URLs). In addition to $q_T$ we also have access to a range of other information, such as the result URLs and the snippet terms, as additional search context. These different types of information (the target query terms, the result URLs, and the result snippet terms) can be used individually or in combination to produce a *stak query* ($S_Q$) as the basis of recommendation.

In what follows we will describe two alternative approaches to using this context information for recommendation. The first is the technique described in [8], which adopts a term-based approach but delivers only limited stak recommendation success; we will use this as a benchmark against which to judge our new technique. The second is a new classification-based approach, which uses machine learning techniques to build a classification mo

## 2.1 Term-Based Recommendation Strategies

The work of [8] described a term-based approach to stak recommendation. Briefly, each stak corresponds to a set of web pages that have been selected, tagged, shared etc. during the searches of stak members. These pages are indexed based on the queries that led to their selections, their associated snippet terms, and any tags that users may have provided. For a given stak, its index can be viewed as single document (*stak summary index*). [8] then proposed a straightforward approach to stak recommendation, by using the stak query $S_Q$ as a retrieval probe against the user's *stak summary index*. Specifically, [8] proposed a TF-IDF approach to retrieval (see [4]), calculating a relevance score for each of the user's staks. These recommended staks are then suggested directly to the user at search time as a reminder to set their appropriate stak context.

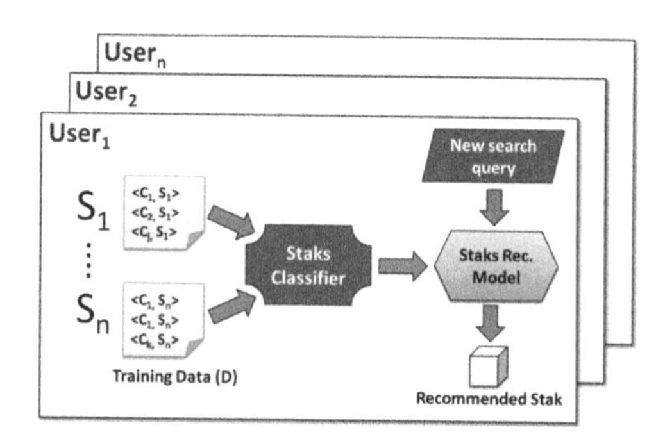


**Fig. 1** Classifier-based approach for stak recommendation

## 2.2 Classification-Based Recommendation Strategies

In this paper we propose the use of a classification-based recommendation approach. In essence, the objective here is to use stak-related search data to train a stak classifier that is capable of classifying future search sessions. For a given user who is a member of $n$ staks $S_1, ..., S_n$ we train a n-class classifier based on the content of these staks. Each stak $S_i$ contributes a set of training data in the form of $\langle C_j, S_i \rangle$ where $C_j$ is the context data of a single search experience and $S_i$ is the correct class for this context, namely the stak identifier. In other words, each training instance corresponds to a single page in $S_i$ and the context data associated with this page are the query terms, snippet terms and tags that the page is indexed under in $S_i$ and the URL of the page. Note, whereas stak information is compiled into a single document for the term-based technique, for this classification-based approach, the original search sessions are used as separate training instances.

In this way we can compare different classifiers trained on different types of context data (query terms vs. snippet terms vs URLs, or combinations thereof) for a direct comparison to the work of [8]. For example, this means that we can train a classifier for a user based on the combination of query, snippets and URLs. In this case, each instance of training data will be drawn from the user's staks and will correspond to a single search experience. It will include the query terms submitted by the user, the snippet terms for results returned by the underlying search engine, and the URLs for said results. These will form the set of training features for this instance. This approach is presented in summary form in Figure 1.

To build our classifiers we also consider the follow three distinct machine learning techniques for the purpose of comparison: (i) Multinominal Naive Bayes (MNB) [5], (ii) Support Vector Machines (SVM) [1] and (iii) Decision Trees (DT) [7]. We use standard reference implementations for these classifiers from the Weka-library [3]. Once a stak classifier has been learned for a given user, it can be used as the basis for stak recommendation at search time: for a new search session the relevant user context data is presented, in the form of a stak query $S_Q$, as new test data to the user's classifier and a ranked list of staks (or a single best stak) is produced based on the classifier's class predictions and confidence scores.

## 3 Evaluation

In this section we adopt a similar methodology to the work of [8] to provide a comparative analysis of the classification-based approaches proposed in this paper. Our dataset is based on HeyStaks query logs for a group of 28 active users, who have each submitted at least 100 queries (including those for the default stak) through Heystaks. In total the dataset consist of 4,343 search queries and 229 unique staks. We focus on queries/sessions where users have actively selected a specific stak during their search session because for supervised learning we need to have training instances that query with other context information and an actual stak selection as a target class. We test the 7 different sources of stak information. In addition to the benchmark term-based technique we test the three classification-based techniques described above to give a total of 28 (7x4) different recommendation strategies. For each of these strategies we compute an average *success rate* by counting the percentage of times that a given recommendation technique correctly predicts the target stak for a user. We perform a 10-fold cross-validation for each user and compute the average overall success rate for each recommendation technique.

### *3.1 Success Rates by Recommendation Strategy*

Figure 2 presents the overall success rates across the 7 different sources of context information ($S_Q$) and for each of the 4 recommendation techniques. The first point to make is that the benchmark term-based strategy (TF-IDF) does not perform as well as the classification-based approaches (MNB, SVM, DT) across any of the context types. For example, we can see that the term-based approaches deliver a success rate of just over 50% when snippet terms are used for the stak query (the *snippet* condition) compared to a corresponding success rate of 70-76% for the classification-based approaches. This pattern is seen across all stak query settings.

Moreover, we can see that among the classification-based approaches MNB consistently outperforms SVM and DT across all conditions.

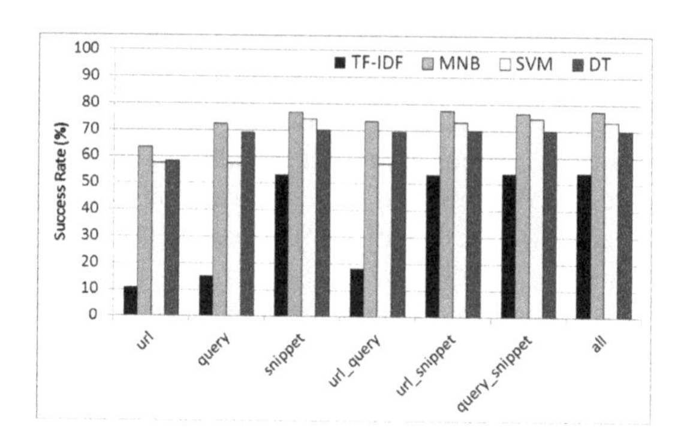


**Fig. 2** Comparison of success rates.

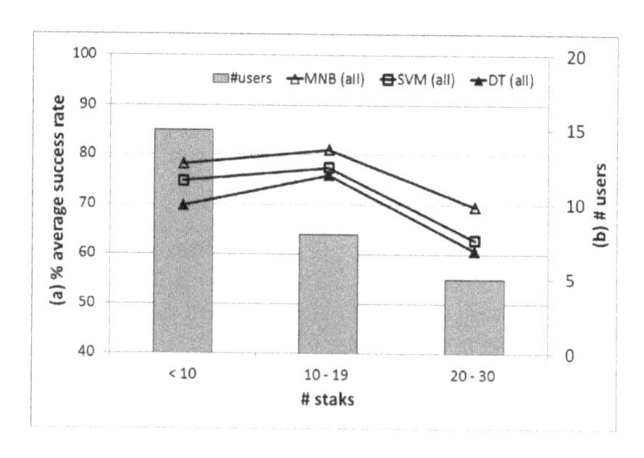


**Fig. 3** An analysis of recommendation success rate based on stak membership.

It is also worth considering the relationship between recommendation accuracy and the number of staks a user is a member of. This is important because in the case of users who are members of fewer staks the task of selecting the right stak should be easier, all other things being equal, because there are fewer staks to choose from. To investigate this we computed the average success rate across three different user groups: 1) users who are members of fewer than 10 staks (the $< 10$ group); 2) users who are members of between 10 and 19 staks inclusive (the *10-19* group); and 3) users who are members of between 20 and 30 staks (the *20-30* group). The results are presented in Figure 3. We can see that for the $< 10$ group (there were 15 users who belonged to less than 10 staks) we have an average success rate of between 70% (DT) and approximately 78% (MNB). These success rates fall off to 60%-70% for the *20-30* group, as expected since all other things being equal there are many more staks to choose from for these users during stak recommendation. Somewhat unexpected is the slight rise in success rates for the *10-19* group compared to the $<$ 10 group. Obviously, all other things are not always equal in terms of the relationship between success rate and the number of staks that a user has joined. For instance, some of the joined staks may be very mature (containing lots of content submitted by many different users) and one might expect these mature staks to benefit from a much stronger recommendation signal than smaller, less mature staks.

## 4 Conclusions

This study provides a preliminary validation of our classification-based approach to stak recommendation. The results demonstrate a significant benefit when compared to the term-based benchmark but the limitations of the study must also be acknowledged. We feel it is reasonable to draw some tentative conclusions at this stage. Principally, there is strong evidence to support the hypothesis that our classification-based approaches represent a significant improvement on the term-based benchmark. Moreover, the evidence suggests that MNB outperforms the other classification-based strategies across all conditions.

**Acknowledgements** This work is supported by Science Foundation Ireland under grant 07/CE/I1147, HeyStaks Technologies Ltd, Ministry of Higher Education Malaysia and Universiti Teknikal Malaysia Melaka.

## References

1. Chang, C.C., Lin, C.J.: LIBSVM: A library for support vector machines. ACM Transactions on Intelligent Systems and Technology **2**, 27:1–27:27 (2011)
2. Evans, B.M., Chi, E.H.: An elaborated model of social search. Information Processing &; Management **46**(6), 656 – 678 (2010)
3. Hall, M., Frank, E., Holmes, G., Pfahringer, B., Reutemann, P., Witten, I.H.: The weka data mining software: an update. SIGKDD Explor. Newsl. **11**(1), 10–18 (2009)
4. Hatcher, E., Gospodnetic, O.: Lucene in action. Manning Publications (2004)
5. Kibriya, A., Frank, E., Pfahringer, B., Holmes, G.: Multinomial naive bayes for text categorization revisited. In: AI 2004 Advances in Artificial Intelligence, *Lecture Notes in Computer Science*, vol. 3339, pp. 235–252. Springer Berlin / Heidelberg (2005)
6. Morris, M., Teevan, J., Panovich, K.: What do people ask their social networks, and why? a survey study of status message q and a behavior. In: Computer Human Interaction (2010)
7. Quinlan, J.R.: Programs for Machine Learning. Morgan Kaufmann Publishers (1993)
8. Saaya, Z., Smyth, B., Coyle, M., Briggs, P.: Recommending case bases: Applications in social web search. In: Proceedings of 19th International Conference on Case-Based Reasoning, ICCBR 2011, pp. 274–288 (2011)
9. Smyth, B., Balfe, E., Freyne, J., Briggs, P., Coyle, M., Boydell, O.: Exploiting query repetition and regularity in an adaptive community-based web search engine. User Model. User-Adapt. Interact. **14**(5), 383–423 (2004)

# A Cooperative Multi-objective Optimization Framework based on Dendritic Cells Migration Dynamics

**N.M.Y. Lee and H.Y.K. Lau**[1]

**Abstract** *Clonal Selection* and *Immune Network Theory* are commonly adopted for resolving optimization problems. Here, the mechanisms of migration and maturation of Dendritic Cells (DCs) is adopted for pursuing pareto optimal solution(s) in complex problems, specifically, the adoption of multiple characters of distinct clones of DCs and the immunological control parameters in the process of signal cascading. Such an unconventional approach, namely, *DC-mediated Signal Cascading Framework* further exploits the intrinsic abilities of DCs, with the added benefit of overcoming some of the limitations of conventional optimization algorithms, such as convergence of the Pareto Front.

## 1 Introduction

In the research of artificial immune systems (AIS), the immuno-mechanism of distinguishing safe and danger signals, namely, *Dendritic Cell Algorithm* (DCA) has been widely demonstrated to be an effective classifiers [1][2] and intrusion detectors [3][4] by Greensmith [5][6]. These studies demonstrated promising results with the adoption of the immunological concepts, namely, antigen sampling and antigen processing. Apart from the distinctive performance demonstrated in these studies, the underlying mechanisms of Dendritic Cells (DCs) mediated immunity also provides an alternative approach in demonstrating the capabilities in cooperation, decision making and optimization, which are established by extensive research conducted by immunologists and biologists over the last few decades.

DCs exhibit a highly dynamic, cooperative and decision control characteristics for maximizing the efficacy of T-cell immunity in the adaptive immune system, which is conceivably (i) diversify and (ii) optimize the presentation (or permutation) of the antigen prior to priming a specific DC mediated T-cell immunity [7][8]. By mimicking these behaviors by the proposed framework, known as, *DC-mediated Signal Cascading Framework*, following novelties are exhibited:

– Solutions (or the Pareto front) are evolved iteratively through the processes of gleaning and cascading multiple signals, in a real time manner.

---

1 Department of Industrial and Manufacturing Systems Engineering, The University of Hong Kong, Pokfulam Road, HKSAR

- Unlike the immune-inspired optimization algorithms [9][10], diversity of solutions are controlled by the control parameters (of the framework) adaptively and iteratively, instead of governed by defined probability functions as conventional approaches such as Genetic Algorithm [11].

Revealing the "unexplored" immune-phenomenon and mechanisms for resolving complex problems like Multi-objective Optimization Problems (MOP) is therefore the focus of the paper. In this respect, the remarkable immunological features and control mechanisms of DCs, and the analogy between DC-mediated immunity and the proposed framework will also be stipulated in the following sections at the level of (i) an individual DC and (ii) a collection of distinctive clone of DCs for problem solving.

## 2 Proposed Dendritic Cell Signal Cascading Framework for Optimization

The proposed framework consists of a number of immunological characteristics that are integrated at the levels of an individual and a collection of DCs,

(i) For each distinctive DC, which inherits the capabilities of selective expression and multi-characteristics of receptor repertoires [12][13], and the capability of multiple-tasking [14] demonstrate imposingly in the progression of maturation, are mimicked in the proposed framework. As for the presented receptor (which is associated to a DC), signal cascading is introduced for a downstream signaling pathway for regulating the migration and maturation properties of DCs (e.g. the survival rate and phenotype of DCs) through the interactions between chemokines and their receptors [15][16][17].

(ii) In a collaborative immune system, a collection of DCs perform various decision-making and cooperation in an efficient manner with a set of quantitative control protocol in signal cascading process, for instances, defining the level of threat [18] and the fate of the cells [19][20][21] as observed by a group of scientists and immunologists. In an engineering prospective, this approach further advances the classical immuno-mechanisms, such as stimulation and suppression [22] and threshold control [23] for Pareto Front generation. In this regards, diverse results will be evolved with respects to the quantified interactions between the candidate solutions and the objective functions.

### *2.1 A Dendritic Cell*

Conceptually, each individual DC acts as a decomposer which "captures" the essential information from the problem domain (analogous to antigen/dangle signals), and the signals are "converted" in terms of the control factors (analogous to chemokines) in the proposed framework, which further control and provide guidance for the Pareto Front generation.

Continuous "capturing" or "recruiting" of DC subsets in the migration and maturation mimics the interactions between the generated candidate solutions and the objective functions. Similar to the immune-features inherent in DCs, the candidate solutions interact with more than one DC-embedded objective function processors. The affinity measure indicates the "Fitness" and "Threat" of the candidate solution to the objectives. The former refers to the antigenic affinity (similarity) between candidate solutions and the member of the Pareto Front, such as Euclidean Distance [24][25]. To further enrich the meaning of "fitness", the magnitude of "threat" is introduced, which is analogous to DCs' capability of differentiating safe and danger signals. In the domains of solving a MOP, the "threat" is measured based on:

- Number of iterations (life cycle) of the candidate solutions
- Number of replications of the candidate solutions
- Number of virulence factors, such as constraints violation

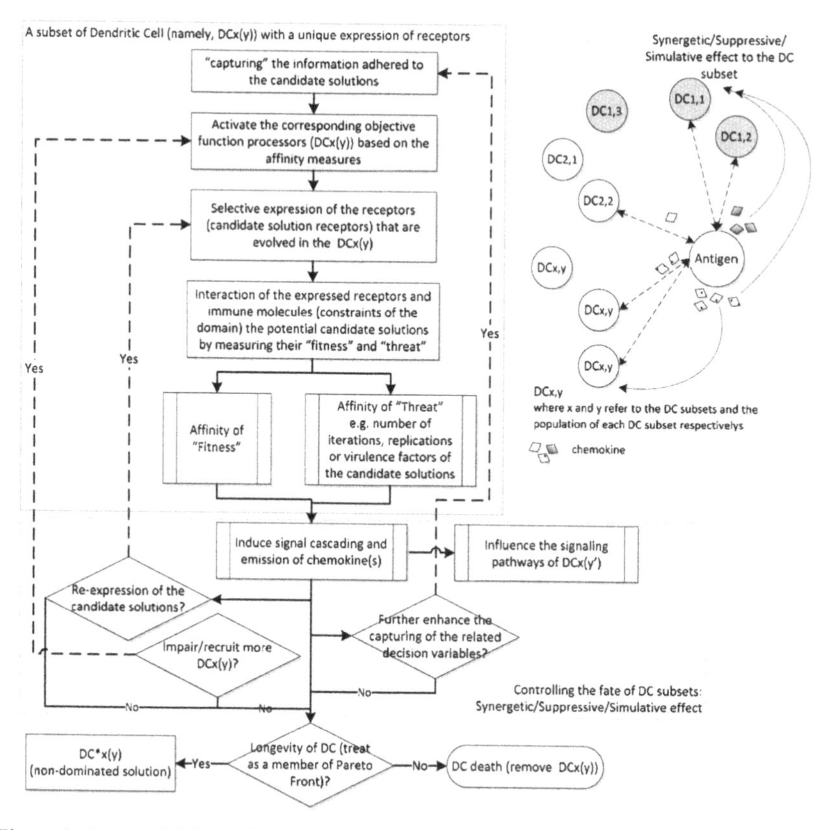


Figure 1. Proposed DC-mediated Signal Cascading framework performed by each DC, and the cooperative control of the collection of DCs pursuing a novel immune-inspired paradigm for solving MOP.

Both "Threat" and "Affinity" take an initiative for diversifying the pool of candidate solutions, and more importantly, in governing the fate of the solutions, and this will result in generating true positive solutions. Each of the decomposed signal will subsequently undergo a series of distinctive signal cascading processes with respects to the given problem constraints, as depicted in Figure 1. The optimal solution(s) are the survivor(s) through collaboration and cooperative interactions between DCs, which will be described in Section 2.2.

## 2.2 Signaling Cascade

As for signal cascading, which mimics the pathway generated in the maturation and migration of DC, which demonstrates the interaction with other immune molecules (which mimics the constraints of the domains), resulted in impairing or amplifying particular functions to control the fate of the candidate solution(s). Notably, this mechanism is not executed by a single binding of a single DC, but multiple bindings of a collection of DCs. The complexity of a network is formed continuously and iteratively. For the sake of simplicity, the effected function (to the optimization algorithm) will be qualified by the summation of the concentration of the emitted signals, namely, the survival, migration speed, the change of the associated receptors of the candidate solution, etc. in the network. To recapitulate briefly, these pathways will be resulted in controlling the fate of the candidate solutions. As for the candidate solution(s), they may either die due to its maximum lifetime or continue in the migration and maturation upon the termination criteria is met. The persisted candidate solution(s) are then treated as the Pareto Front. This paradigm also provides an innovative problem solving (control) approach with salient characteristics highlighted below:

- Apart from inhabitation and stimulation, synergy is introduced. Comparing with the stimulation effect, the effect is further enhanced than triggered by a clone of immune cells.
- Similar to the pathways presented by DC, each induced pathway may provide similar functionality – this will either exhibit or impair the DCs as depicted in Table 1.

The table below summarizes the behaviors and control parameters that are inspired from the migration and maturation of DCs, and they provide innovations to the derivation of novel framework for solving MOP. More importantly, the fate of the candidate solutions can be determined as DCs perform in a stochastic and dynamic environment.

Table 1. The modeling of control parameters in the migration and maturation of DCs and the decision factors in the production of Pareto optimal [19][20]

| The signal pathway in migration and maturation of DCs | The fate of the candidate solutions (Pareto Optimal) in MOP |
|---|---|
| Change in Cytoarchitecture | Changing the phenotype of the candidate solutions |
| Rate of endocytosis | Rate of the phenotypical changes |
| DC Locomotion/ survival | Motility (or longevity) of candidate solutions |
| Recruitment of DC subsets | Increase the interactions (adhesion opportunity) of specific DC subsets |

All in all, the fate of the candidate solution(s) proposed in the *DC-mediated Signal Cascading Framework* is governed by a number of factors as proposed above, namely, (i) "threat" and "affinity", (ii) inhibit, stimulate and synergy, and (iii) decision factors, such as motility rate. The characteristics of dynamic behavior and the diversity of the candidate solutions are distinctive features of the proposed framework, when compared to other immune-inspired optimization algorithm and the classical evolutionary-based optimization algorithm [11].

## 3 Conclusions and Future Work

According to the immunological features and behaviors that are embedded in the migration and maturation program of DCs, (i) antigen capturing and (ii) signaling cascading are implemented in the proposed DC-inspired framework for solving multi-objective optimization. The notions of the proposed framework and DC-inspired capabilities have been described in this paper; yet, further investigation and experiments shall be conducted in details, particularly, benchmarking the evolved Pareto Front with other classical algorithms. Further, better understanding of the performance (behavior) and modeling of these theoretical underpinnings will also be addressed with a view to develop a novel immune-inspired algorithm.

## References

1. Oates, R., Kendall, G. & Garibaldi, J.M., Classifying in the presence of Uncertainty: A DCA Perspective, ICARIS 2010, pp. 75-87 (2010)
2. Oates, R., Greensmith, J., Aickelin, U., Garibaldi, J. & Kemdall, G., The Application of a Dendritic Cell Algorithm to a Robotic Classifier, ICARIS 2007, pp.204-215 (2007)
3. Greensmith, J., Aickelin, U. & Tedesco, G., Information Fusion for Anomaly Detection with the Dendritic Cell Algorithm, Information Fusion, 11, pp.21-34 (2010)
4. Fanelli, R., Further Experimentations with Hybrid Immune Inspired Network Intrusion Detection, ICARIS 2010, pp. 264-275 (2010)

5. Greensmith, J., Aickelin, U. & Cayer, S., Introducing dendritic Cells as a novel immune- inspired algorithm for anomaly detection, ICARIS 2005, pp. 153-167 (2005)
6. Greensmith, J., Aickelin, U. & Tedesco, G., Information Fusion for Anomaly Detection with the Dendritic Cell Algorithm, Information Fusion, 11, pp.21-34 (2010)
7. Martin-Fontecha, A., Lanzavecchia, A. & Sallusto, F., Dendritic Cell Migration to Peripheral Lymph Nodes, In: Lombardi, G. & Riffo-Vasquez (eds.) Dendritic Cells, pp.31- 49, Springer, Heidelberg (2009)
8. Ricart, B.G., John, G., Lee, D., Hunter, C.A. & Hammer, D.A., Dendritic Cells Distinguish Individual Chemokine Signals through CCR7 and CXCR4, J Immunol, 186, pp.53-61
9. Coello Coello, C. A., & Cruz Cortés, N., An Approach to Solve Multiobjective Optimization Problems Based on an Artificial Immune System., ICARIS 2002, pp. 212--221 (2002).
10. Freschi, R., Coello, C.A.C. & Repetto, M., Multiobjective optimization and artificial immune system: a review, retrieved at: http://www.igi-global.com/viewtitlesample.aspx?id=19637 (IGI Global), (2009).
11. Kim, D.H., Abraham, A. & Cho, J.H., A hybid genetic algorithm and bacterial foraging approach for global optimization, Information Sciences, 177(18), pp.3918-3937 (2007).
12. Sozzani, S., Allavena, P. &. & Mantovani, A., Dendritic cells and chemokines, In: Lotze, M.T. & Thomas, A.W. (eds.), Dendritic Cells, pp.203-211, Elseiver Academic Press, London (2001).
13. Sanchez-Sanches, N., Riol-Blanco, L. & Rodriguez-Fernandez, L., The multiple personalities of the chemokine receptor CCR7 in dendritic cells, J. Immunol., 176, pp.5153-5159 (2006).
14. Randolph, G.J., Ochando, J. & Parida-Sanchex, S., Migration of dendritic cell subsets and their precursors, Annu. Rev. Immunol., 26, pp.293-316 (2008)
15. Hochrein, H. & O'Keeffe, M., Dendritic cell subsets and toll-like receptors, In: Bauer, S. & Hartmann, G. (eds.), Toll-like Receptors (TLRs) and Innate Immunity – Handbook of Experimental Pharmacology, pp.153-179 (2008)
16. Van Hasster, P.J. & Devreptes, P.N., Chemotaxis: signaling the way forward, Nature Reviews Molecular Cell Biology, 5, pp. 626-634 (2004).
17. Shi, G.K., Harrison, K., Han, D.B., Moratz, C. & Kehrl, J.H., Toll-like receptor signaling alters the expression of regulator of G protein signaling protein in dendritic cells: implications for G protein-coupled receptor signaling, J. Immunol., 172(9), pp.5175-5144 (2004)
18. Blander, J. M. & Sander, L.E., Beyond pattern recognition: five immune checkpoints for scaling the microbial threat, Nature Reviews Immunology, 12, pp. 215-225 (2012).
19. Murphy, K.M. & Stockinger, B., Effector T-cell plasticity: flexibility in the face of changing circumstances, Nature Immunology, 11(8), pp.674-680 (2010).
20. Goodnow, C.C., Vinuesa, C.G., Randall, K.L., Mackay, F. & Brink, R., Control systems and decision making for antibody production, Nature Immunology, 11(8), pp.681-688 (2010).
21. Pulendran, B., Tang, H. & Manicassamy, S., Programming dendritic cells to induce $T_H2$ and tolerogenic responses, Nature Immunology, 11(8), pp.647-655 (2010).
22. Ko, A., Lau, H.YK. & Lau, T.L., General Suppression Control Framework: Application in Self-balancing Robots, ICARIS 2005, pp.375-388 (2005)
23. Hart, E., Bersini, H. & Santos, F., Tolerance vs Intolerance: How Affinity Defines Topology in an Idiotpypic Network, ICARIS 2006, pp. 109-121 (2006)
24. Timmis, J., Neal, M. & Hunt, J., An artificial immune system for data analysis, Biosystems, 55(3), pp. 143-150 (2000)
25. Coello Coello, C.A. & Cortés, N.C., Solving multiobjective problems using an artificial immune system, Genetic Programming Evolvable Machines, 6, pp. 163-190 (2005).

# R U :-) or :-( ? Character- vs. Word-Gram Feature Selection for Sentiment Classification of OSN Corpora

Ben Blamey and Tom Crick and Giles Oatley

**Abstract** Binary sentiment classification, or sentiment analysis, is the task of computing the sentiment of a document, i.e. whether it contains broadly positive or negative opinions. The topic is well-studied, and the intuitive approach of using words as classification features is the basis of most techniques documented in the literature. The alternative character n-gram language model has been applied successfully to a range of NLP tasks, but its effectiveness at sentiment classification seems to be under-investigated, and results are mixed. We present an investigation of the application of the character n-gram model to text classification of corpora from online social networks, the first such documented study, where text is known to be rich in so-called unnatural language, also introducing a novel corpus of Facebook photo comments. Despite hoping that the flexibility of the character n-gram approach would be well-suited to unnatural language phenomenon, we find little improvement over the baseline algorithms employing the word n-gram language model.

## 1 Introduction

As part of our wider work on developing methods for the selection of important content from a user's social media footprint, we required techniques for sentiment analysis that would be effective for text from online social networks (OSNs). The n-gram model of language used by sentiment analysis can be formulated in two ways: either as a sequence of words (or overlapping sequences of n consecutive words), or more rarely, as a set of all overlapping n-character strings, without consideration of individual words.

Ben Blamey, Tom Crick, Giles Oatley
Cardiff Metropolitan University, Western Avenue, Cardiff, CF5 2YB, United Kingdom, e-mail: {beblamey,tcrick,goatley}@cardiffmet.ac.uk

Our hypothesis is that the character n-gram model will be intrinsically well-suited to the 'unnatural' language common to OSN corpora, and will achieve higher accuracy in the binary sentiment classification task.

For our study, we gathered 3 corpora: movie reviews, Tweets, and a new corpus of Facebook photo comments. We ran traditional word-based classification alongside character n-gram based classification. The aim was to see whether the character n-gram model offers improved accuracy on OSN corpora, with the movie review corpus serving as a non-OSN control.

## 2 Background & Related Work

Text found in social media is rich in 'unnatural' language phenomena, defined as "informal expressions, variations, spelling errors ... irregular proper nouns, emoticons, unknown words" [1]. Existing NLP tools are known to struggle with such language, Ritter et al. have "demonstrated that existing tools for POS tagging, Chunking and Named Entity Recognition perform quite poorly when applied to Tweets" [2, pp. 1532], Brody and Diakopoulos "showed that [lengthening words] is a common phenomenon in Twitter" [3, pp. 569], presenting a problem for lexicon-based approaches. These investigations both employed some form of inexact word matching to overcome the difficulties of unnatural language, we wondered whether the flexibility of the character n-gram language model would make it more appropriate than the word-based language model for sentiment analysis of OSN text.

The character n-gram model is arguably as old as computer science itself [4]. It has been proven successful for tasks within NLP; such as information extraction [5], Chinese word segmentation [6], author attribution [7], language identification [8], and other text-classification tasks [9, 10, 11], and outside it (e.g. compression, cryptography). The word-based model has a number of disadvantages: Peng et al. cite drawbacks of the standard text classification methodology including "language dependence" and "language-specific knowledge", and notes that "In many Asian languages ... identifying words from character sequences is hard" [12, pp. 110].

Research has found character n-grams to perform better than word n-grams: Peng et al. apply the character n-gram model to a number of NLP tasks, including text genre classification observing "state of the art or better performance in each case". [12, pp. 116]. NLP expert Mike Carpenter, has stated that he favours character n-gram models [13].

There are just a few examples of sentiment analysis employing the character n-gram model: Rybina [14] did binary sentiment classification on a selection of German web corpora, mostly product reviews, and finds that character n-grams consistently outperforms word n-grams by 4% on F1 score. This is an extremely interesting result, and our desire to repeat her findings were a key motivation for this work, but some details are unclear: the classifier that was used is closed-source, and it isn't obvious what method was used to label the data. Nevertheless, such a result demands further explanation. Other studies have more mixed findings; Ye et al. [11] classified travel reviews using both character and word n-gram models with LingPipe, and found that neither was consistently better.

Much work has studied sentiment analysis of OSN corpora, especially Twitter, using the word n-gram model. Go et. al [15], in the first sentiment analysis study of the micro-blogging platform, achieved accuracy of around 80%. Pak and Paroubek [16] experimented with exclusion of word n-grams; based on saliency and entropy thresholds, but neither the thresholds themselves nor improvement over original accuracy are quoted. At the state of the art, Bespalov et al. [17] overcome the high-dimensionality of the word n-gram feature-set using a multi-layer perceptron to map words and n-grams into lower-order vector spaces, while retaining meaning. The approach achieves accuracy of more than 90%.

Sentiment analysis studies of Facebook are comparatively rare, in one such study Kramer computed the positivity of status messages using LIWC[1] [18] to create an index of "Gross National Happiness" [19]. To our knowledge, there have been no documented studies of sentiment analysis applying the character n-gram model to online social network text, and none looking at Facebook photo comments using either language model.

## 3 Methods

An emoticon is a sequence of characters with the appearance of a facial expression. For Tweets and Facebook photo comments, we follow the approach of Read [20] of using the presence of emoticons as labels of sentiment (thereby allowing unsupervised machine learning), and them removing them from the text for classification (similarly, the standard approach is to use 'star-ratings' with movie reviews). We distinguish the emoticons from instances where the text resembles something else, such as <3 (a heart). The importance of unnatural language is exemplified by one Facebook photo comment, reading simply: "<3!". A comprehensive list of Western emoticons was compiled from Wikipedia[2]. Around 20,000 Tweets were gathered from the Twitter Search API, using positive and negative emoticons as search terms. With Facebook photo comments, the percentage of comments that contained emoticons was low, (negative emoticons were found in less than 1% of the corpus), so it was necessary to collect a large amount of data. We managed to obtain over 1 million unique Facebook photo comments via the Facebook API.

URLs and Twitter 'mentions' and hashtags were replaced with respective single characters, so their meaning is captured in both word and character n-gram models. Only the documents with exclusively happy {:) :D :-) =) : ) :-D :o) =D} or sad {:( :-( :'( : (} emoticons only were selected, and emoticons were chosen that accurately mean 'happy' and 'sad' - documents with :P ('sticking out tongue') and similar emoticons were excluded, because they tend to be used for jokes and

[1] http://www.liwc.net/

[2] http://en.wikipedia.org/wiki/List_of_emoticons

insults - which might confuse the classifiers. Also excluded were 'winks', e.g. ';-)', incase they were used to indicate flirtatious remarks, which again may disrupt classification, for example: "Nice slippers, hon! (Brother's not bad either... ;-) tee hee!) x". Emoticons are replaced with a period, because they seem to be used as end-of-sentence punctuation, indeed they are commonly suffixed directly onto other words, so it is best to search for them as a substring (technically an application of the character-based model). This yielded labelled corpora of 7000 Tweets and 7000 Facebook photo comments, alongside 1386 movie reviews used in previous studies [21].

For word-based classification, further processing is necessary. Elongated words are squashed to a maximum of 3 repetitions, e.g. '<3<3<3<3<3' becomes '<3<3<3'. Go et al. [15] handle single repeated characters only, shortening to two reptitions. By shortening to three, we preserved the elongated variations as separate features, as word length is known to indicate sentiment strength [3]. We followed the approach of Das and Chen [22] to handle negation (as in "I am not happy") by labelling words in a negative context, yielding a small improvement to classification, consistent with other studies.

For the text classification itself, we used 4 feature-sets: word unigrams, bigrams, and the union of both, and the union of the sets of character n-grams where $n \in \{1,..,8\}$. Note that we do not trim low-frequency features, as it is generally discouraged except where necessary for performance. Three standard classifiers were used in our experiment:

- Naive Bayes (based on feature frequency), with 'plus-one' smoothing.
- Maximum Entropy (i.e. 'loglinear discrimitive') of the Stanford Classifier[3], with default settings (a quadratic prior with $\sigma = 1$).
- SVM$^{light}$ classifier[4] [23], with default settings.
- For character n-grams only, the LingPipe[5] [24] DynamicLMClassifier.

The classification accuracies are shown below in Table 1.

## 4 Conclusions

Our results look a lot like those of Ye et al. [11], with neither word- nor character-grams yielding consistently higher accuracy. Therefore, the findings contradict some existing studies: inconsistency with the results of Rybina [14] (a consistent 4% improvement with character n-grams). Her results are hard to explain – language is a slight possibility (her corpus was German).

Looking more closely at our data, we can see that character n-grams consistently beat word unigrams, which is understandable, as 8 characters will often be enough

[3] http://nlp.stanford.edu/software/classifier.shtml

[4] http://svmlight.joachims.org/

[5] http://alias-i.com/lingpipe/

to contain more than one word, and including word bigrams has often given better accuracy than unigrams alone.

Our hypothesis that character n-grams will be intrinsically well-suited to the 'unnatural' language common to OSN corpora was false: there doesn't seem to be a significant performance difference between the OSN and non-OSN corpora, for social network text and unnatural language – but the language of the social web is a pressing challenge for NLP; and as discussed, many of the existing tools struggle with it. The size of our corpus may have been an issue, to reap the full benefits of the character n-gram model more training data might be needed - LingPipe is designed to scale character n-gram data to the order of gigabytes [25].

Putting the issue of unnatural language aside, proponents of character n-gram models have a point: studies (including this one) have repeatedly shown that the character n-gram can perform as well as simple word n-gram models – whilst being considerably simpler to implement, especially when tokenization is hard, such as in Asian languages. There is no one 'right' way to do tokenization, negation, word squashing, stemming, and precise details are often thought too tedious for publication. Experiments involving word-grams can sometimes be difficult to repeat perfectly for these reasons, and greater use of character n-gram based algorithms would eliminate these inconsistencies between work. Our tendency to automatically adopt the word-based model may suggest some degree of human-centric bias in our research thinking, or perhaps too strong a focus on English and other Western languages, within sentiment analysis research.

**Table 1** 3-fold cross-validated accuracies. The best performing configuration of feature-set and classifier for each corpus is shown in bold.

| Corpus | Features | Bayes | MaxEnt | SVM | LingPipe |
|---|---|---|---|---|---|
| IMDB Movie Reviews | Word Unigrams | 80.5% | 82.7% | 82.8% | |
| | Word Bigrams | 80.5% | 79.4% | 76.7% | |
| | Word Unigrams+Bigrams | 81.4% | 83.5% | 82.2% | |
| | Character n-grams | 81.9% | **84.6%** | 82.6% | 75.9% |
| Tweets | Word Unigrams | 88.7% | 91.2% | 88.8% | |
| | Word Bigrams | 89.5% | 91.4% | 90.5% | |
| | Word Unigrams+Bigrams | 90.6% | 91.6% | 90.8% | |
| | Character n-grams | 90.8% | 91.9% | 90.6% | **92.0%** |
| Facebook Photo Comments | Word Unigrams | 80.2% | 79.8% | 78.6% | |
| | Word Bigrams | 75.8% | 75.8% | 72.5% | |
| | Word Unigrams+Bigrams | **80.5%** | 80.0% | 78.3% | |
| | Character n-grams | 80.4% | 80.1% | 80.0% | 75.9% |

## References


1. M. Hagiwara. Unnatural Language Processing Contest 2nd will be held at NLP2011 (2010). URL http://bit.ly/dGvUnR
2. A. Ritter, S. Clark, Mausam, O. Etzioni, in *Proceedings of the 2011 Conference on Empirical Methods in Natural Language Processing* (ACL, Edinburgh, Scotland, UK., 2011), pp. 1524–1534
3. S. Brody, N. Diakopoulos, in *Proceedings of the 2011 Conference on Empirical Methods in Natural Language Processing* (ACL, Edinburgh, Scotland, UK., 2011), pp. 562–570
4. C. Shannon, Bell System Technical Journal (27), 379 (1948)
5. D. Klein, J. Smarr, H. Nguyen, C. Manning, in *Proceedings of the seventh conference on Natural language learning at HLT-NAACL 2003 - Volume 4* (Association for Computational Linguistics, Stroudsburg, PA, USA, 2003), CONLL '03, pp. 180–183. DOI 10.3115/1119176.1119204
6. N. Xue, Computational Linguistics and Chinese Language Processing **8**, 29 (2003)
7. F. Peng, D. Schuurmans, S. Wang, V. Keselj, in *Proceedings of the tenth conference on European chapter of the Association for Computational Linguistics - Volume 1* (Association for Computational Linguistics, Stroudsburg, PA, USA, 2003), EACL '03, pp. 267–274. DOI 10.3115/1067807.1067843
8. W.B. Cavnar, J.M. Trenkle, in *Proceedings of SDAIR-94, 3rd Annual Symposium on Document Analysis and Information Retrieval* (1994), pp. 161–175
9. S. Raaijmakers, W. Kraaij, in *ICWSM* (2008)
10. R. Pon, A. Cárdenas, D. Buttler, T. Critchlow, in *Computational Intelligence and Data Mining, 2007. CIDM 2007. IEEE Symposium on* (2007), pp. 354–361. DOI 10.1109/CIDM.2007.368896. URL http://dx.doi.org/10.1109/CIDM.2007.368896
11. Q. Ye, Z. Zhang, R. Law, Expert Syst. Appl. **36**(3), 6527 (2009). DOI 10.1016/j.eswa.2008.07.035
12. F. Peng, D. Schuurmans, S. Wang, in *Proc. of HLT-NAACL 03* (2003), pp. 110–117
13. B. Carpenter. Yahoo group message discussion (2010). URL http://tech.dir.groups.yahoo.com/group/LingPipe/message/917
14. K. Rybina, Sentiment analysis of contexts around query terms in documents. Master's thesis (2012)
15. A. Go, R. Bhayani, L. Huang, Processing **150**(12), 1 (2009)
16. A. Pak, P. Paroubek, in *Proceedings of the Seventh conference on International Language Resources and Evaluation (LREC'10)* (ELRA, Valletta, Malta, 2010)
17. D. Bespalov, B. Bai, Y. Qi, A. Shokoufandeh, in *Proceedings of the 20th ACM international conference on Information and knowledge management* (ACM, New York, NY, USA, 2011), CIKM '11, pp. 375–382
18. F.M..B.R. Pennebaker, J.W. Linguistic inquiry and word count: Liwc2001 (2001)
19. A.D.I. Kramer. Facebook gross national happiness application (2010). URL http://www.facebook.com/gnh/
20. J. Read, Proceedings of the ACL Student Research Workshop on ACL 05 **43**(June), 43 (2005)
21. B. Pang, L. Lee, S. Vaithyanathan, in *Proceedings of the 2002 Conference on Empirical Methods in Natural Language Processing (EMNLP)* (Association for Computational Linguistics, 2002), pp. 79–86
22. S. Das, M. Chen, in *Asia Pacific Finance Assc. Annual Conf. (APFA)* (2001)
23. T. Joachims, *Making large-scale SVM learning practical* (MIT press, 1999)
24. Alias-i. Lingpipe 4.1.0 (2008). URL http://alias-i.com/lingpipe
25. B. Carpenter, in *Proceedings of the Workshop on Software* (Association for Computational Linguistics, Stroudsburg, PA, USA, 2005), Software '05, pp. 86–99

# Predicting Multi-class Customer Profiles Based on Transactions: a Case Study in Food Sales

Edward Apeh, Indrė Žliobaitė, Mykola Pechenizkiy and Bogdan Gabrys

**Abstract** Predicting the class of customer profiles is a key task in marketing, which enables businesses to approach the customers in a right way to satisfy the customer's evolving needs. However, due to costs, privacy and/or data protection, only the business' owned transactional data is typically available for constructing customer profiles. We present a new approach that is designed to efficiently and accurately handle the multi-class classification of customer profiles built using sparse and skewed transactional data. Our approach first bins the customer profiles on the basis of the number of items transacted. The discovered bins are then partitioned and prototypes within each of the discovered bins selected to build the multi-class classifier models. The results obtained from using four multi-class classifiers on real-world transactional data consistently show the critical numbers of items at which the predictive performance of customer profiles can be substantially improved.

## 1 Introduction

Customer profiles which encapsulate the detailed knowledge of the customer, provide a means by which predictive models can be built so as to effectively recognize the status and preference of each individual customer. Such predictive models can be incorporated into the company's market segmentation, customer targeting and channelling decisions with the goal of maximizing the total customer lifetime profit.

Behavioural customer profiles, which can be derived from transactional data, typically have much stronger predictive power than factual customer profiles. Transac-

---

Edward Apeh, Indrė Žliobaitė and Bogdan Gabrys
Smart Technology Research Centre, Bournemouth University, UK
e-mail: (eapeh,izliobaite,bgabrys)@bournemouth.ac.uk

Mykola Pechenizkiy
Eindhoven University of Technology, Eindhoven, the Netherlands
e-mail: m.pechenizkiy@tue.nl

tional data can be electronically collected and readily made available for data mining in plenty quantity at minimum extra costs. Transactional data is however, inherently sparse and skewed which adversely affects the performance of ad-hoc predictive models built using transactional data based customer profiles.

We introduce a new approach for customer profile prediction via multi-class classification. We address the challenge of transactional data sparsity and skewness by modelling customer profiles which have been locally specialized by first binning them into homogeneous groups, instead of building a global model of all diverse customer profiles. Prototypes of customer profiles are then extracted from the discovered bins and used in building multi-class classifier models. The learned models can then be used to predict the class (e.g. cafeterias, schools, supermarkets, etc.) of customer profiles based on their purchases. The approach is validated on the case study of a food retail and food service company operating in the Dutch food and beverages market.

This study presents two major contributions. First, we propose a new algorithm for predicting the class of customer profiles based on their transactions that can handle sparse multi-class datasets, as exhibited by real life applications. Second, we present an extensive case study in the food sales domain which validates our proposed approach and illustrates the set of practical challenges in customer profile prediction. An extended version of this study can be found in our technical report[1].

The paper is organized as follows. Section 2 discusses background and related work. Section 3 presents the approach. Section 4 presents our experimental case study. Section 5 summarizes the implications to a customer-centric business.

## 2 Background and Related Work

Transactional data are time-stamped data collected over time at no particular frequency. Examples of transactional data may include: point of sales (POS) data, call centre data, trading data, etc. The inherently largeness, skewness and sparsity of transactional data makes modelling and performing prediction on the entire dataset available prohibitive in terms of computational time and costs. Also, conventional data reduction techniques may not work well due to the concentration of the transactional data around the point where a vast majority of the customers buy only a few items.

Our approach to customer profile class prediction is closely related to user modelling in textual data. Furthermore, the proposed binning approach technically relates to contextual learning (e.g. [1, 3]), where specialized predictive models are used in making predictions based on the current context during operation.

---

[1] Available at: http://alturl.com/f7wg7

# 3 Predicting Customer Profiles from Transactional Data

Our approach builds upon the algorithm proposed in [2]. We introduce two principal extensions: (1) the algorithm is redesigned from binary to a multi-class setting that is required by realistic application tasks; (2) we introduce k-means prototyping to be able to efficiently build multi-class classifier models using large sparse datasets.

Formally, given a set of $M$ customer profiles $P = \{\mathbf{p_1}, \mathbf{p_2}, \ldots, \mathbf{p_M}\}$, with each profile $\mathbf{p_i}$ having its aggregated $d$-dimensional transaction, our goal is to build models that *accurately* assign the $i$th customer $\mathbf{p_i}$ to the $j$th class label $y_j \in \omega_A, \ldots, \omega_K$.

Customer profiles may vary in magnitude and normalizing the customer profiles may occlude some predictive features, we partition the customer profiles into homogeneous bins based on the number of items purchased. We then use K-Means to cluster the customer profiles in each bin and select the instances whose Silhouette Statistics is greater than a user defined threshold (0.6 in our case) as prototypes to form our training set. Then we train classifiers on the discovered prototypes. Algorithm 1 formally describes the procedure for training the predictors.

**Algorithm 1:** Train predictors

**Input**: Training set $P$
**Output**: Predictors $C_B$, bins cluster centres $\mu_k^B$
**Initialize**: $S = 0.6$

```
1  Bin P into B bins based on the number of items per transaction.;
2  for b ← 1 to B do
3      Cluster into K_Classes groups using K-means, where K_Classes is the number of classes in T;
4      Record bin cluster centre μ_k^b;
5      for each instance x_i in bin b do
6          for each group k ∈ K_classes do
7              Compute the Silhouette Statistics sw_i^k;
8              if sw_i^k > S then
9                  include x_i into the set of prototypes P*_kb
10     Train a predictor C_b on instances in P*_kb;
```

New customer profiles are predicted by first determining their closest bin based on the number of transactions and then using the classifier trained within that bin for the prediction as formally described in Algorithm 2.

**Algorithm 2:** Predict class of new customer profile

**Input**: new customer profile $x_i$, predictors $C_B$, bins cluster centres $\mu_k^B$
**Output**: predicted class $k_i$

```
1  Assign x_i to the closest bin B_p : p = arg min ||x − μ_k^b||;
2  classify x_i using the predictor C_b;
```

# 4 Case Study in Food Sales Domain

To validate and support the proposed approach this section presents an experimental case study using a real-world transactional data from the food sales domain.

**Data.** The data was provided by Sligro Food Group N.V., which is a group of food retail and food service companies selling to the Dutch market. The provided data consists of 408625 aggregated transactions of which a total 148601 SKU products were transacted by 65 Sligro customer categories over three consecutive years. 148 of the top selling products were used for the analysis in this paper.

**Experiments protocol.** The main goal of our experiments was to validate the proposed approach on a real-world case study. In particular, we were interested in determining the effect of the number of items bought in each category on the classification performance. We compared the performance of our binning-based approach with the random sampling baseline, which randomly sub-samples customer profiles for classification.

Experimental comparison were performed using WEKA's built-in One-Vs-All (OVA) and error-correcting output codes (ECOC) on four base classifiers (Logistic regression, Decision Tree (J48), Naive Bayes and Support Vector Machine) . For each of the selected prototypes 10-fold cross-validation was repeated 10 times.

To compare classifier performance on the entire multi-class transactional dataset, we use the weighted average AUC, where each target class $c_i$ is weighted according to its prevalence thus: $AUC_{weighted} = \sum_{\forall c_i \in C} AUC(c_i) \times p(c_i)$.

## 4.1 Data Analysis

**Table 1:** Identified Data Bins.

| Class | Bins | | | | |
|---|---|---|---|---|---|
| | 1..5 | 6..20 | 21..127 | 128+ | Sum |
| 100 | 551 | 470 | 699 | 1439 | 3159 |
| 190 | 4443 | 3957 | 2677 | 300 | 11377 |
| 230 | 9284 | 7963 | 6358 | 1125 | 24730 |
| 300 | 1454 | 1765 | 2697 | 2676 | 8592 |
| 310 | 972 | 1117 | 2069 | 3083 | 7241 |
| 331 | 970 | 1052 | 1713 | 3235 | 6970 |
| 360 | 898 | 957 | 1064 | 948 | 3867 |
| 380 | 714 | 768 | 1116 | 1037 | 3635 |
| 390 | 974 | 1071 | 1814 | 1593 | 5452 |
| 391 | 789 | 935 | 1140 | 586 | 3450 |
| 590 | 1088 | 1075 | 1334 | 900 | 4397 |
| 620 | 891 | 1104 | 1457 | 1189 | 4641 |
| 800 | 8718 | 8749 | 7904 | 1701 | 27072 |
| 820 | 1518 | 1431 | 1105 | 134 | 4188 |
| 840 | 4051 | 4396 | 4089 | 609 | 13145 |
| 890 | 1443 | 1513 | 1364 | 217 | 4537 |
| 900 | 2941 | 1928 | 1420 | 206 | 6495 |
| Total | 41699 | 40251 | 40020 | 20978 | 142948 |

**Table 2:** Selected Prototypes.

| Class | Bins | | | | |
|---|---|---|---|---|---|
| | 1..5 | 6..20 | 21..127 | 128+ | Sum |
| 100 | 23 | 17 | 12 | 115 | 167 |
| 190 | 105 | 185 | 47 | 31 | 368 |
| 230 | 246 | 381 | 94 | 112 | 833 |
| 300 | 80 | 72 | 82 | 268 | 502 |
| 310 | 55 | 36 | 7 | 305 | 403 |
| 331 | 39 | 39 | 23 | 320 | 421 |
| 360 | 30 | 43 | 12 | 93 | 178 |
| 380 | 24 | 22 | 13 | 103 | 162 |
| 390 | 36 | 42 | 21 | 158 | 257 |
| 391 | 41 | 42 | 16 | 59 | 158 |
| 590 | 42 | 56 | 25 | 90 | 213 |
| 620 | 18 | 60 | 21 | 109 | 208 |
| 800 | 287 | 328 | 101 | 171 | 887 |
| 820 | 38 | 51 | 11 | 14 | 114 |
| 840 | 121 | 199 | 43 | 61 | 424 |
| 890 | 25 | 69 | 21 | 22 | 137 |
| 900 | 72 | 68 | 17 | 21 | 178 |
| Total | 1282 | 1710 | 566 | 2052 | 5610 |

Following the proposed approach outlined in Section 3, the customer profiles were first binned, using the equal-frequency binning algorithm, to obtain the homogeneous groups for each of the categories (i.e. classes) as shown in Table 1. It can be seen that employing the equal-frequency binning algorithm enables the proportions of the categories (i.e. classes) to be maintained across the bins. The K-means prototype selecting algorithm as outlined in Section 3 was then applied to the discovered bins to obtain the prototypes for each class in each bin as shown in Table 2.

**Analysis of the Predictive Performance.** Figure 1 shows the plots of the predictive performance results. It can be seen from all four plots that there is a critical number of items purchased $o^\star$ at which the overall AUC classification performance is higher than that obtained from the baseline approach of random sampling customer profiles for classification.

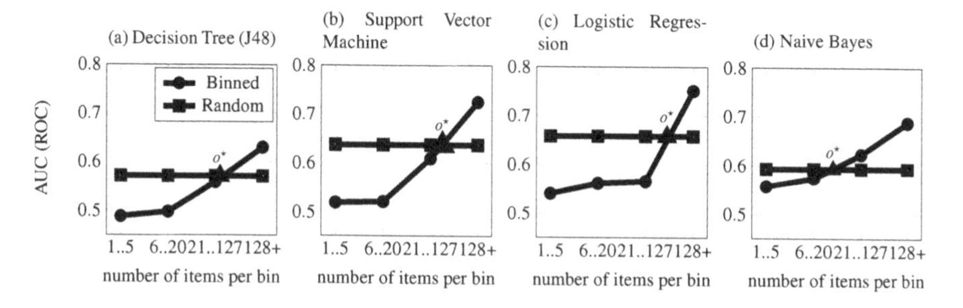


**Fig. 1:** Performance on the selected prototypes of customer profiles.

Identifying the critical point $o^\star$ validates the contribution to predicting multi-class customer profiles based on transactional data, in that, not only does it help in overcoming the challenge of transactional data sparseness and skewness but it also enables practitioners to build specialized classifier models that can more accurately classify customers based on the number of items purchased.

From a business perspective, the customers with profiles whose classification fall above the critical point can be prime candidates for direct interactive/one-to-one marketing campaigns while customers whose profiles fall below the critical point can be candidates for general marketing campaigns.

In addition, the differences in classification performance on individual categories across the bins provide insight that can be valuable for developing better relationship with the customers. As an illustration, the top 10 items for the category code 310 (Cafeteria) across the four bins in Table 3 show that the highlighted product codes[2] have a strong influence on the classification of the customer profiles based on transactional data in that category. These products could be included in product promotions and customer targeting programmes as incentives for product growth and customer retention.

[2] 184532 drink, 192653 Coca-Cola, 882627 bread, 882627 cigarettes

**Table 3:** Top 10 products of the *Cafeteria* category.

| Bin1 | | Bin2 | | Bin3 | | Bin4 | |
|---|---|---|---|---|---|---|---|
| Product Code | Total | Product Code | Total | Product Code | Total | Product Code | Total |
| 184532 | 111 | 184532 | 119 | 297296 | 146 | 192653 | 5062 |
| 93456 | 60 | 93456 | 70 | 93456 | 92 | 184532 | 4782 |
| 81491 | 8 | 64787 | 65 | 882627 | 63 | 190902 | 3689 |
| 394548 | 2 | 652378 | 36 | 936251 | 60 | 882627 | 2329 |
| 637263 | 1 | 190855 | 26 | 24224 | 54 | 265943 | 2130 |
| 192653 | 1 | 936251 | 24 | 736177 | 13 | 432257 | 1868 |
| 269117 | 1 | 591881 | 20 | 591881 | 11 | 255710 | 1578 |
| 476997 | 1 | 87353 | 9 | 282241 | 8 | 516140 | 1569 |
| 736177 | 1 | 432257 | 4 | 900360 | 7 | 231708 | 1483 |
| 882627 | 0 | 882627 | 3 | 192653 | 6 | 401963 | 1450 |

## 5 Conclusion

We presented an investigation into the classification of multi-class customer profiles using real-world transactional data in the food sales domain. We proposed and validated a new approach for predicting customer profiling that can deal with sparse realistic transactional data using the 'divide-and-conquer' principle. The approach first partitions data into homogeneous bins, then crystallizes each bin by extracting the most representative prototypes to be used for training the predictive models.

The experimental case study validates the approach on a difficult real world problem. The predictive performance is consistent across different base classifiers. The overall accuracy improves with the number of items purchased. The analysis demonstrates that it is possible to find a critical number of items to be purchased to ensure accurate classification. Knowing this point allows for the filtering of customers and for focused marketing activities to be undertaken on the ones where better predictive accuracy can be expected. Our case study also illustrated that the proposed approach can be used not only for the prediction of new customer profile classes, but also for business analysis, as closer insights can be gleaned from the predicted customer profiles thereby enabling better understanding of the customers of the business.

**Acknowledgements** We thank Sligro Food Group N.V. for providing the data and the domain knowledge for the case study.

## References

1. Adomavicius, G., Tuzhilin, A.: Context-aware recommender systems. In: F. Ricci, L. Rokach, B. Shapira, P.B. Kantor (eds.) Context-Aware Recommender Systems, pp. 217–253. Springer (2011)
2. Apeh, E., Gabrys, B., Schierz, A.: Customer profile classification using transactional data. In: Proceedings of the Third World Congress on Nature and Biologically Inspired Computing (NaBIC2011) (2011)
3. Žliobaitė, I., Bakker, J., Pechenizkiy, M.: Beating the baseline prediction in food sales: How intelligent an intelligent predictor is? Expert Syst. Appl. **39**(1), 806–815 (2012)

# Controlling Anytime Scheduling of Observation Tasks

J W Baxter, J Hargreaves, N Hawes and R Stolkin

**Abstract** This paper describes how multiple independent observation tasks can be scheduled for an autonomous vehicle. Presented with large numbers of tasks, of differing reward levels, a vehicle has to evaluate the best schedule to execute given a limited time to both plan and act. A meta-management framework acting on top of an anytime scheduler analyses the problem and the progress made in generating solutions to identify when to stop planning and start executing. We compare a probabilistic management technique with active monitoring of the current execution reward and conclude that in this case detecting a local maxima in the predicted reward is the most effective policy.

## 1 Introduction

Anytime algorithms provide a useful technique for trading off planning time against execution. In situations where fast but well-supported decisions are needed they provide a way of providing bounded optimality [5]. This requires a stopping condition for the algorithm which provides the best response given the time and computation limits faced by an agent. Our problem domain consists of planning surveillance tours where a vehicle has to plan a tour of a number of locations making observations at each one. The observation tasks have a predicted duration and reward value and the aim is to maximise the expected reward from a tour. This is therefore a prize-gathering orienteering problem. These sorts of problems arise in many different situations such as unmanned aerial vehicle mission planning, robotic security patrols and the gathering of scientific data remotely. We have applied an anytime

J W Baxter
University of Birmingham, Poynting Institute and QinetiQ Ltd. BT15 2TT, e-mail: j.baxter@poynting.bham.ac.uk

J Hargreaves, N Hawes, R Stolkin
University of Birmingham, BT15 2TT e-mail: {jxh576, n.a.hawes , r.stolkin} @cs.bham.ac.uk

algorithm based upon weighted A* search [7] which can plan observation schedules in an anytime fashion and can provide the optimal solution if given sufficient time and memory. In initial trial work in a UAV mission planning domain we applied a simple fixed run time limit for this anytime scheduler. This paper compares the performance of different meta-management techniques for deciding when to stop planning and start executing.

## 2 Related Work

Team orienteering problems and the related prize gathering travelling salesman problem have been covered in the operations research literature [1, 2] where exact solutions for problems with up to 102 tasks have been found using integer programming. Some problems could be solved very rapidly but others took minutes or hours of CPU time, especially as the length of tours increased (roughly equivalent to longer total execution time in our problem formulation). Shi et al. [6] propose an ant colony optimisation method for solving these problems but don't give data on the computational cost. We are not aware of other work describing an anytime planning approach to these problems. Anytime planners have a long history and anytime heuristic search is analysed in depth in [3]. More recent work [7] has looked in detail at the effect of using different heuristic weights on the speed with which bounded optimal solutions can be found and the effect of combining admissible and non-admissible heuristics. Zilberstein and Hansen [4] propose a range of different, and increasingly complex, meta-management schemes for anytime algorithms. They focus on a specific problem type (the 20 city travelling salesman problem) without considering if this data generalises to different sizes of problem.

## 3 Meta-Management

In our UAV planning domain we compared two different techniques for stopping an anytime algorithm in order to get the best result: the 'myopic' stopping criteria proposed in [4] which uses statistics gathered over previous problem instances, and a 'loss limiting' policy which executes the plan as soon as the benefit of executing the current plan falls below a set threshold of the best (predicted) execution reward seen so far.

Developing the meta-management system consist of two steps, gathering data to characterise generic performance and analysing a specific problem instance to predict how it is likely to perform. Unlike the approach in [4] we assert that the cost of time is not independent of the problem instance. In our case we represent time as a lost chance to gain rewards. Therefore we need to be able to estimate from a problem what the expected reward is and how much reward a delay is likely to cost.

### *3.1 Characterising performance*

To characterise the performance of the scheduler we used a random problem generator and ran a series of 100 test runs recording the reward of the best plan found against time. The reward level was compared against the best result found after 60s of runtime (on a 2.3Ghz Intel 2 Core Duo with 2Gb of RAM) or the optimal if the algorithm terminated before that time having proved its solution was optimal. In order to better represent real world problems we used both fully random positioning of tasks and a clustering generator where tasks were more likely to be close to other tasks than spread uniformly through the space. The algorithm is highly effective at finding high quality solutions very quickly in single vehicle cases and the curve showing the average plan quality (as a proportion of the best found in 60s) can be seen in Figure 1. This data is used to predict future improvements for the 'myopic' scheme and to predict how close a solution is likely to be to the optimal. The reward lost by executing a plan later is estimated by calculating a linear expected reward per unit time from the estimated optimal reward divided by the time allowed for execution.

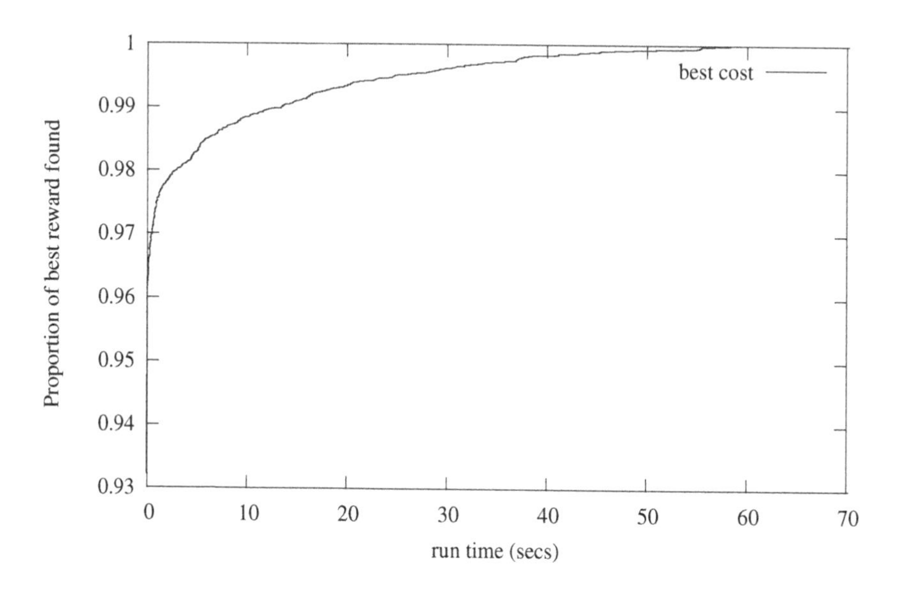


**Fig. 1** Reward level of best plan found against time (averaged over 100 runs)

### 3.2 Online decision making

The 'myopic' management policy uses the (estimated) current quality level of a solution and a probability table derived by sampling 100 test runs to predict if the expected gain in reward through planning will exceed the reward lost by delaying execution. The 'loss limiting' policy attempts to execute just after the estimated execution reward has peaked (at the maxima of the curves shown in Figure 2 ). It does this by recording the maximum execution reward found so far and executing if the current estimated reward ever falls below 99.5% of this maximum. The tolerance allows for short periods in which no improvement is found.

## 4 Results

To analyse the effect of the meta-management layer we examined the effect of *time pressure*, i.e. how the system responds to changes in the balance between the speed of computation and the speed of execution. To do this we vary the relative rates of computation and execution so that the meta-management layer has to adjust the execution decision point. Figure 2 shows the result of scaling the times in the problem by rates between 1 and 0.1. This is equivalent to either reducing the computational speed or reducing the travel time, task durations and total execution time by the rate factor. To keep the same time axis, the graph uses the scaled task times, giving the same computation time but with the task times and distances between them reduced. The actual execution reward is calculated by assuming the output schedule is run but only those tasks which can be fully completed provide a reward. As more time is spent running the scheduler it is more likely that one or more tasks at the end of the schedule will not be completed reducing the actual reward. The graph shows the proportion of the maximal reward obtainable after 60 seconds of planning (at the base rate) if execution is started at the time on the x axis (averaged over 100 test instances). As can be seen at the base level of time pressure planning can continue for 20 seconds before the benefit of improved plans is outweighed by the lost rewards due to the delay. As the time pressure increases, the total reward peaks much earlier and decreases swiftly, showing that the benefit of additional planning is rapidly lost.

To identify the performance of the meta-management layer, Table 1 shows the average result of 100 runs using different policies to decide when to stop planning and start executing. The values are the proportion of the reward of the best possible plan found after 60 seconds when stopping planning according to the policy and starting to execute. The first three policies use a fixed planning time with no meta-management. The final policy, hindsight, shows the maximum possible result which could be obtained given the scheduler's performance by looking back over every decision point of a 60 second run for each problem to find the maximum.

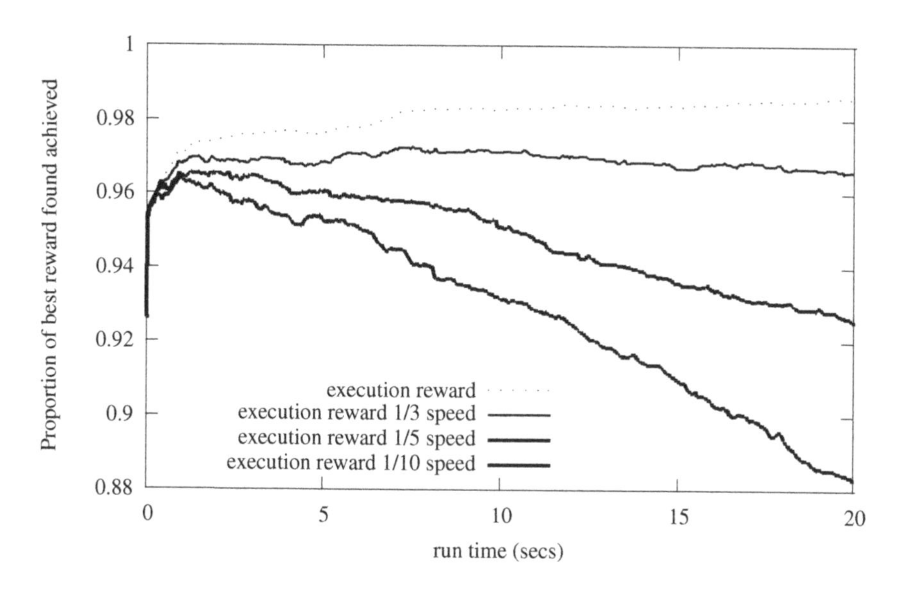


**Fig. 2** Actual reward achieved at execution for increasing levels of time pressure (averaged over 100 runs)

**Table 1** The resulting reward as a proportion of the maximum possible averaged over 100 runs for each policy with varying processing speeds

| Policy | Rate<br>1 | 0.33 | 0.2 | 0.1 |
|---|---|---|---|---|
| fixed 1s | 96.7679 | 96.3957 | **96.2939** | 95.5168 |
| fixed 10s | **97.0843** | 95.1745 | 93.1818 | 88.1618 |
| fixed 20s | 96.5668 | 92.5464 | 88.1677 | 77.0558 |
| myopic | 96.6338 | 96.2636 | 95.987 | 95.5899 |
| loss limiting | 96.9283 | **96.5873** | 96.2759 | **95.7427** |
| hindsight | 98.8795 | 97.8987 | 97.6254 | 97.2579 |

## 5 Conclusions

The anytime scheduler we are using is highly efficient for these problems, getting on average 93% of the reward found after a minute within the first step (which typically takes 33ms) and rapidly improving the solution.

The best result over all policies is highlighted in bold. This shows that the fixed time policies produce good results if they are matched to the problem but that they cannot match the performance of the managed policies across the different levels of time pressure. The 'loss limiting' policy outperforms the 'myopic' forward looking policy and is the best or second best in all cases. The 'myopic' forward looking strategy is limited because its performance is dependent on the quality of the improvement probability table as well as the estimates of the maximal reward and cost of time. As its future value prediction is based on averages over large numbers of runs it tends to perform badly in cases where the current problem is non typical and big improvements are made much earlier or later than the average. The 'loss limiting' policy is more robust to early gains (which skew the estimates of the optimal reward) but still terminates early in cases where the scheduler finds large improvements relatively late. The 99.5% loss limiting parameter was found by trial and error, improved versions could make use of information about the variability of previous problems to set this more appropriately.

## 6 Further Work

We hope to extend this work to looking at problems with multiple vehicles and time windows on tasks. The cost of time could be better estimated by examining the slack in the schedule. A schedule with a few seconds of 'spare' time at the end has effectively zero cost of time and this could be used to continue planning. In the longer term we are interested in applying the meta-management to more complex problems, and planning and execution strategies. In particular we would like to analyse the ability of meta-management to cope with more dynamic environments where tasks and travel times may change and multiple decisions on planning and re-planning are necessary.

## References

1. de Arago, M.P., Viana, H., Uchoa, E.: The team orienteering problem: Formulations and branch-cut and price. In: T. Erlebach, M.E. Lbbecke (eds.) ATMOS, *OASICS*, vol. 14, pp. 142–155. Schloss Dagstuhl - Leibniz-Zentrum fuer Informatik, Germany (2010). URL http://dblp.uni-trier.de/db/conf/atmos/atmos2010.html
2. Boussier, S., Feillet, D., Gendreau, M.: An exact algorithm for team orienteering problems. 4OR: A Quarterly Journal of Operations Research **5**, 211–230 (2007). URL http://dx.doi.org/10.1007/s10288-006-0009-1. 10.1007/s10288-006-0009-1
3. Hansen, E.A., Zhou, R.: Anytime heuristic search. Journal of Artificial Intelligence Research (JAIR **28**, 267–297 (2007)
4. Hansen, E.A., Zilberstein, S.: Monitoring and control of anytime algorithms: A dynamic programming approach. Artificial Intelligence **126**, 139–157 (2001)
5. Russell, S., Wefald, E.: Do the right thing. In: Studies in Limited Rationality. MIT Press (1991)
6. Shi, X., Wang, L., Zhou, Y., Liang, Y.: An ant colony optimization method for prize-collecting traveling salesman problem with time windows. In: Proceedings of the 2008 Fourth International Conference on Natural Computation - Volume 07, ICNC '08, pp. 480–484. IEEE Computer Society, Washington, DC, USA (2008). DOI 10.1109/ICNC.2008.470. URL http://dx.doi.org/10.1109/ICNC.2008.470
7. Thayer, J.T., Ruml, W.: Faster than weighted a*: An optimistic approach to bounded suboptimal search. In: proceedings of the international conference on planning and scheduling (2008)

# Applications and Innovations in Intelligent Systems XX

# BEST APPLICATION PAPER

# Swing Up and Balance Control of the Acrobot Solved by Genetic Programming

Dimitris C. Dracopoulos and Barry D. Nichols

**Abstract** The evolution of controllers using genetic programming is described for the continuous, limited torque minimum time swing-up and inverted balance problems of the acrobot. The best swing-up controller found is able to swing the acrobot up to a position very close to the inverted 'handstand' position in a very short time, which is comparable to the results which have been achieved by other methods using similar parameters for the dynamic system. The balance controller is successful at keeping the acrobot in the unstable, inverted position when starting from the inverted position.

## 1 Introduction

Genetic Programming (GP) has been shown to perform well in some difficult control problems, even those where traditional control methods have failed [4]. However, there are surprisingly few applications of GP to difficult control problems. The swing up and balance of the acrobot has been considered such a difficult control problem. The discrete, half-swing-up variant of the acrobot problem (a simpler version of the problem than the one considered here) was one of the six problems included in the 2009 Reinforcement Learning Competition [11].

In the literature, there are only two GP approaches to the acrobot problem that we are aware of [3, 7]. Neither attempts the balance task, one only attempts the simplified half-swing-up [3] and the other uses GP to solve the three section variant of the acrobot problem [7]. Thus this is the first application of GP to solve the standard acrobot swing-up and balance tasks as proposed by [12]. Through GP, two computer programs are evolved successfully here: one which is able to swing up the

Dimitris C. Dracopoulos, Barry D. Nichols
School of Electronics and Computer Science, University of Westminster, 115 New Cavendish Street, London, W1W 6UW, e-mail: {d.dracopoulos,b.nichols1}@westminster.ac.uk

acrobot in time comparable to that achieved by other methods and another which can successfully balance the acrobot when starting from the inverted position.

## 2 The Acrobot Problem

The acrobot is a two link robot, based on a human acrobat. The links are connected by an actuated joint and one end is connected to a bar by an unactuated joint [12], see Fig. 1. The state $\mathbf{x} = [\theta_1, \theta_2, \dot{\theta}_1, \dot{\theta}_2]^T$ comprises the first angle $\theta_1$, second angle $\theta_2$ and the two angular velocities $\dot{\theta}_1$ and $\dot{\theta}_2$. The control variable is the torque $\tau$ to be applied to the actuated joint.

The acrobot is a difficult control task as it is a four dimensional, highly nonlinear, under-actuated control problem [11, 12, 13].

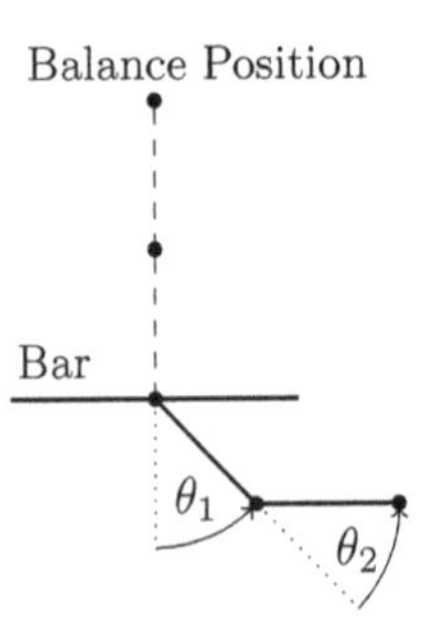


**Fig. 1** Diagram of the acrobot

The task of the acrobot is to swing up, in the minimum time, from the initial state $\mathbf{x}_0 = [0,0,0,0]^T$ to the inverted, unstable, target state $\mathbf{x}_t = [\pi,0,0,0]^T$ and then to remain balanced in that position. This could be split into two tasks: swing-up and balance, with two distinct controllers which are switched at a given point. The controller switching point can be either found through trial and error [12] or by applying other techniques to find the most appropriate point to switch [18].

### 2.1 Other Approaches

There have been many approaches to the acrobot problem, and its many variations; however, usually unbounded, or very large, torque values are permitted [5, 10, 12, 16, 18]. Also, the task is often simplified to perform only the swing-up [1] or half-swing-up, where the task is to swing the end of the acrobot one link length above the bar [14, 17], commonly with a discrete set of allowable torque values. Another simplification which is often used in computational intelligence approaches, and occasionally in other approaches, is to use a linear quadratic regulator (LQR) [6] for the balance phase, hence, solving only the swing-up task [7, 8, 15].

There have only been two GP approaches to the acrobot, to our knowledge: one attempts the half-swing-up task using a discrete action set [3], and the other addresses the different, albeit related, task of swinging up a three section version of the acrobot which has two actuated joints, but applies an LQR controller to the balance task [7].

## 2.2 The Dynamic System

The results presented here are based on computer simulations of the acrobot using the equations of motion (1) described in [14]. The parameter values used are shown in Table 1. The angular velocities were limited to $-4\pi \leq \dot{\theta}_1 \leq 4\pi$ and $-9\pi \leq \dot{\theta}_2 \leq 9\pi$. Continuous values were allowed for the control input in the range $\tau \in [-2,2]$. The acrobot state $\mathbf{x}$ was updated every 0.05 simulated seconds and the controller selected the $\tau$ to apply to the actuated joint every 0.2 seconds.

$$\begin{aligned}
\ddot{\theta}_1 &= -d_1^{-1}\left(d_2\ddot{\theta}_2 + \phi_1\right) \\
\ddot{\theta}_2 &= \left(m_2 l_{c2}^2 + I_2 - \frac{d_2^2}{d_1}\right)^{-1}\left(\tau + \frac{d_2}{d_1}\phi_1 - m_2 l_1 l_{c2}\dot{\theta}_1^2 \sin(\theta_2) - \phi_2\right) \\
d_1 &= m_1 l_{c1}^2 + m_2\left(l_1^2 + l_{c2}^2 + 2 l_1 l_{c2}\cos(\theta_2)\right) + I_1 + I_2 \\
d_2 &= m_2\left(l_{c2}^2 + l_1 l_{c2} cos(\theta_2)\right) + I_2 \\
\phi_1 &= -m_2 l_1 l_{c2}\dot{\theta}_2^2 \sin(\theta_2) - 2 m_2 l_1 l_{c2}\dot{\theta}_2\dot{\theta}_1 \sin(\theta_2) \\
&\quad + (m_1 l_{c1} + m_2 l_1)\, g\cos\left(\theta_1 - \frac{\pi}{2}\right) + \phi_2 \\
\phi_2 &= m_2 l_{c2}\, g\cos\left(\theta_1 + \theta_2 - \frac{\pi}{2}\right)
\end{aligned} \tag{1}$$

**Table 1** Acrobot parameter values

| Parameter | Symbol | Value |
|---|---|---|
| Mass of link 1 | $m_1$ | 1 |
| Mass of link 2 | $m_2$ | 1 |
| Length of link 1 | $l_1$ | 1 |
| Length of link 2 | $l_2$ | 1 |
| Length to centre of mass of link 1 | $l_{c1}$ | 0.5 |
| Length to centre of mass of link 2 | $l_{c2}$ | 0.5 |
| Inertia of link 1 | $I_1$ | 1 |
| Inertia of link 2 | $I_2$ | 1 |
| Gravitational constant | $g$ | 9.8 |

## 3 The GP Approach

GP was applied to the acrobot task by including **x** and a randomly generated constant in the terminal nodes, see Table 5. The fitness of each individual was calculated by running a simulation of the acrobot where the GP program was presented with the current state of the dynamic system **x** and returned the value of $\tau$, which was limited to the allowable range using:

$$\tau = \begin{cases} \tau_{MIN}, & \text{if } \tau < \tau_{MIN} \\ \tau_{MAX}, & \text{if } \tau > \tau_{MAX} \\ \tau, & \text{otherwise} \end{cases} \tag{2}$$

Two experiments were conducted: one for the swing-up control and one for the balance control. Both used the same GP parameters, (Table 3), but different function sets (Tables 4 and 6). The parameters and function sets were selected through experimentation with several different values and combinations on each problem. The other main difference between the two approaches was the method of calculating the raw fitness $f(i)$ of individual $i$.

However, once $f(i)$ was calculated, the adjusted fitness $f_a(i)$ was computed using equation (3) to exaggerate the difference between the best performing individuals [9].

$$f_a(i) = \frac{1}{1+f(i)} \tag{3}$$

After the population had been generated, and $f_a(i)$ was computed for all individuals in the population, the normalised fitness $f_n(i)$ was calculated for each of the individuals, see equation (4). The normalised fitness is required to apply over-selection, which was used for both experiments.

$$f_n(i) = \frac{f_a(i)}{\sum_{j=1}^{M} f_a(j)} \tag{4}$$

### 3.1 Swing-up

For the swing-up task GP was applied in order to evolve a controller to swing the acrobot up in minimum time. At the same time, it also had to be ensured that the angular velocities $\dot{\theta}_1$ and $\dot{\theta}_2$ were small, to allow the balance controller to stabilise it when the inverted position is reached. The fitness function was defined to ensure the fitness of an individual would be proportional to its performance in the aforementioned control task:

$$f(i) = t_{up}(i) \tag{5}$$

where $t_{up}$ is the time to reach the inverted position and the individual was considered to have reached the inverted position when the following evaluates to true:

$$\cos(\theta_1) \leq -0.95 \wedge \cos(\theta_2) \geq 0.95 \wedge |\dot{\theta}_1| \leq 0.5 \wedge |\dot{\theta}_2| \leq 0.5$$

The performance of the best individual was observed after 50 runs of the program, a single run consisted of 100 generations. The best two individuals from each generation were directly copied into the next generation and there was a limit to the size of the GP tree $D_c$. If this limit was exceeded by one of the child GP trees as a result of crossover one of the parents was reproduced directly, if both children exceeded $D_c$ then both parents were directly reproduced in the next generation.

## 3.2 Balance

In order to 'catch' the acrobot from the swing-up and balance it, the balance controller must be able to stabilise the acrobot from angles slightly off of the inverted position; therefore eight two dimensional angle offset vectors $\mathbf{o}_j$ were defined: $\mathbf{o}_0 = [0,0]^T$ and $\mathbf{o}_j, j = 1,2,\ldots,7$ were randomly generated, but were kept constant throughout the experiment, see Table 2. To calculate the fitness of an individual the acrobot simulation was run eight times, each time the corresponding offset vector was added to the target angles $[\pi,0]^T$ to use as the initial acrobot angles. The fitness of an individual was, therefore, an accumulated measure of how far the acrobot deviated from the balance position at each time step before falling down added to the difference between the maximum balance time and that which was achieved (normalised to the same magnitude as the accumulated fitness). This was then summed over all eight offsets (Equation (6)), where $\theta_{n,t}$ is the $n$-th angle at time $t$ of the current trial and $t_{down,j}(i)$ is the time that it took for individual $i$ to fall down when starting from $[\pi,0]^T + \mathbf{o}_j$. An individual was considered to have fallen down when the following evaluates to true:

$$\cos(\theta_1) > -0.9 \vee \cos(\theta_2) < 0.9$$

Each of the offset simulations was only run for a maximum time of 30 seconds, but was stopped as soon as the acrobot fell down. However, when producing the results of the best of run for each of the 50 runs, a 60 second simulation was applied to validate the controller over a longer period of time from each of the offset positions.

$$f(i) = \sum_{j=0}^{7} \left\{ \frac{(T - t_{down,j}(i))}{2.5} + \sum_{t=0}^{t_{down,j}(i)} \left( (1+\cos(\theta_{1,t}))^2 + (1-\cos(\theta_{2,t}))^2 \right) \right\} \quad (6)$$

**Table 2** Offset Vectors

| Vector Number | $\theta_1$ Offset | $\theta_2$ Offset |
|---|---|---|
| $\mathbf{o}_0$ | 0 | 0 |
| $\mathbf{o}_1$ | -0.091 | -0.048 |
| $\mathbf{o}_2$ | 0.08 | 0.106 |
| $\mathbf{o}_3$ | 0.056 | -0.015 |
| $\mathbf{o}_4$ | -0.121 | -0.022 |
| $\mathbf{o}_5$ | -0.034 | 0.04 |
| $\mathbf{o}_6$ | -0.113 | -0.07 |
| $\mathbf{o}_7$ | 0.046 | -0.121 |

**Table 3** Genetic Programming parameters

| Parameter | Value |
|---|---|
| Probability of crossover ($P_c$) | 90% |
| Probability of reproduction ($P_r$) | 10% |
| Probability of mutation ($P_m$) | 0% |
| Population size ($M$) | 2000 |
| Number of generations (swing-up) ($G$) | 100 |
| Probability of choosing internal points for crossover ($P_{ip}$) | 90% |
| Maximum size of initial GP trees ($D_i$) | 6 |
| Maximum size of GP trees created during run ($D_c$) | 17 |
| Over-selection cumulative percentage ($c$) | 16% |
| Initialisation method | Grow |

**Table 4** Swing-up Function set

| Symbol | Function | Formula |
|---|---|---|
| + | add | $arg_1 + arg_2$ |
| − | subtract | $arg_1 - arg_2$ |
| * | multiply | $arg_1 \times arg_2$ |
| / | protected divide | $\begin{cases} arg_1/arg_2, & \text{if } arg_2 \neq 0 \\ 0, & \text{otherwise} \end{cases}$ |
| tanh | hyperbolic tangent | $\tanh(arg_1)$ |
| abs | absolute value | $\lvert arg_1 \rvert$ |
| iflt | if less than | $\begin{cases} arg_3, & \text{if } arg_1 < arg_2 \\ arg_4, & \text{otherwise} \end{cases}$ |

## 4 Results

The results of the two experiments are discussed separately, first the swing-up, then the balance. The progression of fitness of the best individual and the average of all individuals of the run is shown in Fig. 2 for the swing-up task and Fig. 3 for the balance task. The values were taken from an average performing run of the 50 from each task.

**Table 5** Terminal set

| Terminal | Description |
|---|---|
| $\theta_1$ | angle one |
| $\theta_1$ | angle two |
| $\dot{\theta}_1$ | angular velocity at angle one |
| $\dot{\theta}_2$ | angular velocity at angle two |
| $R$ | randomly generated constant terminal swing-up: $\in [-10, 10]$, balance: $\in [-100, 100]$ |

**Table 6** Balance Function set

| Symbol | Function | Formula |
|---|---|---|
| + | add | $arg_1 + arg_2$ |
| − | subtract | $arg_1 - arg_2$ |
| * | multiply | $arg_1 \times arg_2$ |
| / | protected divide | $\begin{cases} arg_1/arg_2, & \text{if } arg_2 \neq 0 \\ 0, & \text{otherwise} \end{cases}$ |
| sin | sine | $\sin(arg_1)$ |
| cos | cosine | $\cos(arg_1)$ |
| tanh | hyperbolic tangent | $\tanh(arg_1)$ |
| exp | exponential | $\exp(arg_1)$ |
| log | natural logarithm | $\begin{cases} \log(arg_1), & \text{if } arg_1 > 0 \\ 0, & \text{otherwise} \end{cases}$ |
| sqrt | square root | $\sqrt{\lvert arg_1 \rvert}$ |
| gt | greater than | $\begin{cases} 1, & \text{if } arg_1 > arg_2 \\ -1, & \text{otherwise} \end{cases}$ |
| abs | absolute value | $\lvert arg_1 \rvert$ |
| sgn | sign | $\begin{cases} 1, & \text{if } arg_1 > 0 \\ -1, & \text{if } arg_1 < 0 \\ 0, & \text{otherwise} \end{cases}$ |
| square | raise to the power of two | $arg_1^2$ |
| cube | raise to the third power | $arg_1^3$ |
| ifltz | if less than zero | $\begin{cases} arg_2, & \text{if } arg_1 < 0 \\ arg_3, & \text{otherwise} \end{cases}$ |
| iflt | if less than | $\begin{cases} arg_3, & \text{if } arg_1 < arg_2 \\ arg_4, & \text{otherwise} \end{cases}$ |

## *4.1 Swing-up*

Each of the 50 runs produced a GP controller able to swing up the acrobot with an average swing-up time of 11.81 seconds. The best controller of the experiment was capable of reaching the position $\mathbf{x} = [-3.07184, 6.03887, 0.0238457, 0.393939]$ in only 8.8 simulated seconds, Fig. 4. This performance is comparable to other approaches which use the same parameter values and torque limits. Moreover, it is

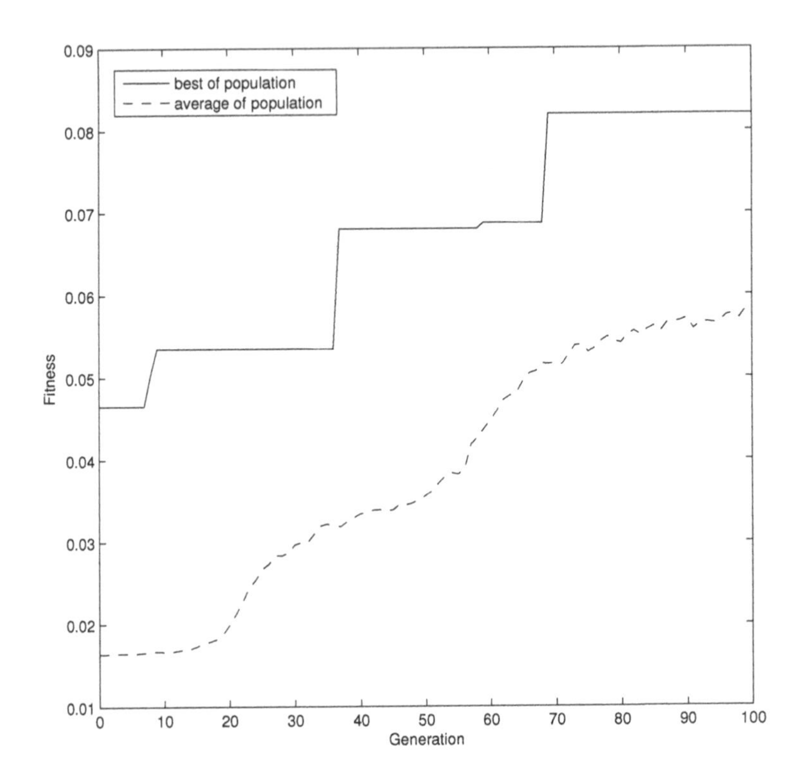


**Fig. 2** Fitness of the Best Individual and Average of Population Taken From an Average Run of the Swing-up Task

an improvement on that of a reinforcement learning method which uses the same torque limits [2], and took approximately 11 seconds to swing up. Also, this was only took slightly longer than a method which allowed torque values of more than 20Nm, which took almost 8 seconds to reach a similar position [10].

## 4.2 Balance

Nine of the 50 runs generated controllers which were able to successfully balance the acrobot when starting from $\mathbf{x}_t$ for the full 60 seconds, and some could also balance from one of the other offsets, $\mathbf{o}_4$. However, as the fitness is only calculated from the first 30 seconds, there were some individuals which scored a high fitness but were unable to balance, even starting from $\mathbf{x}_t$, for the full 60 seconds.

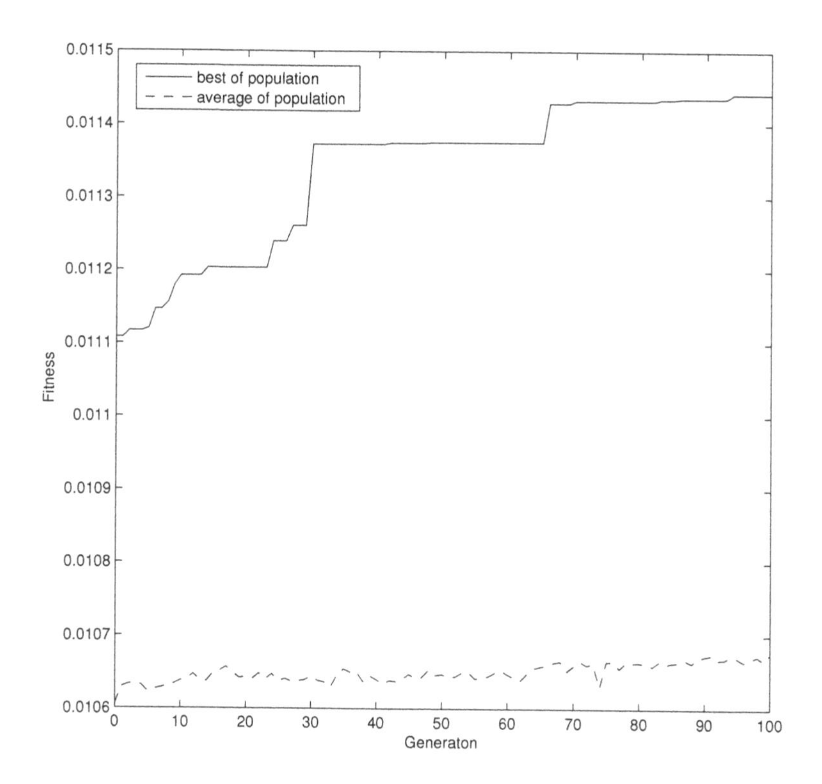


**Fig. 3** Fitness of the Best Individual and Average of Population Taken From an Average Run of the Balance Task

The best balance controller of the experiment was able to balance the acrobot for the full 60 seconds when starting from $\mathbf{x}_t$, see Fig. 5, and from $[\pi, 0]^T + \mathbf{o}_4$, see Fig. 6, but was unable to balance for even 2 seconds from any of the other offsets.

It is worth noting, that as far as the authors understand, the previous published approaches by other researchers to the problem (Section 2.1) do not attempt to generalise, i.e. the start state(s) used for testing was (were) also included in the set of start state(s) to derive a controller during training. The exceptions are [2] and [18], which start from random initial states for training and, hence, do not ensure the balance position is included in the training set, although they do not preclude it either. However, the only generalisation initial condition case which these 2 approaches test is the balance position.

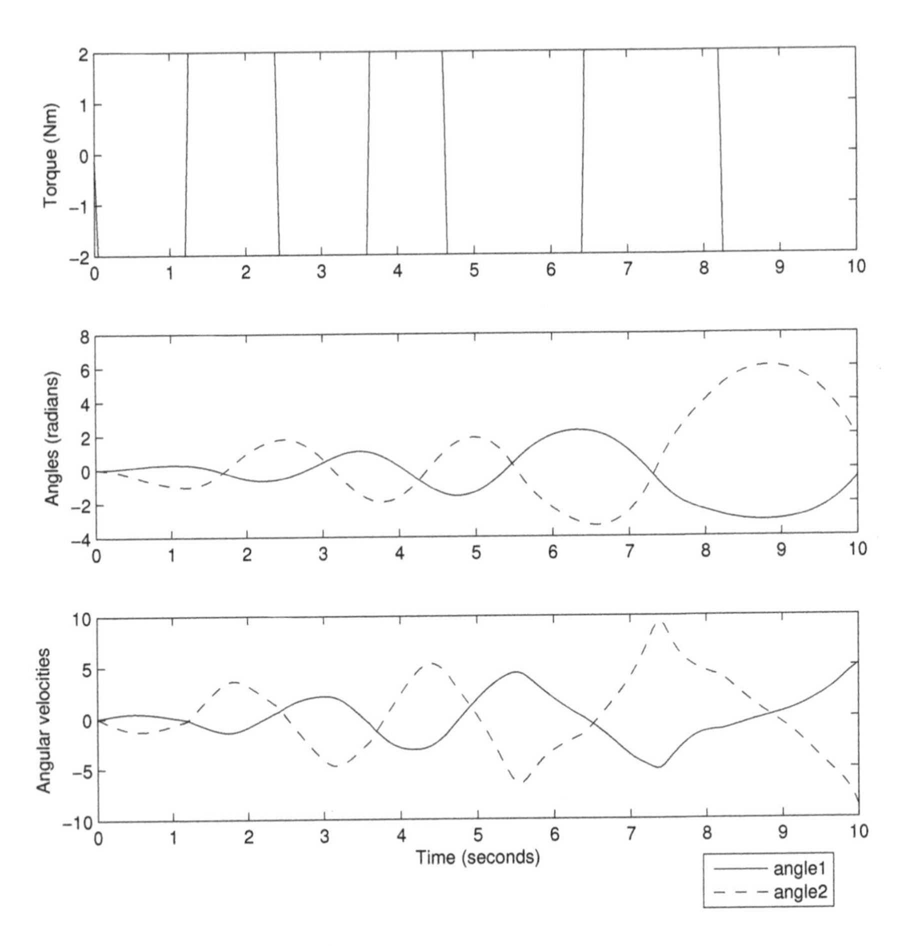


**Fig. 4** Best GP Swing-up Controller

## 5 Conclusions and Future Work

GP was applied to the swing-up and balance tasks of the acrobot and achieved good results, particularly in the swing-up task, which swings up the acrobot very close to the target position within a short time, comparable to that of other methods. The balance controller was successful in balancing when starting from very close to the inverted position, but couldn't be used to 'catch' the acrobot after a swing-up controller was applied as it was unable to stabilise the acrobot when starting even small angles from the balance position. The approach described here is the first which uses GP to tackle both the swing-up and the balance acrobot problems and additionally there are only a few other approaches which have been attempted on both problems.

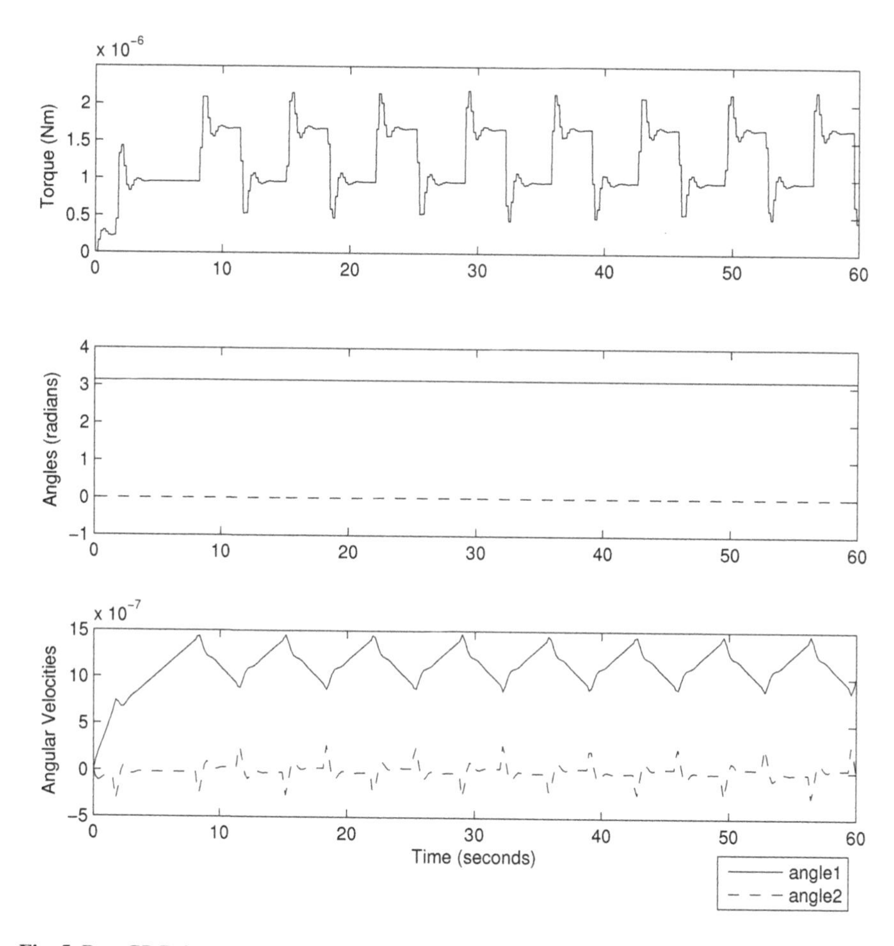


**Fig. 5** Best GP Balance Controller $\mathbf{o}_0$. The diagrams illustrate the torques, the two angles and the two angular velocities respectively as a function of time.

Current work in progress attempts to improve the angles from which the balance controllers are able to stabilise the acrobot. Additionally, the possibility of a single control program evolved using GP is investigated, which is able to both swing-up and balance the acrobot.

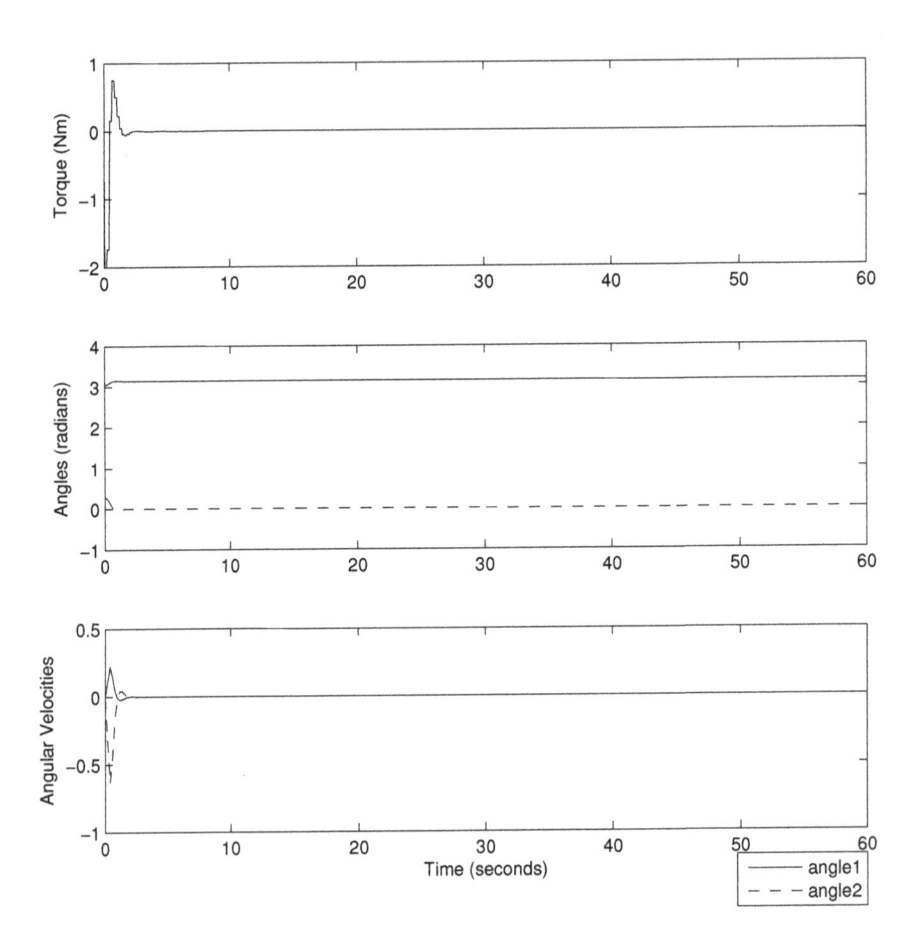


**Fig. 6** Best GP Balance Controller $\mathbf{o}_4$. The diagrams illustrate the torques, the two angles and the two angular velocities respectively as a function of time.

## Appendix

### *Best Swing-up Controller*

(iflt (* (velocity1) (abs (abs (+ (/ 5.92578 (abs (angle1))) (* (velocity1) (+ (iflt (3.24332) (velocity2) (0.431627) (/ (tanh (-7.66423)) (/ (5.92578) (angle1)))) (abs (angle1)))))))) (angle1) (4.89173) (-4.45234))

## *Best Balance Controller*

```
(* (- (velocity2) (angle1)) (+ (/ (velocity1) (sqrt (abs (exp (cos (tanh (+ (tanh (velocity1)) (ifltz (ifltz (+ (velocity2) (tanh (tanh (cos (angle2))))) (tanh (tanh (log (cos (tanh (angle2)))))) (log (cos (angle2)))) (ifltz (+ (velocity2) (tanh (tanh (cos (angle2))))) (log (cos (tanh (angle2)))) (tanh (ifltz (tanh (velocity1)) (+ (velocity2) (+ (velocity2) (tanh (angle2)))) (log (cos (angle2)))))) (ifltz (tanh (cos (angle2))) (+ (velocity2) (tanh (angle2))) (ifltz (tanh (velocity1)) (+ (velocity2) (+ (velocity2) (tanh (angle2)))) (tanh (log (cos (angle2))))))))))))) (+ (velocity2) (* (angle1) (angle2)))))
```

# References


1. Boone, G.: Minimum-time control of the acrobot. In: Robotics and Automation, 1997. Proceedings., 1997 IEEE International Conference on, vol. 4, pp. 3281–3287 (1997)
2. Coulom, R.: High-accuracy value-function approximation with neural networks. In: European Symposium on Artificial Neural Networks (2004)
3. Doucette, J., Heywood, M.I.: Revisiting the acrobot 'height' task: An example of efficient evolutionary policy search under an episodic goal seeking task. In: Evolutionary Computation (CEC), 2011 IEEE Congress on, pp. 468 –475 (2011)
4. Dracopoulos, D.C.: Genetic evolution of controllers for challenging control problems. Journal of Computational Methods in Science and Engineering **11**(4), 227–242 (2011)
5. Duong, S., Kinjo, H., Uezato, E., Yamamoto, T.: On the continuous control of the acrobot via computational intelligence. In: B.C. Chien, T.P. Hong, S.M. Chen, M. Ali (eds.) Next-Generation Applied Intelligence, *Lecture Notes in Computer Science*, vol. 5579, pp. 231–241. Springer Berlin / Heidelberg (2009)
6. Franklin, G.F., Powell, J.D., Emami-Naeini, A.: Feedback Control of Dynamic Systems, 4 edn. Prentice Hall, New Jersey (2002)
7. Fukushima, R., Uezato, E.: Swing-up control of a 3-dof acrobot using an evolutionary approach. Artificial Life and Robotics **14**, 160–163 (2009)
8. Jung, T., Polani, D., Stone, P.: Empowerment for continuous agent-environment systems. Adaptive Behavior - Animals, Animats, Software Agents, Robots, Adaptive Systems **19**, 16–39 (2011)
9. Koza, J.R.: Genetic Programming: On the Programming of Computers by Means of Natural Selection. MIT Press, Cambridge, MA, USA (1992)
10. Lai, X.Z., She, J.H., Yang, S.X., Wu, M.: Comprehensive unified control strategy for underactuated two-link manipulators. Systems, Man, and Cybernetics, Part B: Cybernetics, IEEE Transactions on **39**(2), 389–398 (2009)
11. RLC: Reinforcement learning competition. http://www.rl-competition.org (2009)
12. Spong, M.W.: Swing up control of the acrobot. In: Robotics and Automation, 1994. Proceedings., 1994 IEEE International Conference on, vol. 3, pp. 2356–2361 (1994)
13. Spong, M.W.: The swing up control problem for the acrobot. Control Systems, IEEE **15**(1), 49–55 (1995)
14. Sutton, R.S., Barto, A.G.: Reinforcement Learning: An Introduction. MIT Press, Cambridge, Massachusetts (1998)
15. Wiklendt, L., Chalup, S., Middleton, R.: A small spiking neural network with lqr control applied to the acrobot. Neural Computing & Applications **18**, 369–375 (2009)
16. Willson, S., Mullhaupt, P., Bonvin, D.: Quotient method for controlling the acrobot. In: Decision and Control, 2009 held jointly with the 2009 28th Chinese Control Conference. CDC/CCC 2009. Proceedings of the 48th IEEE Conference on, pp. 1770–1775 (2009)
17. Xu, X., Hu, D., Lu, X.: Kernel-based least squares policy iteration for reinforcement learning. Neural Networks, IEEE Transactions on **18**(4), 973–992 (2007)

18. Yoshimoto, J., Nishimura, M., Tokita, Y., Ishii, S.: Acrobot control by learning the switching of multiple controllers. Artificial Life and Robotics **9**, 67–71 (2005)

# LANGUAGE AND CLASSIFICATION

# Biologically inspired Continuous Arabic Speech Recognition

**N. Hmad**[1] **and T. Allen**[2]

**Abstract** Despite many years of research into speech recognition systems, there are limited research publications available covering Arabic speech recognition. Although statistical techniques have been the most applied techniques for such classification problems, Neural Networks have also recorded successful results in speech recognition. In this research three different biologically inspired Continuous Arabic Speech Recognition neural network system structures are presented. An Arabic phoneme database (APD) of six male speakers was constructed manually from the King Abdulaziz Arabic Phonetics Database (KAPD). The Mel-Frequency Cepstrum Coefficients (MFCCs) algorithm was used to extract the phoneme features from the speech signals of this database. The normalized dataset was used to train and test three different architectures of Multilayer Perceptron (MLP) neural network identification systems.

## 1 Introduction

Speech is natural communication method between humans and is the fastest form of data input method between humans and technology. In addition, as speech capture devices are built into many modern devices (mobile phones, PCs etc.) capturing the data input requires no extra peripherals. Consequently, applications with speech interfaces are desirable.

Automatic speech recognition is a process that converts an acoustic signal, captured by the microphone, to a set of textual words. The output of the recognizer can be used for several applications such as command and control or real-time communication between human and computer, etc. Speech recognition can be basically classified into two modes. The first is isolated word recognition; where the words in this mode are surrounded by clear silence i.e. well known boundaries.

1 Nottingham Trent University, NG1 4BU, UK
Nadiaf_hammad@yahoo.com

2 Nottingham Trent University, NG1 4BU, UK
Tony.allen@ntu.ac.uk

The second is continuous speech recognition. This second mode is more difficult than isolated word recognition because word boundaries are difficult to detect due to the variability of speakers in terms of the speed of speech and word pronunciation. In addition, all speech recognition systems are affected by variability between speakers and variability between environments. A word may be uttered differently from speaker to another; also it might be uttered in different way by the same speaker as a result of emotion and illness. Moreover, the environment may introduce corruption into the speech signal because of background noise, microphone characteristics, and transmission channels [1].

## *1.1 General Speech Recognition Techniques*

Hidden Markov models (HMM) are the most successful speech recognition method developed to date and have been extensively used for continuous speech recognition [2]. For example, Al-Ghamdi [3] used the SPHINX toolkit (which is based on HMM methods) on the Arabic Phonetics Database (King Abdulalaziz Phonetic Database KAPD). Artificial Neural Network (ANN) algorithms such as Radial Basis Function (RBF), Recurrent Neural Networks (RNN), and Multi-Layer Perceptron (MLP) have also been widely used for such applications [4-6]. Some approaches use a hybrid of these techniques to produce high performance systems [1].

## *1.2 Arabic Speech Recognition*

Arabic is one of the oldest living languages and is one of the oldest Semitic languages in the world. It is also the fifth most generally used language. The Arabic language is the mother tongue for roughly 200 million people and classical Arabic is the language used to speak most of the Islamic literature. Although, there are numerous research publications reporting on speech recognition for many languages, there is only limited work that has been produced for the Arabic language.

### 1.2.1 Arabic Phoneme Implementation

Phonemes are generally considered as the smallest meaningful part of speech compared with any other speech parts (syllable, word, phrase, etc.). In several standard corpora, the word or phoneme segmentation has been prepared manually by experts [3]. An Arabic equivalent is not available. Therefore, in this work, a database of 31 Arabic phonemes was manually created from the King Abdulalaziz Phonetic Database [7].

Phonemes in any language can be categorised into two main groups (consonants and vowels). Vowels and consonants are the main categories of all languages. In Arabic languages, consonants are further categorised into four classes. These are: voiced and unvoiced stops class, voiced and unvoiced fricatives class, nasal class and the trill and lateral class. Long and short vowels then make up a fifth class as shown in Fig. 1 [8].

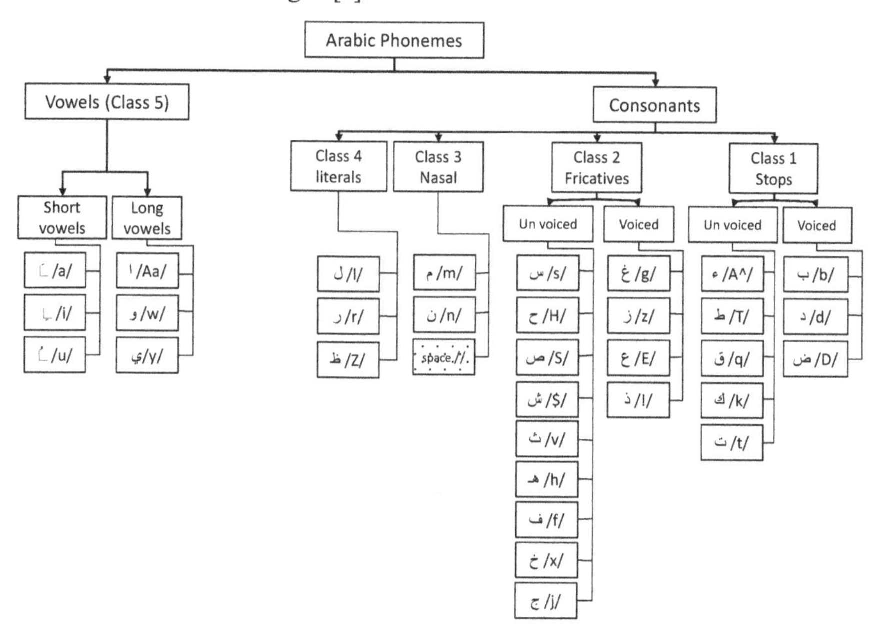


**Fig. 1.** Arabic Phoneme categorisation.

### 1.2.2 Arabic Phoneme Segmentation Techniques

Speech phoneme segmentation is a real challenge for continuous speech recognition systems. In limited vocabulary isolated word recognition the problem can be solved by determining the correct boundary of the isolated words and rejecting the artifacts of speech such as noise and intra-word stops. With regard to large vocabulary continuous speech boundary detection, the problem becomes much more difficult than isolated word boundary detection because of the intra-word silences and other artifacts. These problems can be reduced by applying the following pre-processing steps.

Voiced/Unvoiced detection algorithms can be applied on pre-emphasised speech signals to detect syllable boundaries. The most common methods used for end point detection are Energy profile and Zero Crossing Rate (ZCR) [9] and Entropic contrast [10]. Typically, in the Energy profile and Zero Crossing Rate

(ZCR) algorithms an adaptive threshold is applied based on the characteristics of the energy profile, in order to differentiate between the background noise and the speech segments. However, this algorithm is very sensitive to the amplitude of the speech signal, meaning that, the energy of the signal will affect the classification results. This is especially a problem in noisy environments.

Recently, a new end point detection algorithm has been proposed that uses Entropic contrast [10]. This algorithm uses features of the entropy profile of the speech signal rather than the energy profile of the signal. Thus the proposed algorithm uses the entropy of the speech for boundary detection. The calculation of the entropy is applied in the time domain. Crucially, this profile is less sensitivity to the changes in amplitude of the speech signal.

Despite the use of the above techniques, current speech segmentation techniques do still introduce errors into the segmentation process. Therefore, many modern speech recognition systems perform the recognition process without prior segmentation. These systems tend to be based on the use of features extraction techniques such as MFCC [6, 11-14], FFT [15-17], HFCC [18], Linear Predictive Coefficient (LPC) [8] ... etc.

### 1.2.3 Arabic Phoneme Recognition

Although statistical approaches using HMMs have been the most successful for speech recognition, neural networks have also been applied to speech classification problems [6, 8, 12, 15, 17-20]. A couple of research projects have shown the capability of using neural networks for Arabic phoneme recognition.

This paper is organized as follows: In section two, general speech recognition system stages are presented. The three proposed structures of MLP are also detailed in the same section. The efficiency of the proposed algorithms is evaluated in section three where results are presented showing experimentation applied on the proposed database. In section four, a conclusion of the work done is presented.

## 2 Speech Recognition System

The overall speech recognition system is shown in Fig. 2

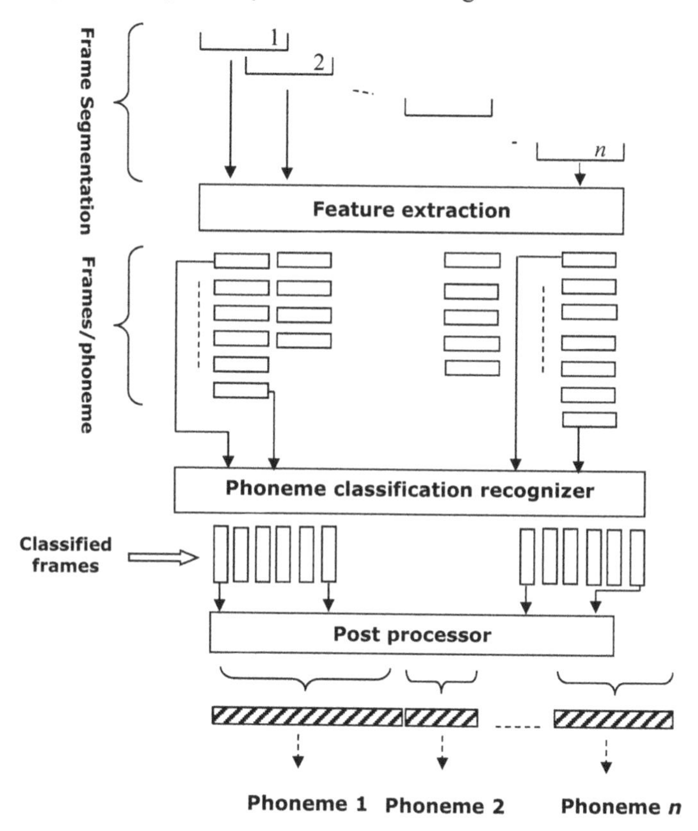


**Fig. 2.** Speech recognition system.

### *2.1 Proposed Arabic Phonemes Database*

KAPD is the King Abdulaziz Arabic Phonetics Database, recorded at the King Abdulaziz City for Science and Technology (KACST) in Saudi Arabia. This database consists of 7 male speakers; where each speaker has recorded the 340 semi-Arabic words that were artificially created to construct the (KAPD). Whilst we recognize that this is a small population of speakers, against which to test out algorithm, it was felt important to use a 'standardized' database as the basis for our research in order that others could then compare their results with ours. All 340 words were recorded 7 times in different environments by the seven speakers (2380 words). Only two groups of words (680 words) for each of the seven speakers, recorded in a quiet environment, were selected for use in this project. The speech was recorded over a microphone at a sampling frequency of 10 kHz and 16 bits.

In preparation for investigating neural network techniques for Arabic phoneme recognition, 2307 pure Arabic phoneme samples were manually extracted in the frequency domain to represent the 33 different Arabic phonemes. These phoneme samples were randomly chosen from the 340 Arabic words in the KAPD. 28 phonemes representing the essential 28 Arabic letters plus 3 phonemes to represent the short vowels (a, i, u) were extracted. In addition, as the letter A in the Arabic language has two different pronunciations, an extra phoneme was also used for the long form of letter A to represent this sound. Also, the silence between phonemes and words is represented as a phoneme. A total of 2307 phonemes were therefore constructed to build an Arabic phoneme database.

## *2.2 Feature Extraction*

Three sub-classes of dataset were extracted from the single 2307 phonemes dataset to produce three 2307 datasets respectively representing phonemes sampled using frames of length 50, 100, and 180 samples respectively; with 50% overlap. Mel-Frequency Cepstrum Coefficients (MFCCs) features were extracted from the speech signal frames for each pattern of all three datasets. More than one frame for each phoneme were extracted from the phoneme database, and each frame was characterized by 12 cepstral coefficients excluding the c0 (frame energy). The c0 coefficient was eliminated from this dataset because it represents the power of speech (the energy of speaker) and this characteristic is not needed in speech recognition. The maximum frame length used for each dataset was chosen to be less than the minimum phoneme length in that dataset. After the feature extraction stage each frame was normalized using (1).

$$\text{Normalized feature} = \frac{\text{feature- minimum cepstral coefficient}}{(\text{maximum cepstral coefficient- minimum cepstral coefficient})\,\textit{in dataset}} \quad (1)$$

## *2.3 Phoneme Classifiers*

The three normalized and optimised datasets, described above, were used to train and test three different architectures for the Multilayer Perceptron (MLP) Neural Network (NN) phoneme recognition system. The structure of the MLP NN was thus optimised by empirical experimentation over many structures of MLP NN.

### 2.3.1 The Initial Architecture

The first structure of MLP NN selected was a single three layer (input, hidden and output layer) network with 11 nodes for the input layer, 20 nodes for the hidden layer and 33 nodes for the output layer- see Fig. 3.

Five speaker dependent neural networks were trained using the three datasets of each speaker separately. A summary of the results from testing these five networks is shown in Table 1.

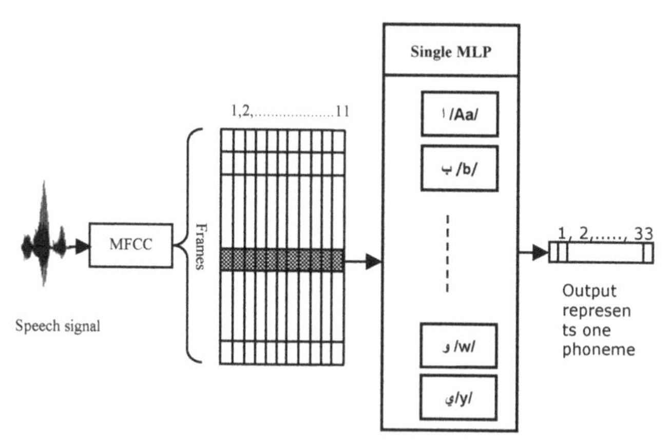


**Fig. 3.** The proposed single MLP phoneme classifier structure.

### 2.3.2 Category-Based Phoneme Recognition System

As several phonemes are very similar in pronunciation (i.e. nasals phonemes 'm' and 'n', short vowel phonemes 'a', 'i' and 'o' and long vowels phonemes 'A', 'y' and 'w'), the recognition process has difficulty in differentiating these phonemes. Therefore, a recognition principle based on dividing the phonemes into several groups, depending on the similarity in pronunciation, could make the recognition problem easier to solve [8]. Fig. 4 shows the architecture of the proposed technique.

In this phoneme classifier structure, the Arabic phoneme database was divided into five categories depending on the similarities between phonemes in pronunciation. The first dataset contains the 8 stops phonemes class:'ء,^ ', 'ب,b', 'د,d', 'ت,t', 'ض,D', 'ط,T', 'ق,q', 'ك,k', the second dataset contains the 13 fricatives phonemes class:'ث,v', 'ج,j', 'ح,H', 'خ,x', 'ذ,!', 'ز,z', 'س,s', 'ش,$', 'ص,S', 'غ,g', 'ف,f', 'ه,h', 'ع,E', the third dataset is nasals and space phonemes class: 'م,m', 'ن,n', 'فراغ,'' ', the fourth dataset contains: 'ر,r', 'ظ, Z', 'ل,l', and the fifth dataset represent the long and short vowels phonemes class: 'ا,A', 'و,w', 'ي,y', 'َ,a', 'ِ,I', 'ُ,u'.

In this structure, each class phoneme dataset was used to train and test a single Multilayer Perceptron (MLP) Neural Network (NN) phoneme recognition system (see Fig. 4). Five MLP NN were used to construct an individual identification system; each with the same structure: 11 nodes in the input layer, 50 nodes in the hidden layer, and 33 nodes in the output layer.

A complete identification system was built from the five phoneme class identification systems. The main point of this structure was to take a final decision for the frame recognition system. Therefore, each frame was tested on the five phoneme classification networks and then the maximum likelihood decision was taken form the five outputs produced

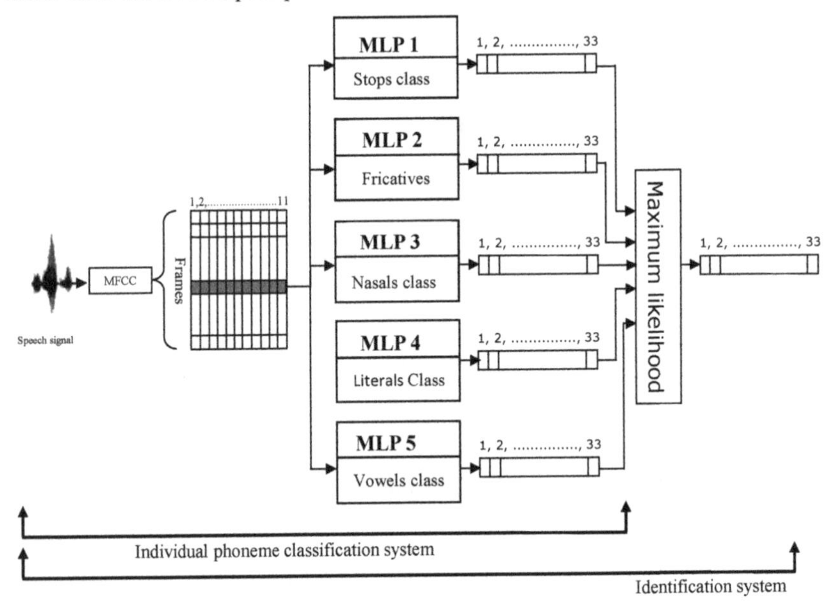


**Fig. 4.** Category-based phonemes identification classifier system.

### 2.3.3 Individual Phoneme Classifier System

Computational complexity is one of the most difficult issues for large vocabulary continuous speech recognition [21]. This computational complexity could be decreased by designing small networks to classify each phoneme in separate network.

In this technique, 33 individual phoneme datasets were used to train 33 separate MLP NNs. Each MLP NN was trained to recognize one phoneme from all other

Arabic phonemes (see Fig. 5). The structure of each MLP NN was the same for all 33 NNs, 11 input nodes, 7 hidden nodes, and one output node.

The performance measure, described by Equation 2 was used to optimise the MLP structures. Each phoneme dataset was trained on the same structures of three layers MLP NN, with 11 features in input layer, from 0 to 15 nodes in the hidden layer and one output in the output layer.

The decision for selecting the number of nodes in the hidden layer was taken experimentally according to the Minimum Average Error Rate (MAER) for each output for each structure. Seven was found to be the optimum number of hidden nodes.

$$\text{Performance}_{phoneme\ i}\ (\%) = \left[1 - MAER_{phoneme\ i}\right] * 100 \tag{2}$$

$$(\text{MAER})_{phoneme\ i} = \frac{\text{FRR}_{phoneme\ i} + \text{FAR}_{phoneme\ i}}{2} \tag{3}$$

Where:

FRR: False Reject Rate.
FAR: False Accept Rate.

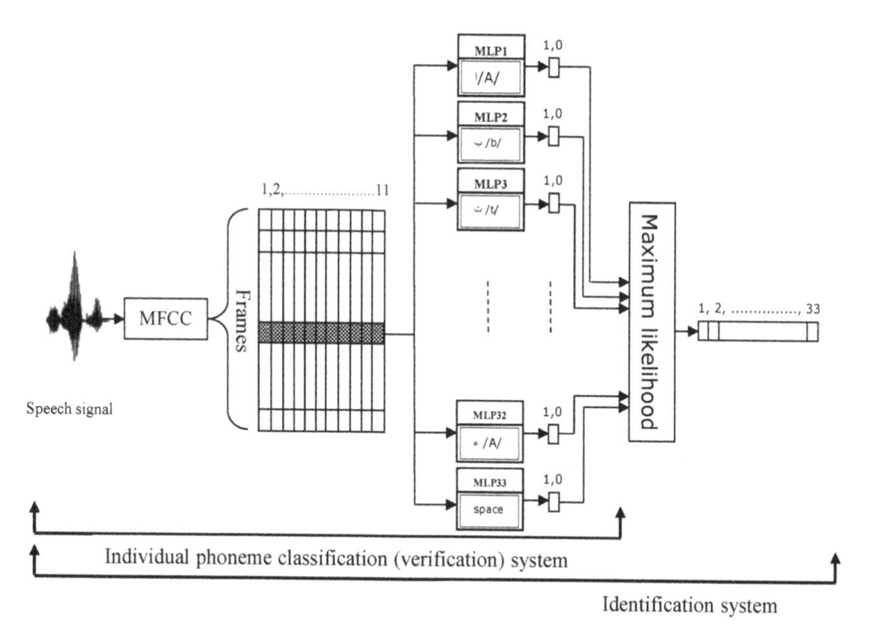


**Fig. 5.** Individual phoneme classifier identification system.

An identification system was built from the 33 phoneme classifier. The main point of this structure is to take a final decision for the frame recognition system. Therefore, each frame was tested on the 33 phoneme classifiers and then the maximum likelihood decision was taken for the 33 results produced.

# 3 Experimental Results

## *3.1 Speaker dependant system classifier*

The dataset was build by extracting phonemes from six speakers. Table 1 shows the results obtained for a speaker dependent system using architecture 1. These show the average classification accuracy, for each speaker, for all the phonemes in their individual datasets and the individual frames within all the phonemes in the same datasets. Post processing enhancement was done within the phonemes itself, where the phoneme classification of all frames within a phoneme was selected as that given by the majority of frames within that phoneme. The main reason for including this step was to overcome the effect of distorted/noisy frames within the phoneme.

**Table 1.** The average results of frames and phonemes recognition, where training dataset equal testing dataset. The frame size is 60 samples per frame.

| Speaker No | No of phonemes | | Classification accuracy of phoneme after the post-processing | Classification accuracy of frames per phonemes after the post-processing |
|---|---|---|---|---|
| | Train | Test | | |
| 1 | 229 | 310 | 72.85 | 74.09 |
| 2 | 408 | 408 | 85.67 | 87.82 |
| 3 | 317 | 317 | 85.88 | 88.32 |
| 4 | 315 | 315 | 79.20 | 81.72 |
| 5 | 322 | 322 | 81.75 | 83.47 |
| 6 | 170 | 170 | 95.27 | 96.51 |

## *3.2 Speaker Independent Classifier*

The full multi-speaker Arabic phonemes database was constructed from these datasets. However, in order to provide a true speaker independent test, speakers 1, 3, 4 and 5 were selected to construct the Arabic phoneme training and testing databases, and speakers 6 and 7 were kept aside for a true speaker independent test.

### 3.2.1 Individual Classifier System Results

A comparison between the frame classification results of the three architectures is shown in Table 2. The results shown are all without the post-processing enhancement as this requires knowledge of the phoneme boundaries which may not be available for unseen speech samples.

**Table 2.** Shows a comparison between the results of the average classification accuracy of frames per phoneme for the three architectures of the MLP NN for individual systems. The results are shown without post processing, and frame size is 64 samples per frame.

| No of phonemes | | The accuracy of frames per phonemes | | |
|---|---|---|---|---|
| Train | Test | Architecture 1[1] | Architecture 2[2] | Architecture 3[3] |
| 634 | 710 | 47.36 | 64.58 | 87.90 |

#### 3.2.1.1 Architecture1: The single MLP NN system

The average result of the phoneme recognition for the single MLP NN system was 48.55% and 66.09% before and after the post-processing enhancement respectively. The frame recognition accuracy for this system was also 47.36% and 72.23% before and after the post-processing enhancement respectively.

#### 3.2.1.2 Architecture2: The Category-based phonemes recognition system

Table 3 shows the average results of frames and phonemes for each class. The average result of the phoneme recognition for the Category-based phonemes

[1] Architecture 1: Single MLP NN.
[2] Architecture 2: Category-based phonemes recognition system.
[3] Architecture 3: Individual Phoneme classifier system.

recognition system was 68.78% and 74.37% respectively before and after the post-processing enhancement. The frame recognition accuracy for this system was also 64.58% and 78.83% before and after the post-processing enhancement respectively.

**Table 3.** The average results of frames and phonemes recognition. These datasets does not include transition areas for the phonemes and the frame size is 64 samples per frame.

| class No | No of phonemes | | Classification accuracy of phoneme after the post-processing | Classification accuracy of frames per phonemes after the post-processing |
|---|---|---|---|---|
| | Train | Test | | |
| 1 | 129 | 158 | 61.02 | 72.25 |
| 2 | 233 | 264 | 80.61 | 81.00 |
| 3 | 130 | 138 | 62.57 | 65.84 |
| 4 | 61 | 66 | 90.00 | 89.55 |
| 5 | 81 | 84 | 88.35 | 96.95 |

3.2.1.3 Architecture3: Individual Phoneme classifier system

The average individual phoneme classification performance of this system gives 87.90% frame recognition.

### 3.2.2 Overall Identification System Results

As indicated earlier, architectures 2 & 3 can be converted into identification systems by the inclusion of a maximum likelihood stage following the classifier output. Table 4 gives a compassion between the results of the three identification systems.

**Table 4.** Shows a comparison between the results of the average accuracy of frames per phoneme for the three architectures of MLP NN for identification system, and the results are shown without post-processing. The frame size is 64 samples per frame.

| No of phonemes | | The accuracy of frames per phonemes | | |
|---|---|---|---|---|
| Train | Test | Architecture1 | Architecture2 | Architecture3 |
| 634 | 710 | 47.36 | 44.58 | 46.63 |

It can be seen from the results, indicate, that the results for all three identification systems are similar.

## 4 Conclusion

This paper presents three different biologically inspired neural network identification system structures, using the Mel-Frequency Cepstrum Coefficients (MFCCs) features, for Continuous Arabic Speech Recognition.

The results of the three neural networks identification systems, trained on an Arabic phoneme database (APD) extracted manually from the King Abdulaziz Arabic Phonetics Database (KAPD), is as follows: the three phoneme classification systems (single MLP identification system, category-based phonemes recognition system, and individual Phoneme classifier system) have comparable performance as complete identification systems. The performances of these systems were 47.52%, 44.58%, and 46.63% frame recognition respectively. On the other hand, the best frame classification performance was with the individual Phoneme classifier architecture which gives 87.90% frame recognition. Therefore, these results indicate that the individual phoneme classification architecture (single MLP NN structure) is the optimum choice for Arabic phonemes verification/identification. This is because it has the best frame classification performance and comparable identification performance.

However, it must be acknowledged that the performance of even the best system is still short of an equivalent GMM based phoneme recognition system. One reason for this disparity may be due to the limitation of MLP systems when being used to classify time varying signals. To address this, our future work will involve applying Echo State Networks as frame/phoneme recognisers within our system. Their proven ability to act as time-series pattern classifiers should help improve the recognition rate of the more variable Arabic phoneme classes.

Finally then, it can also be concluded that, like all speech recognition research, Arabic speech recognition is a big challenge due to the similarity between phoneme pronunciation and the effect of time varying frequency patterns for different phonemes.

**Acknowledgments** We wish to thank Tariq Tashan for his assistance and constructive suggestions during this work.

## References

1. Yuk D.: Robust speech recognition using neural networks and hidden markov models-adaptations using non-linear transformations [dissertation]. New Jersey: The State University of New Jersey; 1999.
2. Tursun, N., Silamu, W. In: Large vocabulary continuous speech recognition in uyghur: Data preparation and experimental results. Network computing and information security (NCIS), international conference; China. ; 2011. p. 197-200.

3. Al-manie, M.A., Alkanhal, M.I., Al-ghamdi, M.M. In: Automatic speech segmentation using the arabic phonetic database. Proceedings of the 10th WSEAS international conference on AUTOMATION & INFORMATION; ; 2006. p. 76-9.
4. Renals, S.: Radial Basis Function Network For Speech Pattern Classification. [Internet]. 1989;25(7):437; 439. Available from: internal-pdf://18-0638587909/18.pdf.
5. Maheswari, N.U., Kabilan, A.P., Venkatesh R.: Speaker independent phoneme recognition using neural networks. Journal of Theoretical and Applied Information Technology (JATIT). 2009:230-5.
6. Chen, W., Chen, S., Lin, C.: A speech recognition method based on the sequential multi-layer perceptrons. Neural Networks. 1996;9(4):655-69.
7. Alghmadi, M.M.: KACST arabic phonetics database. Congress of Phonetics Sci. 2003;15:3109-12.
8. Mosa, G.S., Ali, A.A.: Arabic phoneme recognition using hierarchical neural fuzzy petri net and LPC feature extraction. Signal Processing: An International Journal (SPIJ). 2009;3(5):161-71.
9. Anwar, M.J., Awais, M.M., Masud, S., Shamail, S.: Automatic arabic speech segmentation system. International Journal of Information Technology. 2006;12(6):102-11.
10. Waheed, K., Weaver, K., Salam, F.M.: A robust algorithm for detecting speech segments using an entropic contrast. In proc. of the IEEE Midwest Symposium on Circuits and Systems. Lida Ray Technologies Inc. 2002;45.
11. Hong, L., Yanmin, Q., Jia, L.: English speech recognition system on chip. Tsinghua Science & Technology. 2011;16(1):95-9.
12. Reynolds, T.J., Antoniou, C.A.: Experiments in speech recognition using a modular MLP architecture for acoustic modelling. Information Sciences. 2003;156(1-2):39-54.
13. Jou, S., Schultz, T., Walliczek, M., Kraft, F., Waibel, A.: Towards continuous speech recognition using surface electromyography. Interspeech. 2006:573-6.
14. Kirchhoff, K., Vergyri, D.: Cross-dialectal data sharing for acoustic modeling in arabic speech recognition. Speech Communication. 2005;46(1):37-51.
15. Szczurowska, I., Kuniszyk-Jóźkowiak, W., Smołka, E.: The application of kohonen and multilayer perceptron networks in the speech nonfluecy analysis. Archives of Acoustics. 2006;31(4):205-10.
16. Nakamura, M., Tsuda, K., Aoe, J.: A new approach to phoneme recognition by phoneme filter neural networks. Information Sciences. 1996;90(1-4):109-19.
17. Koizumi, T., Mori, M., Taniguchi, S., Maruya, M. In: Recurrent neural networks for phoneme recognition. Department of information science, fourth international conference; 3-6 Oct 1996; Fukui University, Fukui, Japan. Spoken Language, ICSLP 96; 1996. p. 326-9.
18. Skowronski, M.D., Harris, J.G.: Automatic speech recognition using a predictive echo state network classifier. Neural Networks. 2007;20(3):414-23.
19. Ismail, S., Bin Ahmad, A.M. In: Recurrent neural network with backpropagation through time algorithm for arabic recognition. ; 2004.
20. Bengio, Y.: Neural Networks For Speech And Sequence Recognition. first edition ed. International Thomson Computer Press; 1996.
21. Sweeney, L., Thompson, P.: Speech perception using real-time phoneme detection: The BeBe system. Laboratory for Computer Science, Massachusetts Institute of Technology, Cambridge, MA 02139, USA; 1997.

# A cross-study of Sentiment Classification on Arabic corpora

**A. Mountassir, H. Benbrahim and I. Berrada**[1]

**Abstract** Sentiment Analysis is a research area where the studies focus on processing and analyzing the opinions available on the web. Several interesting and advanced works were performed on English. In contrast, very few works were conducted on Arabic. This paper presents the study we have carried out to investigate supervised sentiment classification in an Arabic context. We use two Arabic Corpora which are different in many aspects. We use three common classifiers known by their effectiveness, namely Naïve Bayes, Support Vector Machines and k-Nearest Neighbor. We investigate some settings to identify those that allow achieving the best results. These settings are about stemming type, term frequency thresholding, term weighting and n-gram words. We show that Naïve Bayes and Support Vector Machines are competitively effective; however k-Nearest Neighbor's effectiveness depends on the corpus. Through this study, we recommend to use light-stemming rather than stemming, to remove terms that occur once, to combine unigram and bigram words and to use presence-based weighting rather than frequency-based one. Our results show also that classification performance may be influenced by documents length, documents homogeneity and the nature of document authors. However, the size of data sets does not have an impact on classification results.

## 1 Introduction

Nowadays, the web is no longer just a source of information for internet users; it represents also a space where simple users can provide information. With the emergence of social media (such as social networking sites, online news sites, online web forums, personal blogs and online review sites), internet users are more and more invited to express their opinions, post comments or share experiences about any topic. Therefore, the online opinion has become an important currency for many researches especially in the field of Opinion Mining (OM) and Sentiment Analysis (SA).

---

[1] ENSIAS, Mohamed 5 University, Rabat, Morocco
asmaa.mountassir@gmail.com, benbrahim@ensias.ma, iberrada@ensias.ma

As a definition, Sentiment Analysis, or Opinion Mining, is a research area that consists of mining in the web opinions expressed by internet users about a given subject. The comments that contain these opinions are collected in order to analyze them and perform a number of tasks on them. Among these tasks we find Subjectivity Analysis [1], Polarity Classification or Sentiment Classification [2], identification of attitudes and opinions intensity [3], and Summarization [4]. We stress that there are three levels of analysis; namely document, sentence and term-level. Working on document-level means that we are interested in identifying the polarity (positive or negative) of the whole document [2]. But if the analysis is conducted on sentence-level, the goal is to determine the polarity of each sentence in the document [1]. The work on term-level is known as Semantic Orientation [5]. It focuses on identifying the polarity of each term (word) in the document. Note that we mean by document a text unit of analysis. In our study, documents are comments or posts written by internet users.

Most of research works performed in this field have been conducted on some European languages (especially English) and Asian ones (Japanese and Chinese). Nevertheless, very few works were performed on Morphologically-Rich Languages (such as Arabic, Hebrew and Czech) [6]. This is due to two major factors; the problem related to the lack of content available on the web on one hand, and the problem related to the complexity of processing these languages on the other hand. As a definition, a Morphologically-Rich Language (MRL) is a language in which significant information concerning syntactic units and relations is expressed at word-level [6]. The Arabic language presents an important case of MRL since it was judged by Internet World Stats as the language with the fastest growth rate, in terms of internet users, in the last 11 years[2]. As Arabic language has many forms [7], we specify that our study focuses on Modern Standard Arabic form (MSA).

The aim of this paper is to report details of the study that we have conducted in order to investigate SA in an Arabic context. We have focused on classification by polarity on document-level, i.e. documents are classified as either positive or negative. We have applied three standard supervised classifiers: Naïve Bayes [8], Support Vector Machines [9] and k-Nearest Neighbor [10]. We have three main goals. First, a comparative study of the three classifiers in classifying sentiments in an Arabic context. Second, the investigation of some settings to identify the best of them. These settings concern stemming type, weighting scheme, term frequency thresholding and n-gram words model. Third, the comparison between results obtained on two different Arabic corpora.

Among works that were performed on sentiment classification on Arabic data sets we find that of Abbasi et al. [11] who worked on document-level by using syntactic and stylistic features and a feature selection algorithm that they developed and named EWGA. Their work was conducted on comments of a Middle Eastern extremist group. Thanks to EWGA algorithm, they achieved high

[2] http://www.internetworldstats.com/stats7.htm

results. Abdul-mageed et al. [1] have worked on sentence-level to determine the document subjectivity (subjective vs. objective) by using Support Vector Machines as a classifier and different types of features (including n-grams). Their study was conducted on the Arabic Tree Bank corpus. The closest work to our study is that of Rushdi-Saleh et al. [12] who worked on the Opinion Corpus for Arabic (OCA), a corpus that they built from movie-reviews. They used Naïve Bayes and Support Vector Machines as classifiers and n-gram words as features to classify OCA documents on document-level. They found that stemming is not recommended in Arabic context.

The remainder of this paper is organized as follows. The second section deals with preprocessing and document representation processes. The third section describes the experimental environment and the performed experiments, then it discusses the obtained results. We finish by a conclusion and future works in the last section.

## 2 Preprocessing and Document Representation

Before we could classify a given document, it is essential to first apply the preprocessing process which consists of cleaning, normalizing and preparing the text document to the classification step. We describe below the most common methods of preprocessing phase.

- *Tokenization:* Basically, this process used to transform a textual document to a sequence of tokens separated by white spaces. If needed, we can remove special characters, numbers and punctuation marks. Nevertheless, in Arabic processing field, this process is called *Segmentation* [1]. It consists of separating a word from its clitics (i.e. proclitics and enclitics) and the determiner Al. As a definition, proclitics and enclitics are a kind of prefixes and suffixes, respectively, which can be associated to Arabic words. These clitics can be prepositions, conjunctions, future markers, etc. As examples, the segmentation of the word "وسنذهب" (and we will go) gives "نذهب" (we go), and the segmentation of the word "كتابهم" (their book) gives "كتاب" (book). In the first example we remove proclitics, while in the second example we remove enclitics.
- *Stemming:* For Arabic language, there are two different morphological analysis techniques; namely stemming and light-stemming. Stemming reduces words to their stems [13; 14]. Light-stemming, in contrast, removes common affixes from words without reducing them to their stems [15]. Stemming would reduce the words « المدرسة » (the school), « المدرس » (the teacher) and « الدراسة » (the study) to one stem « درس » (to study).While light-stemming would reduce the words « الدراسات » (the studies) and « مدرسان » (two teachers) to respectively « دراسة» (a study) and « مدرس » (a teacher). The main idea for using light stemming [13; 14] is that many word variants do not have similar meanings or semantics although these word variants are generated from the same root. For example, the

stemming of the two words « رائع » (wonderful) and « مروع » (horrible) gives the word « روع » (horrify). We can see that the polarity of « رائع » (wonderful) is inversed by stemming. Hence, light-stemming allows retaining words' meanings.

- *Stop Words removal:* Stop Words refer to function words (such as articles, prepositions, conjunctions, and pronouns) which provide structure in language rather than content, they refer also to words that do not have an impact on categories discrimination. Note that the list of stop words (called stoplist) is typically established manually, it is domain and language-specific.
- *Term Frequency Thresholding:* This process used to eliminate words whose frequencies are either above a pre-specified upper threshold or below a pre-specified lower threshold. This process helps to enhance classification performance since terms that rarely appear in a document collection will have little discriminative power and can be eliminated. Likewise, high frequency terms are assumed to be common and thus not to have discriminative power either.

After pre-processing stage, we proceed to the dictionary building. This dictionary consists of all terms that occur in all pre-processed documents. These terms can be words and/or expressions depending on the adopted model. These terms are called features. Let $\{f_1,\ldots,f_m\}$ denote the feature set.

The next step consists of representing our textual documents by the use of the conventional Vector Space Model [16]. We map each document $d_j$ onto its vector $(n_1(d_j),\ldots,n_m(d_j))$; where $n_i(d_j)$ denotes the weight of feature $f_i$ with respect to document $d_j$.

## 3 Experiments

In this section, we report details about the experiments that we have carried out. First, we present the data collections that we have used. Second, we describe the experiments' environment, i.e. some details about the classifiers, the validation method and performance measures that we have used. The results of our experiments are reported and discussed at the end of this section.

### *3.1 Data Collections*

#### 3.1.1 ACOM

As mentioned in the introduction, Arabic suffers from lack in digital content on the web. Indeed, we could not find Arabic sites that invite internet users to let or express their opinions such as IMDb (www.imdb.com) or Epinions (www.epinions.com). Moreover, and after mining in several Arabic web forums such as Maktoob (www.maktoob.com) and Koora (www.koora.com), we concluded that these forums could not be a good data source for our study for two reasons. On one hand, the Arabic used by internet users in these forums is generally dialectal; this could not help us as we are interested in MSA. On the other hand, we noted that for discussions, just the first post has an evaluable content, otherwise the replies are just greeting expressions like for instance « مشكووووور » (Thaaaanks) and « بارك الله فيك أخي » (May God bless you brother).

The source that we judged suitable for our study was the online forums of Aljazeera's site[3]. In order to leave a comment, the user has to write in MSA according to the participation terms. We collected our data from Aljazeera's polls and forums.

We called our corpus ACOM (Arabic Corpus for Opinion Mining). It consists of two data sets; each one is domain-specific. The first data set (called DS1) falls within movie-review domain, it consists of 594 documents. While the second data set (called DS2) is sport-specific data set, it consists of 1492 comments. We point out that, though DS1 pertains at the origin to the movie-review domain, it cannot be comparable to the common movie-review data sets since the participants in this discussion have deviated from the original goal of the discussion which was the reviewing of a famous series. The comments posted by internet users in this discussion were mostly about religious issues. This will affect our classification tasks as we will see in the following.

Since our study deals with supervised learning, we had to label manually our data sets. The categories to consider are POSITIVE, NEGATIVE, NEUTRAL and DIALECTAL. POSITIVE category includes all comments that reflect a positive opinion regardless of the opinion object. It is the same for the NEGATIVE category. In contrast, the NEUTRAL category includes non-opinionated comments (as example a comment that reports just a quranic verse), comments reflecting neutral opinions (as the expression « والله لا أدري الصواب من الخطأ » (I swear that I cannot distinguish what is right from what is wrong)) and comments comprising a mixture of positive and negative opinions (as in « هذا المسلسل رائع لكن لا اؤيد تجسيد الصحابة » (This series is wonderful, but I cannot stand the incarnation of Prophet companions)). Finally, we classified as DIALECTAL each comment that contains expressions in dialectal Arabic. This category used to clean our data sets from comments not written in MSA.

It is of worth to note that, during the annotation phase, we noticed that an opinion (positive or negative) is not often expressed explicitly, but can be expressed in a subtle manner. This is why the task of sentiment classification is more complicated than the other categorization types. We give as examples these two comments that have been classified as negative « لا حول ولا قوة إلا بالله » (There is no power but from God) and « حسبي الله ونعم الوكيل » (God is Sufficient for me; most Excellent is He in Whom I trust). We note also that, for DS2, we found that a large number of negative comments were ironic. If we do not take in account the discussion context while reading these comments, we would consider them as

[3] http://www.aljazeera.net

positive. But in fact they carry a negative sense. We can illustrate this case by this example « هذا هو رايي المتواضع في عبقريتك وشكرا لإتحافنا بعبقريتك أيها البروفيسور » (This is my humble opinion about your genius; I thank you dear professor because we benefited from your intelligence).

In table 1, we present the number of comments per category for each data set of ACOM corpus. As we can observe from this table, the dialectal comments of each data set represent less than 7% of the whole data set. This poor percentage proves that Aljazeera's site is a good data source for our study. However, we can see that, for DS1, the negative comments are much larger than the positive ones. The reason of the negative comments predomination is perhaps the fact that Aljazeera's topics are usually related to the Arab-Muslim world's problems in different domains. This represents a drawback of Aljazeera's comments.

Table 1. Number of comments per category for each data set of ACOM

| Data set | POSITIVE | NEGATIVE | NEUTRAL | DIALECTAL | Total |
|---|---|---|---|---|---|
| **DS1** | 184 | 284 | 106 | 20 | 594 |
| **DS2** | 486 | 517 | 391 | 98 | 1492 |

As our study deals with a binary classification where the categories are POSITIVE vs. NEGATIVE, we have eliminated (for this study) comments of NEUTRAL and DIALECTAL categories. The neutral documents will be introduced in next studies as they represent an important percentage (up to 26.2%) of the whole data set. Afterward, we proceeded to data set balancing with respect to categories distribution. We have eliminated a number of negative comments from each data set in a way to equalize the number of documents for each category. Note that our data sets should be (even approximately) balanced so as to not bias the classifiers' learning. Table 2 gives more details about the resulted data sets; it shows for each data set and for each category the total number of documents and the average of tokens number per document. It also shows the total number of documents for each data set.

Table 2. Statistics on each data set of ACOM

| | POSITIVE | | NEGATIVE | | Total Doc |
|---|---|---|---|---|---|
| | Nb Doc | AVG Tok | Nb Doc | AVG Tok | |
| **DS1** | 184 | 60 | 184 | 62 | 368 |
| **DS2** | 486 | 57 | 514 | 66 | 1000 |

### 3.1.2 OCA

OCA (Opinion Corpus for Arabic) is a corpus built by Rushdi-Saleh et al. [12]. It consists of 500 movie-reviews collected from several Arabic blog sites and web pages. These documents were annotated automatically on the basis of the ratings associated to the reviews. Table 3 gives some statistics about the OCA corpus.

Table 3. Statistics on OCA

| | POSITIVE | | NEGATIVE | | Total Doc |
|---|---|---|---|---|---|
| | Nb Doc | AVG Tok | Nb Doc | AVG Tok | |
| **OCA** | 250 | 485 | 250 | 378 | 500 |

As we can see from tables 2 and 3, DS1, DS2 and OCA have different sizes. Moreover, we conclude that OCA comments are much longer than ACOM's ones. So, in this study, we will have the opportunity to deal with different data sets with respect to document number and document length.

## *3.2 Classification and Evaluation*

In this study, we are interested in a binary classification, where each document is assigned to one of the two categories POSITIVE vs. NEGATIVE.

In order to perform our tasks of preprocessing and classification, we have used the data mining package Weka[4] [17]. We have used three standard classifiers; namely Naïve Bayes (NB), Support Vector Machines (SVM) and k-Nearest Neighbor (k-NN). Each of these classifiers was shown as effective for several classification tasks [2; 18; 19; 20]. For the NB classifier, we have used a kernel estimator, rather than a normal distribution, for numeric attributes [21]. Concerning SVM classifier, we have used a normalized polynomial kernel with a Sequential Minimum Optimization (SMO) [22]. Furthermore, the k-NN classifier that we have used is based on a linear search with a cosine-based distance [23].

As a validation method, we have used 10-fold cross-validation [8]. The metric we have used to evaluate our results is the Macro-averaged F1 measure [20].

## *3.3 Experimental Design*

The present paper has three major goals. The first is about identifying the best settings for sentiment classification on Arabic corpora. These settings are about stemming type, weighting scheme, term frequency thresholding and feature type (n-gram words). The second goal is related to a comparison, in an Arabic context, between three common classifiers known by their effectiveness in text categorization. The third is about comparing the classification effectiveness

[4] http://www. cs.waikato.ac.nz/ml/weka

obtained on three different sized Arabic data sets. We note that, for all tested settings, stop words were removed.

As we advanced in section 2, there are two types of stemming in Arabic, namely stemming and light-stemming. In this paper, we investigate the contribution of each type in enhancing the classification results.

The weighting feature types that we studied are three:

- *Presence:* This weighting (called also binary weighting) focuses on presence rather than frequency term regarding a given document. It consists of assigning 0 as weight to a term if it does not occur in the document, otherwise it is assigned 1. This weighting was introduced originally in sentiment analysis by Pang et al. in [2].
- *Normalized Frequency:* This weighting derives from frequency weighting, where each term is given as weight its frequency (number of occurrences) regarding a given document. We normalize these weightings by dividing them by the length of the document. We mean by length of document the number of its terms. We use normalization to avoid problems that can result from differences in documents size. Indeed, a term can occur many times in a long document, but few times in a short one. The importance of this term can be the same regarding the two documents, while, if we consider a frequency weighting, this term would not have the same importance regarding the two documents.
- *TFIDF:* This term refers to Term Frequency/Inverse Document Frequency [24]. For a term t, the TFIDF weight is assigned the value: $TFIDF(t) = TF(t) \times IDF(t)$; where TF represents the frequency of the term t (number of occurrences in the data set), and IDF measures the significance of this term. It is obtained by the following formula: $IDF(t)=\log(n/n(t))$; where n represents the total number of documents in the data set, and n(t) denotes the number of documents where term t occurs. The higher is the value of TFIDF for a term, the more it is considered significant.

For term frequency thresholding, we were interested in the minimum frequency. The values we considered were 1, 2 and 3. The idea is to eliminate from dictionary the term whose frequency is strictly less than the specified threshold. The objective was to find the best threshold for the three classifiers.

Concerning feature types, we focused on n-gram words [25]. In this study, we first consider the use of unigrams alone. Afterward, we investigate the adding of bigrams. Finally, we examine the combination of unigrams, bigrams and trigrams. We specify that n-gram extraction is performed before stemming and stop words removal.

In order to investigate these four settings (parameters), we change each time the value of a given parameter and we keep fixed values for the other parameters.

## 3.4 Results

In figures 1, 2 and 3, we present the obtained results (in terms of macro-averaged F1) of the three classifiers for the different settings for respectively DS1, DS2 and OCA data sets. We specify that, in these figures, "Stem" and "LStem" refer respectively to stemming and light-stemming processes. "Pres", "NormFreq" and "TFIDF" denote the three weighting schemes that we have tested. "FreqN" means that we apply a thresholding of minimum frequency; N is the value of the threshold. "Uni", "Bi" and "Tri" correspond respectively to unigrams, bigrams and trigrams.

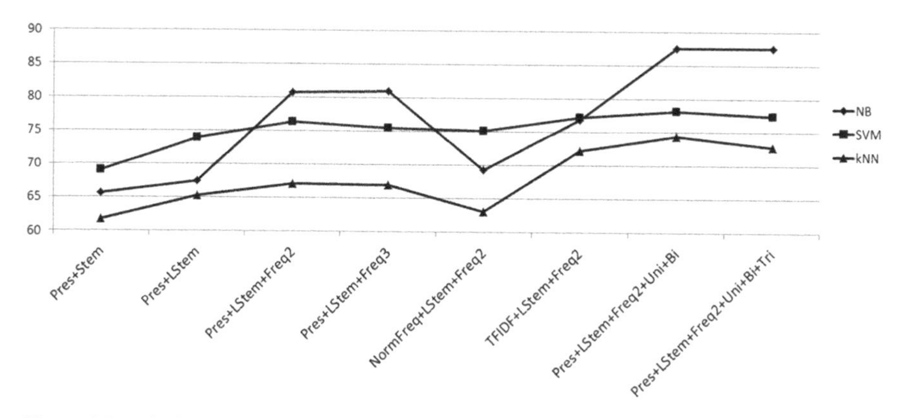


Figure 1 Results in F1 measure for DS1.

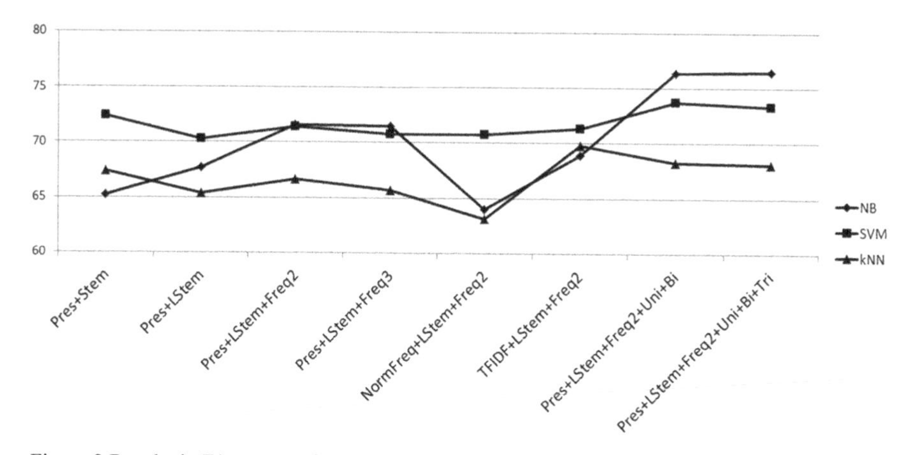


Figure 2 Results in F1 measure for DS2.

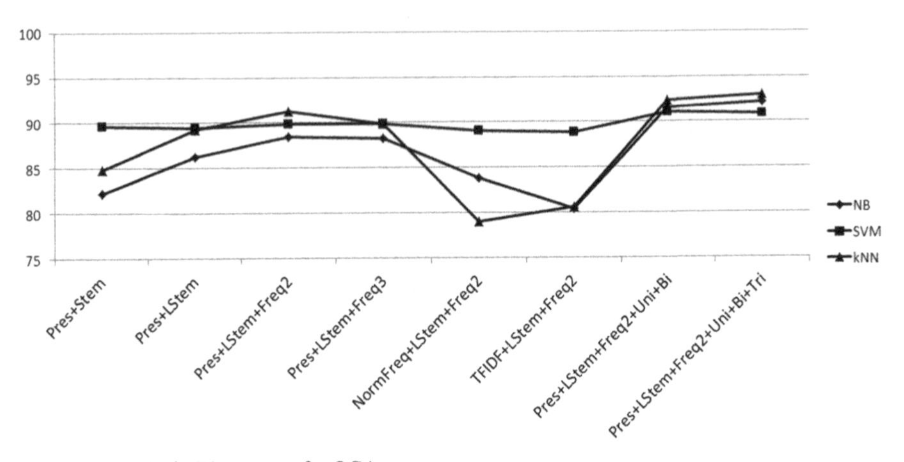


Figure 3 Results in F1 measure for OCA.

The first test is about stemming. According to figures 1, 2 and 3, we can see that, in most of times, light-stemming gives better results than stemming. This result is logic since light stemming allows retaining words meanings. Hence, it is predictable that light-stemming performs well than stemming. This is why we keep applying light-stemming for the following tests. We note that, to our best knowledge, there was no comparative study performed between stemming and light-stemming in the context of Sentiment Analysis.

The second test deals with thresholding; we remove terms that occur once in the data set (Freq2), then we remove those that occur twice (Freq3). We can observe, through figures 1, 2 and 3, that the elimination of terms that occur once (these terms are called hapaxes) improves clearly the results in comparison with the non-application of thresholding. However, the removal of terms that occur twice does not enhance the results. This can be explained by the fact that removing hapaxes means that we eliminate, among others, mistakes made by the authors who are just simple internet users. Indeed, these mistakes can be a source of noise for classification. Nevertheless, the elimination of terms that occur twice may remove significant features. This is why we keep applying the thresholding with a threshold equal to 2 for the rest of tests.

The following test concerns the weighting schemes. As we can see from figures 1, 2 and 3, presence weighting performs well for all classifiers. We can also see that TFIDF weighting allows to SVM and k-NN achieving good results. However, we note that weighting based on normalized frequency gives poor results for all classifiers. Hence, we keep using the presence weighting for next tests. It is of worth to note that this result is consistent with that reported by Pang et al. [2].

The last test has as goal to investigate the use of n-gram words. Figures 1, 2 and 3 show that adding bigrams to unigrams improves considerably the results in comparison with the use of unigrams only. However, and when we add trigrams, the performance is not enhanced for DS1 and DS2, but for OCA it is slightly improved. As the combination of unigrams, bigrams and trigrams is expensive and as it does not give considerable improvement, we keep the combination of unigrams and bigrams as feature model. We can interpret the contribution of bigrams to enhancing the results by the fact that, in contrast to unigrams, bigrams allow the management of negation. Indeed, by considering this example « لا أوافق » (I do not agree) which has a negative sense, if we consider unigrams only we will have just « أوافق » (I agree) since the word negation « لا » is a stop word. We can see that, as a consequence, the polarity is inversed. But if we use bigrams, the expression « لا أوافق » (I do not agree) will be kept as it is, and so the polarity is maintained.

We think that, in the context of Sentiment Classification, the negation treatment is the only contribution of bigram adding in comparison with the use of unigrams only. Indeed, to identify and classify sentiments, we do not need to identify some specific key words such as "Text Mining" and "Computer Science" that could be helpful in Text Categorization. This finding, about the combination of unigrams and bigrams, is different from the result reported by Pang et al. [2] since they found that the use of unigrams gives the best results. Our finding is also different from that reported by Rushdi-Saleh et al. [12] who stated that the obtained results with bigrams and trigrams were very similar to unigrams.

So far, the best setting identified for all classifiers and all data sets is the application of light-stemming, the elimination of hapaxes, the combination of unigrams and bigrams with the use of a presence-based weighting.

As a comparison between the three classifiers, we can observe through figures 1, 2 and 3 that the results regarding the classifiers performance are not the same. For OCA, the three classifiers seem to be competitive in most of the performed tests. The best results achieved are 91% by SVM, 92.2 by NB and 93% by k-NN. However, for ACOM (DS1 and DS2), just NB and SVM are competitively effective; k-NN seems to be less effective since it yielded poor results for almost all tests. The best results achieved, for ACOM, are 87.5% by NB, 78.2% by SVM, and 74.4% by k-NN. Hence, we can conclude that NB and SVM are the most effective. But we cannot draw a conclusion about k-NN performance since its behavior was not the same regarding the two corpora. Yet, we can give an interpretation to this observation. K-NN yielded high results on OCA perhaps because this data set is homogeneous. In other words, the documents of this data set belong purely to the movie-review domain; so the documents of each category can be very similar. As the k-NN classification is based on similarity, we can understand why this classifier performed well on this corpus. However, we can say that k-NN did not give good results on DS1 because the documents of this data set are not homogeneous. In fact, these documents were collected from a discussion about a historical film that has generated much noise in the Arab-Muslim world. So the documents domain is not purely movie-review but also religion and history. It is the same for DS2 since its documents were collected from discussions about 18 topics of sport.

To compare between results obtained on the three data sets (DS1, DS2 and OCA), we can say that the classifiers yielded better results on OCA in comparison with ACOM. The best results were achieved on OCA (up to 93% with k-NN). DS1 was in the second rank with up to 87.5% (a result achieved with NB). Finally, DS2 had the worst results with up to 76.4% (a result obtained by NB). We can explain this difference in results by authors' nature for each corpus. In fact, documents of OCA are produced either by bloggers or professional writers, while documents of ACOM are, generally, posted by mere internet users. So it it is obvious that documents of OCA are well written in comparison with those of ACOM which contain a large number of mistakes. Furthermore, documents of OCA are much longer than ACOM's ones (see tables 2 and 3). So we can draw the conclusion that the more the documents are long, the more the classification is effective. This finding is to be verified in future studies. There is another different aspect, between the two corpora, that concerns annotation. We recall that documents of ACOM were manually annotated by one annotator, while documents of OCA were automatically annotated on the basis of rating systems [12]. We are not sure that annotation method could have an impact on classification performance. This detail is also to be further studied in future works. Finally, we recall that a great part of negative documents of DS2 were written in an ironic manner, i.e. they seem to be positive while, in fact, they carry a negative sense. This type of documents represents a major challenge for classification. This may explain why this data set (DS2) had the worst results.

Concerning the impact of data set size on classification, we can say that the effectiveness of classification does not depend on the number of documents in the data sets. The size of DS2 (1000 documents) is the double of that of OCA (500 documents), and the size of OCA is larger than that of DS1 (368 documents), however the best results were obtained on OCA, then the ones on DS1; DS2 was the worst in terms of classification results.

We can say that our results (up to 93% in F1 measure) is competitive with those reported in the literature. Abbasi et al. [11] report a result of 93.6% in accuracy. Recall that they have used a feature selection algorithm that helps to eliminate non-informative features and hence improves considerably the results. Abdul-mageed et al. [1] achieved a result of 95.52% in F1 measure, but their work cannot be directly compared to ours since their work was on sentence-level, while ours was on document-level. Moreover, their corpus consists of newswire articles, while ours consists of comments written by simple internet users. Rushdi-Saleh et al. [12] report a result of 90.73% in F1 measure, recall that their experiments were carried out on OCA.

## 4 Conclusion and future works

The study that we have reported in this paper had as objective to investigate sentiment classification in an Arabic context. We carried out our study on two Arabic corpora with different sizes, namely ACOM and OCA. ACOM is a corpus that we have developed (from Aljazeera's site) and annotated manually with two main categories: POSITIVE and NEGATIVE. It consists of two data sets; DS1 which is a collection of 368 comments about a series reviewing and DS2 which is a collection of 1000 comments from sport domain. OCA is a collection of 500 movie-reviews collected by Rushdi-Saleh et al. [12]. We aimed to, first, investigate some settings to find those that yield the best results. The considered settings were about stemming type, term frequency thresholding, term weighting, and n-gram words model. We used for classification three common classifiers known by their effectiveness, namely Naïve Bayes, Support Vector Machines and k-Nearest Neighbor. The second goal of this study was to compare the effectiveness of these three classifiers in an Arabic context. The third goal that we chase was about the comparison between the results yielded on the two corpora. The obtained results show that the best setting for almost all classifiers on all data sets was the application of light-stemming, the elimination of hapaxes (as thresholding), the combination of unigrams and bigrams, and the use of a presence-based weighting. We note that the TFIDF-based weighting is also a suitable weighting for both SVM and k-NN. To compare between the three classifiers, results show that they did not have the same behavior toward the two corpora. Indeed, for ACOM, NB and SVM were competitively effective (up to 87.5%), but k-NN was less effective. Nevertheless, for OCA, the three classifiers showed a high effectiveness and were competitive along the performed tests (up to 92.2% for NB, 91% for SVM and 93% for k-NN). As a comparison between the two studied corpora, we found that the best results were obtained on OCA. This is due perhaps to the differences between the documents of the two corpora, namely the nature of document authors, document length and documents homogeneity. This hypothesis is to be further studied in future works. We show also that the size of data sets does not influence the performance of classification.

As future works, we look forward to further studying and verifying the different findings reported in this paper. These findings are about the factors that can influence classification performance, namely documents length, documents homogeneity, the nature of authors and annotation method. We will also verify the assumption about the non-influence of data set's size on classification effectiveness. We will also use features to handle negation; this might help us to avoid the use of bigrams since they are expensive.

## References

1. Abdul-Mageed, M., Diab, M.T., Korayem, M.: Subjectivity and Sentiment Analysis of Modern Standard Arabic. In Proc. ACL (Short Papers), pp.587-591 (2011).
2. Pang, B., Lee, L., Vaithyanathain, S.: Thumbs up? Sentiment classification using machine learning techniques. In Proceedings of the Conference on Empirical Methods in Natural Language Processing, pp.79-86 (2002).

3. Wilson, T.A., Wiebe, J., Hwa. R.: Recognizing strong and weak opinion clauses. In Computational Intelligence, 22(2):73–99 (2006).
4. Zhuang, L., Jing, F., Zhu, X.: Movie Review Mining and Summarization. In CIKM'06, Virginia, USA (2006).
5. Turney, P.: Thumbs up or thumbs down? Semantic orientation applied to unsupervised classification of reviews. In ACL'02, pp. 417–424 (2002).
6. Tsarfaty, R., Seddah, D., Goldberg, Y., Kuebler, S., Versley, Y., Candito, M., Foster, J., Rehbein, I., Tounsi, L.: Statistical parsing of morphologically rich languages (spmrl) what, how and whither. In Proc. NAACL HLT 2010 First Workshop on Statistical Parsing of Morphologically-Rich Languages, Los Angeles, CA, (2010).
7. Saad, M.K., Ashour, W.: OSAC: Open Source Arabic Corpora. In 6th ArchEng Int. Symposiums, EEECS'10 the 6th Int. Symposium on Electrical and Electronics Engineering and Computer Science, European University of Lefke, Cyprus, (2010).
8. Mitchell, T.: Machine Learning. McCraw Hill (1996).
9. Vapnik, V.: The Nature of Statistical Learning. Springer-Verlag (1995).
10. Dasarathy, B.V.: Nearest Neighbor (NN) Norms: NN Pattern Classification Techniques. McGraw-Hill Computer Science Series. Las Alamitos, California: IEEE Computer Society Press (1991).
11. Abbasi, A., Chen, H., Salem, A.: Sentiment analysis in multiple languages: Feature selection for opinion classification in web forums. ACM Trans. Inf. Syst., 26, pp.1–34 (2008).
12. Rushdi-Saleh, M., Mrtin-Valdivia, M.T., Urena-Lopez, L.A., Perea-Ortega, J.M.: Bilingual Experiments with an Arabic-English Corpus for Opinion Mining. In Proc. Of Recent Advances in Natural Language Processing, Hissar, Bulgaria, pp.740-745 (2011).
13. Duwairi, R., Al-Refai, M., Khasawneh, N.: Feature reduction techniques for Arabic text categorization. Journal of the American Society for Information Science. Volume 60 Issue 11, pp. 2347-2352 (2009).
14. Duwairi, R., Al-Refai, M., Khasawneh, N.: Stemming Versus Light Stemming as Feature Selection Techniques for Arabic Text Categorization. 4th Int. Conf. on Innovations in Information Technology. IIT'07. Pp. 446-450 (2007).
15. Khoja, S., Garside, R.: Stemming Arabic text. Computer Science Department, Lancaster University, Lancaster, UK (1999).
16. Salton, G.: Automatic Text Processing: The Transformation, Analysis, and Retrieval of Information by Computer. Reading, Pennsylvania: Addison-Wesley (1989).
17. Witten, I.H., Frank, E.: Data Mining: Practical machine learning tools and techniques. In 2nd Edition, Morgan Kaufmann, San Francisco, California (2005).
18. Abbasi, A., Chen, H.: Identification and comparison of extremist-group web forum messages using authorship analysis. In IEEE Intelligent Systems 20, 5, pp.67-75 (2005).
19. Zheng, R., Li, J., Huang, Z. Chen, H.: A framework for authorship analysis of online messages: Writing-style features and techniques. In Journal of the American Society for Information Science and Technology 57, 3, pp.378-393 (2006).
20. Yang, Y.: An evaluation of statistical approaches to text categorization. Inform. Retr. 1, 1–2, pp. 69–90 (1999).
21. John, G.H., Langley, P.: Estimating Continuous Distributions in Bayesian Classifiers. In: Eleventh Conference on Uncertainty in Artificial Intelligence, San Mateo, pp.338-345 (1995).
22. Platt, J.: Fast training on SVMs using sequential minimal optimization. In Scholkopf, B., Burges, C., and Smola, A. (Ed.), Advances in Kernel Methods: Support Vector Learning, MIT Press, Cambridge, MA, pp.185-208 (1999).
23. Salton, G., McGill, M.: Modern Information Retrieval. New York: McGraw-Hill (1983).
24. Sebastiani, F.: Machine learning in automated text categorization. In ACM Comput. Surv., Volume 34, Number 1, pp.1-47 (2002).
25. Shannon, C.: A mathematical theory of communication. In Bell System Technical Journal, 27, Bell System Technical Journal (1948).

# Towards the Profiling of Twitter Users for Topic-Based Filtering

Sandra Garcia Esparza, Michael P. O'Mahony and Barry Smyth

**Abstract** There is no doubting the incredible impact of Twitter on how we communicate, access and share information online. Currently users can follow other users or hashtags in order to benefit from a stream of data from people they trust or on topics that matter to them. However at the moment the *following* granularity of Twitter means that users cannot limit their information streams to a set of topics *by* a given user. Thus, even the most carefully curated information streams can quickly become polluted with extraneous content. In this paper we describe our initial steps to improve this situation by proposing a profiling approach that can be used for information filtering purposes as well as recommendation purposes. First, we demonstrate that it is feasible to automatically profile the interests of users by using machine learning techniques to classify the pages that they share via their tweets. We then go on to describe how this profiling mechanism can be used to organise and filter Twitter information streams. In particular we present a system that provides for a more fine-grained way to follow users on specific topics and thereby refine the standard Twitter timeline based on a user's core topical interests.

## 1 Introduction

Twitter has had an incredible impact on the world's information flow and communication landscape. Its power as an information sharing platform is only surpassed by its unique simplicity of form and function. Today Twitter's 140m users generate about 340 million tweets a day[1], 25% of which contain URLs [2]. However, notwith-

---

Sandra Garcia Esparza, Michael P. O'Mahony and Barry Smyth
CLARITY: Centre for Sensor Web Technologies, School of Computer Science and Informatics,
e-mail: {sandra.garcia-esparza,michael.omahony,barry.smyth}@ucd.ie

[1] http://blog.twitter.com/2012/03/twitter-turns-six.html
[2] http://techcrunch.com/2010/09/14/twitter-seeing-90-million-tweets-per-day/

standing its success to date, we believe that there are a number of key challenges that quickly become apparent to new users and that ultimately may limit the long-term efficacy of Twitter as a platform.

Twitter allows users to share short text messages (tweets) with their *followers*. These tweets can contain raw text, tags (hashtags) and URLs. Users participate in a complex social network by choosing to follow people they trust or topics (hashtags) that matter to them. The end result is that each Twitter user will see a unique and dynamic stream of tweets that have been posted by the people/topics they follow (their *friends*). If the user has been judicious about the people and topics that they follow then the hope is that this stream of information will contain relevant content. This hope is often far from the reality, however, because as Twitter users accumulate more and more friends their stream quickly becomes polluted by a variety of content not all of which is relevant to their core interests. The reason for this is that a typical user will tweet about a range of topics and, conversely, for any given topic (hashtag) an abundance of users can tweet about it. But since Twitter only allows users to follow other users or hashtags in their entirety, it is inevitable that their information stream will become polluted. For example, user @GadgetMary may follow @TechiePeter because she is interested in his insightful technology related tweets but by doing so she is also subjected to his passion for photographing everything he eats; so called *foodspotting*. @GadgetMary's only option is to either stop following him or put up with his early-morning fry-ups and late-night refrigerator raids.

Would it not be better for all concerned if @GadgetMary could follow user @TechiePeter only in relation to his technology related tweets? This is the challenge we wish to address and it is one that contains two key elements. First, we need a way to automatically profile the interests of Twitter users so that users can follow other users based on a subset of these interests. Second, we need a way to classify new tweets and filter those that are off topic. In this way, for example, we can recognize that @TechiePeter's interests include *technology*, *indie music*, *foodspotting*, and *skateboarding* and when @GadgetMary considers following @TechiePeter she can do so with respect to *technology* only. Thus @GadgetMary's timeline integrity will be preserved by filtering out @TechiePeter's non-technology related tweets.

The main contribution of this paper is two-fold. First, we describe a machine learning approach to profiling the interests of Twitter users based on the URLs that they share through their tweets. Focusing on URLs is important because it provides a richer source of content for the purpose of topic classification. To demonstrate the effectiveness of this approach we describe the results of an analysis based on a curated collection of 480 influential Twitter users provided by the marketing department of a major mobile operator. This dataset provides a useful (although coarse-grained) ground-truth against which to evaluate the profiles we produce. In turn we also describe the results of a further experiment to compare these profiles to a more fine-grained set of third-party user topics. In each case we show that our approach to profiling performs well when compared to these manual alternatives.

Secondly, we describe how this profiling technique can be applied to filter Twitter streams in real-time and so form the basis of a more topic-centric Twitter client that can deliver more relevant streams of information to end-users. In particular we have implemented a system with such filtering capabilities called *CatStream* and performed a preliminary user evaluation. Results from such evaluation show that most users would prefer to use a system that allowed to categorise and filter their streams in comparison to the standard Twitter client.

## 2 Related Work

Existing work has focused on profiling Twitter users using the content of their tweets [2, 7, 13]. However, analyzing tweets poses some challenges due to their short length, their noisy nature (i.e. spelling and grammatical errors, use of abbreviation, etc.) and the heterogeneous nature of their content. To deal with these challenges, external sources of information such as Wikipedia or news websites can be leveraged to enrich and disambiguate tweet content [1, 7, 8]. Similarly, metadata associated with URLs from online forums has also been used to improve the categorisation of blog posts [6]. Instead, in this work we analyze the content of URLs, which are commonly used by twitterers. While many tweets might not reveal much about a user's topical interests due to a lack of useful information (e.g. a large percentage of tweets consist of daily chatter), posted URLs provide a much richer source of information in terms of identifying the preferences of users for particular kinds of content (i.e. topics).

In particular we are interested in categorising URLs posted by users in real-time streams like Twitter and to build a profile made of categories (i.e. high level topics such as movies, sports, politics, etc.). Currently, information stream services lack such multiple categorisation. Instead they provide a user categorisation where a user is assigned a single category they are considered to be influential on. This is the case of Twitter and Pinterest, where such categorisation is used to suggest people to follow in these services. This approach works well for famous personalities (i.e. Obama mostly tweets about politics) and other specialised users (BBCSport just tweets about sports). However, for the rest of common people, this single-category assignation is rather limited since users tend to be have more diverse interests and post about many different topics. In such cases, user suggestions based on single categories tend to be poor.

Twitter user profiling has often been applied to recommendation tasks. For example [1, 2] use topic-based profiles to recommend news URLs, while a content-based recommendation approach has been used in [9]. While [1, 2] focus on fine-grained topics, here we are interested in coarse-grained topics or *categories*. This is important in order to organise content and to filter information at a high level. For example, some users might not be interested in *sports* or *finance* and so we can filter tweets containing URLs relating to these categories from the user Twitter feed.

Filtering can also be applied to recommendation scenarios. For example, we can apply categorisation techniques to determine which tweets can be used as a source of recommendation information for a particular domain (i.e. if we want to recommend movies, tweets containing movie-related URLs can be identified). Such an approach which uses micro-blogging messages as recommendation knowledge was explored in [4]; however, in that scenario the information was already categorised, but this is often not the case and such categorisation is needed.

The second part of this paper describes a system that uses category-based profiles that have been built using URLs to filter Twitter information streams. Categorisation for message filtering has also been applied in [11] to classify Twitter messages into five categories (*news, events, opinions, deals* and *private messages*). Here, we also wish to filter micro-blogs but we apply filtering based on the category of the message rather than on the particular communication intention of the author.

## 3 Methodology

Our approach to user profiling is based on user categorisation, this is, the user profile consists of a list of categories the user is interested in. In particular these categories of interest are extracted by classifying the URLs that are present in users' posted tweets. The reason for this is that many tweets consist of daily chatter which does not contain much useful information. Further, the short length of the tweets makes the classification task harder and so, relying on URLs instead, provides us with a much richer source of information.

In order to profile a user we use a text classifier that returns a single category for each URL. By combining the output of all the classified URLs we can build a user profile which consists of a set of categories of interest. In this section we first discuss how the URL classifier is built, and then how the output produced by the classifier is used to create the user profile.

### *3.1 URL Classification*

The first step to profiling is the classification of the user's URLs contained in tweets. The URL classifier is trained using a balanced dataset containing URLs from a set of M categories, as shown by Figure 1. These categories correspond to broad topics such as music, sports or health, and they are sourced from Twitter. The reason for this is because Twitter posts are often annotated by users using hashtags, many of which correspond to categories (e.g. #music, #sports, #health, etc.).

Then, we can represent each category ($C_i$) as a set of URLs that have been annotated with that category hashtag:

$$C_i = \{URL_1...URL_K\} \tag{1}$$

One of the typical approaches to text categorisation lies in the application of machine learning techniques [10]. In past work, naïve Bayes or Support Vector Machines have shown to be very effective for this task [5, 14]. Here we use naïve Bayes

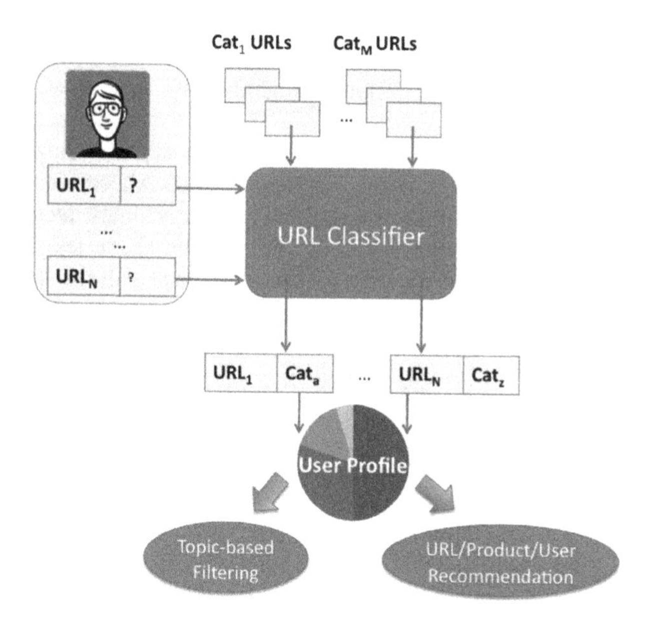


**Fig. 1** System Architecture.

multinomial classifier since in past work with categorisation [3], we found it to be one of the best performing classifiers. We use a single-category classifier, this is, only one category is returned for each classified URL.

We use a bag-of-words approach to represent each URL, where the attributes of the classifier correspond to the terms appearing in the page pointed to by the URL, weighted according to the standard TF-IDF weighting function [12]. We also performed some preprocessing on the data such as removing stop-words, special symbols and digits, but we did not apply feature selection as it did not improve classification performance during testing.

## 3.2 User Profiling

As seen in Figure 1, we can represent a user $U_i$ as a set of URLs that he/she has posted about:

$$U_i = \{URL_1 ... URL_N\} \tag{2}$$

Once the classifier has categorised the user's URLs, we can then compute a weight $w_i$ for each category $C_i$ based on the percentage of URLs that are classified as belonging to that category ($w_i = 0$ if the category doesn't appear in the list). This list of categories and their associated weights conform the user profile P($U_i$).

$$P(U_i) = \{c_1.w_1, ..., c_n.w_N\} \quad (3)$$

Thus, for example, the profile for a user might be 50% *technology*, 30% *travel*, 15% *books* and 5% *art*.

An application of such as profiles would be in information filtering. In the case of Twitter we could just present to the users those tweets whose categories are in their profile. Further, tweets from a category that the user is highly interested should appear more often than tweets that belong to a category the user is less interested in. Section 5 describes and evaluates system we have developed with such functionality.

# 4 Evaluation of User Profiles

In this section we evaluate our approach in three ways. First we evaluate the classification accuracy of the above classifier using a standard 10-fold cross-validation approach. The real test of our profiling approach, however, is whether the profiles produced using this classifier are a real reflection of the tests users' genuine interests. With this in mind we compare the user profiles constructed by our approach to two separate reference sets of user topics.

This work was part of a project to profile users of a large European mobile phone operator. In particular, the operator was interested in profiling a set of 480 influential Twitter users on three high level topics (categories): *technology* (220 users), *music* (193 users), and *sports* (67 users). The operator also provided us with a list of 19 categories they were interested in profiling users by, which are shown in Figure 2.

## *4.1 Categoriser Evaluation*

The first step to evaluate the profiling approach is to build and evaluate the classifier. We collected and downloaded the content of URLs from the 19 different categories which were present in tweets posted between September 2011 and January 2012. We removed duplicate URLs and selected 100 URLs at random for each category, producing a total of 1900 URLs.

We performed 10-fold cross-validation to evaluate classification accuracy. Figure 2 shows the accuracies achieved by each category. Across the 19 categories, the average accuracy achieved by the classifier was 83.6% ($\sigma$ = 6.3%). This relatively high performance allows us to build user profiles based on our approach with a high degree of confidence.

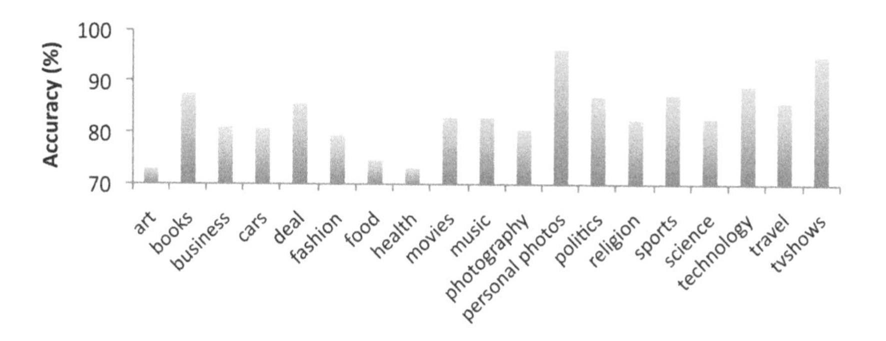


**Fig. 2** Category Accuracies.

## 4.2 User Profiling Evaluation

In order to profile the 480 users provided by the operator, first we downloaded the content of the URLs contained in their tweets. Then, we categorised each of these URLs and created profiles using the approach described in Section 3.2. Each user had on average 200 URLs; the minimum and maximum number of URLs per user was 20 and 2000, respectively. After profiling, each user profile contained on average 11 of the 19 categories considered, indicating that users tend to post about a wide variety of topics. To evaluate the accuracy of our profiling approach, two sources of ground truth are used as described in the following sections.

### 4.2.1 Mobile Operator Categories

We first compared the user profile categories with the categories of influence as specified by the operator[3]. Bear in mind that just because the operator views a user to be influential on a particular category does not mean that this category corresponds to the user's only or primary field of interest. Most probably that is not the case and so an appropriate test of our profiling technique is to look at how frequently our profiles contain each user's category of influence within the top $k$ profiled categories; in this case $k = 10$ was selected as a reasonable cut-off.

Thus, for each group of influential users (*technology*, *music* and *sports*), we ranked the profile categories by their weights and computed, for each group, the percentage of users that had the category of influence contained in the top 10 ranked categories. The results are shown in Figure 3 and indicate that, for the 3 groups of users, our profiling approach is able to find the category of influence within the top 10 categories for at least 80% of users. The best performance was achieved for the

[3] Using a proprietary tool, the operator identified influential users on three categories: *music*, *technology* and *sports*.

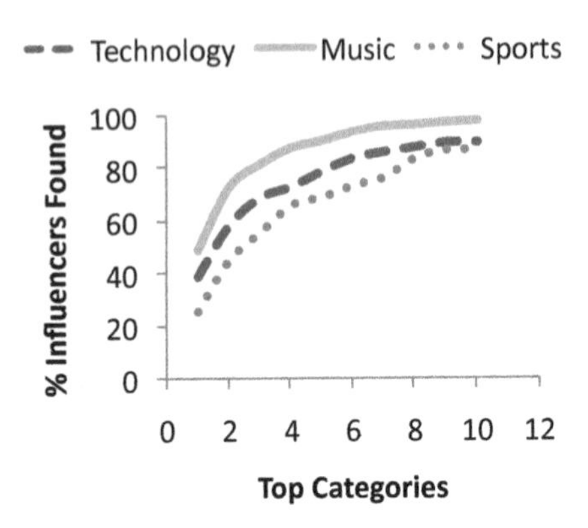


**Fig. 3** Percentage of users that have the category of interest (*technology*, *music* or *sports*) in their top 10 profile categories.

*music* group, where the *music* category was the top-ranked category by our profiling approach in 50% of cases.

### 4.2.2 Klout Topics

The above experiment provides a high-level indication of the performance of our profiling approach with respect to a single category of user interest. In order to evaluate how good it is at finding other interest categories, we used Klout[4] topics as an additional ground truth. Klout is a service that scores users according to their influence in social networks such as Twitter or Facebook. Klout also extracts for each user, a list of topics on which they are considered influential. Some topic examples are *gadgets*, *football* or *Rihanna*. In order to compare our categories with *raw* Klout topics, a manual mapping between each and the 19 categories described in Section 4.1 was performed to produce a mapped set of topics for each user. This set also captures the number of times each (mapped) Klout topic was associated with each user. We ignored those raw Klout topics which were vague, ambiguous or could not be mapped (14% of the total topics). After mapping, the majority of users were represented by about 5 distinct topics; the minimum and maximum number of distinct topics per user was 4 and 8, respectively.

In order to evaluate the performance of our profiling approach, we computed a standard overlap score and the cosine similarity between profile categories and Klout topics. Figure 4(a) shows the overlap between categories and topics. An overlap of 1 indicates that one set is a complete subset of the other set. Recall that, on average, users were represented by 5 and 10 distinct Klout topics and profile categories, respectively. Our analysis indicates that an overlap of 1 was achieved for 50% of users; on average, an overlap of 0.8 across all users was achieved. From Fig-

[4] `http://www.klout.com/`

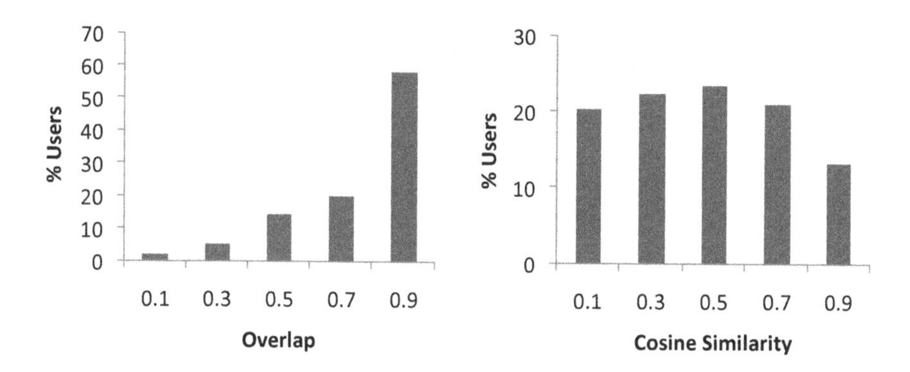


**Fig. 4** Overlap (a) and cosine similarity (b) between profile categories and Klout topics.

ure 4(b), the average cosine similarity between the (weighted) profile categories and (weighted) Klout topics was 0.5; the key difference between the overlap and cosine metrics is that the latter considers the weights associated with the profiles categories and the Klout topics that are associated with users.

Overall, these findings indicate that the majority of Klout topics associated with each user matched those identified by our user profiling technique, which speaks to the validity of the profiling technique at least when it comes to discovering the key topics in which users are considered influential (as determined by Klout), and in which the core interests of users presumably lie. In addition, the user profiling approach discovers, on average, an additional 6 distinct category interests per user compared to Klout, indicating that while users may be considered influential on a few topics, they generally post about a broader set of topics.

## 5 CatStream: A Twitter Interface for Stream Filtering

In the previous section we have demonstrated that it is possible to profile users by the categories extracted from URLs posted in their tweets. Following from these findings we have created a system called *CatStream*, which categorises and filters Twitter information streams based on such category-based profiles. CatStream's primary goal is to allow users to find useful content. For this reason CatStream only focuses on tweets which contain URLs and ignores those which are considered *daily chatter*. The system is currently live at `http://catstream.org`.

### *5.1 System description*

In order to use the system, users need to have a Twitter account. Once the user has logged in, the system retrieves the most recent tweets posted by the user and

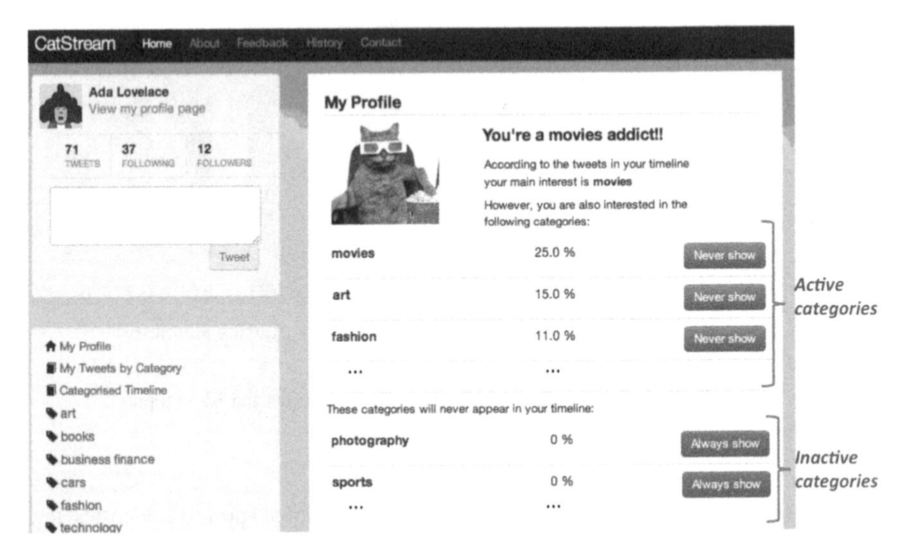


**Fig. 5** A typical *CatStream* profile.

builds (or updates, if the user is returning) a profile following the methodology described in section 3. In the current version of CatStream we only retrieve the last 200 tweets, which in a preliminary evaluation it has shown to be enough in order to create accurate profiles. If the user is returning to the system, then new tweets will be downloaded and categorised in order to keep the profile updated. A screenshot of such a profile can be seen in Figure 5. Here, the user is provided with a list of 19 categories and their associated weight (i.e. the percentage of URLs predicted as belonging to that category). For space reasons here we only show a subset of those.

By default, categories with a percentage higher than 0 are active, this is, tweets belonging to that category will be visible in the user's information stream, also called timeline. Categories with a percentage equals to 0 (i.e. the user has never posted URLs about these categories) are inactive by default, this is, tweets from that category will not be visible in the user's information stream. Some active categories in Figure 5 are movies, art and fashion, while photography and sports are inactive categories. However, the user is free to set any category as inactive or active by clicking on the "never show" or "always show" buttons, respectively. Note that, an inactive category is not necessarily a category the user is not interested in. For instance, a user could be interested in it but not have posted an URL about it and so by default is inactive. Moreover, a user may decide to inactivate a category because, despite being interested in it, they do not wish to receive tweets belonging to that category in a particular time. For example, a user might be interested in "sports" news only during the Olympic Games but not after that. Further, users can examine the category assigned to each of their posted tweets (i.e. tweets containing a URLs) from the "My Tweets by Category" page on the left menu.

The main feature in CatStream is the categorised timeline which can be seen in Figure 6. This is very similar to the standard Twitter timeline, where users can read

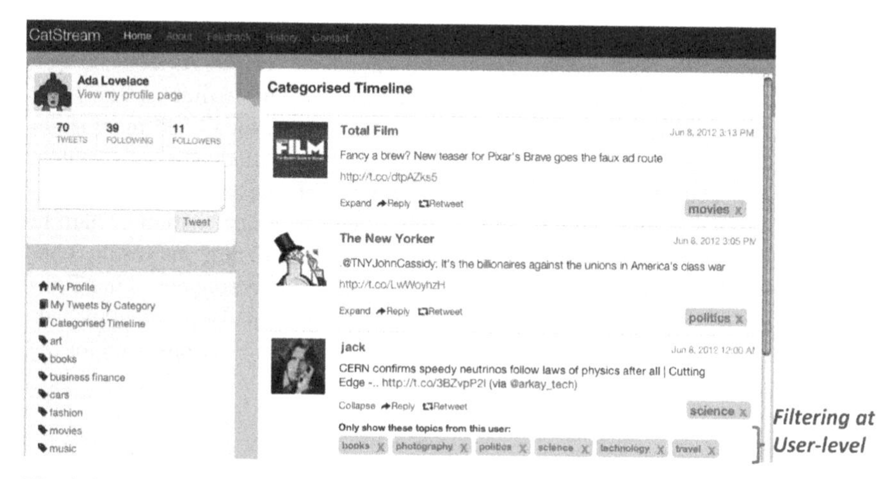


**Fig. 6** CatStream's categorised timeline.

about the tweets posted by the people they follow. However, CatStream's timeline only displays those tweets that belong to the active categories. Category annotations are also visible in the timeline next to the tweet.

Further filtering can also be applied at a user level. For example, we might follow a person who works in the technology field to keep up to date with the area; however we might not be interested in their tweets about movies or politics, despite also having an interest in such topics. To address this, CatStream also allows to refine filtering by category at a user level. We can do this from the timeline, by clicking on the user image and selecting on the categories we want to remove for that user. This action is illustrated on the bottom tweet of Figure 6.

Finally, other current CatStream features also include tweeting, re-tweeting and replying to messages. In the next section we perform a preliminary user study to evaluate the CatStream features described above.

## *5.2 Preliminary User Evaluation*

We ran a preliminary user evaluation among 15 people over the period of a week. Out of those participants, 60% were from the university environment and 40% outside. 67% were male and 33% female and their ages ranged between 24 and 35 years old. The participants were asked to log into CatStream and examine their category profile and the categorisation of their posted tweets and to interact with the categorised timeline. After, they completed a survey on their experience with Twitter and CatStream. Most of the answers were given as a likert scale, which is a 5 rating scale (from 1 to 5) where 1 is equivalent to strongly disagree, 3 is the neutral rating and 5 is equivalent to strongly agree.

In the first part of the survey, we asked users whether they were familiar with Twitter and what they liked and disliked the most about the service. As it can be seen in Figure 7 (a), 37.5% of the participants used Twitter multiple times a day while 13% rarely/never used Twitter. When we asked users what they liked most about Twitter answers included "information in real-time", "short and concise", "simple interface" and "ability to follow experts on subjects". When asking about the worst features in Twitter, answers frequently included "too much information", "hard to know what's important", "noisy data", and "waste of time finding interesting content". In general we found that while users provided diverse reasons for liking Twitter, those for disliking were more similar and often focused on the problem of finding relevant information. We believe these answers provide a clear rationale for the kind of system proposed in this paper.

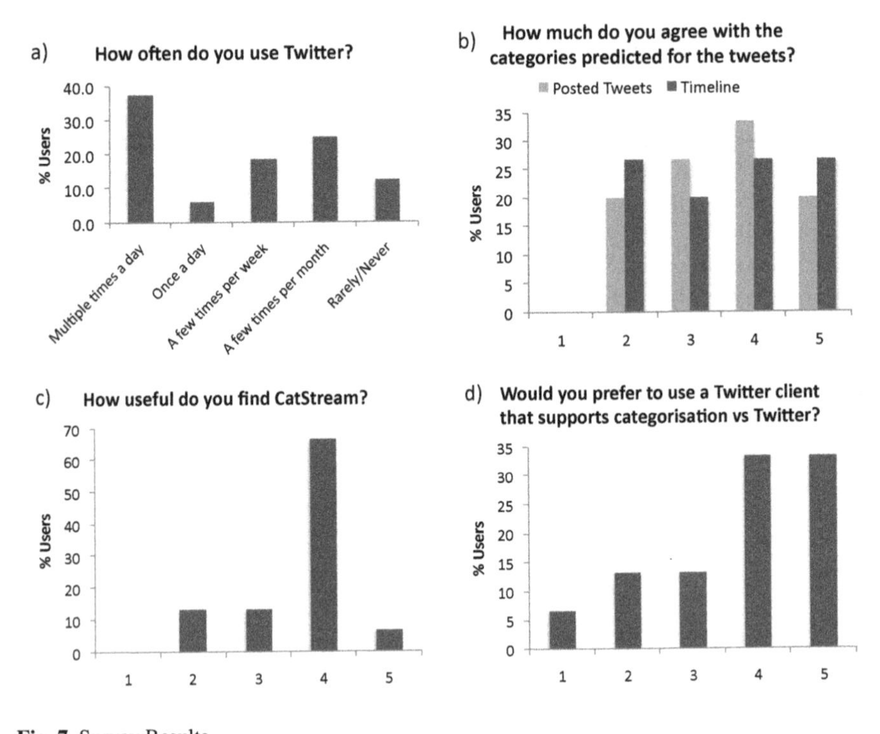


**Fig. 7** Survey Results.

The second part of the survey focused on the evaluation of CatStream. Firstly, we wanted to know how accurate users found the categories predicted by the classifier. For this, we asked users how much they agreed with the categories predicted for their posted tweets and the tweets in their timeline. In Figure 7 (b) it can be seen that about 53% of users were either satisfied or very satisfied with the predicted categories of both posted tweets and timeline tweets, in comparison with 23% and 26% of users that were not satisfied with the categories of their posted tweets and timeline tweets, respectively. This shows that the classifier's accuracy could be improved, which we plan to do on future work by studying other techniques and extending the set of categories, which currently is limited.

On the other hand, as shown in Figure 7 (c), when participants answered to the question "How useful do you find CatStream?", a high percentage of users (73.3%) responded either positively or very positively. This is an interesting and encouraging finding as it suggests that despite the categorisation of tweets was not entirely correct, they still found the system useful.

Moreover, as seen in Fig 7 (d) when asked if they would prefer to use a Twitter client that supports categorisation such CatStream in comparison with a traditional Twitter client, the majority of participants strongly agreed or just agreed (66.6%). This confirms the fact that users feel a need for more organised and filtered information streams. We believe these preliminary results are a high indication that work in this line of research should be continued.

## 6 Conclusions and Future Work

In this paper we have presented a user profiling approach where users' interests on categories are predicted using the content of URLs posted in tweets rather than the actual textual content of tweets. A first evaluation based on ground truth established from commercial mobile operator user-influence data and data derived from Klout topics indicates that our profiling approach is successful at predicting the categories a user is influential on, and thus likely to be interested in. In fact, results show that topics of influence usually form only a subset of a usually much larger set of topics of interest.

One application of our user profiling technique lies in a *topic-centric* approach to information filtering, in contrast to the *user-centric* approach that is typically available to users of social networking sites. For example, topic-centric information streams can be constructed where information is organised according to user preference for different topics. In this paper we have built CatStream, a system with a similar interface to Twitter but which also provides such automatic topic filtering provided by categorisation of URLs. In particular, the timeline in CatStream displays only tweets about topics the user is interested in; further, the ability to manually select which topical tweets to display from each user in the timeline is also possible. A preliminary evaluation of the system has shown that users often feel there is too much information on the real-time web and that such systems would benefit from a categorisation and filtering layer.

In this work we found that 19 categories was a limited set due to the diverse nature of Twitter. In future work we will include additional categories in order to produce better classifications. Further, we will extend the profiling approach by performing topic detection and entity extraction from URLs in order to provide for a more refined filtering. For instance, a user might want to receive tweets about the

Olympic Games in particular, but not all tweets about sports. Finally we will also perform a more extensive evaluation of CatStream with a bigger number of participants.

**Acknowledgements** Based on work supported by Science Foundation Ireland, Grant No. 07/CE/I1147.

## References

1. F. Abel, Q. Gao, G.-J. Houben, and K. Tao. Analyzing user modeling on twitter for personalized news recommendations. In *Proceedings of the 19th international conference on User modeling, adaption, and personalization*, UMAP'11, Berlin, Heidelberg, 2011.
2. J. Chen, R. Nairn, L. Nelson, M. Bernstein, and E. Chi. Short and tweet: Experiments on recommending content from information streams. In *Proceedings of the 28th international conference on Human factors in computing systems*, CHI '10, New York, NY, USA, 2010.
3. S. Garcia Esparza, M. O'Mahony, and B. Smyth. Towards tagging and categorization for micro-blogs. *21st National Conference on Artificial Intelligence and Cognitive Science, AICS'10*, 2010.
4. S. Garcia Esparza, M. O'Mahony, and B. Smyth. Effective product recommendation using the real-time web, Mar. 22 2012. US Patent 20,120,072,427.
5. T. Joachims. Text categorization with support vector machines: learning with many relevant features. In *ECML '98: European Conference on Machine Learning*, London, UK, 1998. Springer-Verlag.
6. S. Kinsella, A. Passant, and J. G. Breslin. Topic classification in social media using metadata from hyperlinked objects. In *Proceedings of the 33rd European conference on Advances in information retrieval*, ECIR'11, pages 201–206, Berlin, Heidelberg, 2011. Springer-Verlag.
7. M. Michelson and S. A. Macskassy. Discovering users' topics of interest on twitter: a first look. In *Proceedings of the fourth workshop on Analytics for noisy unstructured text data*, AND '10, New York, NY, USA, 2010.
8. X.-H. Phan, L.-M. Nguyen, and S. Horiguchi. Learning to classify short and sparse text & web with hidden topics from large-scale data collections. In *Proceedings of the 17th international conference on World Wide Web*, WWW '08, pages 91–100, New York, NY, USA, 2008. ACM.
9. O. Phelan, K. McCarthy, and B. Smyth. Using twitter to recommend real-time topical news. In *Proceedings of the third ACM conference on Recommender systems*, RecSys '09, New York, NY, USA, 2009.
10. F. Sebastiani. Machine learning in automated text categorization. *ACM Computing Surveys*, 34(1):1–47, 2002.
11. B. Sriram, D. Fuhry, E. Demir, H. Ferhatosmanoglu, and M. Demirbas. Short text classification in twitter to improve information filtering. In *Proceedings of the 33rd international ACM SIGIR conference on Research and development in information retrieval*, SIGIR '10, New York, NY, USA, 2010.
12. C. J. van Rijsbergen. *Information Retrieval*. Butterworth-Heinemann, Newton, MA, USA, 1979.
13. J. Weng, E.-P. Lim, J. Jiang, and Q. He. Twitterrank: finding topic-sensitive influential twitterers. In *Proceedings of the third ACM international conference on Web search and data mining*, WSDM '10, New York, NY, USA, 2010.
14. Y. Yang and X. Liu. A re-examination of text categorization methods. In *SIGIR '99: Proceedings of the 22nd annual international ACM SIGIR conference on Research and development in information retrieval*, pages 42–49, New York, NY, USA, 1999. ACM.

# RECOMMENDATION

# Content vs. Tags for Friend Recommendation*

John Hannon, Kevin McCarthy, Barry Smyth

**Abstract** Recently, friend recommendation has become an important application in a variety of social networking contexts, whether as part of in-house enterprise networks or as part of public networks like Twitter and Facebook. The value of these social networks is based, in part at least, on connecting the right people. But friend recommendation is challenging and many systems do little to help users make these valuable connections. In this paper, we build on previous work to consider new strategies for friend recommendation on Twitter. In particular, we compare strategies based on the content of users tweets, recommending users who tweet about similar things, to strategies based on Twitter-list tags by recommending users who are members of lists on similar topics. We describe a comprehensive evaluation to highlight the different benefits of these complementary strategies. We also discuss the most appropriate ways to evaluate their recommendations.

## 1 Introduction

Social networks have become an increasingly important part of our online lives. They help us to keep in touch with family and friends. They also play an increasingly important role when it comes to maintaining and developing our professional networks. Today, social networking services like Twitter operate as powerful information diffusion networks, making it possible for users to keep up to date with relevant breaking news and helping users to discover new information that may be of interest. The value of these networks is derived in part from the social connections that users form. By connecting to the right people, users can benefit from the

John Hannon, Kevin McCarthy, Barry Smyth
CLARITY: Centre for Sensor Web Technologies, School of Computer Science & Informatics, University College Dublin, Ireland. e-mail: {john.hannon, kevin.mccarthy, barry.smyth}@ucd.ie

* This work is supported by Science Foundation Ireland grant 07/CE/I1147 and Amdocs Inc.

social fabric of influential users on a range of topics that matter to them. This is perhaps most evident in Twitter as a user's timeline will be made up of tweets which originate mainly from the people that they follow. By following the right people our user will enjoy an informative and relevant stream of tweets and content but, follow the wrong people, and this stream quickly becomes polluted with noise.

Given the importance of social connections in public and private social networking services it is not surprising that most networks have explored ways to help users find connections to new people. For example, Twitter and Facebook allow users to find new friends via simple search functions. However search-based approaches have their limitations and add some friction by relying on explicit user activity. Twitter and Facebook also both support the creation of communities of users that share common interests (*groups* on Facebook and *lists* on Twitter) as a way for users to engage with topics of interest. For example, Twitter introduced user lists in late 2009, allowing users to be grouped according to user-defined topics or themes. A Twitter user might create a list on the topic of "Artificial Intelligence" for example, and include users they view as authorities on this topic. Other users can subscribe to this list to benefit from member tweets. Lists on Twitter have become an important way for users to curate content on topics or themes that are of interest to them and they have been widely adopted by media outlets, news agencies, and marketing departments as a way to better organise content streams and connect with communities. To complement these approaches both Facebook and Twitter also use simple recommendation techniques to suggest new people to connect with by suggesting users that are friends of their friends. However, these approaches tend to be limited because they do not seem to make these recommendations in a way that reflects your fine-grained interests whether short or long-term.

It is perhaps not surprising then to see that friend recommendation has become an important task for recommender systems in a social network context [1] in an effort to make more targeted recommendations that are capable of enhancing the quality of each user's network. In this paper, we build on recent work in this regard by comparing two different approaches to friend recommendation. In particular, we compare a content-based approach in which users are represented by the content of their tweets (c.f. [5]) to a novel tag-based approach that represents users by the set of tags that are associated with any lists that they are members of. We consider the ability of each recommendation approach to make accurate recommendations. Moreover we propose a novel approach for evaluating precision in a social networking context in order to gain a better understanding of recommendation quality.

## 2 Related Work

Social networks have proven to be a fertile ground for recommender systems research, with researchers exploring a number of important research questions. This includes work on content analysis, network structure analysis, and user modeling, for example, all in the pursuit of helping users to better understand and utilise their evolving social networks. In this section we will review a sample of relevant background research as context for our own work. We will pay particular attention to research that has focused on Twitter for a number of reasons. First, this body of research is reasonably representative of the broader set of social network research. Second, a considerable portion of research in social network analysis has been carried out on Twitter, in large part because of the scale and openness of the Twitter eco-system. Third, the focus of this paper is on people recommendation in Twitter.

To begin with it's worth highlighting some general research on Twitter as a social information network. For example, the seminal work of Kwak et al. [8] considered a sample of more than 41 million users and 106 million tweets in order to better understand the dynamics of the Twitter ecosystem, the structure of its social graph, the role of influential users in information diffusion across the network, and the emergence of trending topics. This work remains an important starting point for much of the research that has followed.

One common research theme relates to the role of Twitter as an information filter for users. For example, early work by Shamma et al. [14] studied the microblogging activity that emerged during the 2008 US presidential debates to demonstrate how frequently occurring terms reflected trending debate topics, although vocabulary complexity did tend to obfuscate automatic topic identification. More recently the work of Phelan et al. [12] demonstrated Twitter as a powerful news aggregator and filter via the *Buzzer* social news service. Buzzer extracts stories from a user's RSS-feeds based on the topics being discussed by the user's Twitter friends/followers. The content of tweets (the user's own tweets and the tweets of their friends/followers) is used as a term-based interest profile for the user and matched against incoming RSS-feeds to select and rank matching stories. Related work by Garcia et al. [4] considered the microblogging activity of users (on Twitter & Twitter-like services such as Blippr) as a source of product information to show how recommendations could be adapted to drive product recommendation based on products that a user had liked or disliked.

The publicity surrounding Twitter in the recent past has stemmed from its ability to detect the occurrence of major events and happenings around the world. For example, the important role of Twitter in recent events such as the 'Arab Spring' or the 'Fukushima Disaster' has received considerable attention in mainstream media; although the precise role that Twitter and similar services have played is still to be fully explored. Twitter is changing the face of news coverage and leading to the emergence of a new type of news service that sources on-the-ground news content from curated communities on Twitter and other social networks. For instance, *storyful.com* is one example of such a service actively curating lists of users as a means to monitor and filter newsworthy content. Certainly there is evidence to support the ability of Twitter to help recognise and amplify real world events. For example, the work of Nichols et al. [11] has examined using Twitter as a means to summarize sporting events, by using tweet frequency and content as a signal for 'event spikes'. Related work [6] demonstrated how such techniques could be used to create real-time, video summaries of sporting events.

In this paper, we are primarily interested with the people finding recommendation task. Simply put, for a given user, who are the other users that might be of interest to follow? This is an important gap in services like Twitter and Facebook, which currently provide only very limited recommendation offerings, even though the value of these services depends largely on helping users to connect with the right people. Daly et al. [3] study the network effects of recommending social connections in the context of an enterprise social networking service in IBM. In particular, they study how different recommendation algorithms deliver different types of beneficial network effects (e.g. popularity, long-range connectivity, etc.) to users. Lo et al. [9] describe a graph-based algorithm for friend recommendation, tested on a small-scale Brazilian social network. This approach is based on real message interactions between network members on the assumption that more interactions between users reflect a stronger relationship. Recently the work of Hannon et al. [5] has studied the friend recommendation task on Twitter by comparing a variety of content-based and collaborative filtering style approaches. In the former, users are modeled based on their tweets (and/or the tweets of their friends/followers). In the latter, user profiles simply contain friend/follower ids. In summary, they found that content and collaborative approaches were able to suggest high-quality recommendations and the combination of both approaches can be used as an effective *ensemble* recommender.

In this paper, we will build on these content-based recommendation approaches [5]. In particular, we will explore an alternative content-based strategy. Instead of relying on the content of tweets directly, which is diverse and noisy, we will consider the use of independent tags, which can be associated with users, as a source of recommendation data. Freeform tags have emerged as a useful form of meta-data and have played a role in recent recommender systems and user profiling research with the objective of connecting users to items; see for example, [7, 13]. In this work we harness a novel source of user tags based around Twitter lists and show how they can be used to connect users with other users as part of friend recommendation.

## 3 Recommending Twitter Users

In this section, we introduce two complementary approaches for friend recommendation on Twitter. Both involve using a profile of user interests as the basis for recommendation: new users are chosen based on their sharing of interests with the target user. In one approach we use the content of tweets to represent user interests, based on previous work of [5]. In the second approach we instead profile users by a set of tags that have been curated independently. The general approach is captured in the architecture shown in Figure 1. Briefly, each user is profiled according to a set of tweets or tags and we then construct a content-based or tag-based index for the set of users. Given a new target user, the recommender system matches this user's profile (content-based or tag-based) against the relevant index and retrieves a ranked list of the most similar users ($\{r_1, ..., r_k\}$).

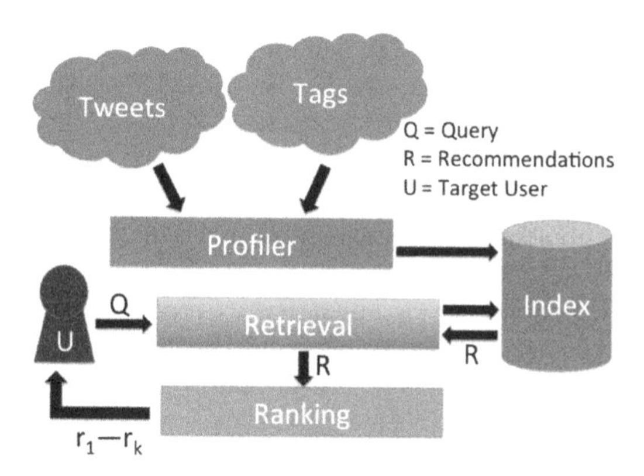


**Fig. 1** Architecture Diagram.

## 3.1 Content Profiling & Recommendation

Hannon et al. [5] proposed using terms extracted from the content of tweets as a way to profile user interests, under the basic assumption that what we talk about on Twitter is a good reflection of what we are likely to be interested in. While this may seem intuitively sound there are a number of important caveats. Firstly, tweet content is a varied and noisy affair: people tweet on a variety of topics and the language used in tweets is often truncated to meet Twitter's strict 140-character limit. In addition, it is not safe to assume that a user's own tweets are a fair reflection of their interests. Many users have interests in topics that they themselves rarely tweet about, and generally they will follow other users to satisfy these interests. For example, given a set of tweets ($\{t_1, ..., t_k\}$) for a target user $U$ we can use term-based weighting techniques such as TF-IDF [2] to build a weighted index of tweets and standard stop word removal to minimize some noise. In effect, we treat the set of tweets for a given user as a document containing the terms of these tweets and when applied across a large index of users, TF-IDF works to weigh the terms for users in a way that helps to distinguish certain terms as being good discriminators of certain users. Then at recommendation time, a target user can be matched against each indexed user using a similarity metric such as cosine [10]. In this work we use Lucene[2] to provide the indexing, weighting, language processing and matching functions for our recommender.

The second issue – whether a user's own tweets are a good representation of their interests – can be addressed by expanding the source of tweets to be used as a user profile. For example, in addition to (or instead of) using the user's own tweets we can profile the user by the tweets of their friends or the tweets of their followers. In the case of the former – profiling using the tweets of their friends – this should provide an effective way of capturing interests that the user has (after all

[2] http://lucene.apache.org/

they chose to follow these friends) but that are not expressed in their own tweets. It should also help to amplify interests that are shared between the tweets of the user and his/her friends. Profiling users by the tweets of their followers is less clear-cut. Some followers may also be friends but others will not be followed by the user directly and so may provide a weaker source of user oriented interests.

In summary our content-based strategies are given by Equations 1 - 5. In each case we will assume to profile users based on for example ($k = 100$) most recent of their own tweets (Equation 1) or the union of the tweets of their friends (Equation 4) or the union of the tweets of their followers (Equation 5).

For a target user, $U_t$, let Equation 1 be the set of the 100 most recent tweets. Extending this, consider $friends(U)$ to be 100 friends of user $U$, and $followers(U)$ to be 100 followers of user $U$. Therefore let Equation 4 & 5 be the union of the status updates of each of the friend/follower users.

$$tweets(U) = \{t_1, ..., t_k\} \tag{1}$$

$$friends(U) = \{f_1, ..., f_m\} \tag{2}$$

$$followers(U) = \{g_1, ..., g_n\} \tag{3}$$

$$friendstweets(U) = \bigcup_{\forall f_i \varepsilon friends(U)} (tweets(f_i)) \tag{4}$$

$$followertweets(U) = \bigcup_{\forall g_i \varepsilon followers(U)} (tweets(g_i)) \tag{5}$$

In this way we can produce 3 separate content-based profiles for users, based on the tweets of the user (*Tweets/User*), friends (*Tweets/Friends*), or followers (*Tweets/Followers*), and we can also produce a fourth profile based on the combination of the above (*Tweets/All*). These tweet based approaches have been recreated from previous work [5] and are intended as an appropriate comparison baseline. In the next section, we will introduce our novel technique for profiling users using tags.

## 3.2 Tag-Based Profiling and Recommendation

Twitter lists provide a way for users to create topical lists of other users, to date users have created millions of lists on a wide variety of topics, from charity and health, to music, entertainment and technology, and a world of topics between. Some lists contain a small handful of Twitter users while others contain much larger cohorts of carefully curated influential users. Some lists are created by regular users while others are created by institutions and media organisations. Some lists are followed by only a few other users while others have amassed followers in their droves.

These lists are interesting for a couple of important reasons. First of all they represent independent opinions about the interests of members. If some user $U_1$ has added $U_2$ to a list on *Gadgets and Technology* then it suggests that $U_2$ is relevant to this topic. In addition lists can be a useful source of tags. Lists themselves have titles and some have descriptions, but more importantly for this work, third parties such as Listorious[3] have created a large database of annotated lists. For example, Listorious has collected numerous lists covering more than 2 million Twitter users. More importantly again, each of these lists has been associated with a set of tags. For example, at the time of writing the *Social Media* list, curated by Pete Cashmore of Mashable[4] includes 102 people and has attracted more than 10,000 followers and it has been tagged with terms such as *twitter, marketing, socialmedia, tech, web*.

Our insight is that these tags can be applied not just to the lists but to the users who are within these lists. Moreover, since users will often be listed in many different lists, sometimes on a range of topics, they will amass a set of tags based on these different memberships. For example, as per Equation 9 we define a set of tags for a user $U$ based on the lists that $U$ is a member of (Equation 6) and the tags associated with these lists (Equation 7). We score these tags (see Equation 8) based on how frequently a user $U$ is associated with a list that has been tagged with a given tag $t$.

$$lists(U) = \{L_1, ..., L_m\} \tag{6}$$

$$tags(L) = \{t_1, ..., t_n\} \tag{7}$$

$$score(t,U) = \sum_{\forall L_i \varepsilon lists(U)} hasTag(L_i, t) \tag{8}$$

$$userTags(U) = \bigcup_{\forall L_i \varepsilon lists(U)} tags(L_i) \tag{9}$$

Extending this to friends and followers we can also generate the set of tags associated with the friends of $U$ as per Equation 10 and, separately, the tags associated with $U$'s followers as per Equation 11; we can also score these tags using a similar scoring function to the ones described above.

$$friendsTags(U) = \bigcup_{\forall f_i \varepsilon friends(U)} userTags(f_i) \tag{10}$$

$$followersTags(U) = \bigcup_{\forall g_i \varepsilon followers(U)} userTags(g_i) \tag{11}$$

In a manner that is analogous to our content-based approach this tag-based approach produces three types of profiles based on the tags of the user (*Tags/User*), friends (*Tags/ Friends*), or followers (*Tags/Followers*) and a fourth profile type based on the combination of the above (*Tags/All*).

[3] http://listorious.com/

[4] http://mashable.com/

# 4 Evaluation

In this section we present a detailed evaluation of the different recommendation approaches described above. For this evaluation we use a set of real-user data and perform a standard leave-one-out style recommendation test. In particular we are interested in understanding the precision characteristics of each recommendation strategy and their ability to contribute complementary recommendations.

## *4.1 Data and Methodology*

The dataset for this study is based on a recent crawl of Listorious. It includes 8,127 users who are members of lists indexed by Listorious such that each user can be associated with at least 10 distinct tags; the average number of distinct tags per user is 24. In addition, we used the Twitter API to extract the necessary user tweet content (up to 100 tweets), friend, and follower information for all of these 8,127 users; note in each case we only used friends and followers that were indexed by Listorious so that we could be guaranteed to be able to produce tags for these users. Summary statistics can be seen in Table 1 for this dataset, note that each user by selection is a member of a Listorious list which can be seen as an indication they are either prolific or well known twitter users, hence the high tweet and follower counts.

**Table 1** Dataset User Summary

| | Tweets | Friends | Followers | Distinct Tags |
|---|---|---|---|---|
| Mean | 7,616 | 1,767 | 2,466 | 24 |
| Median | 2,953 | 1,189 | 1,695 | 19 |

Our evaluation includes 8 different recommendation strategies: the 4 tweet-based and 4 tag-based strategies. In each case, we used a standard leave-one-out style approach to calculate an overall precision for each strategy based on different recommendation list sizes $k$ ($k = 5...20$). Specifically, each user is treated as a test user. His/her profile is matched against the relevant user index and a set of recommendations (user ids) are generated. To measure precision we count the number of these users who are already friends of the test user; these are people the test user has already chosen to follow and the ability to recommend these users is thus a strong indicator of recommendation relevance.

It is worth highlighting that this approach to measuring precision is necessarily conservative. Just because a user does not currently follow a recommendation does not mean it is a bad recommendation, even though this is the way that precision is measured currently. We will return to this issue when describing a modified precision metric that provides for a more fine-grained assessment of precision.

## 4.2 *Precision vs Recommendation List Size*

The overall precision results are presented in Figures 2 - 3. In each case we present a graph of precision versus recommendation list size ($k$) and each graph presents the 4 different recommendation approaches for each one of the 2 strategies.

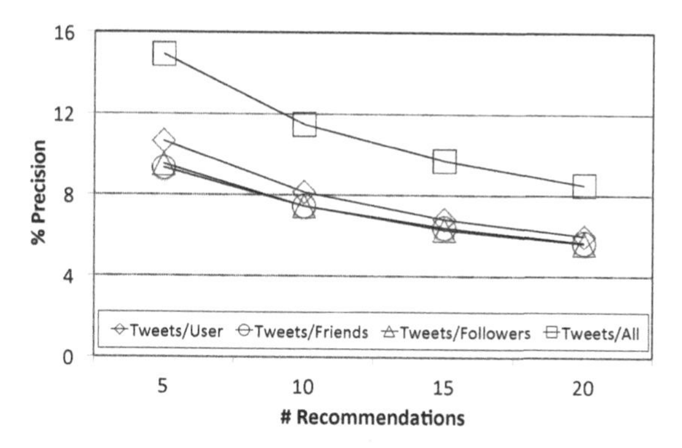


**Fig. 2** Precision vs. Recommendation List Size for Content-Based Approaches.

### 4.2.1 Content-Based Recommendation

Focusing on the results for the content-based strategies first (Figure 2) and following the same evaluation setup as previously [5] albeit on a different dataset of users, we can see that in general the different content-based approaches generate recommendations with precision scores ranging from 5-15%. In each case precision tends to drop off with increasing recommendation list size which indicates relevant recommendations tend to be higher up in the recommendation lists, which is to be desired.

We can see that the *Tweets/User* approach performs better than either of the *Tweets/Friends* or *Tweets/Followers* approaches but the clear winner is the *Tweets/All* with precision scores that are significantly higher than any of the other strategies.

It is interesting that the *Tweets/User* strategy does better than either of the *Tweets/Friends* or *Tweets/Followers* strategies, indicating that the user's own content is a better predictor of friends than the tweets of friends or followers. Earlier we speculated that a user's own tweets might not cover the full range of their interests and his/her choice of friends may be motivated in part to fill in these interest gaps. This certainly seems to be the case because even though *Tweets/User* outperforms *Tweets/Friends* and *Tweets/ Followers*, the combination of all three *Tweets/All* is by far the best overall performer. In other words the combination of user, friends, and follower tweets provide a much richer signal for friend recommendation.

It is also worth noting that both *Tweets/Friends* and *Tweets/Followers* perform very similarly from a precision standpoint. Remember a user does not connect to a follower, a follower connects to him/her, but yet followers turn out to be a useful source of recommendation content when compared to a user's explicit friends.

### 4.2.2 Tag-Based Recommendation

Figure 3 presents the precision versus recommendation list size results for the 4 tag-based recommendation strategies. Broadly speaking we can see a similar range of recommendation precision values when compared to the content-based strategies, and a drop off in precision for increasing recommendation list sizes, but this time the relative ordering of the 4 basic approaches is quite different.

For a start, the *Tags/User* based approach produces far and away the best precision values for this set (from about 9% to 15%) and is inline with the best performer of the content-based strategies. However, the *Tags/All* approach, which combines the tags of users, friends and followers, does relatively poorly in the tag-based approach, with precision scores of only 5% to just under 7%. In this case it is clear that tags of friends and followers do not add any recommendation value to the user's own tags, and in fact result in a significant reduction in precision relative to *Tags/User*.

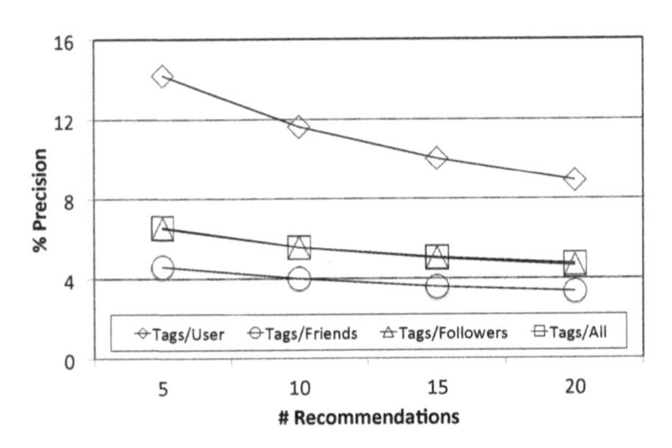


**Fig. 3** Precision vs. Recommendation List Size for Tag-Based Approaches.

We can see that the tags of friends perform worst of all and worse than the tags of followers in particular. This is interesting because, once again, users choose to follow their friends whereas, by and large, they exert little or no control over who follows them and so one would think that a follower list is likely to contain low quality or even spam users. The explanation for this probably depends on a number of factors. First it is important to remember that the tags that we associated with users are actually tags that were associated with *lists* and we have just *transferred* these tags to the members of these lists. Therefore it is likely that some users will be a better fit for some list tags than others; although the fact that we are combining similar tags from multiple lists should help in this regard. More importantly perhaps, at least in order to understand why *Tags/Followers* performs better than expected, is that by definition our tagged users (and tagged followers in particular) are likely to be high quality users to begin with. Why? Because of the fact that someone has chosen to include them in a list. Most users do not find their way into lists and certainly one would expect list curators to do a decent job when it comes to ensure that malicious users are excluded from any lists they create. Hence, the better than expected performance by *Tags/Followers*.

### 4.2.3 Summary Results

A summary of the above results is presented in Figure 4 as a bar chart of mean average precision (MAP) for each of the recommendation approaches. The mean average precision for each technique is simply the average precision for the technique in question across all values of $k$ (recommendation list size). In Figure 4 we show the MAP for each of the 4 recommendation techniques within a given approach as well as a background bar to indicate the average of the MAPs for these 4 individual techniques. As per the above discussion we can clearly see that individual techniques within the content-based and tag-based approaches perform best (*Tweets/All* and *Tags/User* both achieve a MAP of just over 11%). Overall, the content-based approach performs best with an average MAP of about 8% compared to just over 6% for the tag-based approaches.

## 4.3 *Friend-of-a-friend Precision*

Earlier we highlighted the conservative nature of our friend-based precision technique. Clearly it does not allow for the obvious possibility of a good recommendation who is not yet followed by the target user. The concern is that this may skew the performance results in a particular direction and not provide a true insight into practical performance in the field. To address this, in this section we would like to propose an alternative precision metric that provides for a more fine-grained measure of precision and one that can be adjusted in terms of its inherent strictness.

$$FoF(U) = \bigcup_{\forall f_i \varepsilon friends(U)} (friends(f_i)) \tag{12}$$

To do this we make the observation that if we wish to look beyond a user's friends as a source of recommendation test targets then it makes sense to consider the friends of the user's friends or $FoF(U)$ as per Equation 12. Obviously this is typically a much larger set of users than $friends(U)$, often by many orders of magnitude. For example, in our set of test users the average size of $friends(U)$ is about 16 and the average $FoF(U)$ set size is in the region of 500; bear in mind that we are using smaller friend/follower sets for these users than their true friend/follower sets because of the need to ensure that all users are indexed by Listorious. However, it would not be reasonable to consider such a large collection of users as legitimate recommendation targets. Instead, for the purpose of this experiment, we score these *FoF* users by the number of unique tags that they have, as a rough indicator of influence, and then select the top $n$ *FoF* users sorted in descending order of tag count as

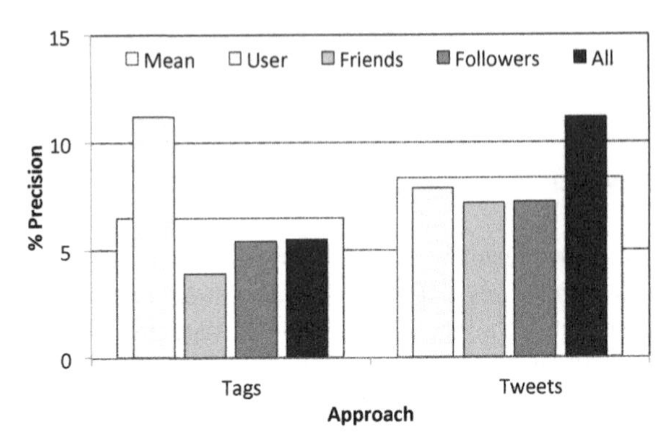


**Fig. 4** Summary of Mean Average Precision for Content, Tag Approaches.

per Equation 13. In this way, we can produce a set of $n$ users who are not friends of $U$ but who are reasonable candidates based on their influence and friendship with the $U$'s friends. Then we calculate a precision score for $U$'s recommendations by comparing them to $FoF(U,n)$. Each user has their own $FoF$ set and can relax/tighten the strictness of the metric by increasing or decreasing $n$.

$$FoF(U,n) = Top_n(Sort_{|user-tags(U_{fof})|}FoF(U)) \tag{13}$$

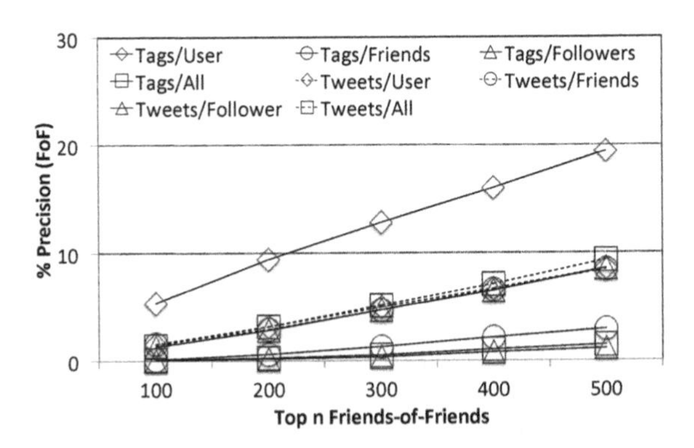


**Fig. 5** Content, Tag-Based Precision using the *FoF* Precision Metric.

To test this idea we re-evaluated the content/tag-based recommendation approaches. We ran each technique to produce a set of 5, 10, 15 and 20 recommendations, then averaged their performance over each set size and evaluated the precision of these recommendations for different values of $n$, the size of the $FoF$ set. The results are shown in Figure 5 as $FoF$ average precision versus $n$. As expected they show a linear increase in precision with increasing values of $n$; obviously if we relax the precision metric to allow for more target test users then precision should increase in line with this. More interesting perhaps is the difference between the relative performance of certain techniques compared to their previous friend-based precision performance. For example, we can see that the *Tags/User* technique significantly outperforms all other techniques, achieving precision scores of 5% - 20%. This was one of the top performers in the previous tests but it was comparable with the *Tweets/All* technique. Now we can see that the *Tweets/All* technique only achieves precision scores of 2% - 9%.

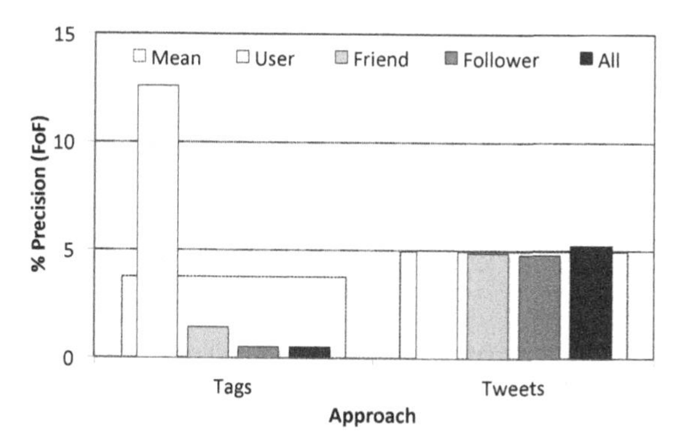


**Fig. 6** Mean Average FoF Precision for Content, Tag-Based Approaches.

These results are clearer in Figure 6 where we chart the mean average precision (averaged over the different values of $n$). *Tags/User* achieves a mean average precision of just over 12.5% compared to about 5% for *Tweets/All*. Once again we can see that despite the high performance of *Tags/User* the other tag-based techniques all perform poorly compared to the content-based approaches.

In summary, this alternative precision metric has revealed some interesting performance differences that were not visible using a more conservative precision metric. Of course this does not mean that the new metric is measuring something useful – it could be that we are no more likely to follow FoF users than any randomly chosen user, but this seems unlikely – and certainly further research is necessary to compare this and our more conservative precision metric to a gold standard such as live-user behaviours. This is beyond the scope of this paper however and as a result we will leave this as a matter for future work.

## 5 Conclusions

In this paper we have described and evaluated a variety of approaches to recommending friends on Twitter. In particular we have proposed a number of novel tag-based approaches using tags extracted from a large collection of manually annotated, curated Twitter lists. And we have compared these approaches to benchmark content-based techniques. We have also proposed a novel approach to measure recommendation precision that suits this particular task and that provides for a more tunable precision metric. During our evaluation this alternative precision metric revealed a number of interesting performance differences compared to a more traditional and conservative precision approach.

## References


1. J. Chen, W. Geyer, C. Dugan, M. Muller, and I. Guy. Make new friends, but keep the old: recommending people on social networking sites. In *Proceedings of the 27th international conference on Human factors in computing systems*, CHI '09, pages 201–210, New York, NY, USA, 2009. ACM.
2. G. Chowdhury. *Introduction to Modern Information Retrieval, Third Edition.* Facet Publishing, 3rd edition, 2010.
3. E. M. Daly, W. Geyer, and D. R. Millen. The network effects of recommending social connections. In *Proceedings of the fourth ACM conference on Recommender systems*, RecSys '10, pages 301–304, New York, NY, USA, 2010. ACM.
4. S. Garcia Esparza, M. P. O'Mahony, and B. Smyth. On the real-time web as a source of recommendation knowledge. In *Proceedings of the fourth ACM conference on Recommender systems*, RecSys '10, pages 305–308, New York, NY, USA, 2010. ACM.
5. J. Hannon, M. Bennett, and B. Smyth. Recommending twitter users to follow using content and collaborative filtering approaches. In *Proceedings of the fourth ACM conference on Recommender systems*, RecSys '10, pages 199–206, New York, NY, USA, 2010. ACM.
6. J. Hannon, K. McCarthy, J. Lynch, and B. Smyth. Personalized and automatic social summarization of events in video. In *Proceedings of the 16th international conference on Intelligent user interfaces*, IUI '11, pages 335–338, New York, NY, USA, 2011. ACM.
7. J. Hannon, K. McCarthy, M. P. O'Mahony, and B. Smyth. A multi-faceted user model for twitter. In *Proceedings of the 20th conference on User Modeling, Adaptation, and Personalization*, UMAP '12. ACM, 2012.
8. H. Kwak, C. Lee, H. Park, and S. Moon. What is twitter, a social network or a news media? In *Proceedings of the 19th international conference on World wide web*, WWW '10, pages 591–600, New York, NY, USA, 2010. ACM.
9. S. Lo and C. Lin. Wmr–a graph-based algorithm for friend recommendation. In *Proceedings of the 2006 IEEE/WIC/ACM International Conference on Web Intelligence*, WI '06, pages 121–128, Washington, DC, USA, 2006. IEEE Computer Society.
10. R. Mihalcea, C. Corley, and C. Strapparava. Corpus-based and knowledge-based measures of text semantic similarity. In *Proceedings of the 21st national conference on Artificial intelligence - Volume 1*, AAAI'06, pages 775–780. AAAI Press, 2006.
11. J. Nichols, J. Mahmud, and C. Drews. Summarizing sporting events using twitter. In *Proceedings of the 2012 ACM international conference on Intelligent User Interfaces*, IUI '12, pages 189–198, New York, NY, USA, 2012. ACM.
12. O. Phelan, K. McCarthy, M. Bennett, and B. Smyth. On using the real-time web for news recommendation and discovery. In *Proceedings of the 20th international conference companion on World wide web*, WWW '11, pages 103–104, New York, NY, USA, 2011. ACM.
13. A. I. Schein, A. Popescul, L. H. Ungar, and D. M. Pennock. Methods and metrics for cold-start recommendations. In *Proceedings of the 25th international ACM conference on Research and development in information retrieval*, SIGIR '02, New York, NY, USA, 2002. ACM.
14. D. A. Shamma, L. Kennedy, and E. F. Churchill. Tweet the debates: understanding community annotation of uncollected sources. In *Proceedings of the first SIGMM workshop on Social media*, WSM '09, pages 3–10, New York, NY, USA, 2009. ACM.

# Collaborative Filtering For Recommendation In Online Social Networks*

Steven Bourke, Michael P O'Mahony, Rachael Rafter, Kevin McCarthy, Barry Smyth

**Abstract** In the past recommender systems have relied heavily on the availability of ratings data as the raw material for recommendation. Moreover, popular collaborative filtering approaches generate recommendations by drawing on the interests of users who share similar ratings patterns. This is set to change because of the *unbundling* of social networks (via open APIs), providing a richer world of recommendation data. For example, we now have access to a richer source of ratings and preference data, across many item types. In addition, we also have access to mature social graphs, which means we can explore different ways of creating recommendations, often based on explicit social links and friendships. In this paper we evaluate a conventional collaborative filtering framework in the context of this richer source of social data and clarify some important new opportunities for improved recommendation performance.

## 1 Introduction

Today recommender systems provide a mature approach for addressing the information discovery challenge facing online users and offer service providers a considerable advantage when it comes to promoting content, products, and services directly to users based on their preferences. Much of the success of recommender systems can be traced back to collaborative filtering techniques [1] that rely on large quantities of ratings or transactional data.

Recently researchers have begun to question some of the basic assumptions that inform collaborative filtering. For example, [3, 23] studied the importance of neigh-

Steven Bourke, Michael P O'Mahony, Rachael Rafter, Kevin McCarthy, Barry Smyth
Clarity: Center for Sensor Web Technologies, University College Dublin, Dublin 4, Ireland e-mail: {steven.bourke, michael.omahony, rachael.rafter, kevin.mccarthy, barry.smyth}@ucd.ie

* Supported by Science Foundation Ireland under Grant No. 07/CE/11147 CLARITY

bours when making a ratings prediction to conclude that in fact neighbours can play a relatively minor role in prediction accuracy or user satisfaction. Elsewhere researchers have started to consider other sources of recommendation knowledge to compliment ratings-based similarity data. For example, with the rise of the social web there has been considerable interest in modelling the reputation of users to bias future recommendations/predictions with respect to users who are both relevant *and* reputable; see for example [11, 18]. On the topic of leveraging social web data for recommendation, research such as [8] explores the use of microblogging services like Twitter as a new source of product data and user opinions, showing how even this noisy information can be used to make reliable recommendations. And finally, the social web has proven to be a fertile ground for new recommendation tasks whether recommending helpful product reviews to users [20, 22], or tags [24], or even suggesting connections and friends in social networks [7, 9].

In this paper, rather that rush to develop a new type of recommender system, we felt it worthwhile to return to recommendation basics by evaluating conventional collaborative filtering in the context of new sources of social recommendation data. This is important if it helps to establish a baseline for future research as well as providing an opportunity to reconsider some basic recommender systems assumptions in the context of the availability of new types of social data. These new data sources have been made available as major social networks such as Facebook and Twitter have provided API access to user data. This data is interesting on two fronts. First it can provide access to large quantities of ratings-like preference data, across a variety of item types. Secondly it can provide access to mature social graphs based on explicit social links and real-world friendships. The work of [2, 3, 15], for example, concluded that people are more likely to respond well to recommendations from their friends suggesting that recommender systems should take advantage of explicit social connections where possible; see also the work of [11, 17] on a social recommendation technique based on trust propagation in social networks.

The central contribution of this work is an experiment designed to evaluate the effectiveness of collaborative filtering under a number of data conditions. This includes varying the way in which the user neighbourhoods and candidate item sets are formed during collaborative filtering; we compare traditional approaches based purely on ratings similarity across a user population to approaches that rely on more constrained populations of users, such as friends and friends-of-friends. We also vary the items used during profiling and recommendation. The results are interesting. For instance they reveal that significant improvements in recommendation quality can be achieved by using more constrained populations of users, at least across most item types. The remainder of this paper is organised as follows. In the next section we describe the relevant background material, after which we describe the experimental setup in terms of the dataset, algorithms, and methodology. Section 4 will describe and discuss the results and finally we will discuss the implications of this work for future recommender systems research.

## 2 Background

In this section we review relevant background material related to our work. Specifically we look at recommender systems with a particular focus on collaborative based systems. We then move onto recommender systems which use some form of social information in the recommendation process, and then finally some newer social networking based recommender systems. Recommender systems can be divided into two categories, content and collaborative filtering based techniques. A content based recommender system will usually leverage some form of meta data alongside user and item transaction information. For example, meta data could include genre or actor information in a movie recommender. How this information is used in a content based recommendation can vary significantly, for instance, in [21] the authors use vector space models to better represent content when generating recommendations; the authors find the vector space model helps to improve the overall accuracy. In [5] the authors represent all content in a hyper graph so that it can be combined alongside implicit data. This allows the authors to treat content based information the same way as transactional information within their recommender system.

The second style of recommendation is known as collaborative filtering, which avoids the need for meta data, and relies solely on user and item transaction information to expose the underlying preferences users have for items. Collaborative filtering techniques can be divided into two categories, known as model-based and memory based techniques [4]. Memory based techniques initially gained popularity due to the seminal work of the GroupLens research group [13], in which the transaction data is represented as a sparse matrix and then either a user-based [10] or an item-based technique [12] is employed to generate recommendations. User-based techniques work by analysing common user preferences for items in order to generate recommendations. This is done by measuring profile similarity between users to form $k$-nearest neighbourhoods. Then a ranking technique is used to score items from the neighbourhood for recommendation to a particular target user. Alternatively, item-based recommendation analyses the similarity between items, expressed in terms of the shared opinions of users who have consumed the items, in order to generate recommendations. Candidate items are ranked based on their similarity to the target user's previously rated items [12]. Improvements in accuracy via the use of extensions for user-based and item-based collaborative filtering are discussed in [4] and [10].

A model-based collaborative technique generally involves the application of a machine learning technique to the user,item matrix, for example clustering or Bayesian networks [4]. More recently, in the Netflix prize [6] model-based techniques based on matrix factorisation have gained a lot of popularity due to the improvements in accuracy they offer over typical memory based approaches on the Netflix dataset [14]. Matrix factorisation involves reducing the user item matrix into a dense latent feature space for both users and items. The latent features represent how strongly an item relates to a certain feature, and the extent to which a user is partial to that feature. Predicting relevant items for the target user is done by calculating the dot product of a user and item feature vector. However, while matrix factorisation has proved to be quite accurate in the Netflix competition, related research has suggested that it may not always be the most accurate approach. For instance in [16] the authors find that user-based recommendation outperforms matrix factorisation in the context of an online auction environment.

The incorporation of social network information into recommender systems is a relatively new direction in this field, prompted by the *unbundling* of networks such as Facebook and Twitter (via open APIs) which reveal rich social data and mature social graphs. For example [8] and [9] exploit social network information in content-based recommendation scenarios. Alternatively, [25] propose a graph based approach for memory-based collaborative filtering that uses transitive similarity metrics to calculate similarities between users based on their social graph, leading to improvements compared to standard user-based collaborative filtering. For model-based collaborative filtering there are approaches such as that proposed in [11] which looks at incorporating trust from social networks into the recommendation process. The authors report improvement over standard matrix factorisation. Finally, in [17], social information is incorporated into the regularisation process of their model training. In our work we focus specifically on the recommendation challenge presented in finding relevant content for users of online social networks. Our work measures the performance of known recommendation techniques in social environments. We add additional social filtering to the process as well as measure cross domain recommendations. To date we are unaware of anyone that has done a detailed study into the effectiveness of recommendation in these types of online social networks.

## 3 Experimental Setup: Data, Algorithms & Conditions

The data for this study was obtained using the Facebook API from Sept. to Oct. 2011 to collect 42,550 user profiles containing links (URLs), videos and check-ins (location ids) shared by these users. In other words, these are unary profiles containing only a user's *positive* preferences, in the form of items that they have actively shared.

We consider a number of basic collaborative filtering style algorithms for the purpose of this study. Specifically we evaluate a user-based and an item-based recommendation algorithm. For our user-based algorithm we use a standard approach described in [19]. To generate a set of recommendations for a target user, the steps involved are as follows:

- The algorithm first finds neighbours using a similarity thresholding approach where users with non-zero similarity to the target user are selected
- User similarity is computed using the Jaccard index over the ratings vector of each user.
- Next, assuming unary ratings, the score for each of the neighbour items (less those that appear in the target user profile) is calculated from the sum of the similarities of the neighbours that contain that item. The $k$ items with the highest

scores are then recommended to the target user. We will refer to this algorithm as the standard user-based approach.

For our item-based approach our technique is based on the the work proposed in [12]. The algorithm works as follows:

- To generate a set of recommendations firstly an *item-item* similarity matrix is built using the Jaccard index against co-rated items.
- The algorithm identifies the most similar items to the target user's items, these are considered the candidate items for recommendation
- The candidate items are then ranked for recommendation by comparing their overall similarity with the previously rated items in the target user's profile.

Next we define two variations of the standard collaborative filtering configuration used in both the user-based and item-based algorithms, by changing the source of users from whom items can be recommended. In the *FRIEND* algorithm we consider only items from friends of the target user, instead of the full set of all users. This allows us to consider the utility of direct social connections as a means of identifying neighbours, rather than using anonymous users with similar ratings patterns. Second, we define *FoF*, this time only considering items from the target user's friends of friends. When applying the FRIEND or FoF variation to the item-based algorithm, only items that exist in either the FRIEND or FoF respectively are considered as candidates for recommendation, i.e. the co-ratings can only come from users that exist within that subgroup. In the case of the user-based variations, only users that exist in the FRIEND or FoF configuration can be used as neighbours when generating recommendations. In both variations once the neighbours have been selected, the candidate items are ranked as in the standard algorithm.

Finally, we also consider a number of input/output variations by changing the rules regarding the types of items that may appear in a profile (input) or be recommended (output). Using the Facebook data we can consider links, videos, location check-ins or any combination thereof as either input, output, or both. This leads to 16 input/output combinations to test with our 3 algorithms (CF, i.e. standard collaborative filtering, FRIEND and FoF).

## 4 Results

### *4.1 Methodology*

From the 42,550 users in the Facebook dataset we identified a subset of 3,419 users whose profiles contained at least 5 items of *each* type (links, videos, check-ins) as the *target users* (the users for whom we will make recommendations); see Table 1. Then, for any other user, in order to qualify as a potential neighbour when recommending a particular type of item, a user must have at least 5 of those items in their profile (although they may have fewer than 5 of the other types of items). Table 1 reports the dataset statistics for each recommendation task (links, videos, or check-ins).

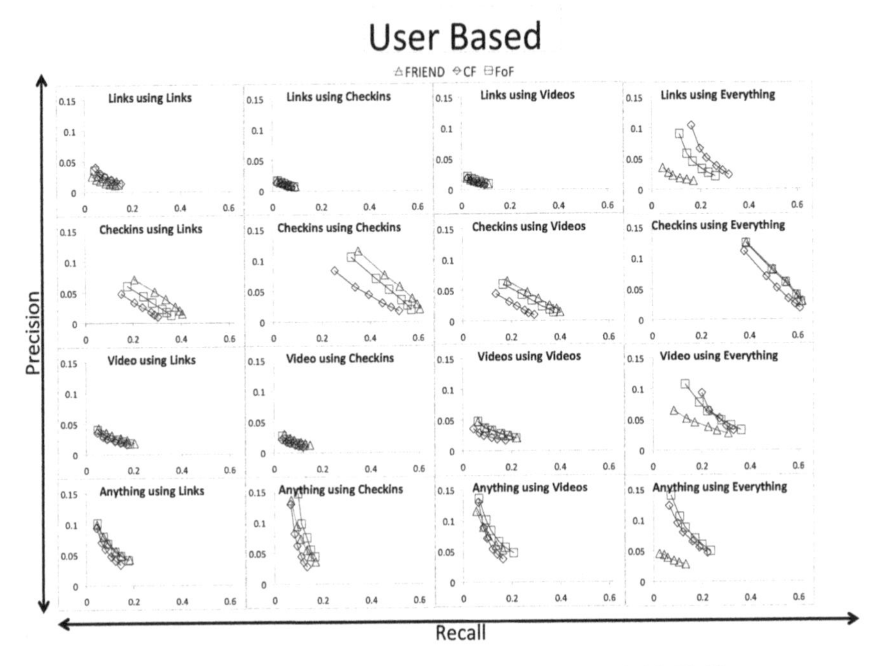


**Fig. 1** Precision vs. Recall for recommendation lists sizes of 5, 10, 15, 25, 35, 50.

**Table 1** Data Statistics

| Type | # Target Users | # Other Users | # Items |
|---|---|---|---|
| Links | 3,419 | 29,517 | 529,257 |
| Video | 3,419 | 23,731 | 171,901 |
| Check-in | 3,419 | 4,776 | 15,259 |

For our experiments we adopted a standard *leave-one-out* testing approach. Each of the 3,419 target users is considered in turn for testing. For each target user, its profile is randomly divided into 20% *test* and 80% *training* data; this is repeated 5 times for each user. Each set of training data is used as the basis for identifying recommendations, which are then compared to the test data items using precision, recall and the F1 measure. We also examine the item coverage in each case. Results are averaged for different length $k$ of recommendation list ($k = 5, 10, 15, 25, 35, 50$).

To measure significance, we firstly analyse the results using Krustal Wallis, then Tukey's is used as the post-hoc analysis to measure for pair-wise significance amongst the different techniques. The p–value is set to 0.05.

The detailed results of these experiment are given in Figs. 1 and 2 (user-based) and Figs. 3 and 4 (item-based). Figs. 1 and 3 show the precision and recall, while Figs. 2 and 4 show the F1 measure, coverage and significance results. All figures contain 16 graphs, one for each of the 16 profile/recommendation variations above. In each graph we indicate the profile/recommendation combination in the graph title; e.g. *Links using Check-ins* indicate links being recommended but using check-in data to select the neighbourhoods. In each graph we present a set of results for the three algorithmic variations. For example, in Figs. 1 & 3 we present the precision and recall results for CF, FRIEND, and FoF for varying recommendation list sizes ($k = 5, 10, 15, 25, 35, 50$). In Figs. 2 & 4 each graph depicts a bar chart of the mean F1 (across $k$) for each of the algorithmic variations, and a line graph showing recommendation *coverage* (the percentage of recommendation trials where at least one recommendation could be made). Finally, statistical significance is indicated by the pattern of the bar in the F1 charts, if the internal pattern for a condition is unique that means that the technique was statistically significant, if it is not unique that represents a non statistically significant result. The F1 charts in Figs. 2 & 4 give us a general performance of each technique. As there are too many charts to discuss individually, we will now highlight what we believe to be the most interesting insights.

## 4.2 User Based Results

From these results we can make some interesting observations. Coverage favours standard CF, which beats FoF and FRIEND in each of the 16 conditions, albeit marginally. This is not surprising since the CF condition enjoys a much larger population of candidate neighbours (some 42,550 users) than FoF (median of 6,500 users) or FRIEND (median of 190 users).

We also see that CF rarely outperforms FRIEND and FoF; in fact only when links are recommended and only then when profiles contain only links, or everything (links, videos, check-ins); Notice too the FRIEND algorithm tends to beat CF and FoF when recommending check-ins (see the $2^{nd}$ row of graphs in Figs. 1 and 2). This makes sense because check-ins are physical locations and thus more likely to be accurately recommended by people who share these locations, and a higher density of these people are likely to be a user's friends. The FoF and FRIEND algorithms outperform CF when profiles contain everything (links, videos, check-ins) and when recommending anything; see the *Anything using Everything* graph in Figs. 1 and 2. It is also noted that when using a richer profile (The *Everything* condition) accuracy is always the best performing approach.

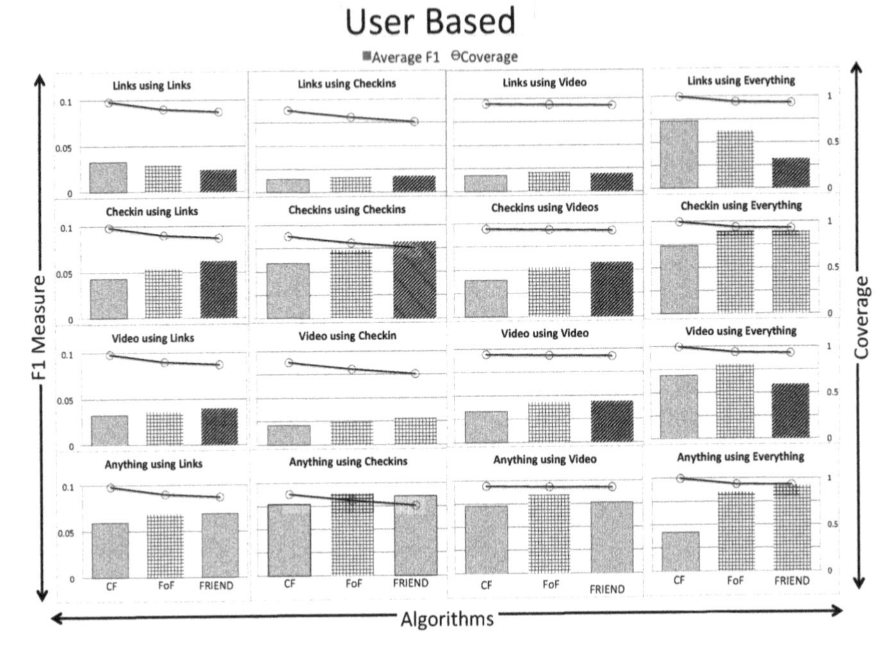


**Fig. 2** User Based: Average F1 and Coverage results.

## 4.3 Item Based Results

In our item-based results we can firstly note that the level of coverage amongst the different techniques and item combinations varies throughout the experiment compared to the user-based coverage. The general trend is that CF has the highest rate of coverage where as FoF and FRIEND drop off in instances where the profile consists of *Links,Videos* or *Checkins*.

We can see that in most cases using FRIEND and FoF outperform CF in terms of accuracy. The only occasion where CF actually outperforms any social based technique is in the case of recommending *Links* and using *Check-ins*. We can note that in general item-based recommendation seems to have a more difficult time accurately predicting recommendations in the *Links* domain under item-based compared to *Videos*, *Checkins* or *Anything*. The highest F1 score is 0.028 when using FoF with *Links* using *Everything* (Fig. 4). Links are difficult for the item-based recommender to accurately predict due to the sheer number of candidate items which reduces the likelihood of two items being co-rated, thus and low levels of similarity between links due to a comparatively low rate of co-rating between items (e.g the number of users who co-rated $item_i$ and $item_j$ is 2). The sparsity of our item-item matrix for links is 0.0037. When recommending check-ins we can see that using FRIEND is the best performing technique in all cases except for *Checkins using Videos* (Fig. 4), but if we look at the overall precision and recall (Fig. 3) we note that while

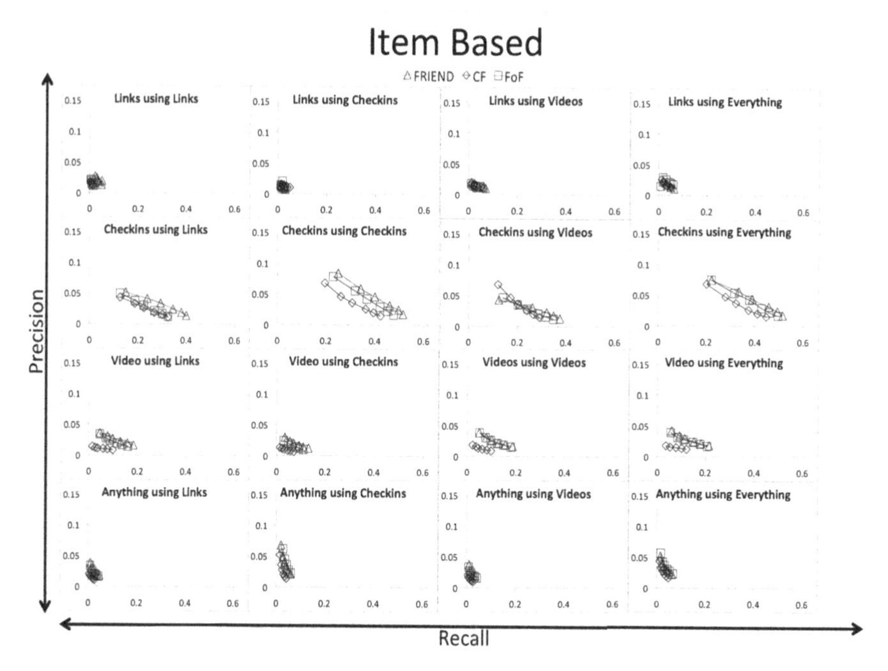


**Fig. 3** Item Based: Precision vs. Recall for recommendation lists sizes of 5, 10, 15, 25, 35, 50.

FRIEND is the best overall approach there are minor differences in overall precision between FRIEND and FoF. Once again we believe that using a persons' friends is more favourable for check-ins because the likelihood of sharing similar physical preferences amongst friends is more likely. When recommending *Anything* using item-based recommendation, we can see that check-ins give a similar performance as to using *Everything*. When we look at precision on its own, Figs. 1 & 3, we can see that the user-based algorithm across different configurations can achieve at least 0.1 in terms of precision where as the item-based algorithm never achieves this.

### 4.4 Summary Analysis

Given the detail in the above charts, it is useful to summarise the results by averaging the results over the different profile and recommendation types. For example, in Fig. 5(a) we present a clustered bar chart of the mean F1 for all list sizes. For each of the three user-based algorithms, results are averaged over each profiling condition (links (L), check-ins (C), videos (V), everything (E)). For each cluster of bars we also show the mean F1 averaged over CF, FRIEND and FoF. In this case we see that CF is consistently beaten by both FRIEND and FoF. It is also clear that there is a benefit to using all three types of profile data (the E condition), with an overall mean F1 of

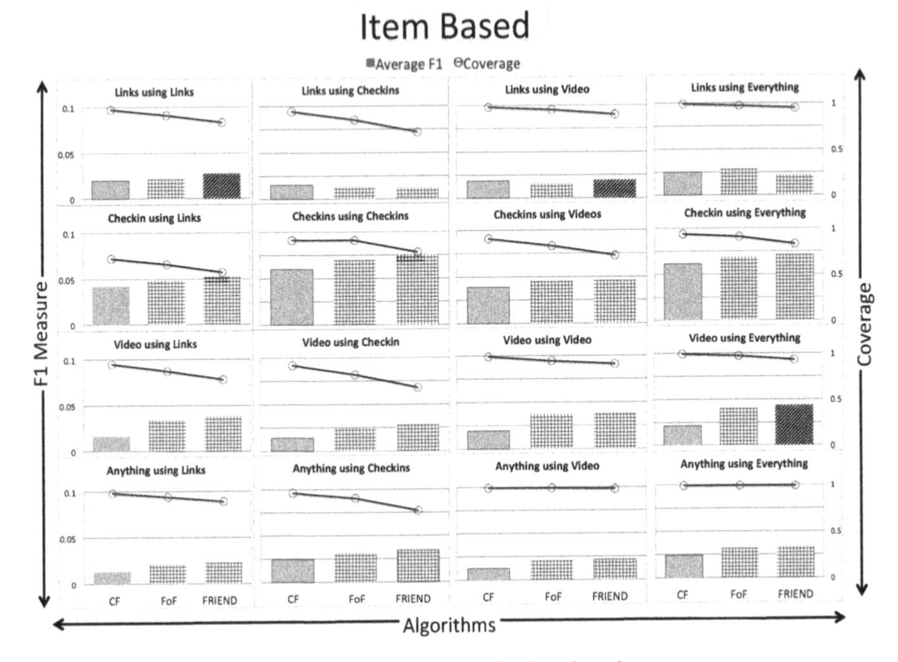


**Fig. 4** Item Based: Average F1 and Coverage results For Item based.

0.07 compared to 0.05 for any individual profiling technique. In general FRIEND and FoF beat CF by about 15% to 20% for each of the L, C, and V conditions, but there is a win for FoF when profiles contain everything (E condition) where we see FoF beating CF and FRIEND by about 25%.

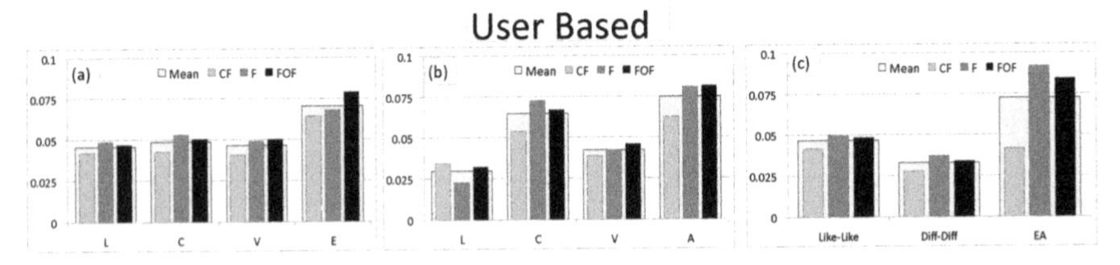


**Fig. 5** (a) F1 when recommending items *using* Links, Check-ins, Videos and Everything; (b) Average F1 when *recommending* Links, Check-ins, Videos or Anything; (c) F1 based on within-domain cross-domain variations.

Summarising the results by recommendation type we obtain an analogous bar graph as shown in Fig. 5(b). This time we see that the best overall performance is achieved when recommending anything (the A condition). Again, both FRIEND and FoF outperform CF, with an average F1 of just over 0.08 compared to just 0.06 for CF, a relative increase of some 30%. Also interesting is the strong performance when recommending check-ins (the C condition); this time the FRIEND algorithm wins overall, beating CF by about 33%. This view of the data also clarifies those conditions where CF produces a higher average F1 than FRIEND or FoF: this only happens in the case where links are recommended (the L condition) where the F1 score for FRIEND is about 30% less than CF.

Next, we consider when the same type of items are used across profiling and recommendation (*Like-Like*) versus when different types of items are used (*Diff-Diff*), and compare these to when we profile using everything and recommend anything (*EA*). The average F1 results are shown in Fig. 5(c). We see superior performance under the *EA* condition. Clearly it is easier to make good recommendations by maximising the availability of profile data without restricting the type of items that can be recommended. While this might not be surprising, what is surprising is the scale of the benefits accruing to the FRIEND and FoF algorithms which deliver more than double the F1 performance of CF. Finally, while cross-domain recommendations are feasible (*Diff-Diff*), they are much less effective than within-domain recommendations (*Like-Like*); on average the F1 score for *Like-Like* is approximately 0.046 compared to 0.032 for *Diff-Diff*, an increase of about 40% for the former. And once again, for both *Like-Like* and *Diff-Diff*, we find that the FRIEND and FoF algorithms outperform CF.

In Figure 6 we performed a similar summary as presented in Figure 5 except that Figure 6 refers to our item-based recommendation results. For the different profiling conditions (Fig. 6(a)) we can see no distinct best performer. We do note that the E condition does slightly outperform the C condition, however the difference is insignificant with E performing at 0.039 and C performing at 0.033. Interestingly we do note that FoF and FRIEND always outperform standard CF. In Fig. 6(b) we see that Checkins(C) is by far the easiest item to recommend for when using a item-based recommender system. For C the F1 being 0.56 while the next best performing approach is Video(V) at 0.029. We do note with the exception of V that CF , FRIEND and FOF all perform to a similar degree, but with CF still being the least effective of the three. Lastly in Fig. 6(c) we can see that item-based recommendation performs best in the typical *Like-Like* configuration, with FRIEND and FoF outperforming CF. In the case of *Diff-Diff* and *EA* we see overall similar results both in terms of accuracy and also in regards to how the CF, FRIEND and FoF approaches perform.

From our summary analysis we can determine that user-based recommendation is generally more preferable in terms of accuracy compared to item-based recommendation. We also note that using either user-based or item-based recommendation we achieve higher accuracy when applying the FRIEND or FOF filters. One difference between user-based and item-based recommendation is that user-based performs well when using the Everything(E) profile (User based: Fig. 5(a), Item-based Fig. 6(a)). We also see that user-based is more accurate when recommending Anything (A) Fig. 5b compared to item-based Fig. 6(b). Finally that when recommending *Everything using Anything* (EA) user-based also more accurate than item-based (Fig. 5(c) & Fig. 6(c)).

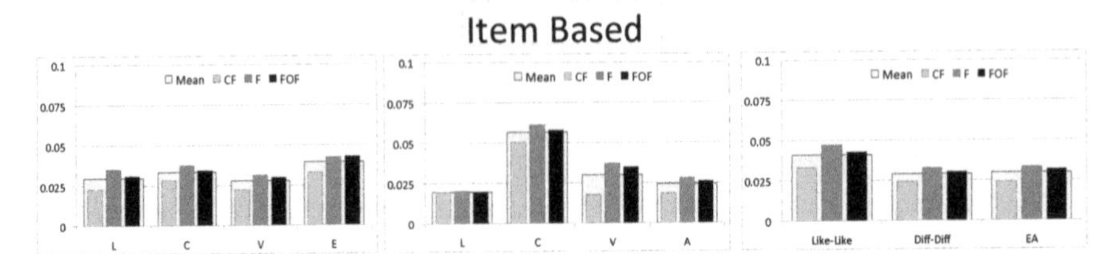


**Fig. 6** (a) F1 when recommending items *using* Links, Checkins, Videos and Everything; (b) Average F1 when *recommending* Links, Checkins, Videos or Anything; (c) F1 based on within-domain and cross-domain variations.

## 5 Conclusions

In this paper we examined user-based and item-based collaborative filtering in the context of the social structures made available from social networks like Facebook. As a result of the large quantities of data being created and shared in these online social networks, users can struggle with the well known challenge of information overload, making it difficult for users to easily find interesting content. To help users with these challenges we performed a detailed evaluation of user and item-based collaborative filtering in comparison to variations that exploited a user's explicit friends or friends-of-friends as alternative sources of information. Furthermore we examined performance when utilising and recommending a variety of item types (links, videos and check-ins). A key result is that the FRIEND and FoF methods tend to outperform a typical collaborative filtering configuration. We also found that recommending check-ins proves to be easier than links or videos, possibly because of the potential for more limited variability and the requirement of a physical relationship to exist between the item and user. Our results confirm higher accuracies for same-domain versus cross-domain recommendations. We also found that in an environment where any type of item can be recommended, building a profile from *check-ins* to perform recommendations performs at a similar level of accuracy as building the profile from *everything*. Finally we have found that user-based collaborative filtering is more accurate than item-based collaborative filtering when recommending information in online social networks. Our work demonstrates that when using collaborative filtering based techniques, the use of friends, or friends of friends, as opposed to typical collaborative filtering returns higher levels of accuracy. Based on these results we conclude that for future work we should look to improve upon results achieved via user based techniques by more extensively leveraging the social graph information while also exploring ways to blend users interests in different item types.

## References

1. Gediminas Adomavicius and Alexander Tuzhilin. Toward the next generation of recommender systems: A survey of the state-of-the-art and possible extensions. *IEEE Trans. on Knowl. and Data Eng.*, 17(6):734–749, June 2005.
2. P. Bonhard and M. A. Sasse. 'knowing me, knowing you' – using profiles and social networking to improve recommender systems. *BT Technology Journal*, 24(3):84–98, July 2006.
3. Steven Bourke, Kevin McCarthy, and Barry Smyth. Power to the people: exploring neighbourhood formations in social recommender system. In *Proceedings of the fifth ACM conference on Recommender systems*, RecSys '11, pages 337–340, New York, NY, USA, 2011. ACM.
4. John S. Breese, David Heckerman, and Carl Kadie. Empirical analysis of predictive algorithms for collaborative filtering. In *Proceedings of the Fourteenth conference on Uncertainty in artificial intelligence*, UAI'98, pages 43–52, San Francisco, CA, USA, 1998. Morgan Kaufmann Publishers Inc.
5. Jiajun Bu, Shulong Tan, Chun Chen, Can Wang, Hao Wu, Lijun Zhang, and Xiaofei He. Music recommendation by unified hypergraph: combining social media information and music content. In *Proceedings of the international conference on Multimedia*, MM '10, pages 391–400, New York, NY, USA, 2010. ACM.
6. Alex Conference Chair-Tuzhilin and Yehuda Conference Chair-Koren. Netflix '08: Proceedings of the 2nd kdd workshop on large-scale recommender systems and the netflix prize competition. August 2008.
7. Elizabeth M. Daly, Werner Geyer, and David R. Millen. The network effects of recommending social connections. In *Proceedings of the fourth ACM conference on Recommender systems*, RecSys '10, pages 301–304, New York, NY, USA, 2010. ACM.
8. Sandra Garcia Esparza, Michael P. O'Mahony, and Barry Smyth. Mining the real-time web: A novel approach to product recommendation. *Knowledge-Based Systems*, 29(0):3 – 11, 2012.
9. John Hannon, Mike Bennett, and Barry Smyth. Recommending twitter users to follow using content and collaborative filtering approaches. In *Proceedings of the fourth ACM conference on Recommender systems*, RecSys '10, pages 199–206, New York, NY, USA, 2010. ACM.
10. Jon Herlocker, Joseph A. Konstan, and John Riedl. An empirical analysis of design choices in neighborhood-based collaborative filtering algorithms. *Inf. Retr.*, 5(4):287–310, October 2002.
11. Mohsen Jamali and Martin Ester. A matrix factorization technique with trust propagation for recommendation in social networks. In *Proceedings of the fourth ACM conference on Recommender systems*, RecSys '10, pages 135–142, New York, NY, USA, 2010. ACM.
12. George Karypis. Evaluation of item-based top-n recommendation algorithms. In *Proceedings of the tenth international conference on Information and knowledge management*, CIKM '01, pages 247–254, New York, NY, USA, 2001. ACM.
13. Joseph A. Konstan, Bradley N. Miller, David Maltz, Jonathan L. Herlocker, Lee R. Gordon, and John Riedl. Grouplens: applying collaborative filtering to usenet news. *Commun. ACM*, 40(3):77–87, March 1997.
14. Yehuda Koren, Robert Bell, and Chris Volinsky. Matrix factorization techniques for recommender systems. *Computer*, 42(8):30–37, August 2009.
15. Danielle H. Lee and Peter Brusilovsky. Social networks and interest similarity: the case of citeulike. In *Proceedings of the 21st ACM conference on Hypertext and hypermedia*, HT '10, pages 151–156, New York, NY, USA, 2010. ACM.
16. Yanen Li, Jia Hu, ChengXiang Zhai, and Ye Chen. Improving one-class collaborative filtering by incorporating rich user information. In *Proceedings of the 19th ACM international conference on Information and knowledge management - CIKM '10*, page 959, New York, New York, USA, October 2010. ACM Press.
17. Hao Ma, Dengyong Zhou, Chao Liu, Michael R. Lyu, and Irwin King. Recommender systems with social regularization. In *Proceedings of the fourth ACM international conference on Web search and data mining*, WSDM '11, pages 287–296, New York, NY, USA, 2011. ACM.

18. Paolo Massa and Paolo Avesani. Trust-aware recommender systems. In *Proceedings of the 2007 ACM conference on Recommender systems*, RecSys '07, pages 17–24, New York, NY, USA, 2007. ACM.
19. Andreas Mild. An improved collaborative filtering approach for predicting cross-category purchases based on binary market basket data. *Journal of Retailing and Consumer Services*, 10(3):123–133, May 2003.
20. Samaneh Moghaddam, Mohsen Jamali, and Martin Ester. Etf: extended tensor factorization model for personalizing prediction of review helpfulness. In *Proceedings of the fifth ACM international conference on Web search and data mining*, WSDM '12, pages 163–172, New York, NY, USA, 2012. ACM.
21. Cataldo Musto. Enhanced vector space models for content-based recommender systems. In *Proceedings of the fourth ACM conference on Recommender systems*, RecSys '10, pages 361–364, New York, NY, USA, 2010. ACM.
22. Michael P. O'Mahony and Barry Smyth. Learning to recommend helpful hotel reviews. In *Proceedings of the third ACM conference on Recommender systems*, RecSys '09, pages 305–308, New York, NY, USA, 2009. ACM.
23. Rachael Rafter, Michael P. O'Mahony, Neil J. Hurley, and Barry Smyth. What have the neighbours ever done for us? a collaborative filtering perspective. In *Proceedings of the 17th International Conference on User Modeling, Adaptation, and Personalization*, UMAP '09, pages 355–360, Berlin, Heidelberg, 2009. Springer-Verlag.
24. Shilad Sen, Jesse Vig, and John Riedl. Tagommenders: connecting users to items through tags. In *Proceedings of the 18th international conference on World wide web*, WWW '09, pages 671–680, New York, NY, USA, 2009. ACM.
25. Panagiotis Symeonidis, Eleftherios Tiakas, and Yannis Manolopoulos. Product recommendation and rating prediction based on multi-modal social networks. In *Proceedings of the fifth ACM conference on Recommender systems*, RecSys '11, pages 61–68, New York, NY, USA, 2011. ACM.

# Unsupervised Topic Extraction for the Reviewer's Assistant

Ruihai Dong, Markus Schaal, Michael P. O'Mahony, Kevin McCarthy, and Barry Smyth

**Abstract** User generated reviews are now a familiar and valuable part of most e-commerce sites since high quality reviews are known to influence purchasing decisions. In this paper we describe work on the *Reviewer's Assistant* (RA), which is a recommendation system that is designed to help users to write better reviews. It does this by suggesting relevant topics that they may wish to discuss based on the product they are reviewing and the content of their review so far. We build on prior work and describe an unsupervised topic extraction module for the RA system that enhances the system's ability to automatically adapt to new content categories and application domains. Our main contribution includes the results of a controlled, live-user study to show that the RA system is capable of supporting users to create reviews that enjoy higher quality ratings than Amazon's own high quality reviews, even without using manually created topic models.

## 1 Introduction

Customer reviews have become an important part of our online (and indeed real-world) shopping experiences. Today the vast majority of e-commerce sites feature customer reviews prominently in their core product and service listings. Indeed this type of user-generated content has become a vital part of the value proposition of services such as Amazon and TripAdvisor, so much so that users will often use these resources and similar as a source of product reviews regardless of whether they wish to make a purchase on the site in question. Online reviews can be helpful and can influence the buying patterns of shoppers. For example, [10] describe the results of one study on the value of online reviews, concluding that consumers do under-

Ruihai Dong, Markus Schaal, Michael P. O'Mahony, Kevin McCarthy, and Barry Smyth
CLARITY: Centre for Sensor Web Technologies, University College Dublin, Ireland e-mail: [Ruihai.Dong,Markus.Schaal,Michael.Omahony,Kevin.Mccarthy,Barry.Smyth]@ucd.ie

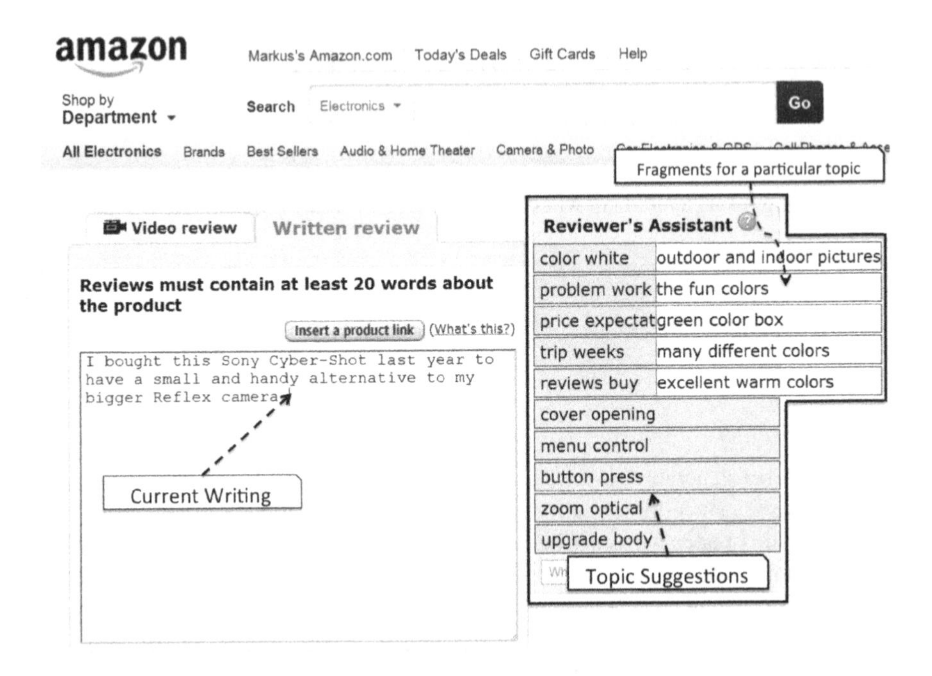


**Fig. 1** The *Reviewer's Assistant* - Browser Plugin

stand the value difference between favorable and unfavorable opinions and respond accordingly. Furthermore, when consumers read online reviews, they pay attention not only to review scores but to other contextual information such as a reviewer's reputation and reviewer exposure. The market responds more favorably to reviews written by reviewers with better reputation and higher exposure. In related work, [15] examine the influence of online reviews on video game sales, indicating that reviews are more influential for less popular games. It is not surprising then that retailers and researchers have started to study different ways to help interested users find high quality reviews for products they are considering. For example, the work of [13, 12, 11] considered different factors such as reviewer reputation, product genre familiarity, and review recency to automatically rank reviews based on their predicted helpfulness.

In this work, we adopt a very different approach when it comes to driving review quality. We focus on the creation of reviews rather than the ranking of existing reviews. We build on prior work on the Reviewer's Assistant (RA) that supports the creation of new reviews by recommending topics to the user while she is writing her review. The aim of RA is to not only produce a better quality of user-generated review, but also to increase the number and diversity of reviews by attracting first-time reviewers who might be initially daunted at the prospect of writing a product review. RA is inspired by the GhostWriter system, first introduced by Bridge et al. [5], as an approach to supporting users to create online adverts for personal goods and items they wish to dispose of and later adapted to support users during the generation of product reviews; Healy and Bridge [9]. Currently GhostWriter 2.0 [9] extracts noun phrases from past product reviews (cases), and suggests these phrases directly to the reviewer/user. Dong et al. adopt a similar approach but compare nouns vs. noun phrases [6] and topics vs. nouns [7] in order to make better suggestions to the user.

In contrast to the work of [7] which relied on costly hand-coded topic models, the central contribution of this work is an automatic approach to topic extraction using the Latent Dirichlet Allocation (LDA) algorithm [4]. This way, new contexts and domains of application can easily and automatically be explored by the RA.

In addition to the technical description of the new technique for automatic topic extraction, we also describe a live-user trial of the RA *auto-topic* system in which we compare the performance of the new variation of RA as compared to the prior versions RA *non-topic* (see [6]) and RA *topic* (see [7]). We do not only report on the perceptions of the users after they use the system but also provide an objective analysis of review quality based on an independent set of review benchmarks. We show that users find the RA system to be useful and helpful and, crucially, the quality of the resulting reviews is statistically superior to high-quality reviews written by users without the RA.

In Section 2 we describe the core architecture and technical features of the Reviewer's Assistant and how this has been incorporated into a browser plugin that can support users as they author Amazon product reviews. Then, in Section 3, we describe the results of a live-user evaluation and a blind-study of review quality before concluding and considering a number of opportunities for future work.

## 2 The Reviewer's Assistant

The Reviewer's Assistant has been developed as a browser plugin so that it can integrate directly with review systems across a wide variety of web sites, cf. [6]. Briefly, the Reviewer's Assistant takes the form of an additional recommendation module that appears on review-creation pages. Figure 1 shows this in the context of an Amazon review page with the RA appearing as a floating set of recommendations which refresh as the user progresses with their review. These recommendations are suggested review topics that have been automatically extracted from a database of reviews (on Digital Cameras in this instance) and selected and ranked according to the content of the user's review so far. At any time the user can even select a topic to see an expanded list of relevant review fragments which is a good aid for the review process.

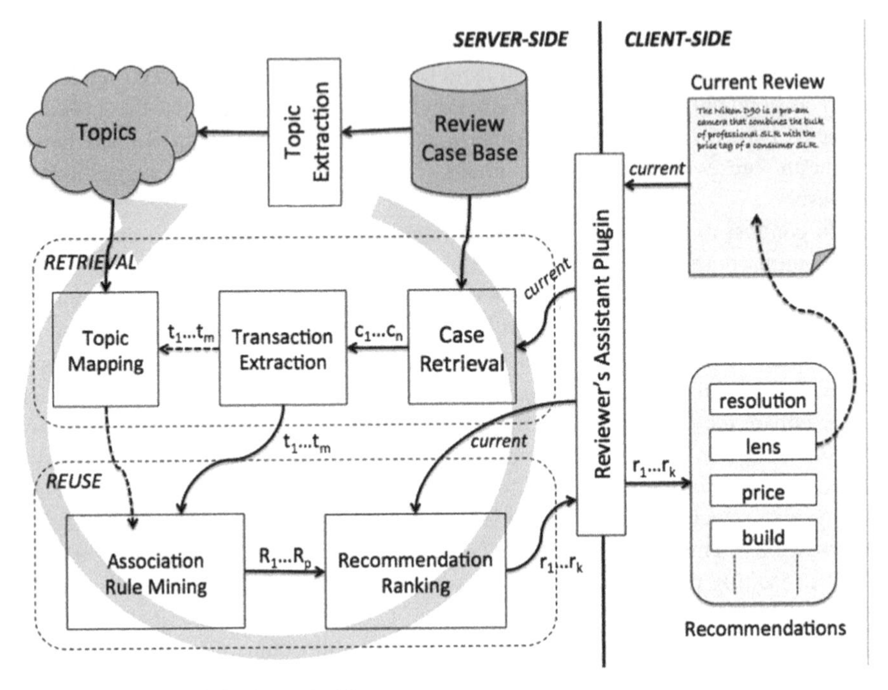


**Fig. 2** The Reviewer's Assistant Architecture

## 2.1 System Architecture

The basic Reviewers Assistant system architecture is depicted in Figure 2. The system has been developed as a case-based recommendation system. Each case corresponds to a previous product review (drawn from an existing review database such as Amazon's) and includes the product id, the review text, and any meta information available, such as the overall review score or helpfulness. These reviews are automatically extracted from the underlying service by using any available API to extract relevant product and review data; typically the RA will extract only high-quality reviews (based on any helpfulness/quality meta data that is available.). As the user writes their review, the review text is periodically (typically on the completion of a new sentence) used as query *current* against the relevant domain case base to retrieve a set of similar reviews, from which term-based transactions are extracted as the basis for association rule mining. In the current implementation we rely on a simple term-based Jaccard similarity metric to retrieve a set of the $n$ review cases that are most similar to *current*.

Each of these $n$ similar reviews is converted into a set of sentence-level transactions and a review-level transaction. This is a straightforward process that starts by identifying the nouns in a review text and then converts each sentence or review into a set of these nouns. If, for example, the review is *The camera takes good pictures. A flash is needed in poor light.* then we would have sentence transactions {*camera,pictures*} and {*flash,light*} and a review transaction {*camera,pictures,flash, light*}. In the earliest version of the RA [6], denoted by *non-topic*, these transactions are directly mined to extract a set of term-based association rules. These rules are then used to map the current review text to a ranked list of concrete topic recommendations for the user.

Since early trials uncovered problems with this approach, arising largely out of the potential for duplication among the recommended items, an improved version of RA [7], denoted by *topic*, subsumed related words under their respective topic labels and thus recommended topics instead of noun words. If, for example, when reviewing a digital camera a user could receive parallel suggestions for *resolution, pixels, picture quality* with RA *non-topic*, these would be replaced by the topic of e.g. *picture quality* when using RA *topic*. Briefly, RA *topic* maps nouns on to a set of *topics* prior to recommendation so that the single noun suggestions can be presented by more intuitive topical suggestions. The current system described in this work goes one step further and proposes a method to generate the topics automatically instead of having them hand-coded.

The details of topic extraction will be described in Section 2.2 and the details of association rule mining are described in Section 2.3

## 2.2 Topic Modeling

The purpose of adopting topics is to afford a level of abstraction (topics vs. nouns) that has the potential to provide a more intuitive set of recommendations based on more meaningful product topics, rather than on the looser vocabulary of user generated reviews. As above describing, there are two topic models supported by our system. The first one is a simple set of synonyms for each of 38 manually identified topics in the Digital Camera space [7]; see Figure 3 for a sample. A second one uses Latent Dirichlet Allocation (LDA) (cf. [3]) and topic (label) discovery (cf. [14]) to automatically extract a set of topics from a collection of reviews, and a sample list of automatically extracted topics is shown in Figure 4.

For the fully automated creation of topic lists, we applied LDA to a set of 9,355 camera reviews. Prior to learning, the input reviews were pre-processed as follows: 1) We removed a list of model and brand names from the reviews. This list was created automatically from the product titles by taking the first four words of each product title. For example, the words *Canon, EOS, Rebel, T2i* were adopted as brand and model names from title *Canon EOS Rebel T2i 18 MP CMOS APS-C Digital SLR Camera with 3.0-Inch LCD and EF-S 18-55mm f/3.5-5.6 IS Lens*. In addition, we considered words consisting of both digits and alphabetic characters and beginning with a character to be model names, e.g. *D90, FZ35, DSC-W530, D7000*, etc. 2) The reviews were tokenized and stemmed with the Porter Stemmer. 3) The standard Google stop word list was also removed. 4) We removed terms occurring in less than 4 reviews and the 5 most common terms.

We used an implementation of Blei's LDA algorithm [4] from the Stanford Topic Detection Toolbox[1], which provides multiple different implementations of Latent Dirichlet Allocation, see [4, 8]. We used 38 topics since this was the number of topics that were manually created. As a result of the LDA training with 1000 iterations, terms (or words) are ranked for each topic according to their frequency.

```
[image] image,photo,picture,photograph
[battery] battery,aa,aaa,lithium,life
[lenses] lens,lenses,mm,af,18-55mm,wide,angle,zoom
[display] display,lcd,screen,display,viewfinder,zoombrowser,...
[video] video,1080p,camcorder,film,hd,720p,movie
[problem] problem,issue,draw,drawback,pros,complaint,...
[purchase] purchase,order,ship,stock,invoice,packaging,bought
```

**Fig. 3** Examples for Manually Created Topics

```
[image quality] image,quality,sensor,comparing,sharpness,...
[battery life] battery,life,charge,charger,extras,spares,hours
[lens lenses] lens,lenses,kit,slr,35mm,primes
[screen lcd] screen,lcd,view,viewfinder,living
[video minutes] video,minutes,capabilities,camcorder,clips,...
[problem work] problem,work,issue,fix,return,repair,warranty
[review buy] review,buy,purchase,money,research,worth,...
```

**Fig. 4** Examples for Automatically Extracted Topics

For further processing, we considered the top 20 words for each topic. We kept each word only in a single topic as we know that the LDA model allows the words belonged to multiple topics. For example, if a word had the highest score in a certain topic, then we would remove that word from all other topics. Finally, we removed non-noun words and presented the top-2 remaining words of each topic as topic suggestions to the user, e.g. we would display *pocket size* as a topic instead of *compact*, see Figure 4.

No matter whether the topics are hand-coded or automatically extracted by LDA, the process of *Topic Mapping* (Figure 2) simply involves the mapping of noun words to their corresponding topic labels prior to presentation. For example, assuming we would process transaction {*1080p, clip, CPU, power*} generated by the sentence {*Viewing the native 1080p clip straight from the camera will need fairly significant CPU power*}, the transaction will become {video, clip, CPU, power} in for the manually created topic list depicted in Figure 3 since *1080p* is one of the words of the topic group *video*.

[1] http://nlp.stanford.edu/software/tmt/tmt-0.4/

## 2.3 Association Rule Mining

In order to generate updating recommendations for the reviewer, RA uses association rule mining techniques to discover patterns of nouns/topics that recur frequently across many reviews. At any point we have a set of transactions (whether *non-topic* or *topic*) as described above, which reflect frequent collections of nouns/topics that occur at the sentence-level or review-level. For example, in the digital camera domain we might have transactions such as $\{image, lens, resolution\}$ and $\{size, price\}$. We apply association rule mining [1] to identify frequently occurring transactions and to generate a set of association rules of the form $\{image, lens\} \rightarrow \{resolution\}$. Following the standard algorithm for association rule mining, we first filter-out rules that fall below a minimum *support* level; that is we keep subsets of transactions that have a pre-defined frequency of occurrence as candidates for rules and their antecedents, so-called item sets.

Finally the resulting rules are ranked in descending order of their confidence, which is basically an estimate of the probability of finding the topic/noun that forms the rule consequent given the occurrence of the antecedent. To generate a set of ranked recommendations we apply each of the extracted rules, in order of confidence, to the current review text. If the current review text triggers a rule of the form $LHS \rightarrow RHS$ then the noun/topic that is the $RHS$ is added to the recommendation list. This process terminates when a set of $k$ recommendations have been generated. In the following sections we will discuss the recommendation and topic extraction algorithms in greater technical detail and show an example of the RA in action.

The whole process used to generate a set of recommendations is presented in Figure 5. Briefly, our approach takes, as input, the set of sentence-level transactions($T_s$), the set of review-level transactions($T_r$) and the current review text by the user (*current*) and outputs a set of $n$ ($n = 10$) suggestions ($S$). We implemented association rule mining (*getARMRecs*) based on the algorithms described in [2]. In our experiments we have set confidence and support thresholds to 3 and 0.25 for sentence-level transactions and to 4 and 0.5 for review level transactions. If association rules do not lead to a set of $n$ recommended topics then additional topics are extracted from reviews based on a simple frequency count as a fallback strategy.

Thus as the user continues to type their review, extracted rules can be triggered leading to updated recommendations. Equally (but not shown here for reasons of brevity) these recommendations fall away as users cover them in their review text.

## 2.4 RA in Action

Figure 6 shows the Reviewer's Assistant in action for our user, who is reviewing a Sony Cybershot DSC-W530 camera on Amazon. The user is presented with the usual Amazon review creation screen and the figure shows the Reviewer's Assistant overlay; the RA widget can be dragged to any suitable location on screen. The RA presents a dynamic set of updating review suggestions (in this case we show

1: **procedure** GETSUGGESTIONS($T_s$,$T_r$, *current*)
2: $S_s = getARMRecs(n, T_s, current, 3, 0.25)$
3: **if** Size of $S_s < n$ **then**
4: $S_r = getARMRecs(n - S_s.size, T_r, current, 4, 0.5)$
5: **end if**
6: $S = S_s \cup S_r$
7: **if** Size of $S < n$ **then**
8: fill-up $S$ by frequency algorithm
9: **end if**
10: return $S$
11: **end procedure**

**Fig. 5** Summary Algorithm for Generating Suggestions

the *auto-topic* version of the RA). Figure 6(a) shows some of the suggestions presented to Joe during the early stages of the review. In this case we see a number of suggestions for some common review topics for this product, including *image quality, battery life* and *photos disappointment.* As shown, the user can view fragments that relate to a particular topic by mousing over the topic. For example, in this case the fragments *"a separate charger"*, *"the battery size"* etc. are displayed for topic *"battery life"*.

In Figure 6(b) we see a snapshot towards the end of the review writing. This time the user is presented with additional topics, many of which are more specialized or not uniquely related to the specific product to provide the reviewer with an opportunity to broaden their review. In Figure 6(b) we can see that the user's review covers a range of topics that have been suggested, including *image quality,* etc.

# 3 Evaluation

We have extended prior work on the Reviewer's Assistant (RA) [6, 7], a browser plugin to support the review writing process, by introducing a non-supervised method for the extraction of topics. This way, not only is the RA able to make suggestions to the reviewer about the type of topics she might consider based on the current review writing, but also we are able to automatically extract topics for various contexts defined by a sufficiently large set of reviews. The novel method (*auto-topic*) will be compared to prior variations of the RA presented in Dong et al. [7]. In this section we will evaluate the new variation as part of a live-user study in order to ascertain whether the expected loss as compared to the *topic* variation with hand-coded topics is acceptable, and how the resulting reviews compare to those created without the help of the RA in order to consider review quality.

The following evaluation has three separate parts and in each, we pay particular attention to performance differences between the *auto-topic* and both the *topic* and *non-topic* variations of the RA, if any. In the first part, we describe the results of a live-user study focusing on how participants used the RA plugin and their feedback

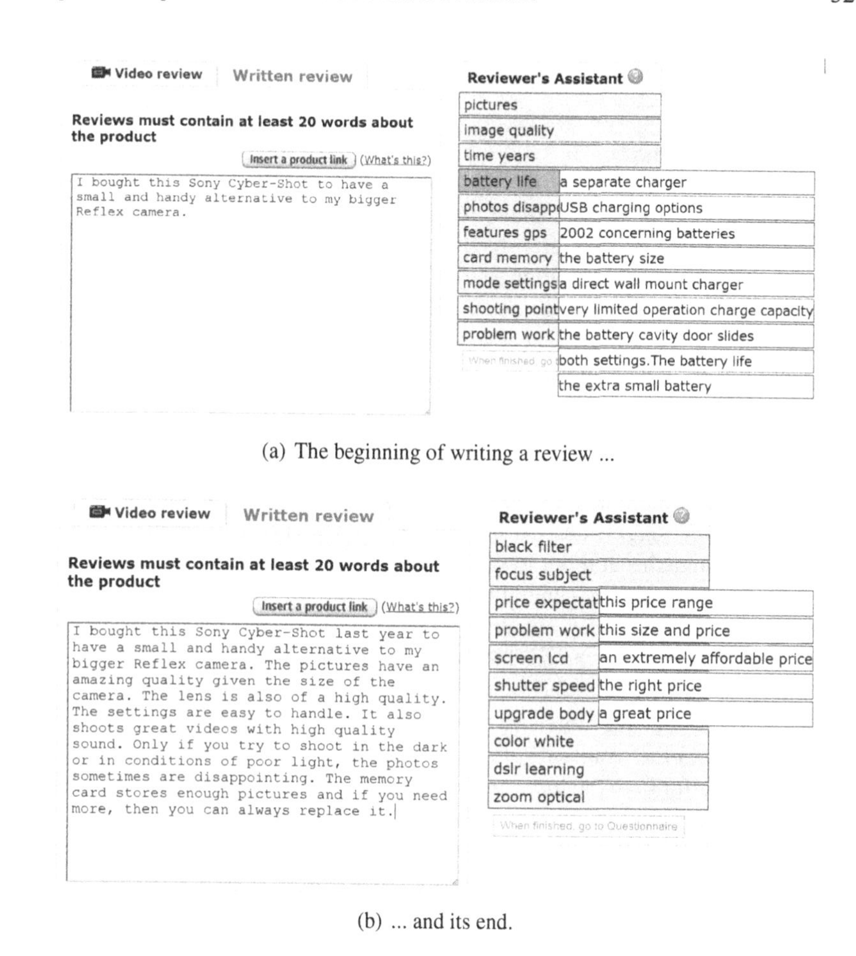


(a) The beginning of writing a review ...

(b) ... and its end.

**Fig. 6** The *Reviewer's Assistant* - Dynamic Behavior

with respect to the utility of the recommendations and their overall satisfaction with the experience. In the second part of the analysis, we perform an objective analysis of the resulting reviews considering the depth and breadth of coverage offered by these reviews with respect to important product features. Finally, we perform a comparison of a subset of the above reviews and a set of comparable Amazon reviews, using a few "expert" users to rate the helpfulness of these different reviews in order to understand if the use of the RA leads to any improvements in review quality.

### 3.1 Usage Analysis

Extending a prior experiment with 40 test users, 26 male and 14 female, for the *non-topic* and *topic* versions of the RA, see Dong et al. [6], we recruited additional 20 test users, 16 male and 4 female, for the *auto-topic* version of the RA. 11 out of the 20 participants in the current study had written at least one online review in the past and the majority had purchased products online through stores like Amazon and iTunes. Exactly in line with our previous experiment, we restricted our target product domain to that of digital cameras on Amazon and configured the RA plugin accordingly. All 20 participants of the new study got access to the *auto-topic* version of the RA. Each user was asked to select a product of interest and to write a review for this product; they were provided with a brief initial tutorial on the RA, the purpose of its suggestions, and how they might avail of them if appropriate.

During the trial user actions were logged as they completed their reviews and availed of the RA suggestions. At the end of the trial each user completed a short post-trial questionnaire in order to rate the RA under four key areas: 1) helpfulness – were the RA suggestions generally helpful? 2) relevance – were the suggestions relevant in the context of the review being written? 3) comprehensiveness – did the suggestions broadly cover the product being reviewed? 4) overall satisfaction – was the participant satisfied with the overall experience provided by the RA?

The results of this questionnaire are shown in Figure 7 and are largely positive. Figure 7(c) shows the results for the new experiment and we also show the corresponding graphs for both the *topic* and *non-topic* variants of the RA for comparison. We can see that overall about 80% – 90% of users found the RA *auto-topic* to be helpful and relevant. The feedback in terms of recommendation comprehensiveness exhibits scores of 65% to 70% and seems to be in line with *topic* users but slightly lower than for the *non-topic* users.

In relation to the post-trial questionnaire we can see that overall there is strong user-support for the RA. Between 70% (*auto-topic*) and 80% (*non-topic*) of users indicated that they were satisfied overall with the system.

### 3.2 Topic Coverage

We now consider the type of reviews that are produced. For example, is there any evidence that the *auto-topic* variation leads to quantitative differences in review quality? In this part of the evaluation we consider review quality in terms of the breadth and depth of topical coverage. For each review, we calculated its *breadth* – the number of topics covered in the review; *depth* – the word length of sentences that cover any topic, divided by the number of topics covered in the review; and *redundancy* – the word length of sentences that do not cover any topic. Finally we also calculated review length based on the number of words in the review.

Thus we analyze the review texts of the 60 reviews produced during both trials and compute their breadth, depth, and redundancy characteristics with reference to

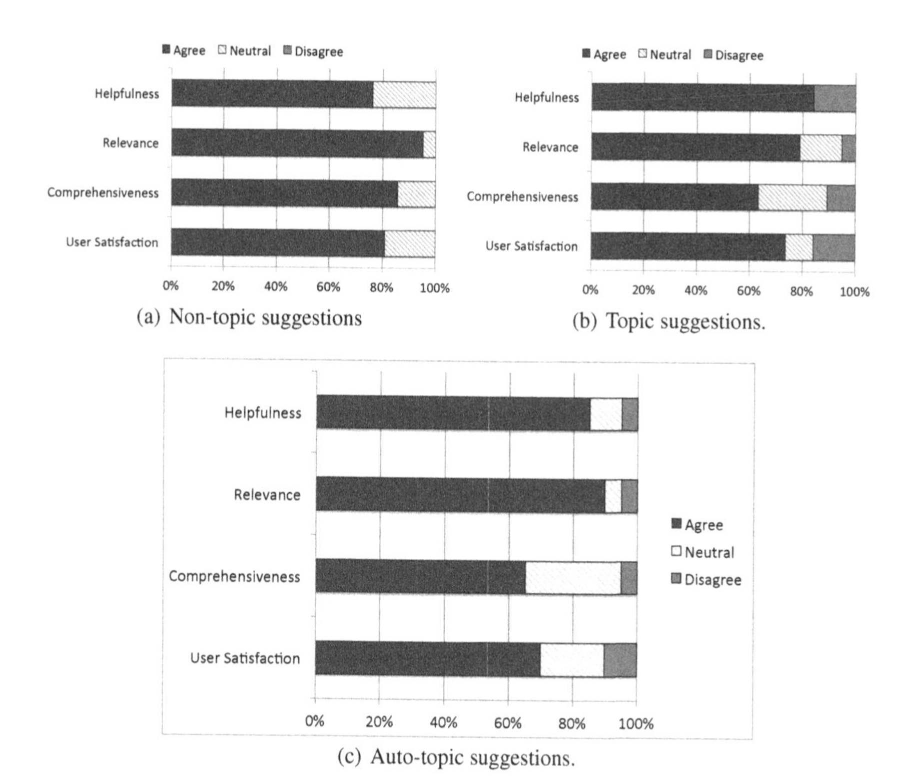


(a) Non-topic suggestions

(b) Topic suggestions.

(c) Auto-topic suggestions.

**Fig. 7** User Feedback.

| | auto-topic | topic | non-topic |
|---|---|---|---|
| Average Breadth* | 9.55 | 10.42 | 7.62 |
| Average Depth | 9.85 | 10.69 | 10.53 |
| Average Redundancy | 10.50 | 9.68 | 10.24 |
| Average Length | 108.80 | 113.58 | 90.43 |

**Table 1** Breadth, Depth, and Redundancy; * indicates a significant difference at 0.1 between *auto-topic* and *non-topic*.

the defined set of product topics used for the digital camera domain. The results are presented in Table 1. We can see that while all techniques perform similarly in terms of review depth (between 9.85 for *auto-topic* and 10.69 for *topic*) and redundancy (between 9.68 for *topic* and 10.50 for *auto-topic*), the reviews produced with the *non-topic* version of RA tend to offer significantly thinner coverage (a breadth of 7.62 for *non-topic* vs. 9.55 and 10.42 for *auto-topic* and *topic* respectively).

Of course this approach provides only a superficial analysis of review quality and it is not clear whether these depth, breadth and redundancy characteristics have any significant bearing on the ultimate perception of review quality or helpfulness. Thus, we analysis performance along these dimensions in the next section.

## 3.3 Review Quality

Ultimately the best test of the RA approach is to consider the quality of the resulting reviews in order to understand whether users find them to be helpful, for example. Even better is if we can compare our test reviews to a benchmark in terms of quality. This is the aim of this final evaluation section.

We report here on the accumulated results of two expert evaluation studies. The first study was conducted to compare RA *non-topic* and RA *topic* with Amazon reviews, see [7]. Later we repeated the exact same study for RA *auto-topic* to obtain comparable results. For each of the two studies we collected sets of reviews with similar lengths. One set was chosen from the reviews written by participants of the RA variants. We collected 10 reviews written using the help of RA *topic*, RA *non-topic*, and RA *auto-topic*, respectively. A second set was selected from two groups of Amazon reviews to serve as a benchmark, for both studies, against which to judge the quality of the RA reviews. One group was chosen at random from among the most helpful Amazon camera reviews. We picked 10 reviews (*Amazon+*) that had a helpfulness score of at least 0.7 (meaning 70% of raters considered them helpful). In fact, the average helpfulness score for these reviews was 0.87 and thus we can view these as examples of very high quality product reviews written without the aid of RA. Next we chose another group of 10 random Amazon reviews (*Amazon-*), but this time we picked reviews that had a helpfulness score of less than 0.7; the average helpfulness score for reviews in this group was 0.40 and thus represent examples of lower quality reviews written without the help of RA.

| | New Study | | | Prior Study | | | |
|---|---|---|---|---|---|---|---|
| Review Set (No. of Reviews) | auto-topic (10) | Amazon+ (10) | Amazon- (10) | topic (10) | non-topic (10) | Amazon+ (10) | Amazon- (10) |
| Helpfulness* | 3.67 | 3.33 | 3.10 | 3.90 | 3.90 | 3.33 | 3.07 |
| Completeness | 3.27 | 2.97 | 2.73 | 3.67 | 3.57 | 2.67 | 2.53 |
| Readability* | 4.20 | 3.77 | 3.50 | 3.60 | 3.60 | 3.80 | 3.33 |
| Avg. Breadth** | 9.00 | 6.00 | 5.10 | 11.30 | 8.60 | 5.90 | 7.20 |
| Avg. Depth** | 11.05 | 16.28 | 21.39 | 11.90 | 11.21 | 15.57 | 12.49 |
| Avg. Redundancy | 8.40 | 20.40 | 23.70 | 5.50 | 14.30 | 22.40 | 21.00 |
| Avg. Helpfulness Ratio | - | 0.87 | 0.40 | - | - | 0.90 | 0.41 |
| Avg. Length | 105.20 | 111.50 | 111.70 | 113.30 | 108.50 | 110.30 | 107.60 |

**Table 2** User Evaluation; * indicates a significant difference at 0.1 (one-sided) between *auto-topic* and both *Amazon-* and *Amazon+*; ** indicates a significant difference at 0.05 between *auto-topic* and both *Amazon-* and *Amazon+*.

Next, we recruited, as in the previous study, 15 reviewer "experts" (with a good understanding of the digital camera space) and asked them to perform a blind review of a random sample of reviews from the three sets (*auto-topic*, *Amazon+*, and *Amazon-*). In each case we asked the experts to rate the reviews on a 5-point scale in terms of 1) helpfulness – how helpful did they think the review would be to others? 2) completeness – did the review provide a reasonably complete account of the product in question? 3) readability – was the review well written and readable? In total each test review was reviewed by 3 different experts. Finally, we calculate the average helpfulness, completeness, and readability ratings across each of the 3 review groups and also calculated their average breadth, depth and redundancy scores based on the approach taken previously.

The results are presented in Table 2 together with the earlier study on RA *topic* and RA *non-topic*. The results justify the usage of automatic topic extraction (*auto-topic*), even though it seems to be slightly weaker than the manually created version (*topic*). The average helpfulness rating of RA reviews (3.67 for RA *auto-topic*) is greater than the helpfulness rating for *Amazon+* (3.33) and *Amazon-* (3.10). Similarly, we can see clear benefits for the RA *auto-topic* in terms of readability with a score of 4.20, when compared to *Amazon+* (3.77) and *Amazon-* (3.50). Both of these helpfulness and readability benefits (RA *auto-topic* versus Amazon) are statistically significant at the 0.1 level; statistically significant differences were not found in terms of review completeness for the RA *auto-topic*, but only for the other two RA variants in the prior study, see Dong et al. [7]. Table 2 also shows how the *auto-topic* version of the RA leads to reviews that have a greater topical breadth when compared to *Amazon+* and *Amazon-*; the *auto-topic* reviews cover 9.0 topics per review compared to only 6.0 and 5.1 topics per review for *Amazon+* and *Amazon-*. In the prior study, tendencies were similar. It is worth noting that we found a significantly lower depth for the *auto-topic* version of the RA vs. both groups of *Amazon* reviews, at 0.05 significance level, a result we reported earlier for the *topic* and *non-topic* versions of the RA, but without significance in the prior study.

## 4 Conclusions

In this paper we have described an extension of the RA that supports its applicability in various contexts and domains. The RA is a recommender system that is designed to help users to write better product reviews by passively making suggestions to reviewers as they write. The core contribution of this work is a novel method to extract topics from a set of reviews to replace the RA's previous reliance on manually created topics. Topic extraction here is based on the unsupervised Latent Dirichlet Allocation (LDA) algorithm, which is applied to a set of reviews. Application-specific pre-processing and post-processing tailors the method for seamless integration with the RA. We have also presented a live-user evaluation to show that users are broadly satisfied with that new variation of the RA and that the loss of quality as compared to the expensive hand-coded topics is acceptable. In particular, when we compare the reviews written using RA to helpful reviews from Amazon we find statistically significant benefits accruing to RA in terms of review helpfulness and readability.

This provides strong support for the utility of this recommendation approach and its application on e-commerce sites such as Amazon but there are many options for future work. Now that we have demonstrated the possibility to automatically extract topics for our recommendation approach, future work will focus on automating the discovery of relationship between topics, features, and categories with a view to providing a richer user experience and pro-active stimulation of review quality.

**Acknowledgements** This work is supported by Science Foundation Ireland under grant 07/CE/I1147.

## References

1. Agrawal, R., Imieliński, T., Swami, A.: Mining Association Rules between Sets of Items in Large Databases. ACM SIGMOD Record **22**(May), 207–216 (1993)
2. Agrawal, R., Srikant, R.: Fast Algorithms for Mining Association Rules in Large Databases. In: Proceedings of the 20th International Conference on Very Large Data Bases, VLDB '94, pp. 487–499. Morgan Kaufmann Publishers Inc., San Francisco, CA, USA (1994)
3. Blei, D.M.: Probabilistic topic models. Commun. ACM **55**(4), 77–84 (2012). DOI 10.1145/2133806.2133826
4. Blei, D.M., Ng, A.Y., Jordan, M.I.: Latent dirichlet allocation. J. Mach. Learn. Res. **3**, 993–1022 (2003)
5. Bridge, D., Waugh, A.: Using Experience on the Read/Write Web: The GhostWriter System. In: D. Bridge, E. Plaza, N. Wiratunga (eds.) Procs. of WebCBR: The Workshop on Reasoning from Experiences on the Web (Workshop Programme of the Eighth International Conference on Case-Based Reasoning), pp. 15–24 (2009)
6. Dong, R., McCarthy, K., O'Mahony, M.P., Schaal, M., Smyth, B.: Towards an Intelligent Reviewer's Assistant: Recommending Topics to Help Users to Write Better Product Reviews. In: Procs. of IUI: 17th International Conference on Intelligent User Interfaces, Lisbon, Portugal, February 14-17, 2012, pp. 159–168 (2012)
7. Dong, R., Schaal, M., O'Mahony, M.P., McCarthy, K., Smyth, B.: Harnessing the Experience Web to Support User-Generated Product Reviews. In: 20th International Conference on Case-Based Reasoning, Lyon, France (2012). To appear.
8. Gretarsson, B., O'Donovan, J., Bostandjiev, S., Höllerer, T., Asuncion, A.U., Newman, D., Smyth, P.: Topicnets: Visual analysis of large text corpora with topic modeling. ACM TIST **3**(2), 23 (2012)
9. Healy, P., Bridge, D.: The GhostWriter-2.0 System: Creating a Virtuous Circle in Web 2.0 Product Reviewing. In: D. Bridge, S.J. Delany, E. Plaza, B. Smyth, N. Wiratunga (eds.) Procs. of WebCBR: The Workshop on Reasoning from Experiences on the Web (Workshop Programme of the 18th International Conference on Case-Based Reasoning), pp. 121–130 (2010)
10. Hu, N., Liu, L., Zhang, J.: Do online reviews affect product sales? the role of reviewer characteristics and temporal effects. Information Technology and Management **9**, 201–214 (2008). 10.1007/s10799-008-0041-2
11. Kim, S.M., Pantel, P., Chklovski, T., Pennacchiotti, M.: Automatically assessing review helpfulness. In: Proceedings of the Conference on Empirical Methods in Natural Language Processing (EMNLP 2006), pp. 423–430. Sydney, Australia (2006)
12. Liu, Y., Huang, X., An, A., Yu, X.: Modeling and predicting the helpfulness of online reviews. In: Proceedings of the 2008 Eighth IEEE International Conference on Data Mining (ICDM 2008), pp. 443–452. IEEE Computer Society, Pisa, Italy (2008)
13. O'Mahony, M.P., Smyth, B.: Learning to recommend helpful hotel reviews. In: Proceedings of the third ACM conference on Recommender Systems, RecSys '09, pp. 305–308. ACM (2009). DOI 10.1145/1639714.1639774
14. Schaal, M., Müller, R.M., Brunzel, M., Spiliopoulou, M.: RELFIN - Topic Discovery for Ontology Enhancement and Annotation. In: ESWC'05, pp. 608–622 (2005)
15. Zhu, F., Zhang, X.M.: Impact of online consumer reviews on sales: The moderating role of product and consumer characteristics. Journal of Marketing **74**(2), 133–148 (2010)

# PRACTICAL APPLICATIONS AND SYSTEMS

# Adapting Bottom-up, Emergent Behaviour for Character-Based AI in Games

**Micah Rosenkind, Graham Winstanley, Andrew Blake**[1]

**Abstract** It is widely acknowledged that there is a demand for alternatives to handcrafted character behaviour in interactive entertainment/video games. This paper investigates a simple agent architecture inspired by the thought experiment "Vehicles: Experiments in Synthetic Psychology" by the cyberneticist and neuroscientist Valentino Braitenberg [1]. It also shows how architectures based on the core principles of bottom-up, sensory driven behaviour controllers can demonstrate emergent behaviour and increase the believability of virtual agents, in particular for application in games.

## 1 Introduction

Creating believable, lifelike characters is as old as the craft of storytelling itself. Be it in performance media such as theatre, audio-visual mediums such as painting, radio or television or in books; creating believable characters is one of the central aspects.

In the comparably recent medium of interactive animations i.e. video games, a demand for novel autonomous lifelike characters has emerged [2, 3, 4, 5, 6]. With the increasing complexity of the virtual worlds presented in video games, it is becoming difficult to artistically and financially justify standard means of populating these worlds with pre-defined 'hand-made' non-player characters (NPC). This has lead to an increasing interest in procedural solutions for both animation and artificial intelligence that offer the promise of generating emergent and complex behaviour, without the need for unfathomable amounts of artist-created assets.

This paper investigates a simple agent architecture inspired by the thought experiment "Vehicles: Experiments in Synthetic Psychology" by the cyberneticist and neuroscientist Valentino Braitenberg [1]. It also shows how architectures based on the core principles of bottom-up, sensory driven behaviour controllers can demonstrate emergent behaviour and increase the believability of virtual agents.

[1] University of Brighton, BN2 4GJ, UK
{m.m.rosenkind, g.winstanley, a.l.blake}@brighton.ac.uk

The paper starts with a definition of believability criteria as set out by eminent authors in the field of believable agents research and artists from the character arts. It then presents the architecture that was used to meet these criteria and details the aspects of the emergent behaviour generated by it that enhance believability. It concludes with an evaluation of the architecture and cites further improvements and scenarios that could be applied.

## 2 Believability Criteria

In attempting to create a believable agent, it is important to explore the main approaches and identify the key factors that make an agent seem believable to an audience. This section cites existing definitions of and criteria for believability in various domains, which can be used to evaluate existing agent architectures and contribute to the development of a new approach.

The term "believability" in context of the work underlying this paper, is regarded as the degree to which an audience of a character's performance is willing to "believe" that the character is alive and thinking. It can be useful to look at this term as "suspension of disbelief", since this notion does not require the audience to truly believe that the character is real, but need only feel inclined (or indeed willing) to do so.

Looking back into the history of character-based arts, there have been several attempts made at defining a set of criteria that can help artists achieve this suspension of disbelief in their audience. One of the earliest remarks on this relationship between the characters, its author and the audience comes from Samuel Taylor Coleridge:

> "...to transfer from our inward nature a human interest and a semblance of truth sufficient to procure for these shadows of imagination that willing suspension of disbelief for the moment, which constitutes poetic faith" [7]

This notion of "poetic faith" has carried much of the work in other forms of character arts. Taking a more scientific approach and attempting to define a set of guidelines and criteria for authors and actors was the goal of Stanislavsky and Popper [8, 9] and Egri [10]. In the early 1980's the team behind Disney's seminal animation work released an extensive compendium of guidelines and criteria for believable animated characters [11]. Chuck Jones from Warner Brothers presents a similar approach in "Chuck Amuck" [12]. These guidelines for classical hand-drawn animation were further developed and adapted to the emerging technology of computer generated imagery (CGI) animation by Pixar's John Lasseter [13].

These works formed the basis for defining a set of criteria for believable interactive characters, which are discussed in a Thesis by Bryan Loyall [14] and summarized by Michael Mateas [15]:

1. Personality
2. Emotion
3. Self–motivation
4. Change (character development)
5. Social relationships
6. *Illusion of Life*

While the first 5 criteria focus on properties for characters that are either imagined by a reader or portrayed by human actors, the final point “Illusion of Life” is taken from Disney’s work on animated characters. Loyall [14] summarizes the set of criteria for artificial characters, which focus less on *what* a character must do and more on *how* a character should present its actions:

1. Appearance of Goals:
2. Concurrent pursuit of Goals and Parallel Action:
3. Reactive and Responsive:
4. Situated
5. Resource Bounded – body and mind:
6. Exist in a Social context:
7. Broadly Capable:
8. Well integrated (capabilities and behaviours)

These requirements extend the fundamental principles for animation to include the notions of real-time responsiveness and the properties of embodied agents. Together with the base requirements for believable characters they form an informative checklist that is suitable for comparing embodied agents. The agent architecture presented in this paper was therefore reviewed against these criteria.

## 3 Believable Agents

Modern interactive media such as videogames require an increasingly infeasible amount of artistic work by animators and character authors. Procedural approaches may offer alternative approaches to creating the vast amounts of characters found in modern open-world games. Since there have been successful applications of procedural approaches in animation [16, 17, 18, 19, 20], the notion of extending the bottom-up, procedural approach to behaviour control follows naturally.

The starting point for bottom-up autonomous agent architectures can be seen in the 1940s when William Grey Walter presented a series of mobile, autonomous robot “tortoises”. These were some of the first mobile autonomous robots and displayed complex, emergent behaviour while having only a small set of simple components. The robots followed light sources in unpredictable paths and were able to move around obstacles using a bump sensor. Walter provided a set of criteria that he says automata like his "Machina Speculatrix" and "Machina Docilis" should meet:

> "Not in looks but in action, the model must resemble an animal. Therefore, it must have these or some measure of these attributes: exploration, curiosity, free-will in the sense of unpredictability, goal-seeking, self-regulation, avoidance of dilemmas, foresight, memory, learning, forgetting, association of ideas, form recognition, and the elements of accommodation." [21, Ch. 5]

One of the most famous examples of agent architectures inspired by Walter's tortoises is Valentino Braitenberg's thought experiment "Vehicles: Experiments in Synthetic Psychology" [1]. The central idea to these experiments was first presented in an essay that describes how a finite set of simple brain connection patterns can create complex, emergent behaviour [22]. Braitenberg's thought experiment is comprised of a series of 14 incremental, evolutionary steps from simple, insect-like agents that simply follow or avoid sources of stimuli, to complex beings with emotional tendencies, and behaviour based on experience (including trauma).

## 4 The Agent Architecture

The basic architecture adopted by Braitenberg in all his vehicles is deliberately simple, forming a physical mapping from perception to actuation. The vehicles themselves are seen to be physical robots with two motors driving two wheels, i.e. a differential drive. Each motor is stimulated by sensors connected to them by 'wires' such that the intensity of stimulation and therefore the wheels' rotational speed, is dependent on the level of perception experienced by the sensors. Figure 1 illustrates this.

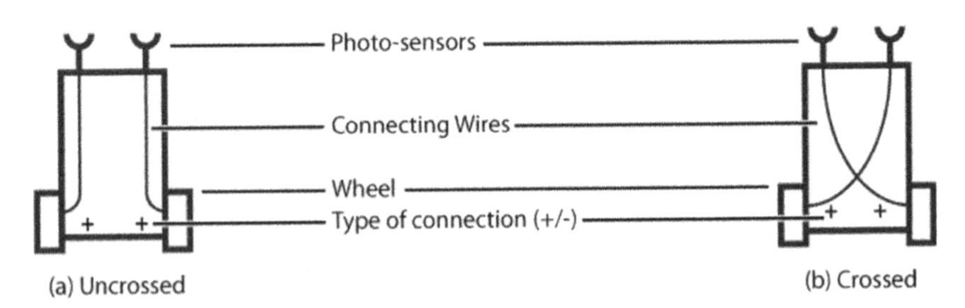


Figure 1 This is the simplest form of dual motor/ sensor vehicle described by Braitenberg, showing the two ways that connections can be made between sensor and motor.

Figure 1(a) shows that each wheel is connected to, and therefore stimulated by, sensors on the same side of the vehicle. Figure 1(b) shows the alternative configuration with crossed connections.

Braitenberg noted that such very simple changes (1(a) vs. 1(b)) would result in radically different perceived behaviour, described along with a human-like mental state to reflect an observer's perception, e.g. love, hate, fear, etc. Hence the term 'Synthetic Psychology.' This "Synthetic Psychology" approach anthropomorphises the agents and stems from the notion that an observer of the complex behaviour exhibited by them will tend to overestimate their internal complexity. Thus these terms, which refer to highly complex phenomena in psychology, can be used to effectively differentiate behaviour patterns and summarize observers' interpretations of them.

This was the starting point for his more advanced thought experiment process and also the work presented here. Table 1 illustrates Braitenberg's base architectures and the mental state labels he used [1, pg. 3-19]. Note that the concept of perception intensity driving the wheels has been 'played with' a little in that a new concept of inhibition has been introduced.

Table 1 Braitenberg's four base architectures for a two-wheeled robot.

| | Architecture | Code | Braitenberg's Labels (1984) |
|---|---|---|---|
| | 1.Parallel-Excitatory | PE | Cowardly (Vehicle 2a) |
| | 2. Crossed-Excitatory | CE | Aggressive (Vehicle 2b) |
| | 3. Parallel-Inhibitive | PI | Loving (Vehicle 3a) |
| | 4. Crossed-Inhibitive | CI | Exploring (Vehicle 3b) |

The definition of the base behaviours is a necessary step because Braitenberg only describes behaviours that drive the agent forward in the early chapters of his work. Backwards motor actions are however referred to in the later Chapter 13 [1, pg. 77] where they are correlated to "fleeing" behaviour and used as an indicator for "bad" situations.

In the model used for this research, the inhibitive forces that are used in two of these base architectures create "backwards" behaviours when attached to motors that can drive both directions. The permutations of these two new behaviours are recorded.

Although he does not explicitly mention it, Braitenberg does infer the presence of an additional passive forward drive on the two motors that is required for the architectures labelled "loving" and "exploring". For these combinations the term "cooperative behaviours" is used, as the four basic architectures and their associated behaviours cooperate with others to form a new set of behaviours. The three passive motor actions used for the experiment were *neutral* (N), *forward* (F), and *backward* (B).

Braitenberg's "Vehicles" thought experiment covers 14 incremental agent architectures. Table 2 below summarizes Braitenberg's descriptions of their behaviour. The "Vehicle" column shows the numbering system used by Braitenberg in his thought experiment to categorize the different architectures. Under the "Components" column, the additions to the properties of the agent model are listed. The "Concept" column is based on the psychological terminology that Braitenberg chose to use to describe the behaviour of the agents in his thought experiment.

Table 2 summarizes the first five agent designs in Braitenberg's thought experiment.

| **Vehicle** | **Components** | **Concept** | **Behaviour** |
|---|---|---|---|
| 1 | 1 Sensor<br>1 Motor<br>Single Wire | ALIVE | Kinesis, Moves in proportion to stimuli. |
| 2a | 2 Sensors<br>2 Motors<br>Uncrossed excitatory connection | COWARD | Turns away from source. Speeds up when near source → Flees |
| 2b | 2 Sensors<br>2 Motors<br>Crossed excitatory connection | AGGRESSIVE | Turns towards source. Speeds up when near source → Attacks |
| 2c | 2 Sensors<br>2 Motors<br>Uncrossed & Crossed excitatory connection | ALIVE | Like Vehicle 1, Moves in proportion to stimuli. |
| 3a | 2 Sensors<br>2 Motors<br>Uncrossed inhibitory connection | LOVE | Turns toward source. Slows down near source. |
| 3b | 2 Sensors<br>2 Motors<br>Crossed inhibitory connection | EXPLORER | Turns away from source. Slows down near source. |
| 3c | Multiple Sensors Cooperating<br>Monotonic Dependences<br>4 Sensors<br>2 Motors | VALUES | Shows COWARD, AGGRESSIVE, LOVE and EXPLORER behaviour towards different stimuli. |

| Vehicle | Components | Concept | Behaviour |
|---|---|---|---|
| 4a | 3c -> with Smooth Non-monotonic dependences | KNOWING INSTINCTS | May circle sources, run between them, approach them to a certain point and turn around Same as 3c, but less predictable |
| 4b | 4a -> Non-monotonic dependencies with Thresholds | DESCISIONS WILL (free?) | Vehicles seem to ponder before acting abruptly |
| 5 | 4b -> with Threshold Devices; some of them networked (counters) | NAMES LOGIC MEMORY | Reacts to specific situations. Counting Elementary (binary) Memory Externalisation of memory through action |

The architecture that is used in the experiments presented here is based on the first 5 Vehicles. The basic architecture that is used for all the experiments is shown in Figure 1. The earlier chapters of Braitenberg's work have been thoroughly explored, usually with a focus on evolving sensory-motor connections using genetic algorithms [e.g. 23, 24] However, the later, more complex architectures present a significant challenge due to their reliance on highly parallel biologically inspired dynamic processing systems. While the work presented in this paper also focuses on the earlier chapters, it presents a set of novel behaviours previously not discusses by the research cited above and describe how they can be used in interactive 'gameplay' scenarios.

## 5 Experimental Design

To test for cooperative behaviours, the base architectures are initially paired with passive motor actions that do not depend on sensory stimulus being present. Combinations of the base behaviours with these passive motor actions will be henceforth referred to as "simple cooperatives".

The extensions of these, the "complex cooperatives" are combinations of base architectures with each other. Braitenberg introduces these as vehicles of type 2 [1 pg. 6] and they are defined as vehicles with a single sensor, but with multiple wire architectures leading to the motors. Braitenberg notes that, of the possible architecture combinations, not all need to be tested and observed exhaustively, because the resulting behaviour can be easily predicted. For example, the architectures with the same wiring, but opposing effect on the motors (CI+CE and PI+PE) cancel out each other's effects on the motors. Architectures with opposing wiring, but the same effect on the motors cancel out each others steering and create a Vehicle type 1 – the equivalent of having only a single sensor and motor, as both motors are driven with equal force.

Table 3 below shows the behaviours that can be predicted and the two combinations of base architectures whose results were unknown prior to the test. These two complex cooperatives (CI+PE and CE+PI) where tested against the three passive motor drives and compared to the 4 base architectures:

Table 3 List of cooperative architectures that needed to be simulated to observe their behaviour.

| **Base** | **Complex co-op** | **Simple co-op** | **Sum Behaviour** |
|---|---|---|---|
| CI | CE | N | *Cancel each other out* |
| CI | PI | N | *Vehicle 1 – wheels driven by the same force* |
| CI | PE | N | } Focus of Experiment |
| CI | PE | F | |
| CI | PE | B | |
| CE | PI | N | |
| CE | PI | F | |
| CE | PI | B | |
| CE | PE | N | *Vehicle 1 – wheels driven by same force* |
| PI | PE | N | *Cancel each other out* |

Figure 2 below is a depiction of the experimental setup and shows one agent facing a single source of stimulus. For each test, the sensor pair is connected to the two motors that drive the left and right wheel, using one of the 4 base architectures presented by Braitenberg [1] and shown in Table 1.

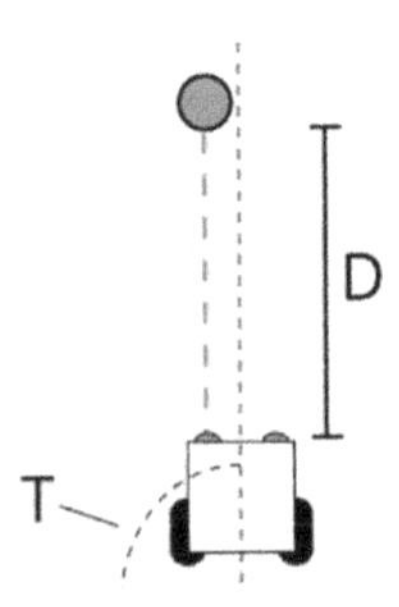


Figure 2 The experimental set-up with the stimulus offset to one side. The changes in the distance to the stimulus D and the angle between the starting orientation and final rotation T were observed while motor activity was recorded.

The reason for this offset is that some types of behaviour may involve turning towards the stimulus – an action that could not be observed if the agent is already directly facing the stimulus. The offset was also deemed a realistic situation for a dynamic environment where objects are expected to be moving (or seem to be moving due to the agent's own movement) and a perfectly symmetrical start position applies to only a limited set of situations, compared to an offset.

This simple, highly controlled setup allowed us to create a series of 'world' situations and to collect data from sensor perception, drive (actuation) and overall movement relative to the source of stimulation. In addition to this, more complex worlds were created in Flash™ for observational studies, i.e. embedded in artificial (virtual) worlds. Each experiment type was found to complement the other and to assist in the identification of 'interesting' emergent behaviour.

## 6 Cooperative Behaviours Test Results

The observations from the *simple cooperatives* test are shown in Table 4. The "Behaviour" column set describes whether the vehicle moved toward or away from the stimulus and whether simultaneously turned toward or away from it. "Easing" of movement is a term borrowed from classical animation, where it traditionally refers to the gradual increase or decrease in animation frames during an animation sequence.

Table 4 lists observations from the simple cooperative behaviour test.

| Name | Configuration | | | Behaviour | | |
|---|---|---|---|---|---|---|
| | **Crossed (C) Parallel (P)** | **Excitatory (E) Inhibitory (I)** | **Passive Motor Drive** | **Distance to Stimulus (D)** | **Angle to stimulus (T)** | **Easing of movement** |
| Aggressive | C | E | N | - | - | Speed up |
| Cowardly | P | E | N | - | + | Speed up |
| Loving | P | I | F | - | - | Slow down |
| Exploring | C | I | F | - | + | Slow down |
| Suspicious | C | E | B | + | -/+ | Slow down |
| Tempting | P | E | B | + | + | Slow down |
| Defensive | P | I | N | + | -/+ | Speed up |
| Revolted | C | I | N | + | + | Speed up |

Figure 3 below is an iconographic depiction of the configurations that constitute the range of basic behaviours possible with a differential drive robot platform. The "simple cooperatives" require a passive drive on the motors (indicated by a grey arrow on the body of the Vehicle) in addition to the drive from the sensory system. For these it is assumed that the passive drive is stronger than the drive from the sensory system.

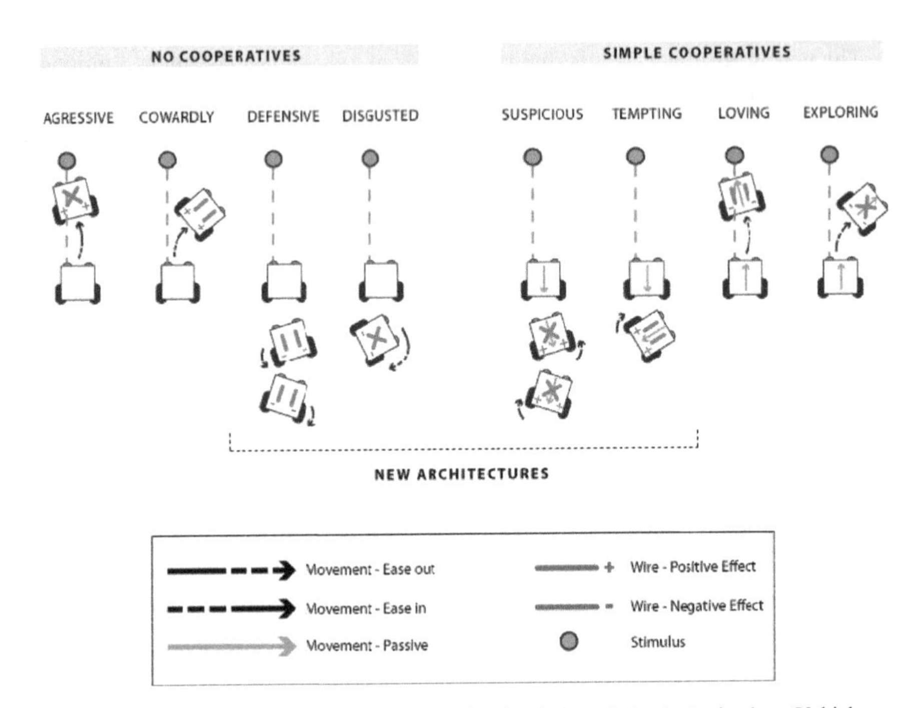


Figure 3 The range of behaviours displayed by the simulation of simple Braitenberg Vehicle architectures. Note that architectures that extend Braitenberg's are identified as "new architectures". Note also that these psychological labels are purely mnemonic and serve as a suggestion to how each behaviour might be interpreted by an observer.

The results for the *complex cooperatives* matched the predictions on Table 3. CI+PI and CE+PE created behaviour where both wheels drive at the same speed, causing zero steering, while CI+CE and PI+PE cancelled each other out completely (no drive on either motors). The observations of the behaviour generated by CE+PI and CI+PE revealed that the resulting behaviour retained their orientation steering, yet cancelled out the directional (forward or backward) drive on the motors. CE+PI turns the agent toward the stimulus while CI+PE turns it away. Based on these results the following table was created. It can be read in the following way:

1. Adding two base architectures ***horizontally*** cancels out the Directional movement (D) in relation to the source of stimulus.
2. Adding two base architectures ***vertically*** cancels out the Turning angle (T) in relation to the source of stimulus.
3. Adding two base architectures ***diagonally*** cancels out both the Turn (T) and the Directional movement (D).

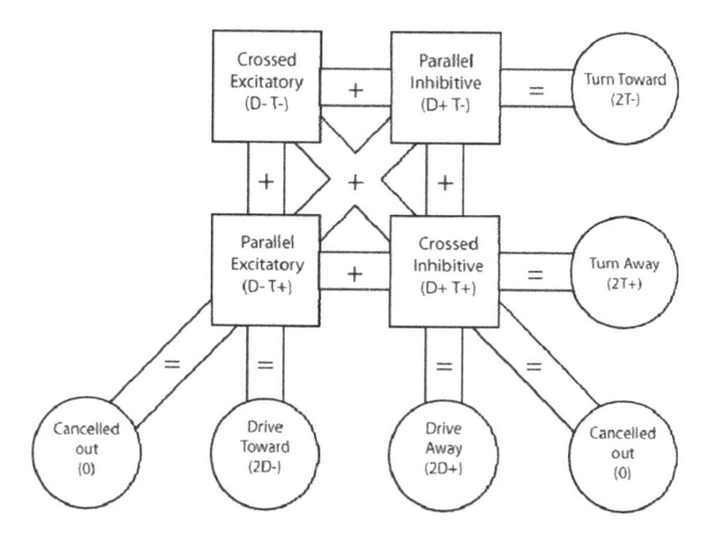


Figure 4 Base architectures combine to form new behaviours or cancel each other out.

So in addition to the 4 base behaviours initially described by Braitenberg, this study has revealed the set of new behaviours listed in Table 5 below.

Table 5 New Behaviours observed in this experiment that were not documented by Braitenberg.

| **Code** | **Label** | **Description** |
|---|---|---|
| *Complex Cooperative Behaviours (combining 2 base architectures)* | | |
| CE+PI | Turn toward | Turns toward the stimulus. Will not move even when stimulus gets close. |
| CI+PE | Turn away | Turns away from the stimulus. Will not move even when stimulus gets close. |
| *Simple Cooperative Behaviours that Braitenberg did not mention* | | |
| PI+N | Defensive | Turns toward the stimulus and will back away when the stimulus gets close. |
| CI+N | Disgusted | Turns away from the stimulus and will back away when the stimulus gets close. |
| PE+B | Tempting/ Taunting | The agent drives backwards and always turns to face way from the stimulus, resulting in an outward spiral path. |
| CE+B | Suspicious | The agent turns toward the stimulus, while driving backwards. |

## 7 Embedded Studies

These were designed to be explicitly visual and to complement analysis of the data elicited from the more formal controlled experiments. There were essentially three situation types used.

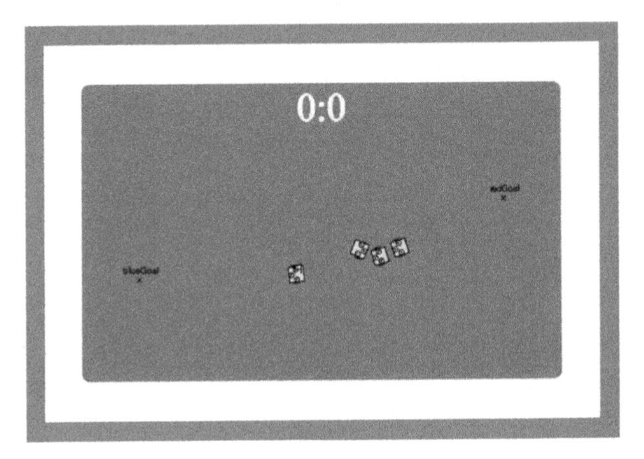


Figure 5 "Capture the flag" Flash game using instinct-driven Braitenberg controller.

Figure 5 shows a simple game of 'capture the flag', to observe some of the simplest behaviour in an adversarial team setting. Rules of the game were imposed, such as what constituted a goal, what happens after a goal is scored, how a flag is actually captured/ visualised, etc., but the way each agent perceives and moves is totally dependent on its architecture. Figure 5 is a screenshot of a game just after initiation. The Agents use a set of 3 competing drives. Initially the 'aggressive' behaviour steers the agents toward the flag. If the agent is the flag carrier, the same behaviour is used for the goal, while simultaneously the 'cowardly' behaviour attempts to avoid enemy agents. A switch is used to tell the other agent in the team that their teammate is carrying the flag. In this case the agent targets nearby enemy agents with the 'aggressive' behaviour in an attempt to block them and protect the flag-carrier.

The world depicted in Figure 6 is not a game situation, but a simple example of how groups of agents can be affected by the behaviour of another. The simulation appears to have elements of group dynamics and illusion of conversation is observed. These "conversational" agents use a balanced set of cooperative instinctual behaviours to create behaviour resembling a group conversation. Groups of agents will face each other and if another agent joins the conversation, will seem to first suspiciously observe and then eventually make space for the newcomer.

Figure 6 “Conversational” agents.

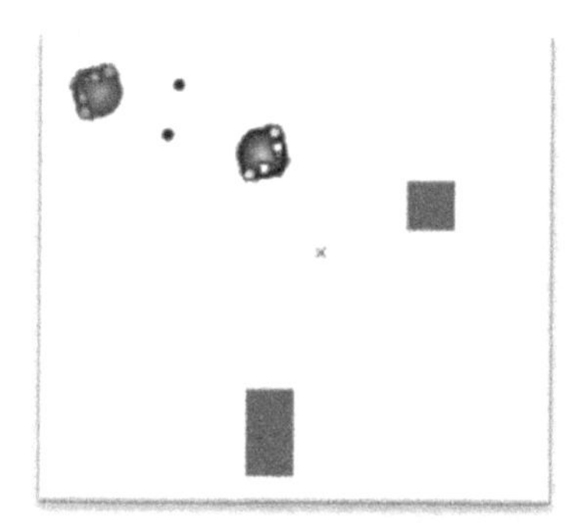

Figure 7 “Combat” shooting game.

The “combat” shooting game shown in Figure 7 involved two agents fighting. As opposed to the standard Braitenberg agent model, they use two separate differential drives to allow for sideways "strafing" movement typical for modern shooting games (2D or 3D). While strafing is mainly used to avoid bullets, the normal differential drive is used to navigate the environment. The aim of this test was to create motion that is different from the tradition 2-wheeled Braitenberg Vehicle.

These animations were designed and implemented to test their impact on a group of researchers/ developers in the field of computer games. By ensuring simplicity in the purpose and goals of these animations, it was easy to gain an appreciation for what was going on very rapidly. They were demonstrated to various attendees of the Paris Game AI Conference in 2011.

## 8 Conclusions

Results obtained, both in terms of sensory and actuator data in time, and through observational studies, indicates that agent (in this case vehicle) behaviour is a simple 2-D world is largely non-deterministic using B’s connectionist approach, and the behaviours emerging from vehicles having essentially the same base architecture can be seen to behave, not only in dramatically different ways, but also in ways that lead to the suspension of disbelief that is the main goal of the work.

During the Paris Game AI conference 2011, these results were presented in the form of a set of demonstrations featuring agents in the Adobe Flash simulation environment interacting with each other using several of these new behaviour combinations.

## References

1. Braitenberg, V. (1984). "Vehicles: Experiments in Synthetic Psychology", The MIT Press. Cambridge, MA, USA
2. Burke, R., Isla, D., Downie, M., Ivanov, Y., and Blumberg, B. (2001). "CreatureSmarts: The art and architecture of a virtual brain". In Proceedings of the Game Developers Conference.
3. Evans, R. (2001). "The future of ai in games: A personal view". In Game Dev. Magazine **8**:46-49. 31.
4. Isla, D., Blumberg, B. (2002). "New Challenges for Character-Based AI for Games". AAAI Spring Symposium on AI and Interactive Entertainment.
5. Yildirim, S., Berg Stene, S. (2008) "A Survey on the Need and Use of AI in Game Agents". In Proc. of Agent Dir. Sim. (ADS'08) within Conf. SpringSim'08. Ottawa, Canada: 124-131
6. Jacobsen, A., Berg Stene, S., Yildirim, S. (2009). "Possibilities for Learning in Game Artificial Intelligence". Presented at NAIS 2009. Available from http://www.tapironline.no/last-ned/257 [Accessed 06.02.12]
7. Coleridge, S. T. (1817). "Biographia Literaria". Project Gutenberg EBook. Available from: http://www.gutenberg.org/dirs/etext04/bioli10.txt [Accessed 12.02.2009]
8. Stanislavsky, C. and H. Popper (1961). "Creating a Role, Theatre Arts Books". Routledge
9. Stanislavski, C. (1968). "Stanislavski's Legacy: A Collection of Comments on a Variety of Aspects of an Actor's Art and Life", Routledge: A Theatre Arts Book.
10. Egri, L. (2004). "The Art of Dramatic Writing: Its Basis in the Creative Interpretation of Human Motives". Simon & Schuster.
11. Thomas, F. and Johnston (1981). "Disney Animation - the illusion of life", Abbeville Press.
12. Jones, C. (1989). "Chuck Amuck: The Life and Times of an Animated Cartoonist". Farrar, Straus and Giroux.
13. Lasseter, J. (1987). "Principles of Traditional Animation Applied to 3D Computer Animation." SIGGRAPH '87: Proceedings of the 14th Annual Conference on Computer Graphics and Interactive Techniques. ACM, New York, NY, USA: 35-44
14. Loyall, B. (1997). "Believable Agents: Building Interactive Personalities." Thesis (PhD). Carnegie Mellon University, Pittsburgh, PA
15. Mateas, M. (1999). "An Oz-Centric Review of Interactive Drama and Believable Agents". Artificial Intelligence Today. Springer Verlag, Berlin, Heidelberg, Germany: 297-328
16. Reynolds, C. (1987). "Flocks, herds and schools: A distributed behavioural model." SIGGRAPH Computer Graphics 21(4): 25-34
17. Naturalmotion. (2005). "Natural Motion". Available from: http://www.naturalmotion.com/products.htm [Accessed 22.05.09]
18. Autodesk. (2008). "Ubisoft Assassin's Creed: Autodesk 3D Pipeline Helps Bring Centuries-Old Tale Into the Modern Age". Available from: http://images.autodesk.com/adsk/files/cust_success_assassin_s_creed_v3.pdf [Accessed 20.03.09]
19. Perlin, K. (2009). "Actor Machine". Available from: http://www.actormachine.com/manifesto.html [Accessed 03.08.09]
20. Regelous, S. (2009). "Massive Software". Available from: http://www.massivesoftware.com/stephen-regelous/ [Accessed 01.02.09]
21. Walter, W. G. (1953). "The Living Brain". Duckworth, London
22. Braitenberg, V. (1965). "Taxis, Kenesis and Decussation." Prog. in Br. Res. **17**: 210-222.
23. Floreano, D. and F. Mondada (1994). "Automatic creation of an autonomous agent: genetic evolution of a neural-network driven robot". From Animals to Animats 3: Proc. of the 3rd Int. Conf. on Sim. of Adaptive Beh.. Brighton, UK. MIT Press, Cambridge, MA, USA : 421-430
24. Seth, A. (1998). "Evolving Action Selection and Selective Attention Without Actions, Attention, or Selection". From Animals to Animats 5: Proc. of the 5th Int. Conf. on Sim. of Adaptive Behaviour. Brighton, UK. MIT Press, Cambridge, MA, USA: 139-146

# Improving WRF-ARW Wind Speed Predictions using Genetic Programming

**Giovanna Martinez-Arellano, Lars Nolle and John Bland[1]**

**Abstract** Numerical weather prediction models can produce wind speed forecasts at a very high space resolution. However, running these models with that amount of precision is time and resource consuming. In this paper, the integration of the Weather Research and Forecasting – Advanced Research WRF (WRF-ARW) mesoscale model with four different downscaling approaches is presented. Three of the proposed methods are mathematical based approaches that need a predefined model to be applied. The fourth approach, based on genetic programming (GP), will implicitly find the optimal model to downscale WRF forecasts, so no previous assumptions about the model need to be made. WRF-ARW forecasts and observations at three different sites of the state of Illinois in the USA are analysed before and after applying the downscaling techniques. Results have shown that GP is able to successfully downscale the wind speed predictions, reducing significantly the inherent error of the numerical models.

## 1 Introduction

As the level of wind power penetration increases, it becomes more difficult for power system operators to maintain the stability of the power grid. This is due to the inherent intermittency and variability of wind. It has become necessary to forecast the amount of wind power that will be produced in order to plan the scheduling of power plants over the next day.

Substantial research has been done and will continue to be carried out for improving the accuracy and robustness of the wind power predictions. Wind power forecasting tools are commonly composed of a combination of Numerical Weather Prediction models (NWP), terrain models from wind farm locations and real time measurements of wind speed and power outputs from wind farms (SCADA systems) [1]. A vast majority of wind power prediction techniques perform two major steps: first, the prediction of the wind speed at a specific height and location, and second, the transformation of these predictions into wind power according to a power curve. This power curve is dependent on the characteristics of the wind turbine.

[1] Nottingham Trent University, NG1 4BU
{N0204624, Lars.Nolle, John.Bland}@ntu.ac.uk

Different methods have different performance depending on the time horizon of the forecast. Forecasts are usually classified in three categories: *very short-term*, that ranges from minutes to couple of hours, *short term* forecasting, which ranges from very-short term up to 48 or 72 hours, and *medium term* forecasting, which ranges from short term limit up to 7 days. Short term forecasting in mainly used for trading in the day-ahead market, which is the main interest in this research. Having an accurate forecast for the next 48 hours will provide the grid operators with the necessary information to take optimal decisions when deciding on the operation schedule of the grid.

A common approach to short term forecasting is to analyse recent meteorological data using statistical models such as autoregressive processes and artificial-intelligence-based models [2]. These techniques are only of use for forecasts of up to 6-10 hours into the future. In order to provide a good short term forecast, it is important to include meteorological models. Here, the atmospheric dynamic is decisive.

Meteorological models represent the atmospheric flow by a set of physical equations, which govern the behavior of the dynamics and thermodynamics of the atmosphere. These models can be classified into three types: global models, e.g. the Global Forecast System (GFS) from the National Oceanic and Atmospheric Administration (NOAA) in the USA, which produce low space resolution forecasts of the entire globe, mesoscale models, which produce forecasts with a space resolution of up to 1km x 1km in a specific region, and local models, with the highest space resolution. Models such as the High Resolution Limited Area Model (HIRLAM) have been used for short term forecasting [3, 4]. In [5], the mesoscale WRF model, initialized with an ensemble of realizations from the North American Regional Reanalysis (NARR) data, is used to quantify wind uncertainty on the day ahead and to study the economic impact of large amounts of wind power in the electric grid. In [6] the use of global physical models together with neural networks and autoregressive models was proposed for short term forecasting, neural networks being the approach that provided better results. In [7] a combination of a global and mesoscale physical model together with neural networks has been used for short term forecasting. The neural network is used to perform the final downscaling from the mesoscale model to the observation sites. However, the neural network approach behaves as a black box, which does not provide information of the model that was found and needs a significant amount of training data to ensure generalisation.

In this paper, the integration of global and mesoscale physical models using genetic programming for wind speed prediction is proposed. The rest of the paper is organised as follows: the next section presents an analysis of the correlation between numerical models and wind speed observations at three different sites in the state of Illinois in the USA [8]. Section 3 shows in detail the proposed approach for wind speed forecasting. Section 4 describes the experiments carried out and presents the results obtained. Finally, section 5 concludes with the discussion and proposed future research.

## 2 Numerical weather prediction models

Numerical weather models have been proved to be relevant for short term forecasting due to their capability to capture the trend of the atmospheric flow [3, 4]. Global models can be used as a first step towards the prediction of wind speed at specific sites. Despite their low resolution, these forecasts can be used as the start point and boundary conditions for a mesoscale model that could improve the space and time resolution of these forecasts. Specifically, the GFS model was used in this research. GFS runs four times a day, at 00Z, 06Z, 12Z and 18Z (UTC time). At each run, it produces low resolution forecasts, this means the entire globe is divided into a grid, of usually 1° x 1°, producing forecasts at each of the intersection points of the grid. Each GFS model execution predicts up to 16 days in advance with a three-hour time step. To make sure that there is a relationship between GFS and real wind speed observations, a correlation analysis has been carried out.

### *2.1 GFS correlation analysis*

As it was previously mentioned, the GFS model runs four times a day, each time producing a forecast of up to 16 days into the future. For short-term prediction purposes, only the first 48 hours of each GFS forecast was taken into account. To analyse the correlation between global forecasts and observations, 50 contiguous runs of the GFS model were used, considering only those values forecasted 48 hours into the future. The reason for this is that values forecasted 48 hours into the future are more likely to have a larger error than the rest of the forecasted hours within the 48-hour horizon.

Wind speeds at ten meters height were extracted from the model forecasts at the closest point in the grid to the observation sites. Figure 1 shows the correlation of GFS with observations from the Wilmington site. The global model is able to capture the trend of local observations despite the low space resolution of the model. It has been found that GFS model tends to over estimate the wind speed. The same behavior occurred in the data from the Cuba and SIUE sites. This evident relationship between the GFS model and the observations proofs the potential of the model for short-term prediction.

### *2.2 WRF correlation analysis*

In order to improve the space and time resolution of GFS forecasts (physical downscaling), a mesoscale model was applied. Usually, mesoscale models can be fed with not only global forecasts, but also local measurements of the terrain and atmospheric conditions to improve the quality of the forecast [7].

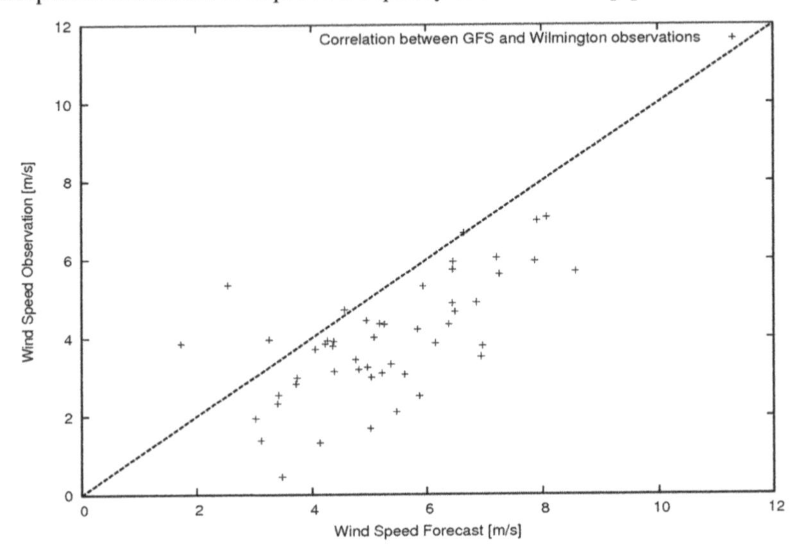


Figure 1. Correlation between GFS forecasts and wind speed observations at Wilmington site.

For this research, the state-of-the-art WRF-ARW model from the National Center of Atmospheric Research (NCAR) in the USA was used. WRF-ARW is a non-hydrostatic limited area model, which, similar to GFS, solves a system of differential equations that represent the dynamics of the atmospheric flow, but does not take into account the ocean-land interactions. This mesoscale model uses a series of parameters to set the initial state and boundary conditions, as well as the parameters to determine how the execution will be carried out. In this research, the initial state was generated using GFS forecasts and terrestrial data. WRF-ARW, as other mesoscale models, allows *nesting*. This means the model can run at different resolutions or domains; one contained into the other, were the inner domains have a higher resolution in a smaller area. Figure 2 shows the domain settings that were used. As shown in the figure, the model was set to run in two domains. The first domain, which covers a major part of the United States, has a resolution of 30km x 30km and results from the first integration of the WRF model from the GFS grid (111km x 78km). The second domain, which is limited to the state of Illinois, the area of interest, has a resolution of 10km x 10km and is obtained by a second model integration that uses the first domain as boundary conditions. A third domain was considered, but the computational cost of running at 3km resolution was considered too high. A similar setup is described in [5].

In order to assess the quality of the model, the 50 runs considered for GFS analysis were used to produce the same 50 48-horizon forecasts but this time improving the space resolution to 10 km and the time resolution to 1 hour. As it was done with the global model, values forecasted 48 hours into the future were taken and compared with real wind speed observations.

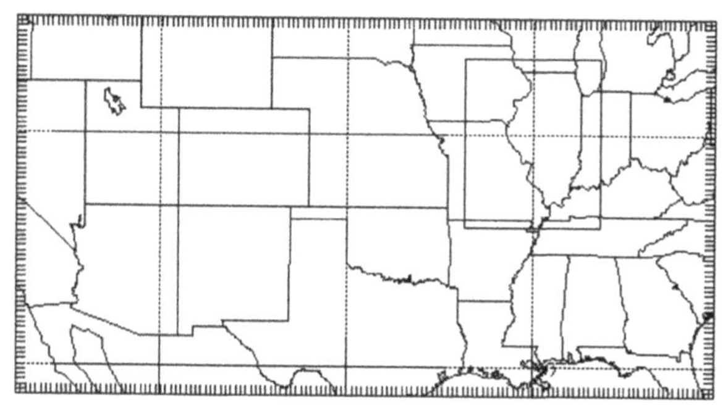

Figure 2. WRF-ARW two domain settings

Figure 3 shows how the mesoscale model still captures the trend of the wind speed observations. However, the quality of the forecast can depend on the local conditions of the site. For the SIUE site, WRF-ARW maintains the level of correlation that was seen on GFS forecasts but for Wilmington, in Figure 4, the correlation shows that there is a decrease in the quality of the forecast. This is reasonable; due to the fact that the mesoscale forecast is done based on another forecast which inherently contained errors.

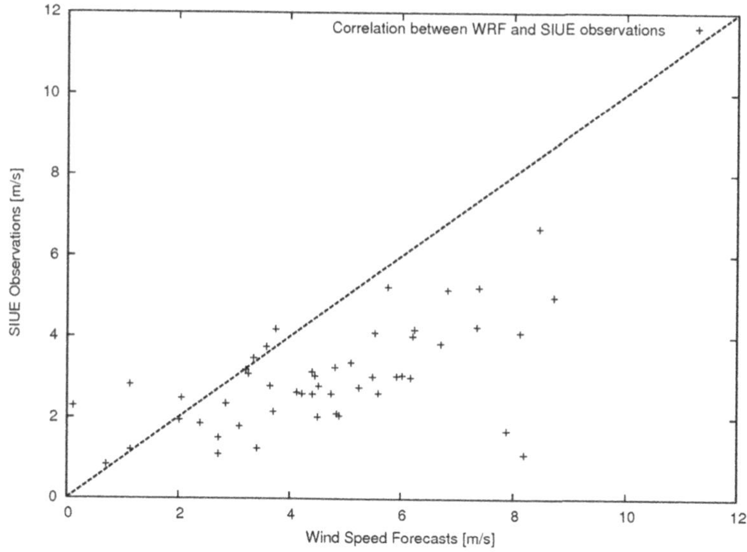


Figure 3. Correlation between WRF-ARW forecasts and wind speed observations at SIUE site.

Despite the improvement in the level of resolution of the forecasts, the mesoscale model it is not accurate enough to be able to forecast the wind speed in specific sites, which are not necessarily close to the one of the intersection points of the model grid. There is a need of a final process that could downscale the forecasts from the grid to the final site. This paper presents four approaches for this downscaling step, three purely mathematical based ones and one based on genetic programming.

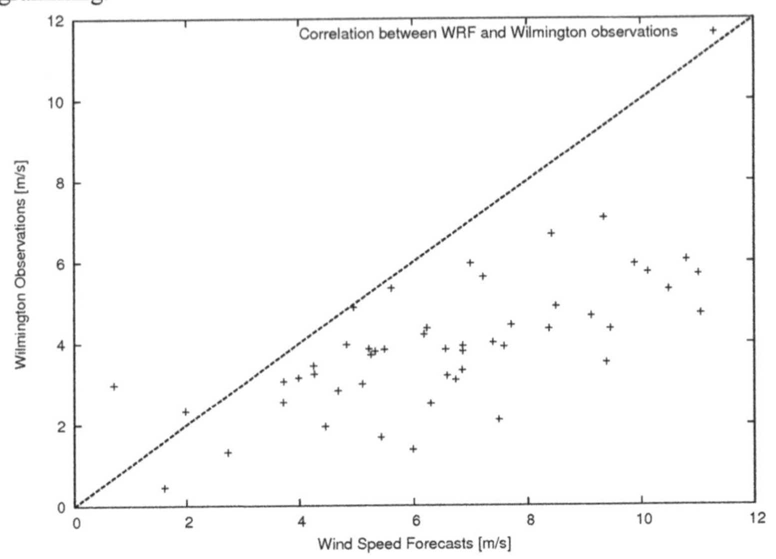


Figure 4. Correlation between WRF-ARW forecasts and wind speed observations at the Wilmington site.

# 3 Hybrid model for wind speed forecasting

As it has been shown, numerical models play an important role in short-term wind speed forecasting. However, the resolution of these models is not high enough to be able to forecast wind speed at specific sites. In this research, an integration of numerical models and statistical downscaling methods is proposed. The layout of the model is shown in Figure 5.

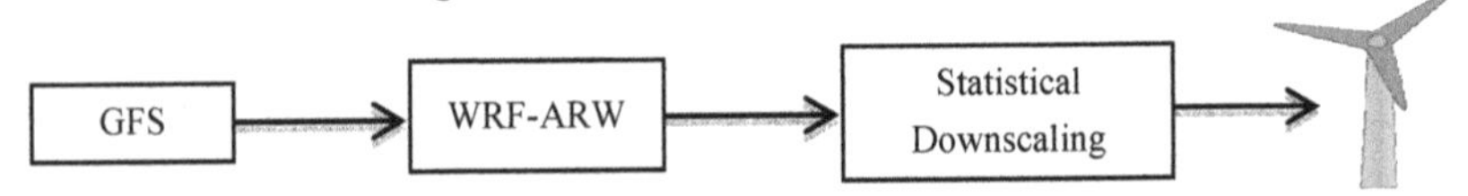


Figure 5. Hybrid forecasting model layout

## *3.1 Mathematical methods and preliminary results*

WRF-ARW outputs a set of atmospheric variables every hour at each of the intersection points of the model grid. Assuming that the observation site is located somewhere between any four points of the grid, as shown in Figure 6, it can be deducted that any of these surrounding wind speeds or a combination of the four, can contribute to the estimation of the wind speed at the desired location. With this idea in mind, the first method that was implemented was the Inverse Distance Weighting (IDW) method [9].

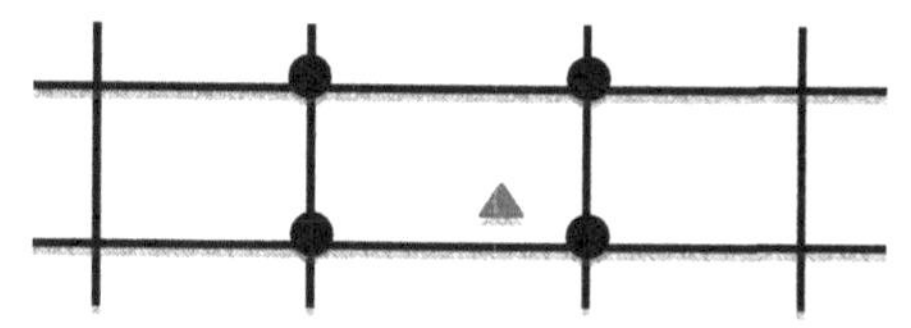

Figure 6. The WRF-ARW grid and a hypothetical location of the observation site.

This method consists of combining the four surrounding forecasts, each one with a specific weight $w_i$ according to its distance to the observation site. IDW assumes that each forecasted point has a local influence that diminishes with distance. The mathematical representation is as follows:

$$v_o = w_1 v_1 + w_2 v_2 + w_3 v_3 + w_4 v_4 \tag{1}$$

where

$$w_i = d_i^{-2} / \sum_{i=0}^{4} d_i^{-2}$$

The variables $v_1$, $v_2$, $v_3$ and $v_4$ represent the four forecasted wind speeds of the mesoscale grid. $v_0$ is the new calculated wind speed at the observation site and $d_i$ is the distance between grid point $i$ and the site. The second method (Best Weights) that was implemented consists of combining the four surrounding wind speeds giving a specific weight to each of them. The mathematical representation is as follows:

$$v_o = w_1 v_1 + w_2 v_2 + w_3 v_3 + w_4 v_4 \tag{2}$$

where, as with the IDW method, the sum of the weights is equal to one. The third mathematical method (Best Coefficients) is similar to the previous one, with the only difference that the sum of the weights, which in this case are called coefficients $c_i$, is not necessarily equal to one.

$$v_o = c_1 v_1 + c_2 v_2 + c_3 v_3 + c_4 v_4 \tag{3}$$

The IDW method will only use the forecasts form the grid to calculate the new wind speed, no past data needs to be available to do the estimation. This is not the case for the other two methods. In order to find the values of the weights, or coefficients for methods two and three, past data is used such that the values of the unknowns will be those that minimise the error between the forecasts and the observations in a given sample of past data. Once the unknowns are found by solving a system of four linear equations, they could be used to calculate new forecasts.

To compare these three methods, the 48-hour ahead values from the 50 forecasts of the WRF-ARW model were used. For IDW, it was just a matter of extracting from the model the four surrounding points, calculate the distance to the observation site and apply equation (1). For the other two methods, the unknowns were found by solving a system of equations that would find the optimal values such that the overall error between the 50 forecasts and the observations are minimised. Once the weights or coefficients were found, they were applied to each of the 50 forecasts (only for values forecasted 48 hours into the future). The mean average error and root mean square error for the three methods on the 50 forecasts sample are shown in Table 1. It can be seen that over the 50 forecast samples, the method that had less error is the third one, based on coefficients. The IDW method can only be as good as the mesoscale model, because no "intelligent" correction is done in order to compensate for the error of the numerical prediction. The other two methods show better results but their quality depends directly on the accuracy of the downscaling model, which in this is case is a simple linear combination of the four wind speeds. Hence, a method that is independent of a specific model was required. For this, genetic programming was used in this research.

Table 1. Mean absolute error (MAE) and root mean square error (RMSE) in meters per second of the three methods over the 50 forecasts sample

| | IDW | | Best Weights | | Best Coefficients | |
|---|---|---|---|---|---|---|
| Site | MAE | RMSE | MAE | RMSE | MAE | RMSE |
| Cuba | 2.84723 | 3.27872 | 1.118538 | 1.386675 | 0.900721 | 1.167497 |
| SIUE | 1.88096 | 2.35104 | 1.630793 | 2.12192747 | 0.799778 | 1.094980 |
| Wilmington | 2.7784 | 3.23304 | 2.530139 | 3.078297 | 0.839859 | 1.059227 |

## 3.2 Genetic Programming

Genetic programming (GP), as many other biologically inspired heuristics, is based on the evolution of individuals over time, through events such as cross over and mutation, where only the fittest individuals survive to next generations [10]. In genetic programming, instead of evolving binary chromosomes, as used in genetic algorithms, programs (in a binary tree layout) are evolved, which represent a set of instructions to solve a specific problem. In terms of statistical downscaling, the GP will be able to perform a symbolic regression to find the model that represents the best combination of the existing forecasts to estimate the new forecast at the observation site with the minimum error. The interesting aspect of GP, compared to neural networks, is that the final model is known, and represented by the best tree that survived. In [11], a GP based strategy was used to perform a symbolic regression in the time series prediction problem, where the genetic program represents a combination of past wind speed data to calculate the wind speed in the near future. In contrast, in this research the genetic program will learn from the error of the past forecasts to be able to compensate the error in future forecasts. A GP tree is formed by a set of terminals and functions. The terminals can be either constants or any of the four wind speed forecasts of the grid, and the functions are the basic operators {+, -, *, /}. The structure of the program is presented in Figure 7. The learning task of this approach consists of minimising the error between the program output and the observations in the training set.

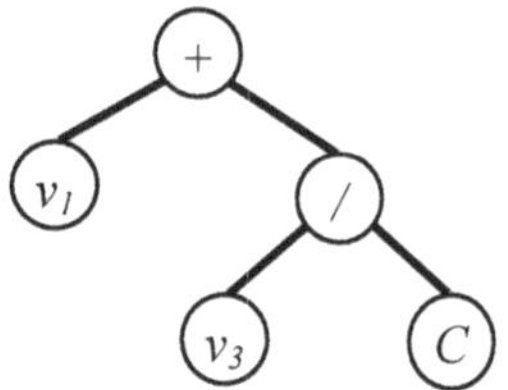


Figure 7. Example of a tree expression of the program $v_0 = v_1 + v_3 / C$

As any other Machine Learning (ML) technique, achieving good generalization is one of the most important goals of the GP approach. Failure to generalize, or overfitting, happens when the solution performs well on the training cases but poorly on the test cases. The most common approach to reduce overfitting is to bias the search towards shorter solutions. Significant research has been done in the use of this strategy. In [12], a parsimony pressure approach is presented. This method consists in computing the complexity of the tree in terms of the Vapnik-Chervonenkis dimension. This complexity factor is used as a penalisation of the fitness of each tree. The mathematical representation of this penalty function is shown on equation 4.

$$f = \frac{1}{s} \sum_{i=0}^{s} e\,(i) + k \left(\frac{t^2 \log_2(t)}{s}\right)^{\frac{1}{2}} \quad (4)$$

The first term of the fitness equation (4) is the sum of the errors between the obtained output (new forecast) and the desired output (observations) in the test set (test set size = s). The second term is the complexity factor, where t is the number of nodes of the GP tested and $k$ is a trade-off weight that allows to control the level of pressure of the complexity factor. A small value of $k$ (e.g. $k$=0.0001) would be translated into low complexity pressure, and higher values of $k$ (e.g. $k$=0.1) will result in a strong pressure to the penalisation.

Other approaches based on Random Sampling Technique (RST) have been proposed to manage overfitting [13, 14]. RST basically consists of using a random subset of the training set instead of the complete set, and calculate the fitness of the programs based only on this subset. The subset can be re-selected every iteration or at every $t$ iterations. The programs that survive through the generations are those performing reasonably good on different subsets. These programs will be capturing the underlying relationship of the data instead of overfitting it. In this research, a combination of complexity penalisation and random sampling was used.

## 4 Experiments and results

With the preliminary results of the mathematical downscaling methods, an exhaustive experimentation phase was done to analyse the potential of genetic programming. The reason for using two overfitting strategies is that, if only the RST strategy is considered, even getting good generalization models, the problem of bloat or code growth can occur [13]. This is due to the concept of *introns*, which are useless pieces of code that do not affect the fitness of the program but contributes to the development of large solutions.

The experiments were conducted as follows. For each of the three observation sites, different sizes of the random training subset (*rss*) were tested. From the 50 forecasts generated using WRF-ARW, the four wind speed predictions of the grid that were closest to the observation sites were extracted, considering only those values forecasted 48 hours into the future. Hence the training set consisted of 50 vectors ($v_1$, $v_2$, $v_3$, $v_4$) each one with its corresponding observation. From this training set, the subset size values that were used were 50, 49, 40, 30, 20, 10 and 1. For each of this subset sizes, three different values of $k$ where used, $k$=0.1, $k$=0.01 and $k$=0.001. Each combination of this two parameters was executed 50 times, keeping track of the best solution at each run and taking an overall average of the best solution. Additional parameters were set to carry out the experiments. These are shown in Table 2.

Table 2. Fixed GP parameters used in the experiments

| Runs | 50 |
|---|---|
| Population | 1000 |
| Generations | 100 and 500 |
| Crossover operator | Standard subtree crossover, probability 0.9 |
| Mutation operator | Standard subtree mutation, probability 0.1, maximum depth of new tree 17 |
| Tree initialization | Ramped Half-and-Half, maximum depth 6 |
| Function set | +, -, *, / |
| Terminal set | Input variables and random constants |
| Selection | Tournament of size 20 |
| Elitism | Best individual always survives |

The first set of experiments were carried out with k=0.1 and trying different subset sizes. Results for the SIUE site are shown in Figure 8. It can be seen from the figure that as the size of the training set increases, the better the average cost of the best solution. However, improvement over training data does not mean generalization for new data. For k=0.1, the generalization is best achieved by subsets of size 40. In Figure 9, with parameter $k$=0.01, again it can be appreciated how the best results over training data are obtained with the larger subset (*rss*=50).

However, the best generalization is achieved with subsets of size 30. Similar results were obtained in the Cuba and Wilmington sites.

Table 3 summarizes the results obtained at the three observation sites with all methods over the complete training data set. In the case of the GP, the result shown is taken from the best run. It can be seen that GP obtains the best results for the three sites. Table 4 shows the results obtained with the best two methods over new 48-hour forecasts. It can be seen that, in case of GP, the best result over new data is not obtained using the complete training set. In general, the results obtained with GP improve the previous methods due to its ability to generalize. A value of $k$=0.01 produced the best results over training and test data and random subsets of size 20 and 49 let to a good level of generalization. Smaller values of $k$ did not produce comparable results due to the low parsimony pressure, which resulted in more complex models.

Table 3. MAE in meters per second of all downscaling methods over the training sample

| Site | IDW | Best Weights | Best Coefficients | GP (*k*, *rss*) |
|---|---|---|---|---|
| Cuba | 2.84723 | 1.118538 | 0.900721 | **0.88917118** (0.01, 49) |
| SIUE | 1.88096 | 1.630793 | 0.799778 | **0.72547812** (0.01, 50) |
| Wilmington | 2.7784 | 2.530139 | 0.839859 | **0.764965**(0.01, 50) |

Table 4. MAE and RMSE in meters per second of the two best downscaling methods over 20 new values forecasted 48 hours ahead. In the case of the GP, the average on 50 runs is reported.

| Site | Best Coefficients | | GP (*k*, *rss*) | |
|---|---|---|---|---|
| | MAE | RMSE | MAE | RMSE |
| Cuba | 1.4668495 | 1.761794 | 1.47404 (0.01, 20) | 1.712134 |
| SIUE | 1.042474 | 1.321985 | 1.01643 (0.01, 20) | 1.229757 |
| Wilmington | 1.3233674 | 1.569485 | 1.16398 (0.01, 49) | 1.38117 |

Finally, figures 10 and 11 show the correlation between the best GP model and observations at the SIUE site, on the training set and on new, previously unseen data. The new predictions correspond to the next model run, after the last sample in the training set. The error obtained in this horizon is comparable to results reported in [15].

## 5 Discussion and Conclusions

In this paper, a genetic programming approach was proposed to improve wind speed predictions from numerical models. It has been shown that numerical weather prediction models play an important role in the short term forecasting due to their ability to capture the atmospheric flow. GFS and WRF-ARW models, with basic parameterization, tend to overestimate the wind speed, depending significantly on the local characteristics of the site. Due to this overestimation and low space resolution of the models, a downscaling method is needed in order to forecast the wind speed in specific locations. Results have shown that the downscaling methods based on a simple combination of the numerical forecasts are not enough to compensate the inherent error of the numerical models. The interesting aspect of using GP as a downscaling technique is that the downscaling model is learned, i.e. not specified by the user. Results have shown that the approach renders promising results compared to the mathematical approaches. The error of the numerical models was reduced by 58% at the Wilmington site, by 48% at the Cuba site and by 42% at the SIUE site.

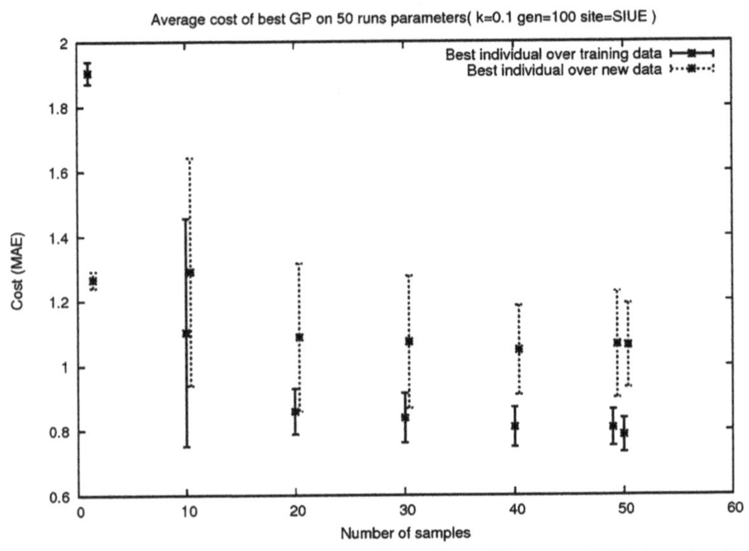


Figure 8. Average cost and standard deviation of best GP on 50 runs at SIUE site using k=0.1 and different subset sizes.

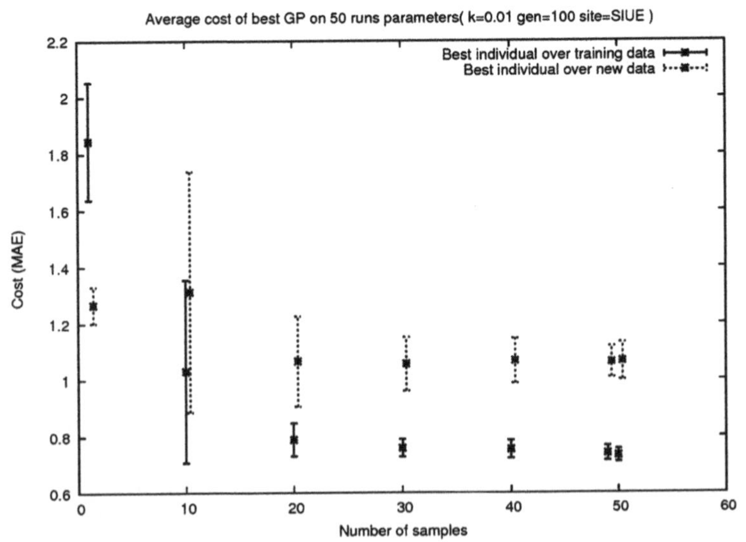


Figure 9. Average cost and standard deviation of best GP on 50 runs at SIUE site using k=0.01 and different subset sizes.

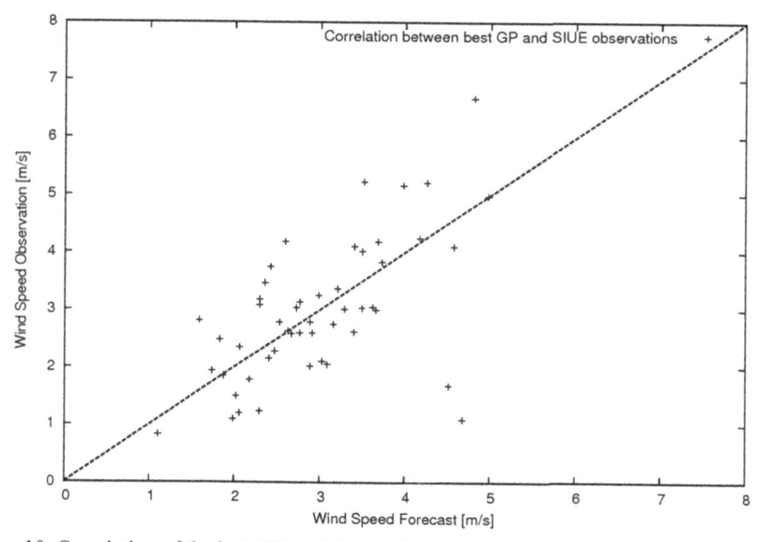

Figure 10. Correlation of the bets GP and observations at SIUE site over the training set.

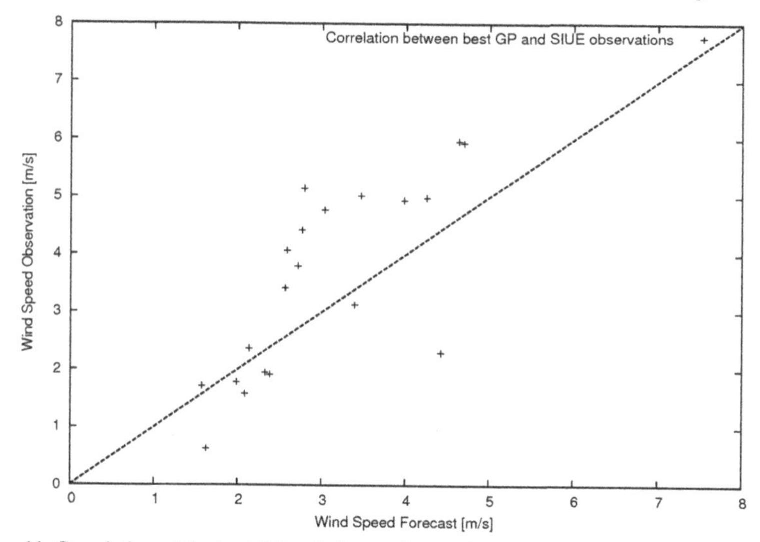

Figure 11. Correlation of the best GP and observations at SIUE over new data

The error at SIUE had less improvement due to the quality of the numerical model in this site. The use of the overfitting techniques revealed the importance of varying the size of the training set, to obtain better generalization models. The results obtained by the GP approach are comparable to the ones reported in [7], where the ANN approach obtains MAE values between 1.45 and 2.2 m/s, which are slightly greater than the MAE values reported in this paper. Similarly, the RMSE obtained with the GP approach ranges between 1.2 and 1.7 m/s, while results reported in [15] range between 0.94 and 2.02 m/s. The terrain complexity plays and important factor in the numerical model performance, hence, a more exhaustive experimentation needs to be performed to ensure the approach is robust in different terrain complexity and in any time period. The execution time of the proposed approach, depending on the input parameters, is between 18 and 59 seconds on average, being almost negligible when compared to the numerical model execution times (2-3 hours). Further research will focus on adding other inputs to the GP, like temperature and wind direction, among other atmospheric variables. These inputs could further improve the generalization of the models. Adding other operators to the function set will also be explored.

## References


1. Monteiro, C., Bessa, R., Miranda, V., Boterrud, A., Wang, J. and Conzelmann, G. Wind Power Forecasting: State-of-the-Art 2009, INESC Porto and Argonne National Laboratory, 2009. Tech. Rep.
2. Bilgili, M., Sahin, B., Yasar, A. Application of artificial neural networks for the wind speed prediction of target stations using reference stations data. *Renewable Energy*, vol. 32, pp. 2350-2360, 2007.
3. Landberg, L. Short-term prediction of the power production from wind farms. *Journal of Wind Engineering and Industrial Aerodynamics*, vol. 80, pp. 207-220, 1999.
4. Landberg, L. Short-term prediction of local wind conditions. *Journal of Wind Engineering and Industrial Aerodynamics,* vol. 89, pp. 235-245, 2001.
5. Constantinescu, E.M., Zavala, E.M., Rocklin, M., Sangmin Lee, Anitescu, M. A computational framework for uncertainty quantification and stochastic optimization in unit commitment with wind power generation. *IEEE Transactions on Power Systems*, vol. 26, pp. 431- 441, 2011.
6. Alexiadis, MC., Dokopoulos, PS., Sahsamanoglou, H., Manousaridis, IM. Short-term forecasting of wind speed and related electrical power. Solar Energy, vol. 63(1), pp. 61-68, 1998.
7. Salcedo-Sanz, S., Perez-Bellido, A.M., Ortiz-Garcia, E.G., Portilla-Figueras, A. Hybridizing the fifth generation mesoscale model with artificial neural networks for short-term wind speed prediction. *Renewable Energy*, vol. 34, pp. 1451-1457, 2009.
8. Illinois Institute of Rural Affaris, 2012. Illinois wind monitoring program, accessed 1 January 2012, <http://www.illinoiswind.org>.
9. Luo, W., Taylor, M. C., Parker, S. R. A comparison of spatial interpolation methods to estimate continuous wind speed surfaces using irregularly distributed data from England and Wales. *International Journal of Climatology,* vol. 28, pp. 947-959, 2008.
10. Poli, R., Langdon, W.B., McPhee, N.F. A field guide to genetic programming, accessed April 2012, <http://www.gp-field-guide.org.uk> (With contributions by J.R. Korza).
11. Flores, J. J., Graff, M., Cadenas, E. Wind prediction using genetic algorithms and gene expression programming. *In Proc. Of the Int'l Conf on Modeling and Simulation in the Enterprises, AMSE 2005.* Morelia, Mexico, 2005.
12. Amil, N.M., Bredeche, N., Gagne, C. A statistical learning perspective of genetic programming. *In Proceedings of the* $12^{th}$ *European Conference on Genetic Programming,* EURO GP'09, vol. 6(104), pp. 327-338, Berlin, Heidelberg, 2009.
13. Liu, Y., Khoshgoftaar, T. Reducing overfitting in genetic programming models for software quality classification. *In Proceedings of the Eighth IEEE International Symposium on High Assurance* Systems *Engineering,* HASE'04, pp. 56-65, 2004.
14. Gonçalves, I., Silva, S. Experiments on Controlling Overfitting in Genetic Programming. $15^{th}$ *Portuguese Conference on Artificial Intelligence (EPIA 2011),* Oct 2011.
15. Sweeney, C., Lynch, P., Nolan, P. Reducing errors of wind speed forecasts by an optimal combination of post-processing methods. *Meteorological Applications,* doi, 10.1002/met.294

# Optimizing Opening Strategies in a Real-time Strategy Game by a Multi-objective Genetic Algorithm

Björn Gmeiner and Gerald Donnert and Harald Köstler

**Abstract** This paper presents modeling, forward simulation, and optimization of different opening strategies in the real-time strategy game Starcraft 2. We implemented an event-driven simulator in C# with graphical user interface. In order to find optimal build orders, we employ a modified version of the multi-objective genetic algorithm NSGA II. Procedural constraints e.g. given by the tech-tree or other game mechanisms, are implicitly encoded into the chromosomes. Additionally, the size of the active part of the chromosomes is not known a priori, and the objectives values have a small diversity. The model was tested on different Tech-Pushes and Rushes, and validated with empirical data of expert Starcraft 2 players.

## 1 Introduction

Electronic sports, i.e. competitive play of video games, attracts more and more people, and currently Starcraft 2[1] is one of the most popular real-time strategy games. Typically, during a match in the BattleNet[2], up to eight human players in two teams compete with each other. The goal is to destroy all enemy bases. Each player chooses one of the three races (Terran, Protoss, or Zerg) that all have different unit types and properties. A match can be divided up into three different phases: early, mid, and late game. In the first phase, usually the first few minutes, the players build up their economy by producing workers to mine minerals or vespin gas and their main base consisting of several building types to enable the construction and

Björn Gmeiner, Harald Köstler
University of Erlangen-Nuremberg,
e-mail: bjoern.gmeiner@cs.fau.de, harald.koestler@cs.fau.de,

Gerald Donnert
e-mail: gerald_donnert@yahoo.de

[1] http://blizzard.com/en-us/games/sc2/
[2] http://eu.battle.net/de/

improvement of simple unit types. In mid game, there are first encounters, new bases are created at resources usually close to the main base, and buildings necessary for advanced unit types are placed. In late game, bigger armies of various unit types try to destroy all enemy units and bases. A whole game takes typically between 10 and 45 minutes.

We restrict ourselves here on early game, where each player has to choose a certain opening strategy. One can focus on worker production and fast expansion to other resource locations in order to increase later income, building production to develop straightly new technology and enable advanced unit types, or unit production to build as many simple units as possible in order to rush the enemy, i.e. to attack very quickly in the first few minutes. The best opening strategy depends on the map, the race, and also the opponents strategy. Our goal in this paper is to evaluate the best opening strategies under certain assumptions

- The races of all players are known
- The player is not interrupted by actions done by other players
- The player either wants to build an advanced unit of one type as fast as possible (Tech push) or to attack at a certain time with as many units of one or two types as possible (Rush)

Therefore, we have implemented a simulator for Starcraft 2 build orders in C#. A build order is an ordered list of buildings, units, and other actions that are performed by the player. For a given build order we are able to check if it is feasible and at which time the single steps like building of a unit are executed in the game.

Since the number of choices is very limited at the beginning, we can for short games up to 2-3 minutes test all possible build orders and find the optimal one with respect to a given strategy directly. For longer games a comparison of the achieved results by random search with strategy lists provided by the community convinced us, that we were far off an optimal solution. Therefore, we implement a genetic algorithm (see e.g. [1, 2, 8, 10]) to find a good strategy.

To optimize several objective functions, we adopted the non-dominated sorting-based multi-objective algorithm II (NSGA-II) by [3, 4]. Other well-known multi-objective optimization evolutionary algorithms can be found e.g. in [6, 11, 13, 16].

We compare our results to existing build orders found by experienced Starcraft 2 players and provide a compact table at the end to help players to decide which build order they can choose based on their opponents and time and unit restrictions. Due to practical reasons, we neglect micro-management of the units because it heavily depends on the player's skills and therefore we cannot predict the chances to win a game choosing a certain strategy accurately. Instead, we provide the income and value of the army to give a rough estimate of the quality of the defined strategies.

Our paper is structured as follows: The next section discusses implementation details of the simulation tool in C# and its features. Here, we explain the different stages of the simulation and all constraints. The third section shows that a random search is not enough to receive an optimal build order. Instead, we adopt a multi-objective genetic algorithm to our problem. Section 4 presents results, which we achieved with our program.

## 2 Forward simulation

An efficient forward simulation is the key for an efficient optimization strategy, since most of the computational time is spent there. However, the price for this is usually a less flexible and harder readable code. In our program we decided to implement a compromise: An object oriented approach with C#, where we loose performance in terms of using high level object oriented software design but gain performance by using an efficient programming language.

In a simulation run, we have a list of virtual players. A single player stores his build order in a strategy list that he either gets from the optimizer or he chooses all building decisions in a random way. Like mentioned before, the build order consists of actions like producing buildings, units, upgrades or do special operations like chrono-boost. Chrono-boost is an ability bound to a building to accelerate its unit production. During the simulation, the resources, i.e. minerals and vespin gas, are updated in a time-discrete manner each second in game time, the remaining actions are event-driven.

A player chooses its next action in advance. If an action is possible depends on two kinds of requirements: *Weak requirements* are delaying the start of the planned action. Weak requirements are e.g. the necessary amount of minerals or vespin gas or if there is a free slot in a production facility to produce a unit. In contrast *hard requirements* have to be fulfilled, otherwise the simulation will not proceed. Hard requirements are given by the tech-tree and the maximal available units that can be built depending on the number of Pylons, a specific building type. When all requirements are fulfilled the player starts the action, which is put in an active action list. After completion, the action is moved from the active action list to a finished action list. Altogether, the number of possibilities increases exponentially with the number of performed actions.

Apart from dependencies due to the tech-tree the following (partially race dependent) rules have to be taken into account:

- Protoss can only build one unit per production building at one time
- Unlike for other actions, it is not enough to decide, whether a Protoss' chrono-boost should be activated or not. An additional decision is required to determine on which type of building it shall be applied. If there are more buildings of one type, we decided to apply the following rules: Buildings producing something are preferred, and secondly the building with the next expiring chrono-boost is chosen.

During a forward simulation, we enter four different stages in each second:

1. *Check possible actions*: If the next action is random, all possible actions are collected. If a strategy list is given, the next action is taken from the list. Additionally it is determined, if a waiting is necessary due to a weak requirement.
2. *Global values* like mineral, vespin, and energy required e.g. for chrono-boost are updated.
3. *Start a production*: If an action is possible (i.e. no waiting state), it is put to the active action list. For Protoss units, a free place in a suitable production facility

is searched and assigned. Furthermore, new objects are generated for the units, buildings, or upgrades.

4. *Check active action list*: Produced objects are removed from the action list and associated operations like increasing income, updating maximum number of units, or clearing production slots are done.

## 3 Optimization

The simplest but most expensive way to find an optimal strategy is to check all possible ones. However, this can take hours on one compute core, e.g. even if there are only five actions per decision possible and we want to simulate all possibilities for ten subsequent actions, we end up with $5^{10} \approx 10^7$ forward simulations. For more complex builds, which easily involve e.g. two different units, six different buildings, four chrono-boost types, and 30 actions (around 6 minutes game-time for many builds), we have no chance to cover the whole search tree and thus apply a genetic algorithm. Nevertheless, it can be helpful to use it for extremely simple and small builds, because here it is ensured to find the optimal solution. These results can be used to evaluate heuristic algorithms.

In general, the optimization algorithm needs to be tailored to three different issues:

- the *search space is extremely large* (e.g. $10^{30}$ possible candidates) in contrast to a *very small objective space* (e.g. one integer between zero and ten),
- strong requirements are *procedural constraints* and have to be fulfilled, while weak constraints, which are resource constraints should be minimized, and
- the number of actions is not known in advance for a certain strategy list.

Next, we explain how we try to cope with these difficulties.

### 3.1 Single-objective optimization

Genetic algorithms are search heuristics from the field of artificial intelligence inspired by natural evolution. Genetic algorithms use populations of strings (chromosomes), which encode candidate solutions (individuals). Initially, all individuals are generated randomly. In a typically genetic algorithm (see fig. 1), the population visits the following steps during one generation:

1. *Forward simulation*: For each individual of the population, a function or simulation is evaluated with the inputs encoded by its chromosomes.
2. *Fitness evaluation*: Results of the forward simulations are evaluated by a fitness value that measure the qualities of the chromosomes.
3. *Selection*: Individuals with a higher fitness are more likely to be selected for the next step, others are withdrawn.

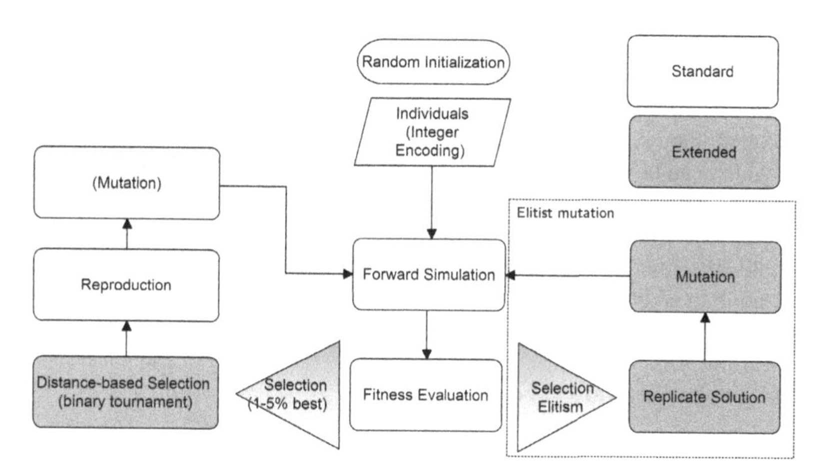


**Fig. 1** Flowchart of a typical genetic algorithm and our extensions.

4. *Reproduction*: New individuals are generated by the combination of two selected individuals. Parts of both parents' chromosomes are mixed together.
5. *Mutation*: Some genes (single entries of the string) of the new chromosomes are randomly changed.
6. *Elitism*: The best few individuals of each generation are directly accepted for the next generation.

In this part we first present the single components of our genetic algorithm: initialization, selection, crossover, and mutation and then discuss their modifications and extensions.

## Encoding

In our strategy lists, we are not only interested in an optimal arrangement of actions, but also to decide which actions and how many of them to put into schedule. In addition, the amount of available resources depends on the number of workers for minerals and vespin gas, and production facilities. In our application it is a quite natural choice to take *strategy lists as individuals*. An entry of a strategy list is equal to an action, which leads to a *integer encoding* of the chromosomes. The length of the chromosomes depends on the simulated time. However, "good" strategy lists tend to be longer. That is due to the reason that it is often beneficial to perform more actions during the simulation time. To give an idea, for five minutes a forward simulation typical needs string lengths of about 20 actions and more. Since we do not know a priori how many action are necessary for its forward simulation, all strategy lists are much longer than they need to be and have equal length.

Initialization

For the initial population we consider random individuals, which are created in such a way that all hard requirements are fulfilled for all actions. Thus, the hard requirements, which can be considered as *constraints are directly fulfilled by the encoding procedure* of the strategy lists. It ensures that the forward simulation will not run into a dead end. For a forward simulation it means that nothing will happen any more until the simulation time ends. Invalid string are not able to provide a good genetic pool. In addition, the user can insert own candidates. There exists a wide list of builds, which usually are encoded similar to our strategy lists. The user is able to add custom lists to the initial population. These lists can help and accelerate the simulation to converge to a good solution.

Selection

During selection all individuals are sorted according to their fitness. Only the best $1-5\%$ individual are chosen. This *very high selection pressure* showed up to be necessary to achieve good results. The reason might be that we have a very large search space compared to the objective space. In our simulation, the objectives are small integers (usually below 10), which means *many solutions have the same objective value*.

Reproduction and distance-based selection

We implemented an adapted uniform crossover. Starting from the first gene of both individuals, each gene is chosen randomly, if the choice does *not violate any hard requirement*. Otherwise the other gene is chosen. Although it is not guaranteed that a valid child is generated, this procedure significantly increases the chance.

We choose the individuals similar to binary tournament selection from the best already selected individuals. Four individuals are randomly chosen. Then the distances between all possible pairs of the chosen individuals are calculated. We define a *distance* $d$, which is similar to a Hamming distance, of two individuals $a_i$ and $b_i$ with a string length of $l$ by

$$d = \sum_{i=1}^{l} (l-i) \cdot (a_i - b_i). \tag{1}$$

The difference of two genes of the strategy lists $a_i - b_i$ is zero, if the genes are equal, and one otherwise. We linearly decrease the effect of different genes w.r.t. the position, and implicitly to the passed simulation time. Investigations showed that the variation of genes at the beginning of the string has a larger effect on the simulation than genes of the end. Genes, which are not processed until the end of the simulation time have no effect. The underlying idea is a similar to e.g. [14], but is placed before the reproduction in a computationally cheap way. We denote this special selection for the reproduction as *distance-based selection*, which helps to retain the diversity. Otherwise the solutions equalize during a few generations due to our aggressive selection.

We also tried to compare strings with other distance metrics, like calculating a dice's coefficient [5] by bigram or trigram, or compare solutions based on the numbers of each object type. However, the first was less effective, the latter similar effective when sorting and selecting the individuals in non-domination fronts but much more expensive.

Elitism and elitist mutation

Out of all solutions with the highest fitness, we choose the following candidates for our *elitism* to keep the possibly most interesting solutions:

- the candidate with the *most workers*,
- the candidates with the *most production buildings* (of each kind),
- and the candidate that *would earliest built an objective unit* after the finished simulation time.

We *apply the mutation operator to copies of elitism solutions* instead after crossover. We fill 20% of a new population pool with such mutated solutions. This is similar to a step of tabu search [7, 15] without storing already visited solution. Those introduced mutated individuals are marked and deleted after evaluation, if they are not a new elitist solution of the next generation. To mutate the individuals after reproduction, as it is done usually, did nearly have no effect.

We tested mutation by flipping two adjacent genes, changing genes, as well as inserting and deleting random genes. The latter mutation performed best and is able to improve the solution, when the population already has converged. The mutation rate was set to 10% for inserting and deleting. Such high mutation rate were shown to be rather effective for small population sizes [9]. One difference might be that we do not perform a reproduction before. We denote this adjustment as *elitist mutation*. Especially for Tech-pushes our mutation is extremely helpful.

## 3.2 Multi-objective optimization

In NSGA II, the population is sorted by non-domination into fronts. An individual dominates another one, if all objective functions are equal or better, and at least one is better than the objective functions of the dominated one. All individuals of the seconds front are dominated by at least one individual of the first front. Any individual of the third front is dominated by at least an individual of the first or second front, and so on. A crowding distance is assigned to each individual, which quantifies the distance to its neighbors. The parents are selected based on the front and crowding distance combined with binary tournament selection.

We apply the genetic operators as described above. However, if we use a standard NSGA-II we encounter the following problems:

1. We do not obtain sufficient diversity preservation by the crowding distance assignment.
2. The maximal population size is limited by the non-dominated sorting algorithm, which still has a complexity of $O(N^2)$.

The first issue arises from the fact that due to our objectives, we have only a very small number of different objectives and many equal tuples. This is not taken into account explicitly by NSGA-II. To relax the second problem, we do not apply the non-domination sorting to all individuals, but only to lists of individuals with the same objective tuples.

### Diversity preservation

The crowding distance calculation ensures that usually at least one individual represents a different object tuple in our problem. Unfortunately, e.g. by the sorting processes, it is even likely that all other chosen individuals have an equal objective tuple. To *penalize* some *individuals with equal objective tuples*, we correct the NSGA-II crowding-distance-assignment algorithm, by calling algorithm 1 afterwards. $I$ is a list of individuals in one front. An objective tuple $m$ corresponds to each individual of a front.

**Algorithm 1** penalize-distance($I$)

```
1:  l = |I| // number of solutions in I
2:
3:  for m = 0..(|m| − 1) do
4:      I[i] = sort(I,m) // stable sorting using each objective value
5:  end for
6:  σ = 0 // initialize a penalty parameter σ
7:
8:  for i = 1..(l − 1) do
9:
10:     if for each m: (I[i − 1].m == I[i].m) then
11:         σ = σ + 1 // increase the penalty for non-changing objective tuples
12:     else
13:         σ = 0 // reset the penalty, when objective tuple changed
14:     end if
15:     I[i]_distance = I[i]_distance − |m| · σ // apply the penalty to the distance
16: end for
```

If several equal objective tuples exists, each additional individual gets a growing negative distance. Please note, algorithm 1 requires a stable sorting. Instead of lines

3-5, one can sort the list of individuals $I$ according to the objective tuples once. An applied penalty is always stronger than a later added (positive) crowding-distance, apart from boundary points.

Non-domination sorting

In a first step all solutions with an equal objective tuple are collected in separate lists. Then, the lists are sorted instead of the single individuals by a (fast) non-domination sorting algorithm. In a typical setup (e.g. the experiment in the next section) one has to sort around 50 lists instead of 50000 single individuals. This modification allowed us to increase the population size beyond 5000 individuals. For problems without such small objective spaces, we refer to a more general approach [12]. Large population sizes are especially important for multi-objective optimization, since we optimize a whole Pareto front of individuals instead of a single individual. In addition to the first front, we take the best individual of each front for our elitism. This is quite cheap, since there are not many different objective values.

# 4 Experimental results

In this section we show results of some (multi-)objective optimization runs.

## *4.1 C# framework: Features*

A major goal is to develop our algorithms in a tool, which is *easy to use*. Users without programming background or knowledge in optimization should be able to create and optimize their builds. A graphical user interface (see fig. 2) guides the user through this process. Here, the user can reduce the search space and adjust simulation time and objective function set. Experienced users can also modify the optimization strategy and its settings. Apart from some special mechanisms (e.g. chrono-boost), properties (e.g. costs or tech-tree requirements) of all buildings, units, and upgrades can be adjusted by text-files.

Although the tool supports all races we restrict ourselves to Protoss for simplicity. We assume that for all builds, one worker does not earn minerals, but is reserved for building. If an Assimilator, the building to enable mining vespin gas, is finished, automatically three of the workers are assigned to it in order to mine. Furthermore, we assume due to in-game measurements an average rate of 0.714 minerals or 0.65 vespin gas per worker and second.

In a first test setup for our algorithms, we try to figure out how many units of the type *Stalkers* can be built in the first 400 seconds of the game. As a first simple approach, a *random search* in Table 1 shows that we were able to find up to four of at

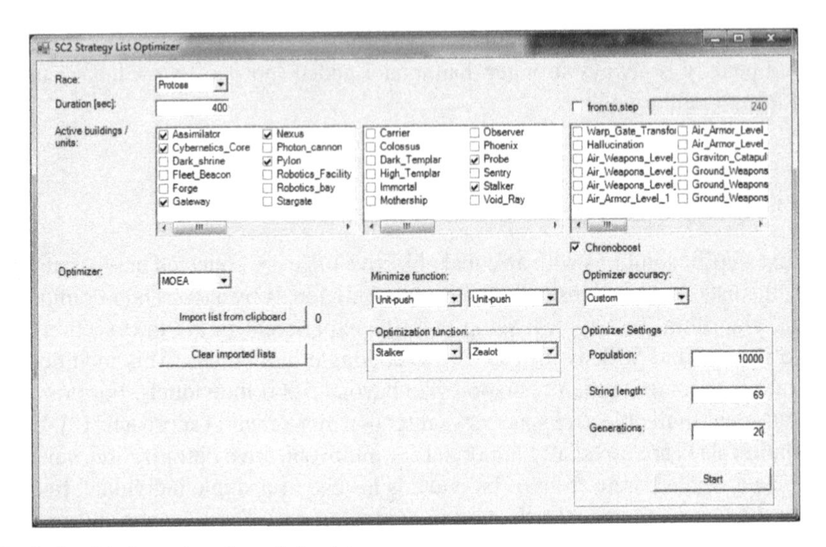


**Fig. 2** Graphical user interface of simulator

least seven possible Stalkers in $10^6$ runs. Here, whenever an action was started, we randomly choose a new possible action and wait until this action can be performed.

The improvement using a plain implementation of a genetic algorithm, with random initialization, forward simulation, fitness evaluation, (aggressive) selection, reproduction, and elitism is shown in *Genetic* in Table 1. The other two variants with distance-based selection and elitist mutation and their effects are discussed in section 3.1. In all test cases we tried to figure out an optimal selection pressure and could choose only 10 generations due to the extreme selection pressure, except from the last case.

**Table 1** Comparison between random search and the genetic algorithm with our extensions. Ten experiments were done for each case and the best one is chosen. Stalkers are the number of units that were obtained after the experimental runs.

| Algorithm | Generations | Individuals / generation | Selection | Stalkers |
|---|---|---|---|---|
| Random search | 1 | 10000 | – | 3 |
| | 1 | 100000 | – | 4 |
| | 1 | 1000000 | – | 4 |
| Genetic | 10 | 1000 | 5% | 4 |
| | 10 | 10000 | 1% | 5 |
| Genetic (distance-based selection) | 10 | 1000 | 5% | 5 |
| | 10 | 10000 | 1% | 6 |
| Genetic (distance-based selection, elitist mutation) | 50 | 10000 | 5% | 7 |

## 4.2 Tech pushes

Next we will cover an example for a Tech-push, where we try to build one advanced unit as fast as possible. Tech-pushes tend to have only view genes compared to the game time, since building up a good economy with e.g. many workers is not so important as for rushes.

For a Void Ray Tech-push, we got by our optimization the strategy list (time to finish: 5:08):

| Gen | Probe | Probe | Probe | Probe | Pylon | Gateway | Probe | Probe | Assimilator |
|---|---|---|---|---|---|---|---|---|---|
| Finished | 0:17 | 0:35 | 0:52 | 1:09 | 1:30 | 2:36 | 1:56 | 2:13 | 2:29 |

| Assimilator | Cybernetics_Core | Probe | Stargate | 3xBoost_Stargate | Void_Ray |
|---|---|---|---|---|---|
| 2:40 | 3:27 | 3:00 | 4:28 | 4:29 | 5:08 |

A population of 10 000 individuals over 200 generations and a maximal simulated time of seven minutes turned out to be sufficient. The longest chain of dependencies (Pylon, Gateway, Cybernetics_Core, Stargate, Void_Ray) gives one theoretical lower bound for the time, which is exactly 4:00 minutes. The remaining objects (Probes, i.e. workers, and Assimilators) have to be built to deliver enough resources. The lack of resources is responsible for the impossibility to build the target unit in a direct way.

Individuals of some generations also try to accelerate the production of workers at the beginning with a boost, but it does not lead to a better minimal time to solution.

## 4.3 Rushes

For the rushes we optimize for two different types of units and apply the multi-objective genetic algorithm. Among the best found strategy lists, we present those which include most workers. We also played some strategy lists in order to validate our forward simulation.

In a rush, players try to get as many as possible of two different kinds of units or a combination of both in a prescribed time. For different times, we do one multi-objective run with 40 000 individuals over 1000 generations to get as close as possible to the Pareto-optimal front. Usually much less individuals and number of generations would be necessary to achieve our fronts. A current compute core needs around four hours for one optimization run.

Figure 3 shows the number of produced Stalkers and Zealots after 240 – 360 seconds. Zealots are cheaper units and they do not need vespin gas and consequently no Assimilator buildings. Hence, Zealots can be build in larger numbers than Stalkers. The builds without Stalkers include zero Assimilators. The fronts with 330 and 360 seconds require two Assimilators for the two builds with most Stalkers each. All remaining strategy lists suggest one Assimilator. Zealots and Stalkers are produced in Gateways. Their optimal number is a trade-off between their costs and the additionally provided production capacities. The number of produced workers lies between

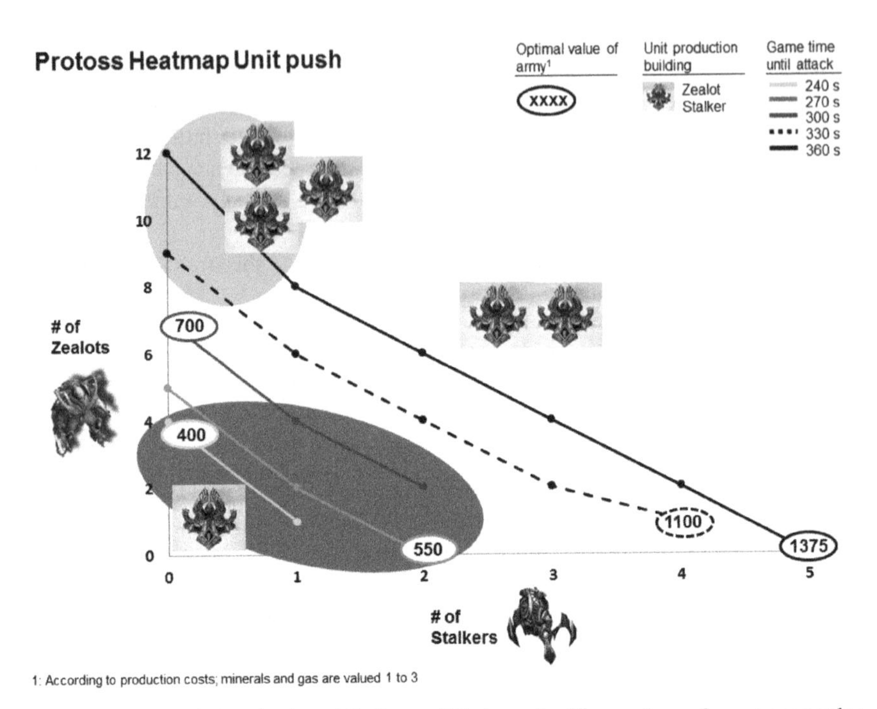


**Fig. 3** Heatmap of the production of Stalker and Zealot units. The numbers of necessary production facilities are marked as regions in the map.

2 and 14. It depends on the simulated time, but not on the produced unit combination in a clear trend. Outstanding points, like $(Zealots, Stalkers) = (12,0), (1,4), (0,2)$ are much more difficult to find, since they only can be build slightly below the given times. Only an extremely small number of candidates allow those objective values.

The two solutions with zero and one Stalker at 279 seconds are found after a few number of generations. However, it is very difficult to find an individual producing two Stalkers only. One reason is that the elitist solution with two Zealots, one Stalker, and one additional Stalker in production that has not been finished is much farther than an elitist candidate without Zealots, one Stalker, and an additional Stalker in production. However, the latter variant is not a member of the first front. Therefore, we take the elitist solutions of all fronts, which we can easily afford due to our few different objective values.

Figure 4 presents the number of produced Stalkers and Immortals after 330 – 420 seconds. The lists of those more advanced builds with around 40 genes and up to 12 build options per gene are much longer than in the last example. A Tech-push showed that one Immortal, that is constructed in a more advanced building called Robotics Facility, can be produced after 325 seconds. We need at least one Gateway in every strategy list, since it is required to be able to build a Robotics Facility. Since most builds (up to pure Stalker builds) require both production facilities, mixed unit builds tend to have better points than the Zealot/Stalker mixtures.

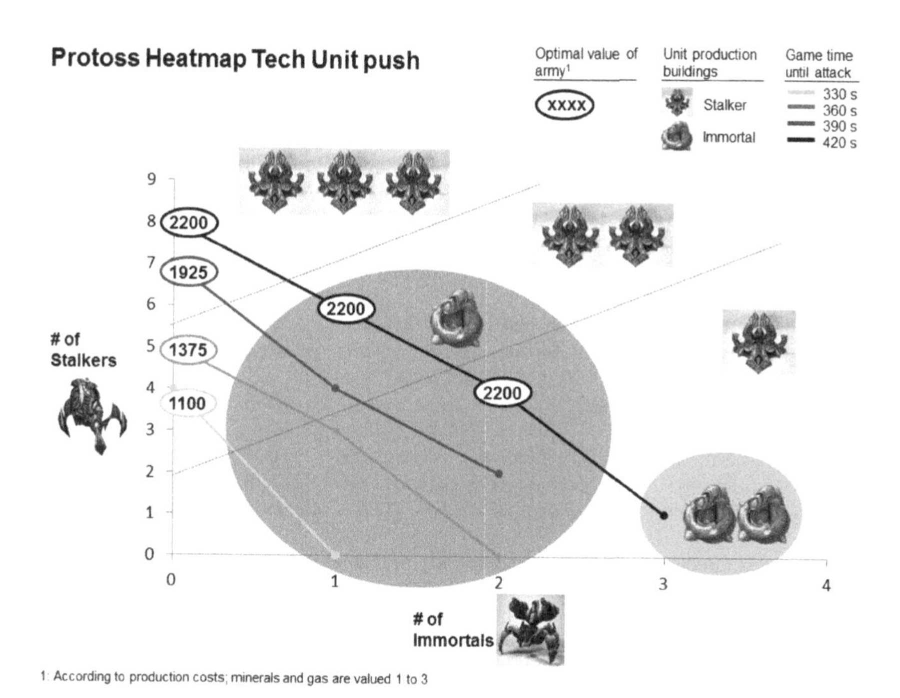


**Fig. 4** Heatmap of the production of Immortal and Stalker units. The numbers of necessary production facilities are marked as regions in the map.

## 5 Conclusions and outlook

We presented a multi-objective genetic algorithm to find the optimal build order for players of real-time strategy games. Most problem dependent algorithmic adjustments help to cope with large search spaces (e.g. $10^{30}$ possible candidates) in contrast to very small objective space (e.g. one integer between zero and ten). We applied the optimization to Starcraft 2 builds with up to 450 seconds simulation time. The resulting build orders of achievable units should feature a valuable tool both for new and experienced players that helps to testify and improve their opening strategies. In the next step, we plan to summarize results for the other two races in Starcraft 2.

## References

1. E. Cantú-Paz. A survey of parallel genetic algorithms. *Calculateurs paralleles, reseaux et systems repartis*, 10(2):141–171, 1998.
2. L. Davis and M. Mitchell. *Handbook of genetic algorithms*. Van Nostrand Reinhold, 1991.
3. K. Deb and T. Goel. Controlled elitist non-dominated sorting genetic algorithms for better convergence. In *Evolutionary Multi-Criterion Optimization*, pages 67–81. Springer, 2001.

4. K. Deb, A. Pratap, S. Agarwal, and T. Meyarivan. A fast and elitist multiobjective genetic algorithm: Nsga-ii. *Evolutionary Computation, IEEE Transactions on*, 6(2):182–197, 2002.
5. L.R. Dice. Measures of the amount of ecologic association between species. *Ecology*, 26(3):297–302, 1945.
6. C.M. Fonseca and P.J. Fleming. Genetic algorithms for multiobjective optimization: Formulation discussion and generalization. In *Proceedings of the 5th International Conference on Genetic Algorithms*, pages 416–423. Morgan Kaufmann Publishers Inc., 1993.
7. F. Glover and M. Laguna. *Tabu search*, volume 1. Springer, 1998.
8. D.E. Goldberg. *Genetic algorithms in search, optimization, and machine learning*. Addison-wesley, 1989.
9. R.L. Haupt. Optimum population size and mutation rate for a simple real genetic algorithm that optimizes array factors. In *Antennas and Propagation Society International Symposium, 2000. IEEE*, volume 2, pages 1034 –1037 vol.2, 2000.
10. R.L. Haupt, S.E. Haupt, and J. Wiley. *Practical genetic algorithms*. Wiley Online Library, 2004.
11. J. Horn, N. Nafpliotis, and D.E. Goldberg. A niched pareto genetic algorithm for multiobjective optimization. In *Evolutionary Computation, 1994. IEEE World Congress on Computational Intelligence., Proceedings of the First IEEE Conference on*, pages 82–87. IEEE, 1994.
12. M.T. Jensen. Reducing the run-time complexity of multiobjective eas: The nsga-ii and other algorithms. *Evolutionary Computation, IEEE Transactions on*, 7(5):503–515, 2003.
13. J.D. Knowles and D.W. Corne. Approximating the nondominated front using the pareto archived evolution strategy. *Evolutionary computation*, 8(2):149–172, 2000.
14. M.L. Mauldin. Maintaining diversity in genetic search. In *Proceedings of the national conference on artificial intelligence*, volume 247, page 250, 1984.
15. A. Misevicius. An improved hybrid genetic algorithm: new results for the quadratic assignment problem. *Knowledge-Based Systems*, 17(2):65–73, 2004.
16. E. Zitzler and L. Thiele. Multiobjective evolutionary algorithms: A comparative case study and the strength pareto approach. *Evolutionary Computation, IEEE Transactions on*, 3(4):257–271, 1999.

# Managing Uncertainties in the Field of Planning and Budgeting – An Interactive Fuzzy Approach

**Peter Rausch[1], Heinrich J. Rommelfanger[2], Michael Stumpf[1], Birgit Jehle[3]**

**Abstract** Despite all effort in last decades, Planning and Budgeting (P&B) is still a challenging task. The corresponding processes for large organizations are usually very resource-intensive, time consuming and costive. One of the major issues is to manage uncertainty. Apart from that, P&B methods have to fulfill other requirements. They have to be consistent with human thinking and should allow realistic modeling. In this paper we analyze whether interactive fuzzy approaches solve these issues and if they could be a new way to fulfill the above mentioned requirements efficiently. For that purpose, prior research related to P&B problems as well as practical issues and challenges will be outlined. Based on a case study, fuzzy interactive solutions will be analyzed. Possible benefits and issues of the applied approach are discussed and ideas about ongoing and further research are given.

## 1 Introduction

Planning and Budgeting (P&B) is applied in many different areas, for instance in governmental and corporate budgeting. Many different budget types can be fixed, for instance budgets for staff expenditures, IT equipment and projects. The budgeting process for large organizations is usually very resource-intensive, time

---

1 Faculty of Computer Science, Georg Simon Ohm University of Applied Sciences, Kesslerplatz 12, 90489 Nuremberg, Germany
{peter.rausch, michael.stumpf}@ohm-hochschule.de

2 Goethe University Frankfurt am Main, Faculty of Economics and Business Administration, Niebergallweg 16, 65824 Schwalbach a. Ts., Germany
rommelfanger@wiwi.uni-frankfurt.de

3 Noris Treuhand Unternehmensberatung GmbH, Virchowstr. 31, 90409 Nuremberg, Germany
b.jehle@noris-treuhand.com

consuming and costive [1]. Due to changes of planning parameters during the planning period and uncertainty about the correct values of parameters, plans can quickly become obsolete [1]. So, because of that plans are of little use in many cases. Additionally, plans and budgets usually contain many assumptions about uncertain conditions, like the macro economical development, the prices of input factors or human assessments. This leads to a limited value of the results of P&B processes. However, without any P&B activities controlling is a very difficult task. Hence, P&B methods have to fulfill specific requirements. They have to be able to cope with uncertainties.

Within the planning periods uncertainties can cause the rescheduling of budgets. So, efficient methods are necessary [9]. In terms of efficiency, it is desirable to have a high level of automation. On the other hand, according to Tate "black box or fully automated solutions are not acceptable in many situations" [14]. In terms of acceptance, methods have to be consistent with human thinking and should allow human interaction. It has to be anticipated that humans are usually not able to overlook the whole solution space and to evaluate all possible solutions for P&B problems at the beginning of the P&B process. In general, humans set aspirations concerning different goals and the satisfaction of constraints [12]. Interactively, they have to explore, what is a "good" solution or not. This process is also influenced by subjective preferences.

Interactive fuzzy approaches could be a new way to fulfill the above mentioned requirements efficiently and to shift the quality of P&B solutions. In this paper, the suitability of interactive fuzzy approaches for solving P&B problems and the related issues will be analyzed. For that purpose, practical issues and challenges of P&B processes will be outlined. Subsequently, prior research and available IT solutions are examined to identify open issues. Based on a case study, fuzzy interactive solutions will be analyzed afterwards. In the subsequent section, possible benefits and issues of the applied approach are discussed. The last section summarizes the paper, provides final conclusions and gives ideas about future work.

## 2 Planning and Budgeting –Issues, Challenges, Related Research and Available Solutions

At the beginning of the 20th century the traditional P&B process as described below came up [15]. Still today, it is used by most of the companies worldwide. Figure 1 shows this process. Especially in large companies or organizations variants of this process are widespread [7].

The P&B process usually starts with the determination of targets and strategic guidelines for the planning period which is derived from the vision of the company and specified from top to bottom afterwards. Different operational objectives, for instance sales, cost, staff or investment targets are defined for the responsible business units. Based on this top-down approach, a bottom-up planning follows. Quite often revisions need to be done to create the final budget. Once the budget is adopted, the actual performance has to be compared with objectives and standards. Finally, necessary actions have to be taken if deviations exceed a certain threshold [4].

**Fig. 1.** Process of P&B in large companies, BBRT [2].

Due to many issues, this approach is not appropriate anymore. Two classes of issues are related with it. *Internal issues* address challenges which can be identified in the internal environment of an enterprise or an organization. The other class, denoted as *external issues*, covers issues originating from external factors. At first, the internal issues will be outlined. Especially in larger companies, big departments need several months each year creating the annual budget. Usually, many employees are involved in the process causing high costs. As the budgets are fixed, there is a lack of incentive for reviewing or even undercutting the budget. On the other hand there are issues which arise due to external influences. Because of today's dynamics, plans with crisp parameters are usually obsolete as soon as they are adopted. Furthermore, they contain various assumptions about uncertain conditions. The budgets are very inflexible and adaptations are complex. Therefore, budgets are of little use for management control. Additionally, linguistic uncertainties can arise. They can be assigned to external issues as well as to internal ones. For e.g. if managers assume that the turnover will increase "slightly" they may think of a range of 1% to 5%. In traditional approaches this is anticipated by using deterministic values and by neglecting variations in this range [13]. Hence, not all solutions of the P&B problem are considered within the planning process.

Oehler [6] points out that nearly every company is using IT to address the above mentioned internal and external issues and categorizes modern software including P&B functions into the following classes: enterprise resources planning systems (e.g. SAP), spreadsheet software (e.g. Excel), OLAP tools (e.g. SAP, Palo, and others) and specific P&B software (e.g. Professional Planner, Corporate Planner, and others). Simple office tools are prevalent in the field [6]. To support the P&B process with IT the following tasks can be covered: data input support, process support, analysis support, support for models and methods, as well as integration aspects [8].

The tasks data input support, process support and integration aspects address internal issues. Data input support is dealing with all issues regarding the inbound information flow to the P&B process, e.g. if data is entered manually and how it is validated or whether data is transferred automatically from other source systems [6]. Process support is dealing with all questions regarding the information flow within the P&B process. The aspect of integration covers the issue of process organization which can be supported by collaboration systems and workflow management systems. This kind of IT support helps a lot to speed up the P&B process, but can't contribute to management of uncertain conditions.

Other IT functions include analysis support which helps to access and analyze P&B data. Also scenarios can be created and visualized. Model and method support is dealing with models which formally describe dependencies between system states, e.g. the dependency between the price of a product and its estimated turnover described by a demand-function [6]. With these models planners are able to do simulations. It could be argued enabling planners to create sophisticated models and to run scenario analyses as well as simulations is a way to handle uncertain conditions. But models, scenario analyses and simulations are not applied interactively within the P&B process and do not allow an explicit management of uncertainty. Instead, they only try to find the "best" deterministic representation of the plan. The resulting plans are fixed like in other approaches, despite the fact that uncertainty has been considered in a most-likely insufficient way. Shim and Siegel [11] point out that software is widely used, not only to speed up the P&B process but also to allow non-financial managers to investigate the effect of changes in budget assumptions and scenarios. Although this is an important step, the above mentioned functions can't help either to solve the issue of managing uncertainty in an efficient way.

## 3 Case Study

To analyze possible solutions for the challenges which were mentioned in the last section we put the focus on a case study concerning an organization which belongs to the public sector. Due to many uncertainties, for instance political influences or unexpected budget cuts, P&B in the public sector is a complex task.

The department of a university which is considered here has to supply an "official" budget plan to the management, once a year. A certain amount of money has to be distributed to different types of sub-budgets. It is also necessary to control expenses during the planning period. If deviations become more and more likely, action should be taken. It is possible to reallocate certain resources during the planning period as long as the complete budget is not exceeded. At the end of the planning period all assigned financial resources have to be spent. Before the details of the P&B problem will be given it is important to motivate and to know about the applied fuzzy approach.

# 4 Suggested approach: FuturePlus

To manage the above mentioned challenges we propose a new framework called FuturePlus (*Fu*zzy *T*ools *U*sing *Re*al world Data for *Pl*anning and B*u*dgeting *S*upport). The basic idea of our approach is to use rolling forecasts to support the generation of plans and to apply fuzzy linear programming to support the planning process. In this paper we focus on the fuzzy model which is the core component of the framework. In section 4.1, we will outline the general concept of FuturePlus. Afterwards, we will introduce the fuzzy linear programming approach, outline the experimental setup and discuss the results.

## *4.1 General concept*

Many applications based on the fuzzy set theory have been developed during the last decades, for instance in the fields of weather analysis [3] or medicine [5]. Fuzzy sets can also help to manage uncertainties and to create more realistic P&B models. As already mentioned in section 2, plans and budgets usually contain a level of assumption about uncertain conditions. Additionally, many planning parameters can change during the planning period. For instance, in our case it is not clear which amount of money is needed for department internal projects or how much is appropriate for hard- and software equipment at the beginning of the planning period. Fuzzy sets allow a realistic modeling of these parameters. As shown in [10], it is possible to describe a fuzzy set very accurate by a membership function $\mu_A\colon X \rightarrow [0, 1]$ which assigns a membership value to each element of a set. However, in practical applications this would cause considerable modeling effort. Rommelfanger discusses different types of membership functions and recommends piecewise linear membership functions [10]. Their accuracy is sufficient for our purpose. They have the advantage that they help minimizing the modeling effort. Figure 2 shows an example of a fuzzy parameter representing an appropriate budget for the sum of all support expenses from a planner's point of view. This membership function gives an account of the budget planner's imagination that a sum between 200,000 € and 210,000 € is absolutely acceptable. But in any case the sum must be greater or equal 180,000 €. Naturally a greater sum is desirable, but maximal a sum of 230,000 € is defensible.

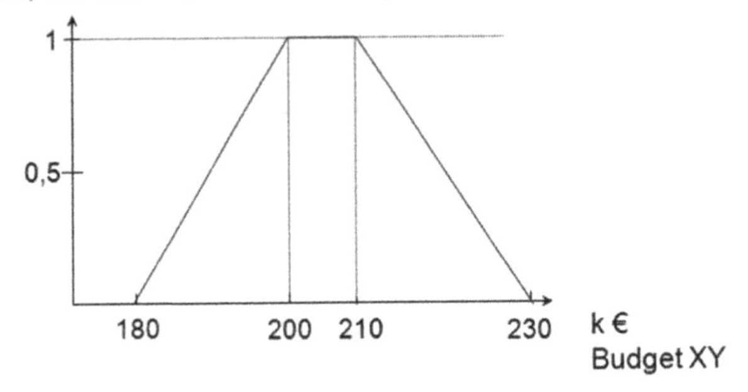


**Fig. 2.** Planning and budgeting with fuzzy parameters.

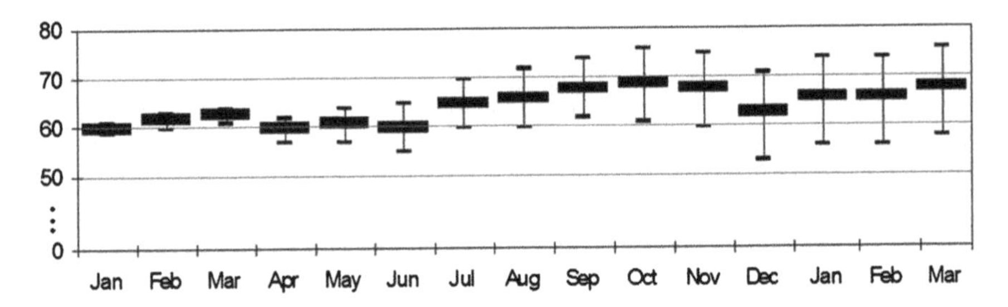


**Fig. 3.** Rolling forecasts with fuzzy parameters.

As in budgeting the whole sum is to be distributed to the different sub-budgets, it seems reasonable to suppose that the allocation is evaluated by a linear weighted addition. Together with simple restrictions this approach leads to a linear optimization model. In the first step, information for the initial model has to be gathered. To provide the parameters and to modify the model in further iterations rolling forecasts are used. Due to the limited space of this paper and its focus on the fuzzy linear approach, just a brief idea concerning the rolling forecast component will be given. In many cases, the longer the forecast horizon is, the more difficult it gets to quantify parameters, exactly. So, like Fig. 3 shows, the user should be allowed to specify fuzzy parameters. The upper and lower bounds represent elements with a membership value which indicates a very low degree of belonging to the considered set. Additionally, one or more elements representing the highest degree of belonging to the considered set are required. The thick black bars denote these intervals. This procedure can be done with an accuracy which depends on the available resources, respectively the costs of information. More precise information can be gathered when the crucial parameters are identified later.

Based on the forecasts, the fuzzy linear model can be set up. To gather all data and to find a solution considering all the methodological requirements which were mentioned before, we use a software called *Fuzzy Linear Programming based on Aspiration Levels* (FULPAL). It is a universal tool and implements the corresponding method. Our approach also reflects the assumption that, usually, human decision makers are not able to overlook and evaluate the whole solution space a priori. To manage this issue, our method is based on the aspiration adaptation theory and interactive planning iterations. In the first step, it allows the user to enter fuzzy parameters for coefficients and constraint borders. For "less than or equal" inequalities constraint borders can be modeled as shown in Fig. 4. The upper bound $b_i$ is undisputed, but in any case the upper bound $b_i + d_i$ should not be exceeded.

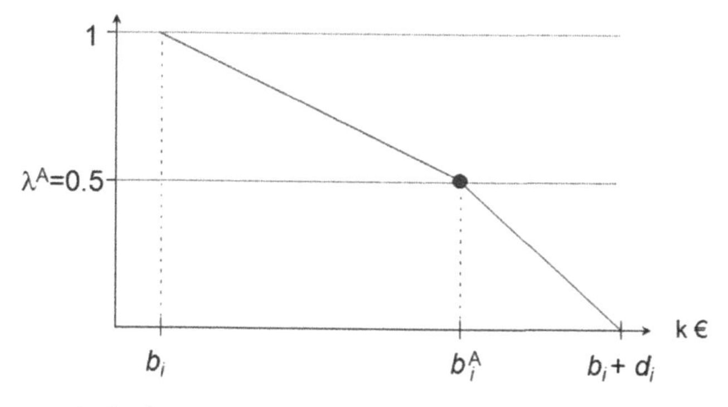


**Fig. 4.** Fuzzy constraint borders.

According to Fig. 5, the minimal and maximal values for the goals are computed when all coefficients and constraint borders are entered. In the next step the decision maker (DM) can specify aspiration levels concerning the restrictions and goals. The aspiration levels are assigned to the membership value $\lambda^A=0.5$. Internally, the software approximates piecewise linear membership functions, computes the optimal degree of satisfaction $\lambda^*$ and determines the unknowns. Details of this process can be found in [10,8]. In case $\lambda^* > \lambda^A$ a solution is found that fulfils all aspiration levels of the decision maker. Therefore, he can accept this satisfying solution or he can improve some aspiration levels, see Fig. 6 and Fig. 7.

If $\lambda^* < \lambda^A$, no acceptable solution can be found the decision maker can modify the aspiration levels. The solution space can be explored iteratively, until an acceptable solution is found or the decision maker terminates the process. At the end of each iteration, it is also possible to check for critical system parameters which are indicated by the lowest $\lambda$ values of the concerning goals or restrictions. For these parameters the decision maker can gather additional information. So, the information procurement process is very efficient.

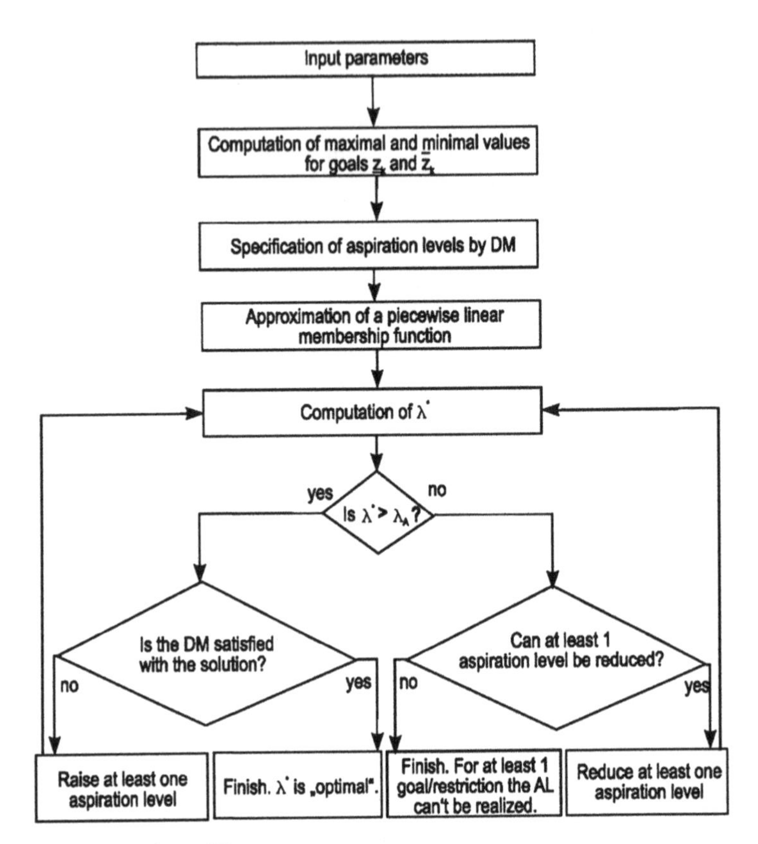


**Fig. 5.** FULPAL at a glance [8].

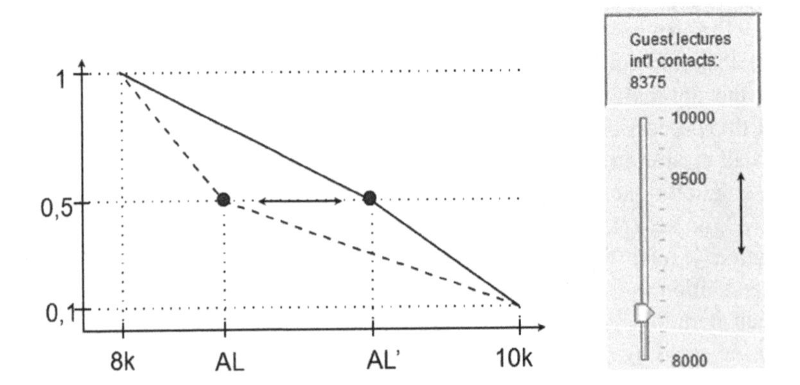


**Fig. 6.** Changing aspiration levels by using the FULPAL GUI (1).

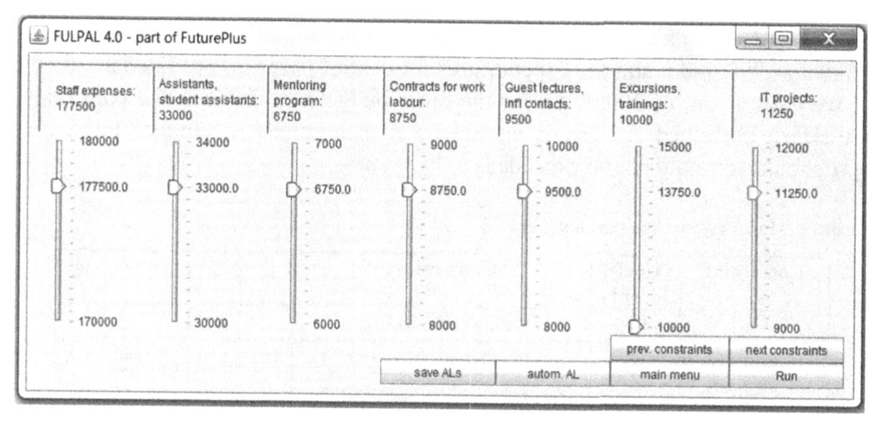


**Fig. 7.** Changing aspiration levels by using the FULPAL GUI (2).

## *4.2 Experiments and Results*

To transfer the ideas of the last section to our case study, we have to setup a model first. The following goals and constraints have to be respected:

1. The available financial resources have to be budgeted completely.
2. A fixed amount is transferred to the head office for central measures during the planning period.
3. Expenses for permanent employees have to be paid.
4. A certain amount is needed for different other expense types: staff expenses (temporary employees), assistants, student assistants, a mentoring program, contracts for temporary work labour, guest lectures, international contacts, excursion and trainings, department internal IT projects, laboratories, library, and equipment for special laboratories.
5. For guest lectures and international contacts a common budget is assigned. Also assistants and student assistants are considered as one category. For all other expense types separate budgets are established.

Some of the listed issues have to be respected strictly. Others represent more or less soft constraints. To illustrate the problem and the proposed solution process we use the following numerical example. As the number of pages is restricted, here we present only a simple model with soft constraints, e.g. the coefficients are crisp and only the constraint borders may be fuzzy. It is no problem to work with more variables, but in our simple case the number of variables is restricted. For simplification we also assume that the lower bounds $c_i$ of the budgets are crisp because of binding arrangements, see Table 1.

The budget for all expenses is 611,600 €. As the regular salary for the staff is known as 293,600 € and the expenditures for central purposes are fixed as 18,000 €, we have a sum of 300,000 € for the other budgets. In Table 1, the constraints are clearly described. As already mentioned, $b_i$ is definitely within the limit, but $b_i + d_i$ should certainly not be exceeded.

**Table 1.** Model constraints (amounts in k €)

| Constraint Nr. | Coef-ficients | Constraints | $c_i$ | $b_i$ | $b_i + d_i$ | $w_i$ |
|---|---|---|---|---|---|---|
| C1 | $x_1$ to $x_{10}$ | Sum of all expenses | | 300 | | |
| C2a/C2b | $x_1$ | Staff expenses (temporary) | 165 | 170 | 180 | 0.6 |
| C3a/C3b | $x_2$ | Assistants, student assistants | 20 | 30 | 34 | 0.7 |
| C4a/C4b | $x_3$ | Mentoring program | 4.5 | 6 | 7 | 0.9 |
| C5a/C5b | $x_4$ | Contracts for work labour | 5 | 8 | 9 | 0.6 |
| C6a/C6b | $x_5$ | Guest lectures and international contacts | 5 | 8 | 10 | 0.9 |
| C7a/C7b | $x_6$ | Excursions and trainings | 5 | 10 | 15 | 0.8 |
| C8a/C8b | $x_7$ | IT-projects | 7 | 9 | 12 | 0.7 |
| C9a/C9c | $x_8$ | Labs (hard- and software) | 8 | 9 | 10 | 0.8 |
| C10a/C10b | $x_9$ | Library and online lecture notes | 25 | 30 | 35 | 0.8 |
| C11a/C11b | $x_{10}$ | Equipment special labs | 23 | 28 | 33 | 0.7 |

As the essential budgets have relatively high lower bounds, the faculty management is interested in shifting more money in national and international contacts and for student requests. After careful consideration the decision maker fixed the weights $w_i$ in Table 1. Then the goal of the budget problem is maximize the objective function

$$z(x_1,x_2,\ldots,x_{10})=0.6x_1+0.7x_2+0.9x_3+0.6x_4+0.9x_5+0.8x_6+0.7x_7+0.8x_8+0.8x_9+0.7x_{10} \quad (1)$$

The requirement that available financial resources have to be budgeted completely is reflected in C1.

$$\sum_{i=1}^{10} x_i = 300{,}000 \quad (2)$$

C2a is also a crisp constraint. It makes sure that the staff expenses are high enough to fulfill the existing contracts.

$$x_1 \geq 165{,}000 \quad (3)$$

The constraints C3a to C11a can be interpreted similarly. They represent crisp lower bounds for budgets. In case of the expenses for temporary staff also constraint C2b has to be fulfilled. Expenses are definitely acceptable up to $b_i$ = 170,000 €, but $(b_i + d_i) \geq$ 180,000 € is certainly beyond the limit. Like Fig. 4

shows, $b_i$ represents the set of valid expense levels very well and the membership value equals 1. In contrast to that $(b_i + d_i)$ represents the set of valid expense levels poorly. Hence, the membership value of $(b_i + d_i)$ is 0 or at least very close to 0. The $\widetilde{\leq}$-symbol denotes a fuzzy less or equal.

$$x_1 \widetilde{\leq} \left[170{,}000;180{,}000\right] \quad (4)$$

The constraints C3b to C11b can be interpreted in the same way. For convenience and to save effort the aspiration levels can be automatically fixed. In our case this option is chosen for the first iteration and all aspiration levels are initially computed according to:

$$g_i^A = b_i + 0.75 d_i \quad (5)$$

As mentioned before, the membership functions will be approximated by piecewise linear functions. By solving the corresponding LP system we get the results shown in the first row of Table 2:[4]

**Table 2.** Experimental results (amounts in k €)

| Iteration | $\lambda^*$ | $x_1$ | $x_2$ | $x_3$ | $x_4$ | $x_5$ | $x_6$ | $x_7$ | $x_8$ | $x_9$ | $x_{10}$ |
|---|---|---|---|---|---|---|---|---|---|---|---|
| 1 | 0.667 | 165.00 | 30.00 | 6.50 | 5.00 | 9.00 | 12.50 | 7.00 | 9.50 | 32.50 | 23.00 |
| 2 | 0.433 | 165.00 | 33.40 | 6.85 | 5.00 | 8.00 | 10.00 | 8.90 | 9.85 | 30.00 | 23.00 |
| 3 | 0.528 | 165.00 | 32.83 | 6.71 | 5.00 | 9.42 | 10.00 | 8.33 | 9.71 | 30.00 | 23.00 |

In iteration 1 $\lambda^*$= 0.667 is greater than $\lambda^A$. The decision maker notices that his aspiration levels can be achieved quite easy. So, he decides to increase some aspiration levels in iteration 2. Based on the results of the first iteration, we assume that the decision maker wants to revise his aspiration levels for constraints c6b (guest lectures and international contacts), c7b (excursions and trainings) and c10b (library and online lecture notes). It is intended to analyze the effect of budget cuts for the mentioned expense types on $\lambda^*$. The corresponding aspiration levels are changed according to Table 3:

**Table 3.** Changes of aspiration levels (AL) (amounts in k €)

| Iteration | AL c6b | AL c7b | AL c10b |
|---|---|---|---|
| 1 | 9.50 | 13.75 | 33.75 |
| 2 | 8.00 | 10.00 | 30.00 |

The corresponding results can be found in row 2 of Table 2. $\lambda^*$= 0.433 is now less than $\lambda^A$. This means that the solution does not correspond to all the preferences of the decision maker. Hence, the decision maker reconsiders his

[4] Details how to convert the fuzzy system into a corresponding LP system can be found in [10].

aspiration levels and relaxes the system by allowing higher expenses for guest lectures and international contacts. The tolerated expenses are reset and the corresponding aspiration level for constraint c6b is changed from 8,000€ to 9,500€. The results of iteration 3 can be found in row 3 of Table 2. Since all aspiration levels can be achieved, the decision maker is completely satisfied and a solution which corresponds to all constraints and his subjective preferences is found. Of course, our example is very simple. For e.g. it could have been also possible that the decision maker decides to gather additional information concerning critical parameters in iteration 2 and to continue with a modified model setup.

## 5 Evaluation

As we have seen, FULPAL allows to process parameters without any artificial accuracy. Discussions how to specify exact limits for expenses can be avoided or at least reduced. It is not necessary to investigate all parameters in detail and information costs can be saved. Decision makers can decide how much effort they want to invest. They are allowed to work with fuzzy parameters. After the first iteration of the process, they can try to get additional information just for crucial parameters. Hence, they can set up a realistic model of their P&B process, efficiently. Finding a solution interactively is consistent with human thinking and allows to include subjective preferences concerning the aspiration levels. This is important in terms of acceptance. Additionally, the approach is very flexible. If a rescheduling is necessary, for e.g. due to new requirements communicated by the university's management, just the relevant parameters have to be adjusted and further iterations can be made.

## 6 Ongoing Work

Due to the limited space of this paper, just a model with a very simple structure was outlined. It represents an aggregated view of the P&B problem and is sufficient to demonstrate improvements compared to the approach we described in section 2. Additionally, an idea how to consider fuzziness in an adequate way could be given. Meanwhile, we are also experimenting with different model setups. It can be useful to refine constraints. For instance, constraints C3a and C3b (the budget for assistants and student assistants) can be detailed by additional constraints limiting the budget for each sub-type of expenses. Additionally, the parameters for the coefficients can be modelled as fuzzy parameters to save information costs and to get a model which is more conform to reality. Another interesting task is to analyze the benefit of fuzzy lower bounds for the budgets. The assumption that the lower bounds of the budgets $c_i$ are crisp because of binding arrangements sounds reasonable at first glance. But due to unexpected events, for e.g. cancellation of guest lectures, also the lower bounds are not always certain.

# 7 Conclusion and Future Work

Despite all progress and promises of software vendors, P&B is still a challenging task. Most of the open issues are caused by uncertainty. In this paper we have presented a new interactive fuzzy approach to manage these issues. The approach is part of a framework called FuturePlus (*Fu*zzy *T*ools *U*sing *Re*al world Data for *Pl*anning and B*u*dgeting *S*upport). It allows decision makers to use fuzzy parameters for P&B. Based on that data and the decision makers' aspiration levels, a first solution can be generated using a fuzzy linear optimization approach. To find an acceptable solution, the solution space can be explored iteratively afterwards. This method is consistent with human thinking and enables the decision maker to manage uncertainties. Experiments were made in the field of an organization which belongs to the public sector. The results were promising. Information costs could be saved and P&B effort was reduced. Additionally, necessary P&B iterations to reschedule budgets could be made much faster in comparison to traditional approaches. The presented approach enables decision makers to investigate the effect of changes in budget assumptions and different scenarios easily. The new framework can be applied in many different areas.

However, there are still many ideas to extend the framework. It could be analyzed whether other instruments and approaches could be combined with FuturePlus and its core component FULPAL. Especially, approaches in the field of complex event processing and event pattern detection could be a useful additional component of the framework. Other technologies and controlling instruments like early warning systems or other business intelligence instruments might also help to improve the quality of P&B. From our point of view it would be also very interesting to have flexible instead of rigid planning cycles. Ideas based on this philosophy are discussed under the term "beyond budgeting" [7]. So, it is important to embed our framework in a new P&B policy which is part of the corporate or the organization's strategy. This could lead to an event triggered P&B approach. Besides, the transfer of the presented ideas to other industries would be very interesting.

## References

1. Barrett, R.: Planning and Budgeting for the Agile Enterprise, Elsevier, Oxford, Burlington (2007).
2. Borck, G, Pflaeging, N., Zeuch A.: Making Performance Management Work. (Available via BetaCodex Network Associates, 2009). http://www.betacodex.org/sites/default/files/paper/3/BetaCodex-PerformanceManagement.pdf. Accessed 14 Aug 2012.
3. Chu, H.-J., Liau, C.-J., Lin, C.-H., Su, B.-S.: Integration of fuzzy cluster analysis and kernel density estimation for tracking typhoon trajectories in the Taiwan region. Expert Systems with Applications, Vol. 39, Issue 10, pp. 9451–9457 (2012).
4. Hansen, D.R., Mowen, M.M., Guan, L.: Cost Management, Accounting & Control, South Western Cengage Learning, Mason (2009).
5. Lupaşcu, C., Tegolo, D.: Stable Automatic Unsupervised Segmentation of Retinal Vessels Using Self-Organizing Maps and a Modified Fuzzy C-Means Clustering. In: Fanelli, A., Pedrycz, W., Petrosino, A. (Eds.), Fuzzy Logic and Applications, Lecture Notes in Computer Science, Vol. 6857, pp. 244–252, Springer, Berlin et al., 2011.
6. Oehler, K.: Unterstützung von Planung, Prognose und Budgetierung durch Informationssysteme. In: Chamoni, P., Gluchowski, P. (Eds.), Analytische Informationssysteme, pp. 359–394, Springer, Berlin et al. 2010.
7. Pfläging N.: Beyond Budgeting, Better Budgeting, Haufe Mediengruppe, Freiburg et al. (2003).
8. Rausch, P.: HIPROFIT - Ein Konzept zur Unterstützung der hierarchischen Produktionsplanung mittels Fuzzy-Clusteranalysen und unscharfer LP-Tools, Peter Lang Verlag, Frankfurt et al. (1999).
9. Remenyi, D.: Stop IT project failures through risk management, Butterworth-Heinemann, Oxford, Woburn (1999).
10. Rommelfanger, H.: Fuzzy Decision Support-Systeme - Entscheiden bei Unschärfe, Springer, 2. Ed., Berlin et al. (1994).
11. Shim, J.K., Siegel, J.G., Shim, A.I.: Budgeting Basics and Beyond, 4th Ed., John Wiley & Sons Inc., Hoboken (2012).
12. Simon, H.A.: The new science of management decision, Prentice Hall PTR, Upper Saddle River, New York, 1977.
13. Strobel, S.: Unternehmensplanung im Spannungsfeld von Ratingnote, Liquidität und Steuerbelastung, Verlag Dr. Kovac, Hamburg (2012).
14. Tate, A.: Intelligible AI Planning - Generating Plans Represented as a Set of Constraints. Artificial Intelligence Applications Institute (2000). http://www.aiai.ed.ac.uk/oplan/documents/2000/00-sges.pdf. Accessed 14 Aug 2012.
15. Weber, J., Linder, S.: Neugestaltung der Budgetierung mit Better und Beyond Budgeting? Wiley-VCH, Weinheim (2008).

# DATA MINING AND MACHINE LEARNING

# Identification of Correlations Between 3D Surfaces Using Data Mining Techniques: Predicting Springback in Sheet Metal Forming

Subhieh El-Salhi, Frans Coenen, Clare Dixon, Muhammad Sulaiman Khan

**Abstract** A classification framework for identifying correlations between 3D surfaces in the context of sheet metal forming, especially Asymmetric Incremental Sheet Forming (AISF), is described. The objective is to predict "springback", the deformation that results as a consequence of the application of a sheet metal forming processes. Central to the framework there are two proposed mechanisms to represent the geometry of 3D surfaces that are compatible with the concept of classification. The first is founded on the concept of a Local Geometry Matrix (LGM) that concisely describes the geometry surrounding a location in a 3D surface. The second, is founded on the concept of a Local Distance Measure (LDM) derived from the observation that springback is greater at locations that are away from edges and corners. The representations have been built into a classification framework directed at the prediction of springback values. The proposed framework and representations have been evaluated using two surfaces, a small and a large flat-topped pyramid, and by considering a variety of classification mechanisms and parameter settings.

## 1 Introduction

In sheet metal forming, especially in Asymmetric Incremental Sheet Forming (AISF), the *springback effect* is a major issue. As a result of springback the actual shape produced by the sheet metal forming process is not the same as the intended (specified) shape. The motivation for the work described in this paper is that if we can predict the springback we can apply a correction to the intended specification so as minimize the springback effect. Springback is caused by a number of factors of which the most significant is the geometry of the intended shape [1]. Further,

Subhieh El Salhi, Frans Coenen, Clare Dixon, Muhammad Sulaiman Khan
University of Liverpool, Department of Computer Science, Ashton Building, Ashton Street, Liverpool L693BX, United Kingdom
e-mail: hsselsal@liv.ac.uk, coenen@liv.ac.uk, cldixon@liv.ac.uk, mskhan@liv.ac.uk

springback is not distributed evenly over a given pshape; in practice springback is more significant over flat surfaces and with respect to some geometries than others. The solution proposed in this paper is founded on the idea of classification whereby a classifier is trained to predict springback at individual locations. Of course, for the classifier to operate correctly a suitable representation of the input data (the 3D surface of interest) is required. Two representations are proposed. The first, the Local Geometry Matrix (LGM) method, is founded on the idea of Local Binary Patterns [7], and can be used to define all possible local geometries. The second, the Local Distance Measure (LDM) method, takes into consideration the proximity of edges and corners with respect to individual points. However, before this second representation can be generated it is first necessary to identify the edges and corners of interest. Thus the contributions of this paper are as follows:

1. A mechanism for describing local 3D geometries using the concept of LGMs.
2. A mechanism for detecting edges and corners in 3D surfaces.
3. A mechanism for describing local 3D geometries using the concept of LDMs
4. A classification framework for the prediction of springback.

The rest of the paper is organized as follows. In section 2 a brief overview of related work is presented. Section 3 introduces the proposed framework for springback prediction, the section focuses on the classifier generation approach adopted. The proposed representations (LGM and LDM), in the context of the framework, are presented and discussed in Sections 3.2 and 3.3. The evaluation of the proposed framework using two surfaces, a small and a large flat topped pyramids, is presented in Section 4 in the context of a number classification paradigms and a variety of parameters. Some conclusions are then presented in Section 5.

## 2 Overview of related work

Asymmetric Incremental Sheet Forming (AISF) is a process for forming sheet metal parts. The potential advantage offered by processes such as AISF is a reduction in manufacturing costs and time. A comprehensive overview of AISF can be found in [9]. However, as already noted in the introduction to this paper, the main limitations of the AISF process is the springback effect. Springback can be defined as the elastic distortion that occurs as a result of the forming process so that the shape produced is not the desired shape. The springback effect is related to both manufacturing parameters and material properties [5, 15, 12]. There has been substantial reported work on springback characterization, analysis and prediction. Numerical and experimental methods have been proposed to predict springback in the context of sheet metal forming processes. The main numerical method used to analyse and predict springback for sheet metal forming is the Finite Element Method (FEM) [3, 14, 18]. Using FEM, the factors that affect the springback may be used to create a simulation model [8]. Although FEM provides a flexible simulation environment (parameters can be easily modified) FEM is an expensive and time-consuming option [17, 5]. Furthermore, FEM is not an accurate prediction method due to the simplification assumptions that must be made [2, 3, 15]. Artificial Neural Network (ANNs) are often quoted as being a good alternative to FEM. ANNs are the most popular experimental method that has been adopted for springback prediction. Some practitioners have used the FEM model to provide the ANN with the required input data to support the training of the network [15, 11]. An alternative approach described in [13] used a genetic algorithm in order to optimize the weights for the nodes of the ANN for the purpose of springback prediction. Nevertheless, computational resources requirements remain the main limitation of ANNs [13, 6]. To the best knowledge of the authors there is no reported work on the application of classification techniques (or data mining techniques in general) for the purpose of predicting springback in the context of AISF.

## 3 Overview Of The Proposed Springback Prediction Framework

The input to proposed springback prediction framework is a *coordinate cloud* describing the desired 3D shape (this is typically extracted from a CAD system). The proposed framework comprises two main components: (i) data preparation and (ii) classification. During the data preparation stage the input cloud is translated into the desired representation. During the classification stage a classifier is applied to the input data to produce springback predictions. The classifier was generated in the standard manner using labelled input data. The classification stage is thus not of specific interest with respect to this paper. The novel elements of this paper are the proposed geometry representation techniques used as input to the classifier. In the context of training the desired classifier the process commences with recasting the input data (the before cloud) into an appropriate format to which further processing can be applied. A grid format is proposed for this purpose. The generation of this grid format is discussed in details in Section 3.1. As noted above classification generation requires a labelled training set, to create a labelled training set we need to compare the before cloud with the cloud produced as a result of applying the AISF (the after cloud)[1]. A description of the mechanism used to calculate the labels (error values) so as to populate a training set is presented in Section 3.1. The next stage is to represent the geometry associated with each grid point in the before coordinate grid. As already noted above, two alternative representations are considered in this paper: the LGM and the LDM methods. These are described in some detail in Sections 3.2 and 3.3, respectively.

[1] This is obtained using the GOM (Gesellschaft fr Optische Messtechnik) optical measuring tool.

## 3.1 Grid Representation

The proposed process, as introduced above, commences with before and after coordinate clouds ($C_{in}$ and $C_{out}$). $C_{in}$ is the point cloud for the desired shape ($T$), while $C_{out}$ is the point cloud for the actual shape ($T'$) produced as a result of application of the AISF process. A cloud point $P_i$ is referenced in terms of a Euclidean coordinate system. Both clouds, $C_{in}$ and $C_{out}$, are translated into a grid representation of the form shown in Figure 1. The size of the grid is defined in terms of a value $D$ which represents the length of a grid square. Each grid square is referenced by the x-y coordinates of its centre point. The z value associated with each grid square is obtained by averaging the z coordinates for all the points located within it. Consequently, each grid square is represented by a central representative point described in terms of x, y, and z coordinates.

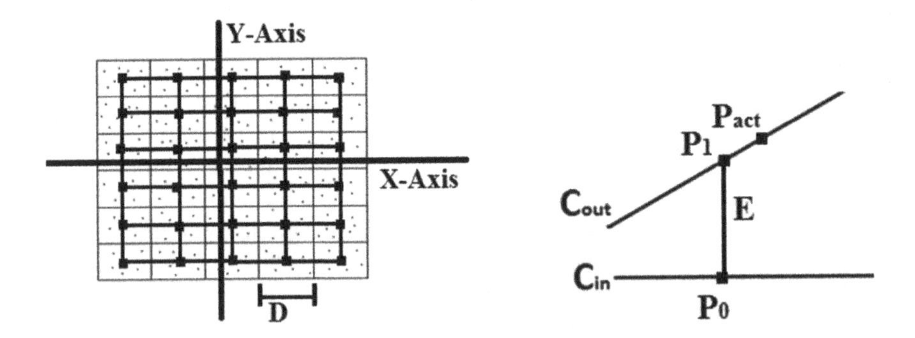


Fig. 1: The grid representation for the $C_{in}$ and $C_{out}$ coordinate clouds

Fig. 2: Springback error calculation

The error label to be associated with each $C_{in}$ grid point is calculated by determining the distance along the normal at each grid point from where it leaves the $C_{in}$ shape to where it intersects the $C_{out}$ shape as shown in Figure 2. The calculation is founded on vectors and plane geometry theory. Equation (1) shows how the distances between $C_{in}$ and $C_{out}$ surfaces are calculated; where $P_0$ and $P_1$ are described in terms of x, y and z coordinates, and the normal to the $C_{in}$ plane is given by the vector $(a,b,c)$.

$$E = \frac{|a(x_1 - x_0) + b(y_1 - y_0) + c(z_1 - z_0)|}{\sqrt{(a^2 + b^2 + c^2)}} \tag{1}$$

Thus two vectors, separated by 90°, adjoining a given centre point to two adjacent points are used to calculate a normal. Given that each point (except at corners and edges) has four neighbours (north, east, south and west) four normals can be calculated. The average of these four normals is used as the normal to a point.

Table 1: Sample training data for the small pyramid when $L = 3$ using LGM method

| Z1 | Z2 | Z3 | Z4 | Z5 | Z6 | Z7 | Z8 | E |
|---|---|---|---|---|---|---|---|---|
| 3 | 6 | 9 | 12 | 15 | 18 | 21 | 24 | 30 |
| 2 | 5 | 8 | 11 | 14 | 17 | 20 | 23 | 28 |
| 3 | 6 | 9 | 12 | 15 | 18 | 21 | 24 | 30 |
| 2 | 5 | 8 | 11 | 14 | 17 | 20 | 23 | 28 |
| 2 | 5 | 8 | 11 | 14 | 17 | 20 | 23 | 30 |
| 1 | 4 | 7 | 10 | 13 | 16 | 19 | 22 | 28 |

Table 2: Sample training data for the small pyramid when $L = 3$ using both the LGM and LDM methods

| Z1 | Z2 | Z3 | Z4 | Z5 | Z6 | Z7 | Z8 | Dist. | E |
|---|---|---|---|---|---|---|---|---|---|
| 1 | 4 | 7 | 10 | 13 | 16 | 19 | 22 | 26 | 28 |
| 1 | 4 | 7 | 10 | 13 | 16 | 19 | 22 | 25 | 28 |
| 2 | 5 | 8 | 11 | 14 | 17 | 20 | 23 | 25 | 28 |
| 3 | 5 | 8 | 12 | 15 | 18 | 21 | 24 | 25 | 29 |
| 3 | 6 | 8 | 12 | 15 | 18 | 21 | 24 | 25 | 28 |
| 2 | 5 | 8 | 11 | 14 | 17 | 20 | 23 | 25 | 28 |

## 3.2 *Local Geometry Matrix (LGM)*

The first shape description mechanism presented in this paper is the Local Geometry Matrix (LGM) method. The concept of LGMs is founded on the idea of Local Binary Patterns (LBPs). A local geometry matric is a $3 \times 3$ grid describing the locations surrounding an individual point (the point of interest is at the centre of the matrix). There are two different options for calculating the values that might be stored in an LGM (Figure 3). The first option is the difference in height ($z$ difference) between the centre point $P_0$ and each of its eight neighbours $P_i$. The second is to store the angle, above or below the horizontal, of the lines connecting $P_0$ and each $P_i$. Whatever the case at the end of the process we have a LGM for each grid point which can be combined with an error value (calculated as described above) so that each grid point forms a record in the training set. Because we are using classifiers that operate with binary valued data we needed to discretise the training data so that each value is replaced by one of $L$ qualitative labels used to describes the nature of the slope in each of the eight directions. An example set of qualitative labels might be $\{negative; level; positive\}$. Using this labelling, and by ordering the matrix elements (grid points) in a clockwise direction from the top left, a record might be described as follows:

$$< positive; positive; positive; level; negative; negative; negative; level; E >$$

Where E is the error value associated with the grid point that the record describes. Table 1 shows a sample set of the training data using LGM for the small pyramid when $L = 3$. The authors experiments with a number of different values for $L$ as reported later in Section 4 of this paper.

## 3.3 *Local Distance Mechanism (LDM)*

The second shape description method presented in this paper is the Local Distance Measure (LDM) method. This is founded on the observation that the springback tends to be greater further from edges. The idea is therefore to describe each grid

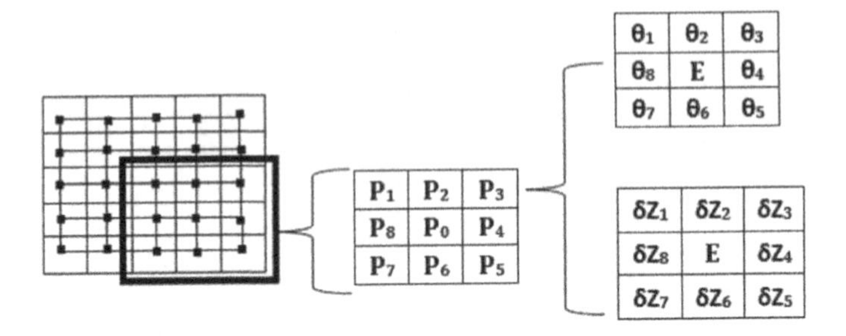


Fig. 3: LGM Representation

square centre in terms of its distance from its nearest edge. This first necessitates identifying the edges in the input grid. Edge detection was conducted by first dividing $C_{in}$ into $K$ regions using the well know K-means algorithm ($K = 4$ was empirically chosen and used with respect to the evaluation described later in this paper). The average angular value for the normals in each region was first determined. The angular difference between the normal at each centre point and the average for the region in which the point was located was then calculated. If the angular difference was greater than some tolerance measure, $\sigma$, the point was considered to be located on an edge. The process is illustrated in Figure 4. Once the edges have been detected, the minimum distance between each grid point and its closest edge was determined simply by adopting a "region growing" process. The result is a set of records each describing a grid square location in terms of its edge distance and its associated error label (springback). Clearly, there are two main factors that affect the process of edge detection:

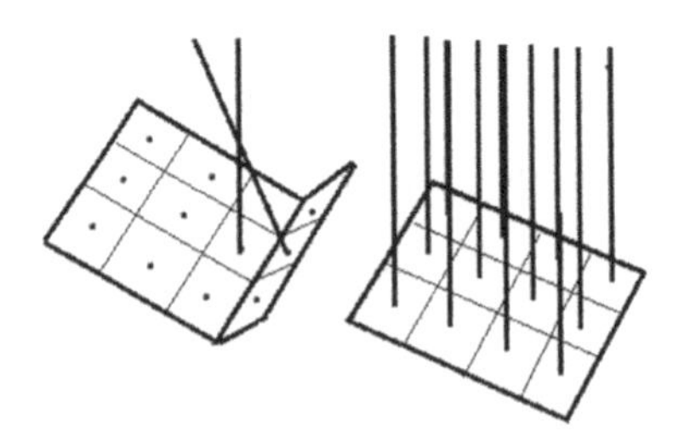

Fig. 4: Edge detection

Table 3: Sample training data for the small pyramid when $L = 3$ using LDM method

| Distance | Error |
|---|---|
| 24 | 30 |
| 23 | 28 |
| 24 | 30 |
| 23 | 28 |
| 24 | 28 |
| 23 | 29 |
| 22 | 30 |
| 22 | 28 |

- **The tolerance value** $\sigma$: As $\sigma$ is increased, the number of points identified as edge points decreases. Conversely as $\sigma$ is decreased, the number of points identified as edge points increases. This is illustrated in Figures 5 and 6 where it can be seen that more points are identified as edge points when $\sigma$=15 than when $\sigma$=5. Therefore, $\sigma$ should be carefully chosen.

- **The grid size** ($D$): The greater the grid size the more difficult it is to detect edges. For a relatively large value of $D$, the grid square will cover a larger area. Therefore the normals are more likely to be parallel than if a small value of $D$ is used. This is illustrated in Figures 7 and 8 where $\sigma$ has been kept constant ($\sigma$=9) and $D$ set at 5 and 15 respectively. In Figure 7 the edges are easily detected as there is an obvious variation in the angular differences. However, in Figure 8 all the points are identified as edges as all the normals are different (by at least a $\sigma$). Table 3 introduces a sample of the training data set for the small pyramid using LDM where each value for each attribute is replaced by one of the qualitative labels for $L = 3$. Table 2 presents another example for the training data set, for the same small pyramid, using the same number of labels ($L = 3$) where LDM and LGM methods are used together.

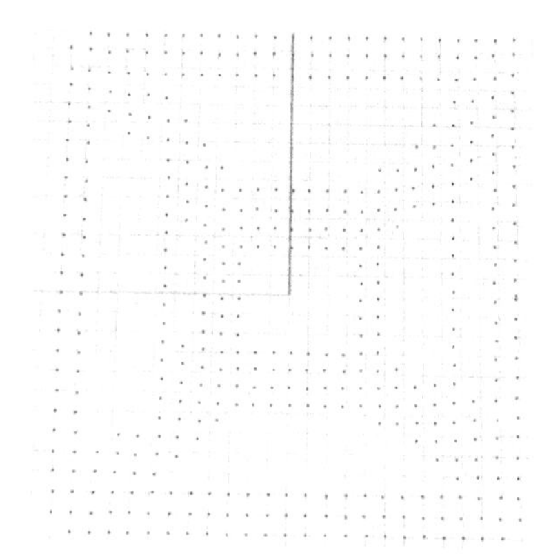

Fig. 5: Large pyramid with D=5 and $\sigma$=5

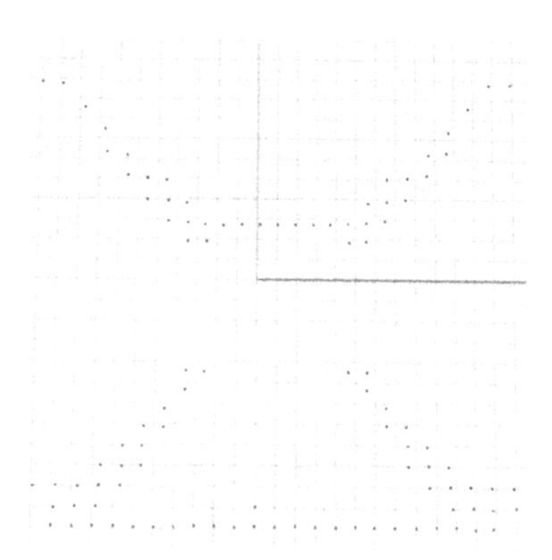

Fig. 6: Large pyramid with D=5 and $\sigma$=15

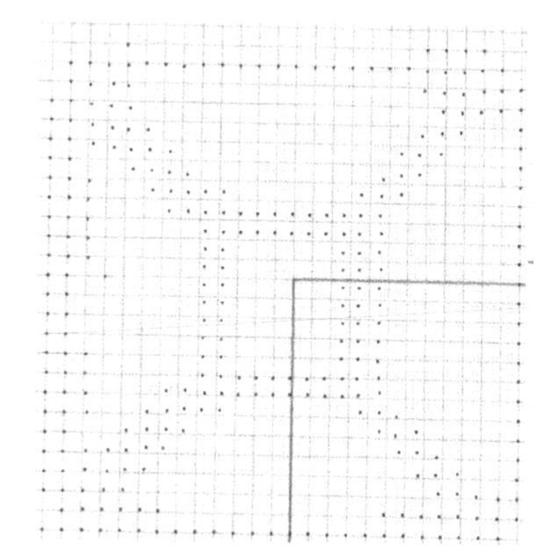

Fig. 7: Small pyramid with D=5 and $\sigma$=9

Fig. 8: Small pyramid with D=15 and $\sigma$=9

# 4 Evaluation

To evaluate the proposed springback prediction framework two sample surfaces were used, a small and a large flat topped pyramid. The small pyramid base size is 140 mm whereas the base size for the large pyramid is 350 mm. Both were defined in terms of before coordinate clouds extracted from a CAD system. These shapes were then manufactured [2] in steel using AISF. After clouds were then generated from the manufactured parts. These before and after clouds were then processed as described above to generate a number of data sets to be used for evaluation purposes. Broadly these data sets can be divided into three categories according to the representation used: (i) LGM, (ii) LDM and (iii) a combination of LGM and LDM. Six sets of experiments were conducted as follows. The first two sets of experiments, reported in Section 4.1 and 4.2, were intended to determine the performance of the proposed approach when applied individually to the small and large pyramids. The next two sets of experiments, reported in Sections 4.3 and 4.4, were designed to determine whether a classifier trained on the small pyramid could be applied to the large pyramid and vice-versa. The significance of these last two experiments was that it would serve to go some way to establishing whether we could produce a generic classifier trained from some appropriate shape that could be applied to other shapes. Some statistics concerning the data sets are presented in Table 4. The results from these four experiments are reported in the following four subsections. Sub-sections 4.5 and 4.6 then analyse the effect of using different values for $D$ (grid size) and $L$ (number of qualitative labels). For the experiments three different classifier generators were considered: (i) Bayes [10], (ii) JRIP [4] and C4.5 [16]. A range of values of $L$ of 3, 5 and 7 was used; and a range of values for $D$ of 5, 10, 15 and 20. For each experiment the results were recorded in terms of percentage accuracy.

Table 4: Number of records generated for the large and small pyramids using different values for $D$

| D | Small Pyramid | Large Pyramid |
|---|---|---|
| 5 | 783 | 4901 |
| 10 | 196 | 1223 |
| 15 | 78 | 528 |
| 20 | 46 | 289 |

[2] The small and the large pyramid were generated purely for research purpose as it is a commonly used shape with respect to AISF related research.

## 4.1 Small pyramid

Table 5 presents the results obtained when applying the proposed techniques to the small pyramid test shape. The reported accuracy values are average values generated using Ten Cross Validation (TCV). From the table we can observe that:

- The highest accuracy obtained was 87.01%, using $D = 15$ and $L = 3$.
- The best performing techniques were the LGM and the combined LGM and LDM technique; the LDM technique on its own did not perform as well as the other two techniques.
- There was no significant distinction between the operation of the three of classification algorithms considered, especially in the context of the LDM technique; however an argument can be made that Bayes and JRIP outperformed C4.5.

Table 5: Accuracy results for the small pyramid

| Method | L | D=5 | | | D=10 | | | D=15 | | | D=20 | | |
|---|---|---|---|---|---|---|---|---|---|---|---|---|---|
| | | Bayes | JRIP | C4.5 | Bayes | JRIP | C4.5 | Bayes | JRIP | C4.5 | Bayes | JRIP | C4.5 |
| LGM | 3 | 61.13 | 84.02 | 84.02 | 71.28 | 85.13 | 86.67 | **87.01** | **87.01** | 85.71 | 57.78 | 75.56 | 75.56 |
| | 5 | 57.80 | 77.88 | 77.49 | 62.05 | 77.95 | 77.44 | 55.84 | 71.43 | 72.73 | 55.56 | 68.89 | 51.11 |
| | 7 | 66.88 | 75.58 | 75.96 | 67.18 | 72.82 | 68.72 | 58.44 | 67.53 | 76.62 | 51.11 | 51.11 | 44.44 |
| LDM | 3 | 84.53 | 84.53 | 84.53 | 84.11 | 84.10 | 84.10 | 75.33 | 75.32 | 75.32 | 75.56 | 75.56 | 75.56 |
| | 5 | 71.87 | 71.87 | 71.87 | 73.85 | 73.85 | 73.85 | 61.04 | 61.04 | 61.03 | 66.67 | 66.67 | 66.67 |
| | 7 | 67.65 | 67.65 | 67.65 | 69.23 | 69.23 | 69.23 | 53.25 | 49.35 | 49.35 | 60.00 | 60.00 | 57.78 |
| LGM and LDM | 3 | 61.13 | 84.27 | 84.02 | 74.36 | 85.13 | 86.67 | **87.01** | **87.01** | 85.71 | 57.78 | 73.33 | 71.11 |
| | 5 | 58.06 | 77.49 | 77.88 | 62.05 | 77.44 | 76.92 | 57.14 | 71.43 | 72.73 | 57.78 | 66.67 | 51.11 |
| | 7 | 66.90 | 75.32 | 75.70 | 67.67 | 72.82 | 69.74 | 59.74 | 62.34 | 76.62 | 55.56 | 53.33 | 44.44 |

From the table we can also observe that the accuracy associated with the LGM and the combination techniques increased gradually as $D$ was increased from 5 to 15, then dropped of when $D = 20$. It was conjectured that this was because when the grid was too small the geometry could not be well defined and when it got too large the geometry was too coarsely defined. It is also interesting to note that the LDM method is more stable in that it features smaller fluctuations in accuracy as $D$ is increased from 5 to 20.

## 4.2 Large pyramid

Table 6 presents the results obtained when the proposed techniques were applied to the large pyramid. Again the results were generated using TCV. From the table it can be seen that:

- The best overall accuracy result obtained was 99.16%, using $D = 5$ and $L = 3$. This is an extremely high accuracy.
- Marginally better results were obtained using LGM and the combination technique, although there was not a significant distinction between the techniques.
- Out of the three classification algorithms considered the best result was obtained using C4.5, but the margin between different data classification techniques was very small, hence it can be argued that there was no significant difference between the classifiers.

Overall the experiments on both the small and large pyramids indicated that very good error predictions could be obtained from application of the proposed framework, especially in the case of the large pyramid (which included more points for the same value of $D$ as indicated in Table 4).

Table 6: Accuracy results for the large pyramid

| Method | L | D=5 | | | D=10 | | | D=15 | | | D=20 | | |
|---|---|---|---|---|---|---|---|---|---|---|---|---|---|
| | | Bayes | JRIP | C4.5 | Bayes | JRIP | C4.5 | Bayes | JRIP | C4.5 | Bayes | JRIP | C4.5 |
| LGM | 3 | 95.12 | 99.10 | **99.16** | 91.65 | 97.95 | 98.04 | 89.94 | 98.29 | 98.29 | 88.19 | 97.22 | 97.22 |
| | 5 | 89.27 | 99.10 | 99.10 | 86.82 | 96.48 | 96.64 | 86.91 | 97.15 | 97.34 | 86.11 | 96.88 | 96.88 |
| | 7 | 86.67 | 98.69 | 98.67 | 86.50 | 95.01 | 95.34 | 85.77 | 95.83 | 96.02 | 81.60 | 95.49 | 95.83 |
| LDM | 3 | 98.94 | 98.94 | 98.94 | 97.71 | 97.71 | 97.71 | 98.29 | 98.29 | 98.29 | 97.22 | 97.22 | 97.22 |
| | 5 | 98.78 | 98.78 | 98.78 | 96.40 | 96.40 | 96.40 | 96.77 | 96.77 | 96.77 | 96.88 | 96.88 | 96.88 |
| | 7 | 98.69 | 98.69 | 98.69 | 95.09 | 95.09 | 95.09 | 96.02 | 96.02 | 96.02 | 95.49 | 95.49 | 95.49 |
| LGM and LDM | 3 | 95.04 | 99.10 | **99.16** | 91.65 | 97.95 | 98.04 | 90.13 | 98.29 | 98.29 | 88.19 | 97.22 | 97.22 |
| | 5 | 89.27 | 99.10 | 99.10 | 86.82 | 96.48 | 96.64 | 87.10 | 97.15 | 97.34 | 86.11 | 96.88 | 96.88 |
| | 7 | 86.67 | 98.69 | 98.67 | 86.58 | 94.84 | 95.50 | 85.77 | 95.83 | 96.02 | 81.60 | 95.49 | 95.83 |

## 4.3 Training on the small pyramid and testing on the large pyramid

The results obtained when building a classifier using the small pyramid and applying it on the large pyramid are presented in Table 7. From the table it can be seen that:

- The best accuracy obtained was 98.94%, using $D = 5$ and $L = 3$. This is again an excellent result.
- The number of labels (3, 5, 7) did not appear to play a significant role although L=3 produced best results (but only with a small margin). However, it should be noted that if we wish to apply corrections to the input cloud we probably need to use a high value for $L$.
- The LDM and the combined technique produced better results than the LGM technique.
- Out of the three classification algorithms considered there was no obvious distinction between their operations.

Table 7: Accuracy results when training on the small and testing on the large pyramid

| Method | L | D=5 | | | D=10 | | | D=15 | | | D=20 | | |
|---|---|---|---|---|---|---|---|---|---|---|---|---|---|
| | | Bayes | JRIP | C4.5 | Bayes | JRIP | C4.5 | Bayes | JRIP | C4.5 | Bayes | JRIP | C4.5 |
| LGM | 3 | 94.90 | 98.65 | 98.65 | 94.60 | 97.71 | 97.71 | 92.22 | 93.36 | 93.36 | 96.18 | 97.22 | 88.68 |
| | 5 | 93.12 | 95.96 | 95.94 | 93.93 | 93.94 | 93.94 | 89.56 | 93.93 | 92.22 | 90.63 | 96.88 | 89.58 |
| | 7 | 94.39 | 96.16 | 96.00 | 90.00 | 90.92 | 93.70 | 95.69 | 94.50 | 91.84 | 89.93 | 95.49 | 94.10 |
| LDM | 3 | **98.94** | **98.94** | **98.94** | 97.71 | 97.71 | 97.71 | 98.29 | 98.30 | 98.30 | 97.22 | 97.22 | 97.22 |
| | 5 | 98.78 | 98.78 | 98.78 | 96.40 | 96.40 | 96.40 | 96.77 | 96.77 | 96.78 | 96.88 | 95.83 | 95.83 |
| | 7 | 98.69 | 98.69 | 98.69 | 95.09 | 95.09 | 95.09 | 96.02 | 96.02 | 96.02 | 95.48 | 95.49 | 95.49 |
| LGM and LDM | 3 | **98.94** | 98.65 | 98.80 | 94.60 | 97.71 | 97.71 | 92.22 | 93.17 | 93.36 | 94.94 | 98.65 | 92.36 |
| | 5 | 93.12 | 95.92 | 95.94 | 94.01 | 93.94 | 94.35 | 90.32 | 93.93 | 92.22 | 91.67 | 95.83 | 89.58 |
| | 7 | 94.39 | 96.16 | 96.04 | 91.57 | 91.00 | 93.70 | 95.88 | 93.36 | 91.84 | 92.71 | 95.49 | 94.10 |

## *4.4 Training on the large pyramid and testing on the small pyramid*

Table 8 shows the results produced when building a classifier using the large pyramid and applying it to the small pyramid. From the table it can be observed that:

- The best accuracy obtained is 84.53%, again using $D = 5$ and $L = 3$.
- There was no significant distinction between the operation of the LGM, LDM and combination techniques, although LDM did produce marginally better results.
- Out of the three data mining techniques considered, there was no significant performance distinction between them.

The results obtained from this experiment, and the previous experiment, indicated that it might be possible to build a generic classifier. It is conjectured that both the small pyramid and the large pyramid contain sufficient examples of all different possible geometries to allow for effective classification, with the larger pyramid producing slightly better performing classifiers than those produced using the small pyramid.

Table 8: Accuracy results when training on the large and testing on the small pyramid

| Method | L | D=5 | | | D=10 | | | D=15 | | | D=20 | | |
|---|---|---|---|---|---|---|---|---|---|---|---|---|---|
| | | Bayes | JRIP | C4.5 | Bayes | JRIP | C4.5 | Bayes | JRIP | C4.5 | Bayes | JRIP | C4.5 |
| LGM | 3 | 59.08 | 84.02 | 84.02 | 29.74 | 84.10 | 84.10 | 36.36 | 55.84 | 55.84 | 46.67 | 75.56 | 75.56 |
| | 5 | 41.18 | 71.61 | 71.36 | 16.92 | 73.85 | 73.85 | 33.77 | 53.25 | 53.25 | 31.11 | 68.89 | 68.89 |
| | 7 | 33.12 | 67.65 | 57.54 | 13.85 | 69.23 | 68.21 | 22.08 | 53.25 | 51.95 | 22.22 | 53.33 | 60.00 |
| LDM | 3 | **84.53** | **84.53** | **84.53** | 84.10 | 84.10 | 84.10 | 75.32 | 75.32 | 75.32 | 75.56 | 75.56 | 75.56 |
| | 5 | 71.87 | 71.87 | 71.87 | 73.85 | 73.85 | 73.85 | 61.04 | 61.04 | 61.04 | 68.89 | 68.89 | 68.89 |
| | 7 | 67.65 | 67.65 | 67.65 | 69.23 | 69.23 | 69.23 | 53.25 | 53.25 | 53.25 | 60.00 | 60.00 | 60.00 |
| LGM and LDM | 3 | 59.08 | 84.02 | 84.02 | 29.74 | 84.10 | 84.10 | 37.67 | 55.84 | 55.84 | 46.67 | 75.56 | 75.56 |
| | 5 | 41.18 | 71.61 | 71.36 | 17.44 | 73.85 | 73.85 | 32.47 | 53.25 | 53.25 | 31.11 | 68.89 | 68.89 |
| | 7 | 33.12 | 67.65 | 57.54 | 13.85 | 69.23 | 68.21 | 22.08 | 53.25 | 51.95 | 22.22 | 53.33 | 60.00 |

## 4.5 The effect of grid size (D)

Clearly, from the foregoing, the grid size parameter, $D$, has an effect. In this subsection a brief analysis of the effect that the value for $D$ has on classification performance is presented. Table 9 shows a comparison between the percentage accuracy and $D$ values for the best performing combinations from the foregoing experiments when $L = 3$. These are identified using the letters A, B, C and D as follows:

A. Training and testing on the large pyramid using, C4.5 and LGM.
B. Training and testing on the small pyramid using JRIP and LGM.
C. Training on the large and testing on the small using JRIP and LDM.
D. Training on the small and testing on the large using Bayes and LDM.

Table 9: The effect of $D$ when $L = 3$

| | D=5 | D=10 | D=15 | D=20 |
|---|---|---|---|---|
| A | 99.16 | 98.04 | 98.29 | 97.22 |
| B | 84.02 | 85.13 | 87.01 | 75.56 |
| C | 84.53 | 84.10 | 75.33 | 75.56 |
| D | 98.94 | 97.71 | 98.29 | 97.22 |

From Table 9 it can be observed that when training on the large pyramid (which produced the better performance) accuracy dropped as $D$ was increased. In the case where we trained and tested on the small pyramid (case B) accuracy peaked at $D =$ 15, while in case C accuracy peaked at $D = 5$. Overall we can conclude that $D = 5$ tends to produce a better result.

## 4.6 The effect of label set size (L)

This section completes the evaluation with a brief analysis of the effect of the label set size ($L$). The analysis was conducted by considering the best performing combinations when $D = 5$ [3]; these are again identified using the letters A, B, C and D as follows:

A. Training and testing on the large pyramid using C4.5 and LGM.
B. Training and testing on the small pyramid using C4.5 and LDM.
C. Training on the large and testing on the small using JRIP and LDM.
D. Training on the small and testing on the large using Bayes and LDM.

[3] It is acknowledged that the best result for the small pyramid was obtained when $D = 15$, however we wished to conduct the analysis concerning $L$ by maintaining $D$ at a constant value.

The results of the analysis are presented in Table 10. In this case we can note that the value of $L$ has little effect with respect to case A and D, while in cases B and C accuracy decreases as the number of labels increases.

Table 10: The effect of $L$ when $D = 5$

| | L=3 | L=5 | L=7 |
|---|---|---|---|
| A | 99.16 | 99.10 | 98.67 |
| B | 84.53 | 71.87 | 67.65 |
| C | 84.53 | 71.87 | 67.65 |
| D | 98.94 | 98.78 | 98.69 |

## 5 Conclusion

This paper has presented a new framework to identify the correlations between 3D surfaces using two different mechanisms for shape description, the first founded on the concept of Local Geometry Matrixes (LGMs) and the second on Local Distance Measures (LDMs). Both mechanisms, and a combination of the two, were evaluated by applying them to two manufactured surfaces, a small and a large flat-topped pyramid. The main finding of the work is that good (very good in some cases) classification accuracy results can be produced not only when the classifier is trained and tested on an identical shape but also when the classifier is trained on one shape and tested on another where every possible pattern that can exist using LGM method is included in the classifier when the pyramid shape considered. The significance of the latter is that this is an indication (further experimentation is clearly required) that we can build a classifier that encompasses all possible geometries by considering a suitable shape, which can then be generally applied. This has implications if we want to build a system that can suggest corrections to be applied to before clouds that can serve to limit the effect of springback. Other significant findings included: (i) confirmation that small grid sizes produce a better performance than large grid sizes, (ii) that both the LGM and LDM techniques worked well and that it could not be argued that there was a significant difference in their operation, and (iii) that the choice of classification algorithm did not make a significant impact. Overall this is a very encouraging result. For future work a new surface representation approach founded on the concept of time series is currently under investigation. The intention is also to conduct further experimentation with a greater variety of surfaces (shapes) and a detailed comparison between the LGM, LDM, the combination of both and the new time series surface representation methods. The ultimate goal is to build an intelligent process model that can predict springback errors and suggest corrections to before coordinate clouds.

**Acknowledgements** This work was partially supported by the EU project Innovative Manufacturing of complex Ti Sheet Components (INMA), grant agreement number 266208.

## References

1. G. Cafuta, N. Mole, and B. Łtok. An enhanced displacement adjustment method: Springback and thinning compensation. *Materials and Design*, 40:476 – 487, 2012.
2. S. Chatti. Effect of the Elasticity Formulation in Finite Strain on Springback Prediction. *Computers and Structures*, 88(1112):796 – 805, 2010.
3. S. Chatti and N. Hermi. The Effect of Non-linear Recovery on Springback Prediction. *Computers and Structures*, 89(13-14):1367 – 1377, 2011.
4. W. Cohen. Fast effective rule induction. In *Twelfth International Conference on Machine Learning*, pages 115–123. Morgan Kaufmann, 1995.
5. M. Firat, B. Kaftanoglu, and O. Eser. Sheet Metal Forming Analyses With An Emphasis On the Springback Deformation. *Journal of Materials Processing Technology*, 196(1-3):135 – 148, 2008.
6. Z. Fu, J. Mo, L. Chen, and W. Chen. Using Genetic Algorithm-Back Propagation Neural Network Prediction and Finite-Element Model Simulation to Optimize the Process of Multiple-Step Incremental Air-Bending Forming of Sheet Metal. *Materials and Design*, 31(1):267 – 277, 2010.
7. Z. Guo, L. Zhang, and D. Zhang. A Completed Modeling of Local Binary Pattern Operator for Texture Classification. *IEEE Transactions on Image Processing*, 19(6):1657–1663, 2010.
8. W. Hao and S. Duncan. Optimization of Tool Trajectory for Incremental Sheet Forming Using Closed Loop Control. In *Automation Science and Engineering (CASE), 2011 IEEE Conference on*, pages 779 –784, 2011.
9. J. Jeswiet, F. Micari, G. Hirt, A. Bramley, J. Duflou, and J. Allwood. Asymmetric Single Point Incremental Forming of Sheet Metal. *CIRP Annals - Manufacturing Technology*, 54(2):88 – 114, 2005.
10. G. John and P. Langley. Estimating Continuous Distributions in Bayesian Classifiers. In *Eleventh Conference on Uncertainty in Artificial Intelligence*, pages 338–345, San Mateo, 1995. Morgan Kaufmann.
11. R. Kazan, M. Firat, and A. Egrisogut Tiryaki. Prediction of Springback in Wipe-Bending Process of Sheet Metal Using Neural Network. *Materials and Design*, 30(2):418 – 423, 2009.
12. W. Liu, Z. Liang, T. Huang, Y. Chen, and J. Lian. Process Optimal Ccontrol of Sheet Metal Forming Springback Based on Evolutionary Strategy. In *Intelligent Control and Automation, 2008. WCICA 2008. 7th World Congress on*, pages 7940 –7945, June 2008.
13. W. Liu, Q. Liu, F. Ruan, Z. Liang, and H. Qiu. Springback Prediction for Sheet Metal Forming Based on ga-ann Technology. *Journal of Materials Processing Technology*, 187-188:227 – 231, 2007.
14. N. Narasimhan and M. Lovell. Predicting Springback in Sheet Metal Forming: An Explicit to Implicit Sequential Solution Procedure. *Finite Elements in Analysis and Design*, 33(1):29 – 42, 1999.
15. V. Nasrollahi and B. Arezoo. Prediction of Springback in Sheet Metal Components With Holes on the Bending Area, Using Experiments, Finite Element and Neural Networks. *Materials and Design*, 36:331 – 336, 2012.
16. R. Quinlan. *C4.5: Programs for Machine Learning*. Morgan Kaufmann Publishers, San Mateo, CA, 1993.
17. M. Tisza. Numerical Modelling and Simulation in Sheet Metal Forming. *Journal of Materials Processing Technology*, 151(1-3):58 – 62, 2004.
18. J. Yoon, F. Pourboghrat, K. Chung, and D. Yang. Springback Prediction For Sheet Metal Forming Process Using a 3d Hybrid Membrane/Shell Method. *International Journal of Mechanical Sciences*, 44(10):2133 – 2153, 2002.

# Towards The Collection of Census Data From Satellite Imagery Using Data Mining: A Study With Respect to the Ethiopian Hinterland

Kwankamon Dittakan, Frans Coenen and Rob Christley

**Abstract** The collection of census data is an important task with respect to providing support for decision makers. However, the collection of census data is also resource intensive. This is especially the case in areas which feature poor communication and transport networks. In this paper a method is proposed for collecting census data by applying classification techniques to relevant satellite imagery. The test site for the work is a collection of villages lying some 300km to the northwest of Addis Ababa in Ethiopia. The idea is to build a classifier that can label households according to "family" size. To this end training data has been obtained, by collecting on ground census data and aligning this up with satellite data. The fundamental idea is to segment satellite images so as to obtain the satellite pixels describing individual households and representing these segmentations using a histogram representation. By pairing each histogram represented household with collated census data, namely family size, a classifier can be constructed to predict household sizes according to the nature of the histograms. This classifier can then be used to provide a quick and easy mechanism for the approximate collection of census data that does not require significant resource.

Kwankamon Dittakan
Department of Computer Science, University of Liverpool, Ashton Building , Aston Street, L69 3BX Liverpool, UK e-mail: dittakan@liverpool.ac.uk

Frans Coenen
Department of Computer Science, University of Liverpool, Ashton Building , Aston Street, L69 3BX Liverpool, UK e-mail: coenen@liverpool.ac.uk

Rob Christley
Institute of Infection and Global Health, University of Liverpool, Leahurst Campus, Chester High Road, CH64 7TE Neston, Cheshire,UK e-mail: robc@liverpool.ac.uk

## 1 Introduction

A census is a collection of information about the nature of a population of a given area. Census collection tends to be undertaken using a questionnaire format. Questionnaires are usually either distributed by post or electronically for self-completion, or completed by field staff. There are many problems associated with the collection of census data, especially in the case of national censuses. The first problem is census budget, the collection of census data requires a considerable resource in terms of money and "manpower". Another problem is the cost of processing the data after it has been collected. A third issue is that there is often a lack of good will on behalf of a population to participate in a census, even if they are legally required to do so, because people are often suspicious of the motivation behind censuses (especially when collected by government organisations) [7]. These problems are compounded in areas where there are poor communication and transport infrastructures; and/or an extensive, but sparsely populated, hinterlands.

The solution proposed in this paper is founded on the idea of constructing classifiers that can predict census information according to the nature of satellite views of households. In some areas, such as inner city areas, this is unlikely to be appropriate; however in sparsely populated rural areas, as will be demonstrated, this can provide a effective and efficient mechanism for collecting census data. The fundamental idea of the proposed research is that, given a set of training data describing households and their geographical locations, we can obtain satellite images of these households and use image classification techniques to construct a classifier that can be applied to the entire region and consequently automatically collect census data at very low cost. Of course this will be an approximation, there is always a trade of between resource reduction and accuracy; and, as already noted, is likely to be more effective in rural and suburban areas, than in city centres and commercial areas.

The rest of this paper is organised as follows. In Section 2 some previous work is presented. Section 3 then provides detail of the proposed census mining framework, including reviews of the image enhancement, image segmentation, feature representation and feature selection mechanisms adopted. Section 4 reports on the evaluation of the framework (conducted using a sequence of experiments applied to data collected from villages lying some 300km to the northwest of Addis Ababa in Ethiopia). Finally, a summary and some conclusions are presented in Section 5.

## 2 Previous work

Satellite image interpretation offers advantages with respect to many applications. Examples include: land usage, regional planning, forest area monitoring, forest mapping, disease and fire detection and wild life studies [4, 13, 11]. The satellite image interpretation technique proposed in this paper is founded on image classification, the process of applying classification techniques to image sets. A typical generic image classification task is to label image data according to a predefined set of classes. In the context of satellite imagery; image classification has been used, for example, for linking urban land cover (obtained from satellite imagery) with urban function characteristics obtained from population census data [14]. However, there is little reported work on the application of image classification techniques to satellite image data for the purpose of census data collection although one example can be found in [19] where classification techniques are applied to satellite images to obtain information such as number of households, total population and urban population at the sub-district level. The distinction between this work and that described in this paper is that the proposed census mining framework operates at a much finer level of granularity.

Image classification, and by extension satellite image classification, requires images to be represented in some form that lends itself to the application of image classification techniques. There are many techniques that have been proposed for representing image features. In general these can be separated into three categories: (i) text based, (ii) semantic based and (iii) content based [10]. In the first category images are represented simply using keywords which describe each image. In the second category researchers have tried to capture the semantic meaning of images. In the third category some general content such as colour, texture and shape, is used to represent entire images or parts of images. The advantage of the third is that this representation can be automatically generated, and thus this is the technique adopted with respect to the work described in this paper where the proposed representation is founded on image colour.

One method of encapsulating image colour is to represent the distribution of colours using histograms [6, 12, 20]. This can be applied to entire images or parts there of. In the case of the satellite imagery of interest we wish to capture the parts of images that represent households. This in turn requires some segmentation process. Image segmentation is the process of partitioning an image so as to identify objects within the given image. Popular segmentation techniques include Threshold segmentation [16, 17], Region-growing segmentation [2, 18], Edge-based segmentation [15] and Texture-Based segmentation [24]. Because the households of interest tended to be rectangular in shape, the Edge-based segmentation technique was adopted for the purpose of identifying (segmenting) households.

## 3 Census Mining Framework

The proposed census framework is described in this section, the framework comprises six stages: (i) coarse segmentation, (ii) image enhancement, (iii) detailed segmentation, (iv) representation, (v) feature section and (vi) classifier generation. The proposed census mining framework takes as input a set of labelled satellite images, feeds then into an appropriately trained classifier and outputs prescribed census data. The nature of the classifier will depend on the nature of the desired census data. Different types of census data will require different types of classifier. The focus of the work describing in this papers is household size in terms of number of people normally resident in a given household. To build such a classifier suitable labeled training data was required. To act as a focus for the work the research team arranged for the manual collection of census data in a rural area lying some 300 km to the northwest of Addis Ababa in Ethiopia (Figure 1), which could then be used to generate an appropriate training data set.

**Fig. 1** The test site location: Harro district in Ethiopia (indicated by arrow)

The collected census data included geographical coordinates so that individual households could be related to satellite imagery. The research team used Google Earth imagery but clearly other forms of satellite imagery could equally well have been used. The identified households could therefore be located and segmented so that a collection of household images could be obtained. The first stage in the framework was thus the coarse segmentation of satellite imagery to isolate individual households or groups of households so that a collection of household sub-images was arrived at. A typical satellite image is presented in Figure 2, satellite images of this forum will thus first be roughly segmented to give $N$ household sub-images. Note that in some cases the sub-images may still contain several households.

The next stage in the process is image enhancement; this is briefly described in Sub-section 3.1. Once suitable enhancement had been applied the third stage was the detailed segmentation of the household sub-images so that individual households could be identified. The proposed segmentation process is described in Sub-section 3.2. The fourth stage was to represent the images in such a way that some image classification technique could be applied. As already noted in Section 2, there are a great many image representation techniques that have been proposed in the literature. The technique used for the proposed census mining was founded on a histogram based representation from which various statistics could be extracted. The acquired histogram data could then be coupled with class data (family size with

**Fig. 2** Example satellite image showing individual households (Harro district in Ethiopia)

respect to the evaluation presented in this paper) and used to generate the desired classifier. The process for generating histograms is presented in Sub-section 3.3. It is commonly accepted that not all features are significant with respect to individual image classification tasks, and that it is desirable (for computational resource saving reasons) to limit the dimensionality of the feature space, hence a feature selection process was also applied. This is presented in Sub-section 3.4. Once a suitable set of features had been identified, expressed in terms of a feature vector, a suitable classifier generator could be applied (stage 6) to produce the final desired classifier.

## *3.1 Image Enhancement*

This sub-section describes the image enhancement processes applied to the input data. An example household sub-image, obtained from a satellite image of the form shown in Figure 2, is presented in Figure 3(a). However, before image enhancement can be applied it was first necessary to register and align the images so that each household (delimited by its surrounding rectangular boundary) was aligned in a north-south direction. The purpose of this registration and alignment was to facilitate future segmentation. The result is a shown in Figure 3(b), where the image shown in Figure 3(a) has been appropriately aligned.

The main issue to be addressed during the image enhancement is that the colours within the collected sub-images are frequently not consistent across the image set. In our work histogram equalisation [11, 9, 8] was applied to the satellite image data

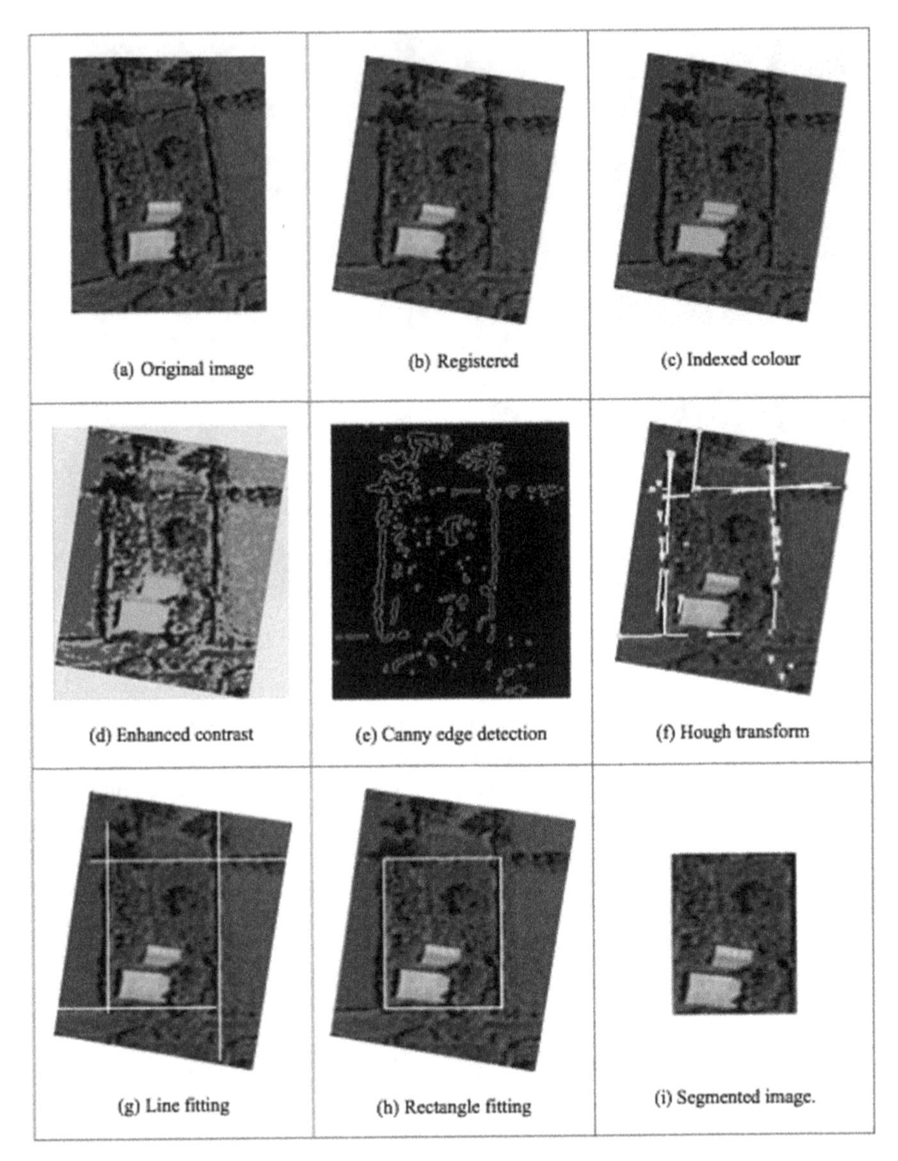


**Fig. 3** Sub-image processing

so as to obtain a consistent colour regime. The first step was to convert each RGB colour represented images into an eight colour indexed image (as shown in Figure 3(c)), and then transform them to grayscale images to which histogram equalisation was then applied. Histogram equalisation is a method for image enhancement directed at ensuring an equal colour distribution (Figure 3(d)). The process commences by selecting a reference image which is then used for normalisation purposes with respect to the remaining images.

## 3.2 Image Segmentation

Once the image enhancement process was complete the next stage was to segment the images so as to isolate individual households. It has already been noted that the households of interest are typically defined by a rough rectangular boundary in which buildings and related objects are located. We wished to segment these images so that these rectangular areas can be clearly isolated, however the boundaries are frequently not well defined in that the edges are not continuous. Thus, for example, region-growing segmentation techniques would be unlikely to perform well. The Canny edge detection algorithm [3] and the Hough Transform [5] were thus applied. Canny edge detection is an edge detection technique for identifying object contours from intensity discontinuities. The Hough transform is a segmentation technique suited for identifying imperfect instances of objects of certain predefined shapes (such as a straight lines). Prior to applying Canny edge detection and the Hough transform, contrast adjustment was applied to the images so that the household boundaries could be more readily distinguished. Canny edge detection was then applied, the result is shown in Figure 3(e) were the detected edges have been highlighted. As a result of applying the Hough transform, we have a collection of "lines" as shown in Figure 3(f). Each line is defined by a start and end point, and a $\rho$ and $\theta$ value (length and direction).

We now need to fit a rectangle to this set of lines. This was achieved by applying a Least squares approach [1] applied to each group of lines approximating to the top, bottom, left and right sides of the rectangle. As results the rectangle surrounding each household was demarcated by a pair of horizontal and a pair of vertical lines (Figure 3(g)). The intersections of the lines can then be found so as to delimit the surrounding rectangle in term of its four corners (Figure 3(h)). The final result is a set of segmented household images as shown in Figure 3(i).

## 3.3 Features Representation

After the households have been segmented each had to be represented using some appropriate mechanisms that serves to: (i) capture the salient features of the object in a concise manner and (ii) permit the ready application of some classification algorithm. As already noted above, a histogram based representation is proposed. Four histograms were generated for each household describing the three RGB colour channels and a grayscale version. Once the histograms had been generated (one per household) two categories of feature could be extracted: histogram features and statistical features. For the histogram features, each of the four histograms was divided into 16 “bins”, giving 64 features in total. Four categories of statistical feature was considered: (i) general features (G), (ii) entropy features (E), (iii) grey-level co-occurrence matrix features (M) [23] and (iv) wavelet transform features (W) [22]. The complete set of statistical features (19 in total) is presented in Table 1, the letter in parenthesis behind each feature type indicates that category of the feature.

**Table 1** Statistical features.

| # | Description | # | Description | # | Description |
|---|---|---|---|---|---|
| 1 | Average red (G) | 9 | Standard deviation (G) | 16 | Average approximation coefficient matrix, $cA$ (W) |
| 2 | Average green (G) | 10 | Entropy (E) | | |
| 3 | Average blue (G) | 11 | Average Local Entropy (E) | 17 | Average horizontal coefficient matrix, $cH$ (W) |
| 4 | Average hue (G) | 12 | Contrast (M) | | |
| 5 | Average saturation (G) | 13 | Correlation (M) | 18 | Average vertical coefficient matrix, $cV$ (W) |
| 6 | Average value (G) | 14 | Energy (M) | | |
| 7 | Average grayscale (G) | 15 | Homogeneity (M) | 19 | Average diagonal coefficient matrix, $cD$ (W) |
| 8 | Mean (G) | | | | |

## 3.4 Feature Selection

The 64 histogram and 19 statistical features identified (as described above) were used to describe each household. Prior to classifier generation, so as to reduce the number of features to be considered to those that best served to distinguish between different classes of household, a feature selection process was applied. Five feature selection strategies were considered: (i) $\chi$-squared, (ii) Gain Ratio, (iii) Information Gain, (iv) One-R and (v) Relief-F. Note that the $\chi$-squared measure evaluates the “worth” of attributes by computing and comparing their $\chi$-squared value. The Gain ratio, as the name suggests, evaluates attributes by measuring gain ratio. Information gain evaluates the attributes by evaluating the entropy of the class, which characterises the purity of an arbitrary. The One-R (One Rule) feature selection strategy comprises a simple classifier that generate a one-level classifier. The Relief-F method is an instance-based technique, which randomises the instances and checks the neighboring instances (records) for the same and different class labels.

# 4 Evaluation

To evaluate the proposed process, as already noted, labeled training data was collected from a rural region 300km to the northwest of Addis Ababa in Ethiopia. More specifically data was collected from the Horro district in the Oromia Region of Ethiopia, an area bounded by the 9.5N and 9.8N parallels of latitude and the

37.0E and 37.5E lines of longitude as shown in Figure 1. Data, including family size, latitude and longitude for each household was collected by University of Liverpool field staff in May 2011. The minimum and maximum family size was 2 and 10 respectively, the mean was 5.97 and the medium 6. Thus, for evaluation purposes the households were divided into two classes: less than 7 people ("Small family") and greater or equal to 7 ("Large family"). The required household satellite image data was extracted from GeoEye satellite images which had a 50 centimetres ground resolution in Google Earth. The high resolution of these images made them well suited with respect to the desired census mining because objects such as buildings and fields could easily be identified. Figure 2 presented a sample satellite image covering part of the Harro district. From the Figure several households can clearly be observed. In total, data for 30 households was processed for evaluation purposes.

An overview of the experimental set up is given by the schematic shown in Figure 4. The data was processed as described in Section 3 above. Data discretisation was then applied to the selected features so that each continuously valued attribute was converted into a set of ranged attributes. Then the classification learning methods were applied. The Waikato Environment for Knowledge Analysis (WEKA) machine learning workbench [21] was used for classifier generation purposes. 10-fold cross-validation was used throughout. The performances of each classifier was recorded in terms of: (i) accuracy, (ii) sensitivity, (iii) specificity, (iv) precision with respect to the small family, (v) precision with respect to the large family, (vi) F-measure and the (vii) ROC area[1].

Extensive evaluation has been conducted with respect to the proposed techniques. This section reports on only most significant results obtained (there is insufficient space to allow for the presentation of all the results obtained). The evaluation presented in this paper were directed at four goals. The first was to determine the effect of classification performance using either histogram features only, statistical features only and a combination of the two (Sub-section 4.1). The second compared the operation of the various suggested feature selection algorithms (Sub-section 4.2). The third was directed at an analysis of the effect that the number of selected attributes, $K$, had on performance (Sub-section 4.3). Finally, the fourth was directed at determining the effect on classification performance as a result or using different learning methods (Sub-section 4.4).

## 4.1 Data Representation (Histogram Attributes v. Statistical attributes)

Three different data sets were generated, to conduct the experiments described in this paper, in order to investigate the effect of different types of data on classification performance. The three data set were as follow: (i) Histogram, comprised

[1] For calculation of the evaluation metrics a confusion matrix with "small family" as the positive class was used.

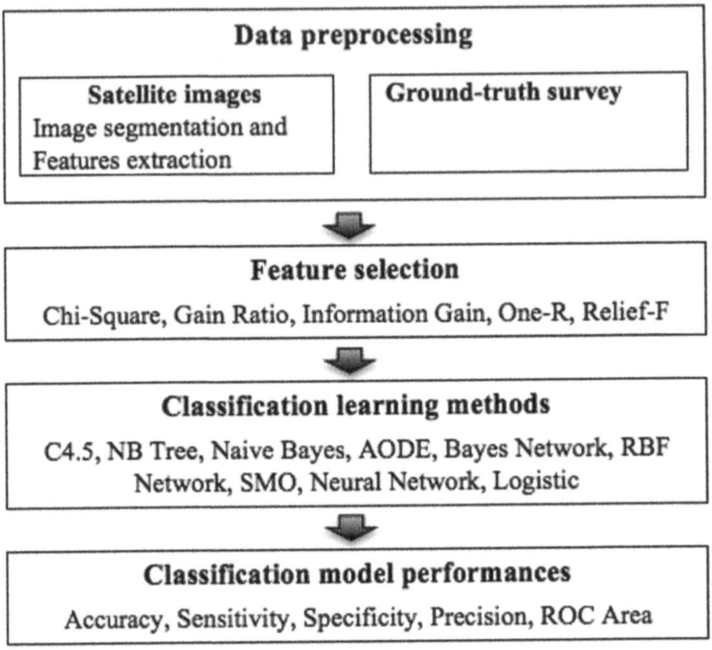

Data preprocessing

Satellite images
Image segmentation and
Features extraction

Ground-truth survey

Feature selection

Chi-Square, Gain Ratio, Information Gain, One-R, Relief-F

Classification learning methods

C4.5, NB Tree, Naive Bayes, AODE, Bayes Network, RBF Network, SMO, Neural Network, Logistic

Classification model performances

Accuracy, Sensitivity, Specificity, Precision, ROC Area

**Fig. 4** Schematic illustration the evaluation set up

of the 64 histogram bin attributes that may be extracted from the four different identified histogram representations; (ii) Statistical, comprising the 19 attributes (including general features, entropy features, grey-level co-occurrence matrix features and wavelet transform measures); and (iii) Combined, which combined the two and hence comprised 83 (64+19) attributes. For the experiments information gain was used as the feature selection method together with $K = 15$ and a Bayesian Network learning method because (as will be seen from the following reported results) it was found these tended to generated the best performance. The value of $K = 15$ was selected because initial experimentation (not report here) indicated that this was the most appropriate. Table 2 shows the classification results produced using the three data sets. Form the table it can be seen that the Combined data produced the best overall classification performance, probably because of the greater quantity of data (number of attributes) used to generate the classifier in this case. It is worth noting that very good results were produced using the Combined data set.

## 4.2 Feature selection

Recall that five kinds of feature selection techniques were considered: (i) $\chi$-squared (ChiSquaredAttributeEval), (ii) Gain ratio (GainRatioAttributeEval), (iii) Informa-

**Table 2** Classification performance using data sets comprising different kinds of feature ($K = 15$, Information gain feature selection and Bayesian Network learning).

| Data Set | Accuracy | Sensitivity | Specificity | Precision "Small" | Precision "Large" | F-Measure | ROC Area |
|---|---|---|---|---|---|---|---|
| Histogram | 0.867 | 0.882 | 0.846 | 0.882 | 0.846 | 0.882 | 0.923 |
| Statistical | 0.600 | 0.706 | 0.462 | 0.632 | 0.545 | 0.667 | 0.688 |
| Combined | 0.933 | 1.000 | 0.846 | 0.895 | 1.000 | 0.944 | 0.959 |

tion gain (InfoGainAttributeEval), (iv) One-R (OneRAttributeEval) and (v) Relief-F (ReliefFAttributeEval). For the experiments used to compare these five methods the Combined data set was used as this had been found to produce the best results as established in the previous sub-section. $K = 15$ was again used together with a Bayesian Network learning method forth the same reasons as before (because they had been found to produce the best results). The results from the experiments are present in Table 3. From the table it can be seen that the $\chi$-square and Information Gain feature selection techniques produced the best results; although, in the context of the ROC area value, information gain outperformed the $\chi$-squared measure, and thus it can be concluded that information gain is the most appropriate measure in the context of the census mining application considered in this paper.

**Table 3** Classification performance using different feature selection techniques (Combined, $K = 15$ and Bayesian Network learning).

| Selection techniques | Accuracy | Sensitivity | Specificity | Precision "Small" | Precision "Large" | F-Measure | ROC Area |
|---|---|---|---|---|---|---|---|
| $\chi$-Square | 0.933 | 1.000 | 0.846 | 0.895 | 1.000 | 0.944 | 0.928 |
| Gain Ratio | 0.900 | 1.000 | 0.769 | 0.850 | 1.000 | 0.919 | 0.950 |
| Information Gain | 0.933 | 1.000 | 0.846 | 0.895 | 1.000 | 0.944 | 0.959 |
| One-R | 0.800 | 0.941 | 0.615 | 0.762 | 0.889 | 0.842 | 0.851 |
| Relief-F | 0.733 | 0.941 | 0.462 | 0.696 | 0.857 | 0.800 | 0.928 |

## *4.3 Numbers of attribute*

In order to investigated the effect on classification performance of the value of $K$ a sequence of experiments was conducted using a rang of values of $K$ from 5 to 25 incrementing in steps of 5. For the experiments the Combined data and information gain feature selection were used, because they had already been shown to produce good performance, together with Bayesian Network learning. Table 4 shows the classification results produced. From the table it can be seen that $K = 10$ and $K = 15$ produced the best performance, although, as in the case of the experiments conducted to find the most appropriate feature selection mechanism, when the ROC area value was considered $K = 15$ produced a slightly better performance than when $K = 10$ was used. Thus it can be concluded that $K = 15$ is the most appropriate measure in the context of census mining, as considered in this paper. It is conjectured that low values of $K$ did not provide a good performance because there was insufficient data to built an effective classifier, while large values of $K$ resulted in overfitting.

**Table 4** The classification performances using different values for $K$ (Combined, Information Gain and Bayesian Network learning).

| Number of attributes | Accuracy | Sensitivity | Specificity | Precision "Small" | Precision "Large" | F-Measure | ROC Area |
|---|---|---|---|---|---|---|---|
| 5 attributes | 0.900 | 1.000 | 0.769 | 0.850 | 1.000 | 0.919 | 0.946 |
| 10 attributes | 0.933 | 1.000 | 0.846 | 0.895 | 1.000 | 0.944 | 0.950 |
| 15 attributes | 0.933 | 1.000 | 0.846 | 0.895 | 1.000 | 0.944 | 0.959 |
| 20 attributes | 0.900 | 0.941 | 0.846 | 0.889 | 0.917 | 0.914 | 0.964 |
| 25 attributes | 0.900 | 0.941 | 0.846 | 0.889 | 0.917 | 0.914 | 0.946 |

### *4.4 Learning methods*

Nine learning methods were considered with respect to the experiments directed at identifying the effect of different learning methods on classification performance: (i) Decision Tree (C4.5, J48 in WEKA), (ii) Naive Bayes Tree (NBTree), (iii) Naive Bayes (NaiveBayes), (iv) Averaged One-Dependence Estimators (AODE), (v) Bayesian Network (BayesNet), (vi) Radial Basis Function Network (RBF Network), (vii) Sequential Minimal Optimisation (SMO), (viii) Neural Network (Multilayerperceptron) and (ix) Logistic Regression (Logistic). In each case the Combined data, $K = 15$ and information gain feature selection was used. The results are presented in Table 5. From the table it can clearly be observed that Bayesian Network learning outperformed all the other classifier generator algorithms considered. The C4.5 decision tree generator produced substantially the worst performance.

## 5 Conclusion

In this paper a framework for remotely collecting census data using satellite imagery and data mining (classification) has been described. The main idea presented in this paper is that classifiers can be built that classify household satellite images to produce census data, provided an appropriate representation is used. The proposed representation is founded on the idea of representing segmented households using a histogram based formalism. The proposed framework was evaluated using differ-

**Table 5** Classification performance using different learning methods (Combined, Information Gain and $K = 15$).

| Learning methods | Accu-racy | Sensi-tivity | Speci-ficity | Precision "Small" | Precision "Large" | F-Measure | ROC Area |
|---|---|---|---|---|---|---|---|
| C4.5 | 0.600 | 0.765 | 0.385 | 0.619 | 0.556 | 0.684 | 0.643 |
| Naive Bayes Tree | 0.900 | 1.000 | 0.769 | 0.850 | 1.000 | 0.919 | 0.946 |
| Naive Bayes | 0.900 | 1.000 | 0.769 | 0.850 | 1.000 | 0.919 | 0.946 |
| AODE | 0.800 | 0.882 | 0.692 | 0.789 | 0.818 | 0.833 | 0.846 |
| Bayesian Network | 0.933 | 1.000 | 0.846 | 0.895 | 1.000 | 0.944 | 0.959 |
| RBF Network | 0.833 | 0.882 | 0.769 | 0.833 | 0.833 | 0.857 | 0.869 |
| SMO | 0.900 | 0.941 | 0.846 | 0.889 | 0.917 | 0.914 | 0.894 |
| Neural Network | 0.900 | 0.941 | 0.846 | 0.889 | 0.917 | 0.914 | 0.964 |
| Logistic Regression | 0.833 | 0.824 | 0.846 | 0.875 | 0.786 | 0.848 | 0.964 |

ent data sets, feature selection mechanisms, numbers of attributes ($K$) and learning methods. The reported results demonstrate that the best performance was produced using: (i) the Combined representation that combined histogram bin data with statistical information derived from the histograms, (ii) information gain as the feature selection method, (iii) $K = 15$ and (iv) Bayesian Network learning. The best results obtained were a sensitivity of 1.000 and a specificity of 0.846; these are extremely good results indicated that accurate census data can be collected using the proposed approach at a significantly reduced overall cost compared to traditional approaches to collecting such data.

## References


1. Otto Bretscher. *Linear Algebra with Applications (3rd Edition)*. Prentice Hall, July 2004.
2. C. R. Brice and C. L. Fennema. Scene Analysis Using Regions. *Artificial Intelligence*, 1(3):205–226, 1970.
3. John Canny. A computational approach to edge detection. *Pattern Analysis and Machine Intelligence, IEEE Transactions on*, PAMI-8(6):679 –698, nov. 1986.
4. Jin-Song DENG, Ke WANG, Jun LI, and Yan-Hua DENG. Urban land use change detection using multisensor satellite images. *Pedosphere*, 19(1):96 – 103, 2009.
5. Richard O. Duda and Peter E. Hart. Use of the hough transformation to detect lines and curves in pictures. *Commun. ACM*, 15(1):11–15, January 1972.
6. J. Faichney and R. Gonzalez. Combined colour and contour representation using anti-aliased histograms. In *Signal Processing, 2002 6th International Conference on*, volume 1, pages 735 – 739 vol.1, aug. 2002.
7. Juan A. X. Fanoe. Lessons from census taking in south africa: Budgeting and accounting experiences. *The African Statistical*, 13(3):82–109, 2011.
8. Rafael C. Gonzalez and Richard E. Woods. *Digital Image Processing (3rd Edition)*. Prentice Hall, 3 edition, August 2007.
9. E. L. Hall. Almost uniform distributions for computer image enhancement. *IEEE Trans. Comput.*, 23(2):207–208, February 1974.

10. Wynne Hsu, S. T. Chua, and H. H. Pung. An integrated color-spatial approach to content-based image retrieval. In *Proceedings of the third ACM international conference on Multimedia*, MULTIMEDIA '95, pages 305–313, New York, NY, USA, 1995. ACM.
11. Jacek Kozak, Christine Estreguil, and Katarzyna Ostapowicz. European forest cover mapping with high resolution satellite data: The carpathians case study. *International Journal of Applied Earth Observation and Geoinformation*, 10(1):44 – 55, 2008.
12. Tzu-Chuen Lu and Chin-Chen Chang. Color image retrieval technique based on color features and image bitmap. *Inf. Process. Manage.*, 43(2):461–472, March 2007.
13. G. Mallinis, I.D. Mitsopoulos, A.P. Dimitrakopoulos, I.Z. Gitas, and M. Karteris. Local-scale fuel-type mapping and fire behavior prediction by employing high-resolution satellite imagery. *Selected Topics in Applied Earth Observations and Remote Sensing, IEEE Journal of*, 1(4):230 –239, dec. 2008.
14. V. Mesev. The use of census data in urban image classification. *PHOTOGRAMMETRIC ENGINEERING AND REMOTE SENSING*, (5):431–438, May 1998.
15. F. O'Gorman and M. B. Clowes. Finding picture edges through collinearity of feature points. *IEEE Trans. Comput.*, 25(4):449–456, April 1976.
16. Nobuyuki Otsu. A Threshold Selection Method from Gray-level Histograms. *IEEE Transactions on Systems, Man and Cybernetics*, 9(1):62–66, 1979.
17. N. Papamarkos and B. Gatos. A new approach for multilevel threshold selection. *CVGIP: Graph. Models Image Process.*, 56(5):357–370, September 1994.
18. T. Pavlidis. *Algorithms for Graphics and Image Processing*. Springer, 1982.
19. K. Roychowdhury, S. Jones, C. Arrowsmith, and K. Reinke. Indian census using satellite images: Can dmsp-ols data be used for small administrative regions? In *Urban Remote Sensing Event (JURSE), 2011 Joint*, pages 153 –156, april 2011.
20. Xiang-Yang Wang, Jun-Feng Wu, and Hong-Ying Yang. Robust image retrieval based on color histogram of local feature regions. *Multimedia Tools Appl.*, 49(2):323–345, August 2010.
21. Ian H. Witten, Eibe Frank, and Mark A. Hall. *Data Mining: Practical Machine Learning Tools and Techniques*. Morgan Kaufmann, Amsterdam, 3. edition, 2011.
22. Yang Zhang, Rongyi He, and Muwei Jian. Comparison of two methods for texture image classification. In *Proceedings of the 2009 Second International Workshop on Computer Science and Engineering - Volume 01*, IWCSE '09, pages 65–68, Washington, DC, USA, 2009. IEEE Computer Society.
23. S. W. Zucker and D. Terzopoulos. Finding structure in co-occurrence matrices for texture analysis. *Computer Graphics and Image Processing*, 12:286–308, 1980.
24. Reyer Zwiggelaar. Texture based segmentation: Automatic selection of co-occurrence matrices. In *Proceedings of the Pattern Recognition, 17th International Conference on (ICPR'04) Volume 1 - Volume 01*, ICPR '04, pages 588–591, Washington, DC, USA, 2004. IEEE Computer Society.

# Challenges in Applying Machine Learning to Media Monitoring

Matti Lyra, Daoud Clarke, Hamish Morgan, Jeremy Reffin, David Weir

**Abstract** The Gorkana Group provides high quality media monitoring services to its clients. This paper describes an ongoing project aimed at increasing the amount of automation in Gorkana Group's workflow through the application of machine learning and language processing technologies. It is important that Gorkana Group's clients should have a very high level of confidence that if an article has been published, that is relevant to one of their briefs, then they will be shown the article. However, delivering this high-quality media monitoring service means that humans are having to read through very large quantities of data, only a small portion of which is typically deemed relevant. The challenge being addressed by the work reported in this paper is how to efficiently achieve such high-quality media monitoring in the face of huge increases in the amount of the data that needs to be monitored. This paper discusses some of the findings that have emerged during the early stages of the project. We show that, while machine learning can be applied successfully to this real world business problem, the distinctive constraints of the task give rise to a number of interesting challenges.

## 1 Introduction

The Gorkana Group provides high quality media monitoring services to its clients. Each client is sent articles that are deemed relevant to some "brief" specified by the client. The sources for potentially relevant articles are typically the latest issues of newspapers, with increasing amounts of content being found from news portals and prominent blogs. This paper describes an ongoing collaboration between Gorkana

Matti Lyra, Hamish Morgan, Jeremy Reffin, David Weir
Department of Informatics, University of Sussex, Brighton, UK. e-mail: {M.Lyra, hamish.morgan, J.P.Reffin, D.J.Weir}@sussex.ac.uk

Daoud Clarke
Gorkana Group, 28–42 Banner Street, London. e-mail: daoud.clarke@gorkana.com

Group and researchers at the University of Sussex's Text Analytics Group with the goal of increasing the amount of automation in the Gorkana Group workflow through the application of machine learning and language processing technologies.

Prior to the implementation of the approach discussed in this paper, determining the relevance of articles was achieved using a workflow summarised in Figure 1. Articles pass through two tiers of filtering. In tier 1 an article is matched against Boolean keyword queries (one per brief) to identify articles that are potentially relevant to each brief. Any article matching at least one brief is passed to a second filtering tier where human readers judge the article's relevance against the brief. Articles deemed to be relevant are sent for further processing and will eventually be dispatched to the client(s).

Gorkana Group provides a very high quality media monitoring service to its clients. In particular, the intention is that their clients should have a very high level of confidence that if an article has been published that is relevant to one of their briefs then they will be shown the article. In order to achieve this, the first tier of filtering is required to achieve very high recall of relevant articles. The problem is that this is achieved at the cost of precision (of the tier 1 filter) — a substantial proportion of the articles marked by the tier 1 filter as being of potential relevance to a brief turn out (when checked by humans) not to be relevant. In other words, the price that is being paid for providing such high-quality media monitoring (with both high precision and high recall) is that humans are having to read through very large quantities of data, only a small portion of which is typically deemed relevant. In extreme cases a relevant article is found only after reading thousands of irrelevant ones (see Table 1). The challenge being addressed by the work reported in this paper is how to efficiently achieve such high-quality media monitoring in the face of huge increases in the amount of the data that needs to be monitored.

This paper discusses some of the findings that have emerged during the early stages of this ongoing collaboration. We show that, while machine learning can be applied successfully to this real world business problem, the distinctive constraints of the task give rise to a number of interesting challenges.

The basis of our approach involves the use of supervised machine learning. As we will see, this turns out to be particularly well-suited to Gorkana Group's scenario described above. In order to improve the efficiency of its workflow, the Gorkana Group has focussed considerable attention on streamlining the manual processing (the tier 2 filter) by building bespoke software designed to facilitate the filtering process. Not only does this provide an efficient way of manually filtering candidate articles for relevance, but, crucially, doubles as a way of efficiently creating labelled training data for machine learning classifiers.

## 2 Outline of Approach

The new workflow is shown in Figure 2. Both tiers of filtering in the original workflow are retained with an additional filtering tier inserted between them (the new tier

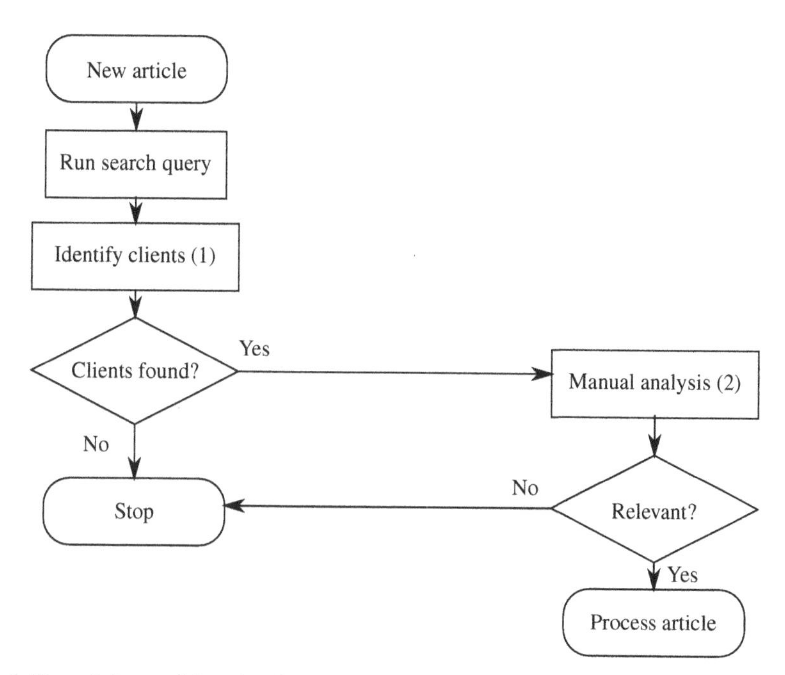


**Fig. 1** The existing workflow that the proposed system would need to fit into. Articles are filtered through two tiers of analysis, indicated by the numbers in brackets.

2). This middle tier comprises a collection of trained classifiers: one for each brief. Once trained, each classifier automatically labels articles as either relevant or not relevant to the brief. Any article marked at the first tier as being potentially relevant to a brief is sent to the tier two classifier for that brief. Only those articles labelled by the classifier as relevant are passed on to the third tier where a final check for relevance is performed by a human reader.

When a new brief is created for a client, deployment of a classifier for that brief within tier 2 is achieved in two phases. In the initial phase the system will undertake processing for that brief according to the old workflow (Figure 1). However, during this initial phase, the relevance judgements made by the human readers are collected for use as training data for the classifier.

As training data is accumulated, the accuracy of a classifier trained on that data can be assessed off-line. The initial phase continues until the performance of the classifier exceeds a pre-specified level of performance, at which point the classifier is enabled and can then be deployed within the second tier of the workflow. This deployment model allows machine learning to be seamlessly integrated into the existing work flow.

We present the results of various empirical studies we have undertaken during the implementation of the approach outlined in Figure 2. Section 3 describes the supervised machine learning framework that has been used during our investigations, and Section 4 presents some of the performance statistics from the data sets that we have been working with.

In the remaining sections of the paper we will focus on three particular issues that arose during the process of deploying this new workflow.

1. As explained above, it is crucial that a high level of recall of relevant documents is achieved. In Section 6 we describe the results of an initial assessment of a method for automatically tuning a hyper-parameter of the classifier in order to satisfy the high relevant recall constraint.
2. Training a classifier for a brief requires the construction of a training set — a collection of hand annotated articles labelled by the tier 1 filter as being potentially relevant for that brief. Section 5 presents an investigation into the impact of using different quantities of training data on the performance of the resulting classifier.
3. After a classifier has been deployed for a brief, we need to be sure that its performance does not degrade over time. Section 7 presents some preliminary evidence that suggests that such degradation will not be a major problem.

## 3 Classification Framework

Many of the data sets we are working with are small, especially with respect to the number of relevant articles. To maximise the savings to the business, in terms of reading reduction, it is important that classifiers can be trained as soon as possible and therefore on as little data as possible. We opted to use a Naïve Bayes classifier, because it has been shown to be an effective approach when using limited quantities of training data, previous work on the same data sets having shown it to yield good results and perform reliably [3]. Naïve Bayes is also simple to implement and the training and classification algorithms have linear time complexity.

Based on the equation shown in 1a, a trained classifier assigns each document a real valued score — the log odds ratio. As shown in equation 1b, this score is used to make a binary classification of the document based on that score being above or below a set threshold $t$. Typically $t$ is set to 0. The relevant article class is taken to be class 1, and the irrelevant article class is class 0.

$$score(\mathbf{x}) = \log \frac{p(C=1)}{p(C=0)} + \sum_{i:x_i=1} \log \frac{p(X_i=1\,|\,C=1)}{p(X_i=1|C=0)} \tag{1a}$$

$$class(\mathbf{x}) = \begin{cases} 1 & \text{if } score(\mathbf{x}) > t \\ 0 & \text{otherwise} \end{cases} \tag{1b}$$

For the purposes of classification, documents are represented as a Boolean feature vector $\mathbf{x} \in \{0,1\}^k$, where $k$ is the number of "significant" features, as determined during training (see below). A document is represented by the vector $(x_1,\ldots,x_k)$ where $x_i$ is 1 if the $i$th feature is in the document, and 0 otherwise.

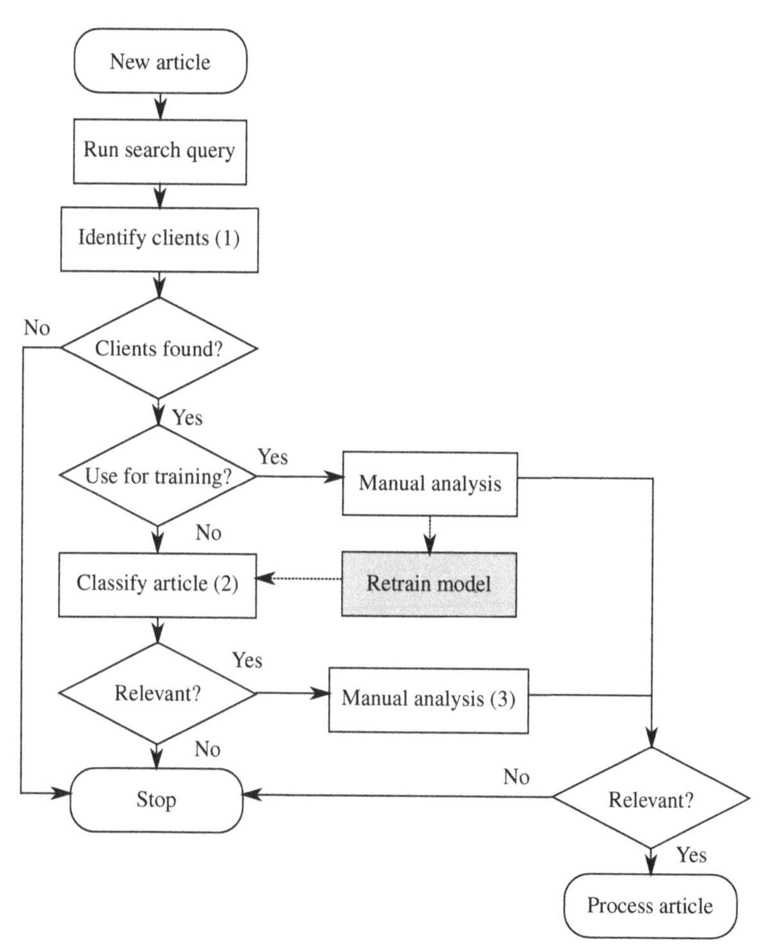


**Fig. 2** The proposed integration of the machine-learning based classification system to the existing workflow. The numbers in brackets indicate the three tiers of filtering applied to articles. Retraining of the model (dashed lines) happens in parallel with classification when new data is available; the retrained model is then used to classify articles arriving subsequently.

The first term in Equation 1a is the log ratio of the prior probabilities of the two classes, and the second term is an estimate of

$$\log \frac{p(\mathbf{X}=\mathbf{x}|C=1)}{p(\mathbf{X}=\mathbf{x}|C=0)}$$

assuming independence of feature occurrences in documents.

Only those features occurring in the training data are considered during scoring; new features occurring in the decoding phase are ignored. All probabilities are de-

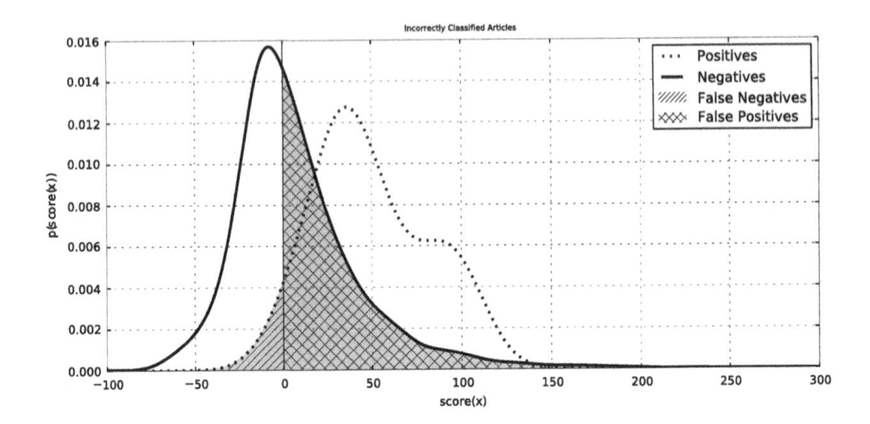


**Fig. 3** The score distributions of the relevant and irrelevant classes for one particular brief. The solid and broken lines show proportion of articles in each class that have been assigned a particular score. The two shaded areas show the two types of errors the classifier is making, relevant articles predicted as irrelevant (False Negatives) and irrelevant articles predicted as relevant (False Positives).

rived from maximum likelihood estimates, except that zero probabilities are avoided by using Laplace smoothing[1].

Not all features seen in the training data are equally useful in making classifications. There is considerable literature on addressing the issue of feature selection; showing that selecting a subset of the entire feature space to use for classification increases predictive performance [1, 4, 6, 9, 10, 11]. After comparing several alternatives we opted to use the Mutual Information (Equation 2) measure as the basis of feature selection. Early experiments with different feature selection methods showed this to perform the best. In the experiments reported below, unless stated otherwise, we used the top 8000 features as ranked by Mutual Information.

$$I(X;C) = \sum_{x\in\{1,0\}} \sum_{c\in\{1,0\}} p(X=x, C=c) \log \frac{p(X=x, C=c)}{p(X=x)p(C=c)} \tag{2}$$

Mutual Information, like other biased feature selection methods, tends to select features from the minority class [11], which in this case is the relevant article class.

The classification threshold $t$ in Equation 1b can be changed to adjust the predictions made by the classifier. Consider Figure 3 showing the score distributions of the two classes. If the classification threshold is at zero then most of the negative articles are classified as relevant because their score is above the threshold. Adjustments to the classification threshold changes the ratio of *false negatives* to *false positives*, and allows us to trade off precision and recall.

[1] Smoothing redistributes the total probability mass over all features so that some mass is deducted from the seen features and assigned to unseen features, thus avoiding zero probabilities [2, 5, 13].

We have adopted a slightly different method for adjusting the tradeoff that is sensitive to the number of non-zero features in the article. We replace Equations 1a and 1b by Equations 3a and 3b, where the hyperparameter $\alpha \in \mathbb{R}_{[0,1]}$ is used to tradeoff precision and recall. An $\alpha$ value of 0.5 is equivalent to using Equations 1a and 1b with threshold $t = 0$. However, since the impact on the score of changing $\alpha$ is sensitive to the number of non-zero features (something that is not the case when making changes to the threshold $t$) the two precision/recall trade-off methods are not equivalent. The intuition behind this approach is that it allows us to dampen the impact of features, increasing the relative impact of the class prior. It also provides a bounded hyperparameter that is more convenient to optimise than the decision threshold.

We conducted statistical significance tests comparing the performance of the two models (1a and 3a) and found that on low amounts of training data (ten or less positives), the model described in 3a is significantly better (1-tailed z-test; $p < 0.05$), yielding a larger amount of correctly rejected irrelevant articles, at an oracle 90% recall level (see below).

$$score(\mathbf{x}) = \log \frac{p(C=1)}{p(C=0)} + \sum_{i:x_i=1} \left[ \log \frac{p(X_i=1|C=1)}{p(X_i=1|C=0)} + \log \frac{\alpha}{(1-\alpha)} \right] \tag{3a}$$

$$class(\mathbf{x}) = \begin{cases} 1 & \text{if } score(\mathbf{x}) > 0 \\ 0 & \text{otherwise} \end{cases} \tag{3b}$$

A final point about the classification framework is related to the way evaluation of the results is performed. One standard measure of performance is **accuracy**; the proportion of articles which are predicted correctly. This measure, however, is inappropriate for our scenario because we are dealing with highly unbalanced data. When deployed in the new workflow, the tier 2 filter being implemented with these classifiers will be processing a far greater number of irrelevant articles than relevant articles. A brief may have as little as 1% relevant articles, in which case a classifier can achieve 99% accuracy by classifying all articles as irrelevant, missing all the relevant articles. One solution to this is to look separately at the accuracy of the classifier on each class, and combine these to get a single measure of accuracy, for example, by using the geometric mean [7]. This approach was applied to building classifiers for sentiment analysis in the presence of highly imbalanced data [3]. However, this metric assumes equal misclassification costs, which is not appropriate in this scenario where the cost of misclassifying a relevant article as irrelevant is far higher than the cost of misclassifying an irrelevant article as relevant. In the empirical investigations reported below, we report **specificity** at a fixed relevant article recall level (e.g. 90%). We will refer to this limit on recall of relevant articles as the **relevant recall constraint**. Specificity is the proportion of irrelevant articles that are correctly identified as irrelevant by the classifier. This gives an indication of the reading reduction we can achieve at the third (manual) processing tier in the workflow, under some specified relevant recall constraint.

## 4 Datasets

The datasets used in the experiments were extracted from the Gorkana Group's system over a period of 11 months. The data consists of online content that is more "mainstream" than typical social media, for example high profile blogs, and articles associated with print newspapers.

The datasets comprise 10 different briefs that vary in the amount of data and class bias. The longest running briefs have data over the entire 11 months, and the shortest ones over a period of 6 months. Table 1 summarises the datasets with duplicate articles and stop words removed.

| Brief | $\#C_1$ | $\#C_0$ | Total | Bias | Avg. length | $\frac{max}{min}$ | ~ Positives per Week |
|---|---|---|---|---|---|---|---|
| A | 938 | 43227 | 44165 | 0.022 | 240 | 218 / 0 | |
| B | 227 | 13633 | 13860 | 0.017 | 242 | 27 / 0 | |
| C | 1115 | 26318 | 27433 | 0.042 | 278 | 64 / 0 | |
| D | 524 | 22471 | 22995 | 0.023 | 259 | 88 / 0 | |
| E | 1995 | 44271 | 46266 | 0.045 | 212 | 171 / 0 | |
| F | 1388 | 54156 | 55544 | 0.026 | 223 | 201 / 0 | |
| G | 289 | 1301 | 1590 | 0.222 | 241 | 53 / 0 | |
| H | 542 | 101 | 643 | 5.366 | 288 | 56 / 0 | |
| I | 1734 | 414 | 2148 | 4.188 | 226 | 115 / 0 | |
| J | 226 | 18821 | 19047 | 0.012 | 208 | 49 / 0 | |

**Table 1** Outline of the datasets. The $\#C_0$ and $\#C_1$ and Total columns show the number of articles in each class and the total number of articles. The Bias column shows the ratio of the two classes $\frac{\#C_1}{\#C_0}$. The Avg. Length shows the mean length of tokenised articles. The sparklines [12] in the last column show the concentration of positives in each dataset per week. The second to last column shows the maximum and minimum values displayed in the sparklines. Note that the sparklines for each brief are scaled independently.

All the data sets were pre-processed by tokenising the articles and removing stop words using the Weka machine learning toolkit[2]. Each tokenised article contains all the token types (features) found from that article, i.e. no feature selection was performed during the pre-processing stage. Feature selection was performed by the classifier during training.

[2] `http://www.cs.waikato.ac.nz/ml/weka/`

## 5 Establishing How Much Training Data Is Required

As discussed previously, it is important to our client that the classifier is trained on as little data as possible. In this section we look at the impact on the quality of classification of varying the amount of data being used to train the classifier. As already mentioned, we require the classifier to produce a pre-defined level of recall of positive articles (relevant recall constraint), achieved by tuning the hyperparameter $\alpha$ during training. In order to eliminate any variation due to inaccurate $\alpha$ estimates for this experiment, we adopt an approach where the value of $\alpha$ is set at a value that meets the relevant recall constraint according to an oracle with access to the data. Results are reported for a relevant recall constraint of 90%. Exact details of how we train the classifier are deferred to section 6, where we look in detail at the problem of tuning the $\alpha$ hyperparameter.

Figure 4 plots specificity at 90% relevant recall, averaged across all 10 briefs, against size of the training data, defined by the number of relevant articles in the data. Two lines are shown. The "macro avg." line weights each brief equally. The "micro avg." line weights each article equally. Both plots show the same trends; as the number of relevant articles used in training increases, performance of the classifier improves. However, this improvement plateaus for more than 40 relevant articles, which is consistent with the result reported in Section 6.

Overall, for the classifier trained on the most data, the results show that at 90% relevant recall the number of irrelevant articles that have to be read is reduced by 30%. At 80% recall this rises to 45%.

The benefit gained from more training data tails off between 10 and 30 relevant articles. The specificity curves in Figure 4 are clearly levelling off after 50 positive articles. This is supported by statistical analysis on the 90% positive recall data; the magnitude of deviation in the specificity scores between different training data amounts are only significant between 10 and 30 relevant articles (1-tailed z-test; $p < 0.05$), and borderline significance for the 30–50 pair (1-tailed z-test;$p = 0.05$).

## 6 Tuning the Hyperparameter

In this section we address the issue of how to determine, given that some quantity of training data is available, a value for the hyperparameter $\alpha$ that will satisfy a desired relevant recall constraint (e.g. 90% recall of relevant articles). Ideally, we want to set $\alpha$ in a way that comes as close as possible to satisfying the recall constraint when applied to unseen test data given the limitations of the training data currently available. Reliable estimation of $\alpha$ is crucial given that the performance of the classifier on the unseen data is highly sensitive to values of $\alpha$, and the importance to the company of satisfying predetermined recall constraints.

We now describe an empirical investigation into how we estimate effective values for $\alpha$ in terms of how well a relevant recall constraint can be satisfied. In a traditional empirical study involving a machine learning framework, a data set would

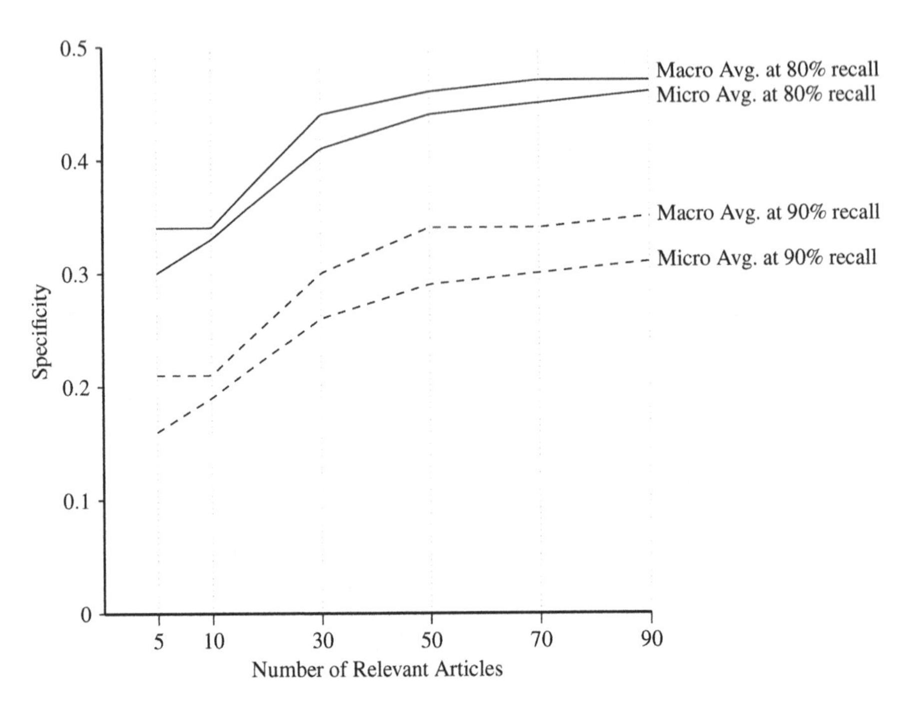


**Fig. 4** Plot showing the micro and macro average of specificity, across all briefs, plateauing as the training size increases. Dataset size is controlled by number of positive articles. Recall of relevant articles is fixed at either 90% or 80% by using an appropriate oracle $\alpha$ value.

be divided into training, hyperparameter optimisation, and evaluation partitions, according to one of the commonly used schemes. For example, one might choose a 60/20/20 split of the whole data set. In the scenario under consideration in this paper, however, we have additional constraints arising from the fact that we are designing, not just an empirical experiment, but a real world implementation, that must meet the client's business objectives. In particular, this means that we need to build classifiers that satisfy the relevant recall constraint with as little training data as possible — more data used to train the classifier means less saving in terms of data annotation effort.

In designing our experiment, we are faced with a scenario where, when a classifier is deployed, it will operate on a stream of data ordered in time. The current point in time divides this stream in two: a "seen" part and an "unseen" part. To simulate this in our experiments, the articles for each brief are first arranged in temporal order. These data are then split in two by selecting a point in time that corresponds to the notional "current time". In our experiments the point where this split is made is set as the earliest point in time where 200 relevant articles have been seen in the data stream.

To calculate what we refer to as the estimated $\alpha$ value we take the average of the values for $\alpha$ produced with 10 repetitions of the following procedure. We take the "seen" portion of the dataset and resample (without replacement) 10 times until a sample with 80 relevant articles is created. For each of these samples we perform $5 \times 2$ cross-validation [8]. This gives a total of 10 repetitions, where for each repetition, we determine the relevant recall scores of a trained classifier across a range of possible $\alpha \in \{0, 0.1, \ldots, 1.0\}$. For each $\alpha$ value we compute the average relevant recall value across all 10 repetitions. We then determine a value of $\alpha$ by calculating a linear least-squares fit and taking the alpha value that gives the required relevant recall constraint (e.g. 90%).

In order to evaluate the accuracy of this $\alpha$ estimate, we determine an optimal "oracle" $\alpha$ value. This is the $\alpha$ value which actually satisfies the required relevant recall constraint on the "unseen" data. To calculate the oracle $\alpha$ value we take the average of the values for $\alpha$ produced with 10 repetitions of the following procedure. We build a classifier using a random sample of the "seen" data ("training sample"). In this and subsequent experiments, the size of the sample is measured in terms of the number of relevant articles that are accumulated during the sampling. The relevant recall scores of this classifier is then evaluated across a range of possible $\alpha \in \{0, 0.1, \ldots, 1.0\}$ against *all* of the "unseen" data. We determine a value of $\alpha$ by calculating a linear least-squares fit and taking the alpha value that gives the required relevant recall constraint (e.g. 90%).

We are now in a position to compare our $\alpha$ estimates with their corresponding oracle $\alpha$ values. Here we undertake this in 4 experimental conditions in which we examine the effect of using different quantities of data to build the classifier on $\alpha$ accuracy by manipulating (1) the number of features used to train the classifier (2000 versus 8000 versus 14000 features); and (2) the number of relevant examples in the "training sample" data used when training the classifier (40 versus 80 relevant articles).

Table 2 summarises the relationship between our $\alpha$ estimates and the oracle values under these experimental conditions. For each experimental condition, the table shows the average $\alpha$ estimate and oracle value across all 10 briefs, with the contribution from each brief weighted equally. We also show the average mean deviation of the estimate from the oracle, with the contribution from each brief again weighted equally.

The results suggest that our method of $\alpha$ estimation generally under-estimates the oracle value, an impression supported by statistical analysis. The $\alpha$ estimations are significantly lower than the oracle values in all cases (1-tailed z-test on signed differences; $p < 0.05$ in all cases).

A higher oracle value is consistent with a classifier performing less accurately against the "unseen" data than against the "seen" data used for $\alpha$ estimation. One possible explanation is that the language is shifting — what counts as "relevant" changes over time (content drift); in Section 7 we investigate further the classifier's accuracy over time since training. Another possibility for error in estimating $\alpha$ is the temporal overlap between data used to train the classifier and data used to optimise $\alpha$. This relies on an assumption that the data carries with it a set of time dependent features and a set of time independent features. The time dependent features are subject to content drift, and too much temporal overlap between the two data sets leads to overfitting the $\alpha$ estimates. Some evidence towards this hypothesis is provided by initial experiments where the "seen data" portion was set to 100 instead of 200 positive articles. Comparing the unsigned difference between the oracle and estimated $\alpha$ values shows a reduction when using 200 instead of 100 positive articles.

Improving the accuracy of the classifier appears to reduce the discrepancy. Table 2 suggests that the number of features used has an impact on the magnitude of this discrepancy, whereas the number of relevant articles does not (over the range used here). A two-way Analysis of Variance of these data confirms that increasing the number of classifier features from 2000 to 8000 narrows the deviation between $\alpha$ estimates and the oracle values (F-statistic $= 7.4$, $p < 0.01$), whilst increasing the number of relevant articles from 40 to 80 has no significant effect (F-statistic $= 0.01$, n.s.), and there is no evidence of any interaction between these factors (F-statistic $= 0.19$, n.s.). Increasing the number of features from 8000 to 14000 however does not cause a significant decrease in the unsigned deviation between the oracle $\alpha$ and the estimated $\alpha$ (Tukey HSD test; $p = 0.99$) compared to (Tukey HSD test; $p < 0.01$) for the 2000 to 8000 case.

Building as accurate a classifier as possible is therefore doubly beneficial. Not only does it improve discrimination, but also it enables the desired discrimination threshold hyperparameter to be determined more accurately. Having an accurate estimate for the classification threshold hyperparameter is very important as the parameter is very sensitive. Table 2 shows the macro average recall across all 10 briefs achieved with the estimated $\alpha$. It shows that very small differences in the $\alpha$ parameter can have a big impact on the positive recall.

The lack of performance improvement between 40 and 80 positive articles suggests that classifier performance may have plateaued at this volume of training data, a result consistent with the results reported in Section 5. Recall that this data set is highly biased (there are on average 27 irrelevant articles for every relevant article), so 40 positive articles is typically equivalent to around 1000 training articles.

## 7 Maintaining the Classifier's Performance

Our final experiment addresses the question of the expected lifespan of trained classifiers, and looks at whether there is evidence for content drift having a negative impact on classifier performance.

The "unseen" data extends over a 9 month period from months 3 to 11 of the data set. This "unseen" data is partitioned into two evaluation sets: the data from months 3–5 and 6–8. The last three months are not usable as only 2 briefs currently extend that far. For each brief, a classifier using 8000 features is trained on a sample of the "seen" data set that contains 50 relevant articles. For each of these classifiers, we use separate, oracle $\alpha$ values for each of the 2 month periods. In particular, we choose a value for $\alpha$ that satisfies a 90% relevant recall constraint on the data for that 3 month period.

| Number of Relevant Articles | Number of Features | Oracle | Estimated | |
|---|---|---|---|---|
| | | $\alpha$ | $\alpha$ | Average of Recalls |
| 40 | 2000 | 0.56 | 0.46 | 75% |
| 40 | 8000 | 0.51 | 0.48 | 79% |
| 40 | 14000 | 0.51 | 0.48 | 78% |
| 80 | 2000 | 0.60 | 0.49 | 71% |
| 80 | 8000 | 0.52 | 0.50 | 80% |
| 80 | 14000 | 0.51 | 0.50 | 79% |

**Table 2** Average Estimates for different classifier settings across 10 briefs. The table shows the oracle $\alpha$ value for each settings pair compared against the estimated $\alpha$ value and their signed macro average difference. The last column shows the macro average recall achieved with the estimated $\alpha$ value.

Table 3 shows the macro and micro average specificity of the classifiers for the two periods and shows no indication of a reduction in performance over time. A 1-tailed z-test on the deviation between the average specificity levels across the bins verifies that there is no significant drop in the performance of the classifier (1-tailed z-test;$p = 0.49$).

| | **Months** | |
|---|---|---|
| | 3–5 | 6–8 |
| Micro | 0.27 | 0.31 |
| Macro | 0.26 | 0.26 |

**Table 3** Mean specificity values across all briefs over time. Because the briefs have varying amounts of data the latter bins have fewer briefs. First two months of data were reserved as training data.

## 8 Conclusion

We have introduced a practical method for increasing the automation in high precision, high recall media monitoring. We find that robust classification models can be developed with as few as 30 relevant articles using the Naïve Bayes framework, and that the performance of these classifiers is stable over many months of time ordered data. Perhaps the greatest challenge is to identify an appropriate evaluation threshold for the classifier that consistently meets pre-defined positive recall constraints and thus guarantees good performance on future data. We find that increasing the size of the classifier's feature space (8000 vs. 2000) yields a significant improvement in identifying the threshold.

Our work is ongoing and presents some questions that have not yet been fully answered. In applying machine learning practitioners face many problems not present in traditional machine learning research. Many of these constraints are set by business requirements and cannot be avoided. We continue to develop the methods described in this paper, in particular looking at methods of setting parameters to deliver performance within tight pre-determined constraints and exploring further the temporal effects prevalent in the data.

## References

1. A.L. Blum and P. Langley. Selection of relevant features and examples in machine learning. *Artificial intelligence*, 97(1-2):245–271, 1997.
2. Stanley F. Chen and Joshua Goodman. An empirical study of smoothing techniques for language modeling. *Computer Speech & Language*, 13(4):359 – 393, 1999.
3. Daoud Clarke, Peter Lane, and Paul Hender. Developing robust models for favourability analysis. In *Proceedings of the 2nd Workshop on Computational Approaches to Subjectivity and Sentiment Analysis (WASSA 2.011)*, pages 44–52, Portland, Oregon, June 2011. Association for Computational Linguistics.
4. G. Forman. An extensive empirical study of feature selection metrics for text classification. *The Journal of Machine Learning Research*, 3:1289–1305, 2003.
5. William A Gale and Kenneth W Church. What's Wrong with Adding One? In Nelleke Oostdijk and Peter de Haan, editors, *Corpus Based Research in Language: In Honour of Jan Aarts*, pages 189–200. Rodopi, Amsterdam, 1994.
6. P.D. Green, P. C. R. Lane, A.W. Rainer, and S. Scholz. Selecting measures in origin analysis. In *Proceedings of the Thirtieth SGAI International Conference on Artificial Intelligence*, 2010.
7. M. Kubat, R.C. Holte, and S. Matwin. Machine learning for the detection of oil spills in satellite radar images. *Machine learning*, 30(2):195–215, 1998.
8. Christopher D. Manning, Prabhakar Raghavan, and Hinrich Schtze. *Introduction to Information Retrieval*. Cambridge University Press, New York, NY, USA, 2008.
9. D. Mladenić. Feature subset selection in text-learning. *Machine Learning: ECML-98*, pages 95–100, 1998.
10. M. Rogati and Y. Yang. High-performing feature selection for text classification. In *Proceedings of the eleventh international conference on Information and knowledge management*, pages 659–661. ACM, 2002.
11. Lei Tang and Huan Liu. Bias analysis in text classification for highly skewed data. In *Proceedings of the Fifth IEEE International Conference on Data Mining*, ICDM '05, pages 781–784, Washington, DC, USA, 2005. IEEE Computer Society.
12. Edward Tufte. Sparkline theory and practice. `http://www.edwardtufte.com/bboard/q-and-a-fetch-msg?msg_id=0001OR&topic_id=1`, May 2004.
13. Chengxiang Zhai and John Lafferty. A study of smoothing methods for language models applied to information retrieval. *ACM Trans. Inf. Syst.*, 22(2):179–214, April 2004.

# SHORT PAPERS

# Comparative Study of One-Class Classifiers for Item-based Filtering

Aristomenis S. Lampropoulos and George A. Tsihrintzis

**Abstract** In this paper, we address the recommendation process as a one-class classification problem. One-class classification is an umbrella term that covers a specific subset of learning problems that try to induce a general function that can discriminate between two classes of interest, given the constraint that training patterns are available only from one class. Usually, users provide ratings only for items that they are interested in and belong to their preferences without to give information for items that they dislike. The problem in one-class classification is to make a description of a target set of items and to detect which items are similar to this training set. We conduct a comparative study of one-class classifiers from density, boundary and reconstruction methods. The experimental results show that one-class classifiers do not only cope with the problem of missing of negative examples but also, succeed to perform efficiently in the recommendation process.

## 1 Introduction

It is well known that users hardly provide explicit feedbacks in recommender systems. More specifically, users tend to provide ratings only for items that they are interested in and belong to their preferences and avoid, to provide feedback in the form of negative examples, i.e. items that they dislike or they are not interested in. As stated in [10], "It has been known for long time in human computer interaction that users are extremely reluctant to perform actions that are not directed towards their immediate goal if they do not receive immediate benefits". However, common recommender systems based on machine learning approaches use classifiers that, in order to learn user interests, require both positive (desired items that users prefer) and negative examples (items that users dislike or are not interested

Aristomenis S. Lampropoulos, George A. Tsihrintzis
University of Piraeus, Department of Informatics, 80 Karaoli & Dimitriou St., 18534 Piraeus, Greece, e-mail: {arislamp,geoatsi}@unipi.gr

in). Additionally, the effort for collecting negative examples is arduous as these examples should uniformly represent the entire set of items, excluding the class of positive items. Manually collecting negative samples could be biased and require additional effort by users. Moreover, especially in web applications, users consider it very difficult to provide personal data and rather avoid to be related with internet sites due to lack of faith in the privacy of modern web sites [10]. Therefore, recommender systems based on demographic data or stereotypes that resulted from such data are very limited since there is a high probability that the user-supplied information suffers from noise induced by the fact that users usually give fake information in many of these applications. Thus, machine learning methods need to be used in recommender systems, that utilize only positive examples provided by users without additional information either in the form of negative examples or in the form of personal information for them.

Collaborative filtering methods assume availability of a range of high and low ratings or multiple classes in the data matrix of Users-Items. One-class collaborative filtering proposed in [9, 8]provides weighting and sampling schemes to handle one-class settings with unconstrained factorizations based on the squared loss. Essentially, the idea is to treat all non-positive user-item pairs as negative examples, but appropriately control their contribution in the objective function via either uniform, user-specific or item-specific weights.

One-class classification is an umbrella term that covers a specific subset of learning problems that try to induce a general function that can discriminate between two classes of interest, given the constraint that training patterns are available only from one class [7, 4, 6, 3]machine learning models based on defining a boundary between the two classes are not applicable. Therefore, a natural choice in order to overcome this problem is building a model that either provides a statistical description for the class of the available patterns or a description concerning the shape / structure of the class that generated the training samples. Otherwise stated, our primary concern is to derive an inductive bias which will form the basis for the classification of new incoming patterns. In the context of building a recommender system, available training patterns correspond to those instances that a particular user identified and assigned to the class of preferable patterns. The recommendation of new items is then performed by utilizing the one-class classifier for assigning the rest of the items either in the class of desirable patterns or in the complementary class.

Complementary class consisted of negative examples and unlabeled positive examples which are mixed together and we are not able to distinguish them. A common solution is to consider all items of complementary class as negative examples. From empirical experiments it was found that this solution works quiet well. The drawback of this approach is that biases recommendation results because a part of complementary data might be positive. A second approach is to handle all data of complementary class as unknown. This solution utilizes only positive examples and ignores all the missing results. In this case, collaborative filtering algorithms are fed only with positive examples and all predictions on missing values are positive examples [9]Other approaches use weighted low rank approximation and assign different weights to the error terms of positive examples (AMAWN) or to use some sampling strategies and to treat some missing values as negative examples (SMAN). These two different approaches try to address one-class collaborative filtering problems and to to balance between the first two strategies of all missing items as negative (AMAN) and all missing as unknown (AMAU) [9].

The paper is organized as follows: in Section 2, we evaluate one-class classifiers for recommendation and present experimental results. Finally, in Section 3, we draw conclusions and point to future related work.

## 2 Experiments

We conducted experiments on the MovieLens dataset publically available at: www.grouplens.org. The data consists of 100,000 ratings on an integer scale from 1 (bad) to 5 (excellent) given to 1642 movies by 943 users. The sparsity of the data set is high, at a value of 93.7%. Users should have stated their opinions for at least 20 movies in order to be included.

The data formulated in the form of a sparse matrix, where rows correspond to users, columns correspond to items and the matrix entries are ratings. Most cells of the matrix are empty, because every user typically rates only a very small subset of all possible items.For our experiments, we considered all ratings with values 1, 2, as negative and ratings 3, 4 and 5 as positive. For all ratings with value 0 which correspond to unseen data we used the strategy AMAN to consider these as negative. We are interested in constructing personalized models for each user, therefore we built one-class classifier for every user.

Feature vectors constructed by ratings of users to a movie. More specifically, for each item (movie) we used a corresponding feature vector coming from the ratings assigned to this by other users. In other words each movie was represented by feature vector of 942 features equal to the number of the rest of 942 users of our dataset, except from the ratings of active user.

In the first step we applied the term frequency/inverse document frequency (tf-idf) measure to transform rating values into weights. Assume that N is the total number of items (herein movies) that can be recommended to users and that user $k_j$ rated $n$ of them. Moreover, assume that $f_{ij}$ is the number of rating value of the user $k_i$ appears in item $d_j$. Then, the tf-idf weight of rating value of the user $k_i$ in item $d_j$, is defined as: $w_{ij} = \log(1+f_{ij}) \cdot \log(\frac{N}{n})$ and the feature vector of an item is transformed into $Vector(d_j) = (w_{1j},...,w_{942j})$.

In the second step we applied dimensionality reduction through Singular Value Decomposition (SVD) method [1]. Singular Value Decomposition is based on a well known matrix factorization method which takes an $m \times n$ matrix $A$, (herein $m$ is the number of users ($m = 942$), $n$ is the number of items ($n = 1682$)), with rank $r$, and decomposes it as follows: $A = U \cdot S \cdot V^T$. Matrix $U$ is an $m \times m$ orthogonal matrix and the first r columns are eigenvectors of $AA^T$ and represents users in a space of lower dimension. While $V$ is an $n \times n$ orthogonal matrix and the first r columns are eigenvectors of $AA^T$ and represents items in a space of lower dimension. Also, $S$ is an $m \times n$ diagonal matrix and the diagonal r entries are non-negative and have the property that $s_1 \geq s_2, ..., s_r$. By retaining the first $k \ll r$, in other words the k largest singular values, and discarding the rests we produce the best low-rank approximation of the original matrix $A$ which is the matrix $A_k$. $A_k$ is defined as $A_k = U_k \cdot S_k \cdot V_k^T$ and corresponds to the closest linear approximation of the original matrix $A$ with reduced rank $k$. For our experiments we set the $k = 10$. Therefore we calculated the matrix $Items = V_k^T \cdot S_k$ which represents n items in the $k$ dimensional feature space.

In our experiments we built one-classifiers only for users who rated at least 80 items, this means that they provided a rating value from the set $\{1,2,3,4,5\}$. Consequently from 943 users we took 410 users. For each of the 410 users, we have trained five one-class classifier using feature vectors from matrix $Items$. We utilized the following five classifiers which are available in the dd-tools toolbox [11]:

Density methods: a) Gaussian model (gausdd), b) Mixture of Gaussians (mogdd), Boundary methods: Support vector data description (incsvdd) and Reconstruction methods: a) Principal Component Analysis (pcadd), b) Auto-Encoders and Diabolo networks (autoencdd).

The one class classifiers examined over different fractions $v \in \{0.1, 0.2, ..., 0.9\}$. Fraction $v$ is a threshold which put such percentage of the training target objects will be rejected and classified as non-positive. For the evaluation of classifiers we followed for each fraction $v$ a 5-fold cross validation in the labels of each user, where the available labels have been randomly split into a training set (80%) and a test set (20%).

Recommender Systems typically show great variability with respect to choice of evaluation measure [2]. In experiments of [8] showed that no single one-class collaborative filtering technique to dominate with respect to all metrics. In this paper, we considered a number of evaluation metrics such as: accuracy and area under ROC curves (AUC-ROC). The experimental results are shown in Figs. 1, 2. As it is presented in these figures one-class classifiers can model efficiently the preferences of users. Particularly, the reconstructions methods have slightly better performance in comparison to other classifiers in terms of AUC values.

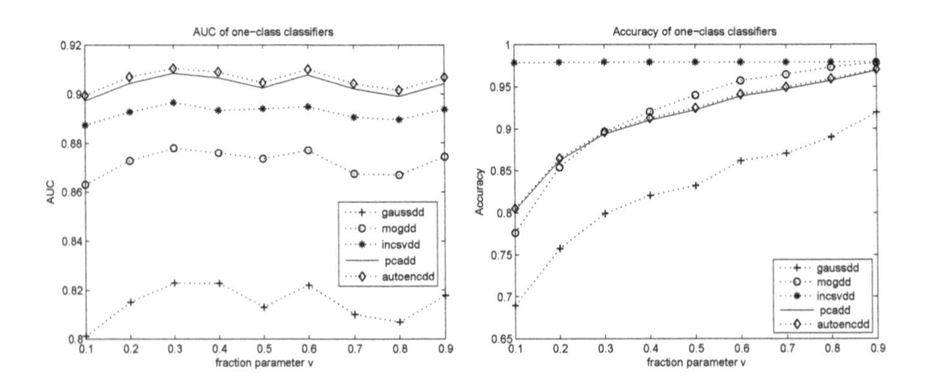


**Fig. 1** Mean Area Under Curve for all users

**Fig. 2** Mean Accuracy for all users

Another important factor that leads toward the selection of one-class learning as a valid paradigm for the problem of recommendation is that the related misclassification costs are analogously unbalanced. The quantities of interest are the false positive rate and the false negative rate. The false positive rate expresses how often a classifier falsely predicts that a specific pattern belongs to the target class of patterns while it originated from the complementary class. The false negative rate expresses how often a classifier falsely predicts that a specific pattern belongs to the complementary class of patterns while it originated from the target class. In the context of designing a recommender system, the cost related to the false positive rate is of greater impact than the cost related to the false negative rate. False positives result in recommending items that a particular user would classify as non-desirable and, thus, effect the quality of recommendation. In contrast, false negatives result in not recommending items that a particular user would classify as desirable. Thus, it is of vital importance to minimize the false positive rate which results in improving the accuracy of recommendation. The following Figs. 3, 4 present the ROC curve and the Cost curve of one user (similar performance observed and for the rest of the 409 users) for the classifier consisted of Mixtures of Gaussian in fraction v=0.2. These two figures illustrate the suitability of one-classifiers for the recommendation problems. In these figures we can observe that for a relative large range of misclassification costs the operating point FP = 0.37, TP = 0.88 will be the optimal one. This is indicated in a cost curve. For a varying cost-ratio between the two classes, the (normalized) expected cost is computed. Each operating point appears as a line in this plot. In Fig. 3 the cost curve for the same user and classifier as of the ROC curve in Fig. 4 is shown. The operating point of Fig. 4 is indicated by the dotted line in Fig. 3. The combination of operating points that form the lower hull is indicated by the thick line, and shows the best operating points over the range of costs.

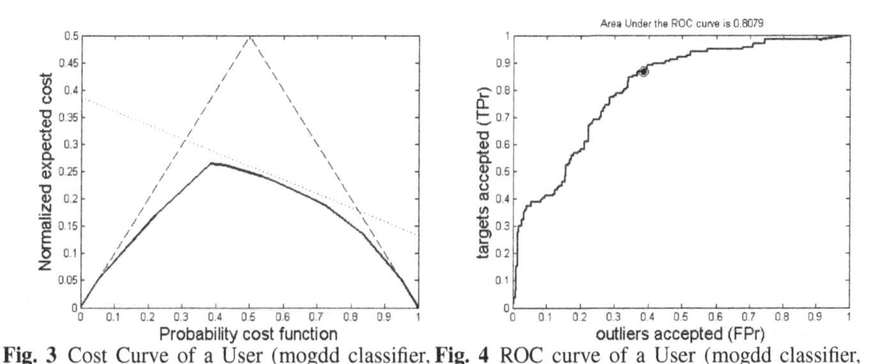


**Fig. 3** Cost Curve of a User (mogdd classifier, fraction v=0.2)

**Fig. 4** ROC curve of a User (mogdd classifier, fraction v=0.2)

## 3 Conclusions and Future Work

We performed a comparative study of one-class classifiers from density, boundary and reconstruction methods for the recommendation problem. Experimental results on publically available dataset of MovieLens revealed that one-class classifiers can model efficiently the preferences of users. Evaluation of classifiers made in terms of AUC and accuracy measures.

Currently, we are in the process of conducting further experiments and improvements to our system, by extending the proposed method into a hybrid cascade recommender system [5]. This and other related research work is currently in progress and will be reported elsewhere in the near future.

## References

1. Billsus, D., Pazzani, M.J.: Learning collaborative information filters. In: Proceedings of the Fifteenth International Conference on Machine Learning, ICML '98, pp. 46–54. Morgan Kaufmann Publishers Inc., San Francisco, CA, USA (1998). URL http://dl.acm.org/citation.cfm?id=645527.657311
2. Gunawardana, A., Shani, G.: A survey of accuracy evaluation metrics of recommendation tasks. J. Mach. Learn. Res. **10**, 2935–2962 (2009). URL http://dl.acm.org/citation.cfm?id=1577069.1755883
3. Japkowicz, N., Stephen, S.: The class imbalance problem: A systematic study. Intell. Data Anal. **6**(5), 429–449 (2002)
4. Juszczak, P.: Learning to recognise. a study on one-class classification and active learning. PhD thesis, Delft University of Technology (2006)
5. Lampropoulos, A.S., Lampropoulou, P.S., Tsihrintzis, G.A.: A cascade-hybrid music recommender system for mobile services based on musical genre classification and personality diagnosis. Multimedia Tools and Applications **59**(1), 241–258 (2012). DOI http://dx.doi.org/10.1007/s11042-011-0742-0
6. Manevitz, L.M., Yousef, M.: One-class svms for document classification. J. Mach. Learn. Res. **2**, 139–154 (2002)
7. Moya, M.R., Koch, M.W., Hostetler, L.D.: One-class classifier networks for target recognition applications. In: Proc. World Congress on NeuralNetworks, pp. 797–801. Portland, USA (1993)
8. Pan, R., Scholz, M.: Mind the gaps: weighting the unknown in large-scale one-class collaborative filtering. In: Proceedings of the 15th ACM SIGKDD international conference on Knowledge discovery and data mining, KDD '09, pp. 667–676. ACM, New York, NY, USA (2009). DOI 10.1145/1557019.1557094. URL http://doi.acm.org/10.1145/1557019.1557094
9. Pan, R., Zhou, Y., Cao, B., Liu, N.N., Lukose, R., Scholz, M., Yang, Q.: One-class collaborative filtering. In: ICDM '08: Proceedings of the 2008 Eighth IEEE International Conference on Data Mining, pp. 502–511. IEEE Computer Society, Washington, DC, USA (2008). DOI http://dx.doi.org/10.1109/ICDM.2008.16
10. Schwab, I., Pohl, W., Koychev, I.: Learning to recommend from positive evidence. In: IUI '00: Proceedings of the 5th international conference on Intelligent user interfaces, pp. 241–247. ACM, New York, NY, USA (2000). DOI http://doi.acm.org/10.1145/325737.325858
11. Tax, D.: Ddtools, the data description toolbox for matlab (2012). Version 1.9.1

# Hybridization of Adaptive Differential Evolution with BFGS

R. A. Khanum and M. A. Jan

**Abstract** *Local search* (LS) methods start from a point and use the gradient or objective function value to guide the search. Such methods are good in searching the neighborhood of a given solution (i.e., they are good at exploitation), but they are poor in exploration. *Evolutionary Algorithms* (EAs) are nature inspired population-based search optimizers. They are good in exploration, but not as good at exploitation as LS methods. Thus, it makes sense to hybridize EAs with LS techniques to arrive at a method which benefits from both and, as a result, have good search ability. *Broydon-Fletcher-Goldfarb-Shanno* (BFGS) method is a gradient-based LS method designed for nonlinear optimization. It is an efficient, but expensive method. *Adaptive Differential Evolution with Optional External Archive* (JADE) is an efficient EA. Nonetheless, its performance decreases with the increase in problem dimension. In this paper, we present a new hybrid algorithm of JADE and BFGS, called *Hybrid of Adaptive Differential Evolution and BFGS*, or DEELS, to solve the unconstrained continuous optimization problems. The performance of DEELS is compared, in terms of the statistics of the function error values with JADE.

## 1 Introduction

"Optimization" is concerned with finding the optimal solution for one or many objective functions, possibly subject to constraints. When no constraints are imposed, the optimization problem can be stated as follows:

---

R. A. Khanum
Department of Mathematical Sciences, University of Essex, Wivenhoe Park, Colchester, Essex, CO4 3SQ, UK. e-mail: rakhan@essex.ac.uk

Muhammad Asif Jan
Department of Mathematics, Kohat University of Science and Technology (KUST), Kohat 26000, Khyber Pakhtoonkhwa, Pakistan. e-mail: janmathpk@gmail.com, majan@kust.edu.pk

$$\text{Minimize } f(\mathbf{x}), \tag{1}$$

where $f(\mathbf{x})$ denotes the objective function, and $\mathbf{x} = (x_1, x_2, ..., x_n)^T \in R^n$ is the decision vector of dimension n.

Though JADE [1], due to its adaptive parameter control strategy, performs better than classic DE on many optimization problems. However, being an EA, its performance worsens with increase in problem dimension. BFGS [2] is a LS technique which has a strong self correcting ability in searching the optimal solution [3], but it is not globally as good as JADE.

The important question is how to reconcile these two different aspects to solve the minimization problem of type 1. A very natural way would be to hybridize these two techniques, JADE and BFGS, together. But, the issue is how to combine them in a way which is easy to understand and implement.

Many hybrid approaches incorporate expensive methods like BFGS to find the best solution. But, DEELS incorporates it in the framework of JADE not only for refining good solutions, but also for locating them in the population during the search process.

The rest of this paper is organized as follows. Section 2 reviews some existing methods that are invoked in the design of the new hybrid algorithm. Section 3 presents the proposed hybrid algorithm, DEELS, in detail. Section 4 reports experimental results and their discussion. Finally, Section 5 presents a brief summary of this paper.

## 2 Relevant Methods

DEELS relies on JADE and BFGS. Thus, this section will first briefly present DE on which JADE is based. It will then introduce JADE and BFGS.

**Classic DE:** DE is a recently developed bio-inspired scheme for finding the global optimum, $\mathbf{x}^*$ of optimization problem 1. In DE, for each individual, $\mathbf{x}_i$, a mating pool of four individuals is formed in which it breeds against three individuals and produces an offspring. To generate an offspring, DE incorporates two genetic operators: mutation and crossover. In mutation, a mutant vector is produced by adding a scaled difference of the two already chosen vectors to the third chosen vector. In crossover, the parameters of parent vector and mutant vector are mixed up and a trial vector is generated, which then competes with the parent vector, based on fitness to survive for next generation. More details about DE can be found in [4–8].

**JADE:** JADE [1] is an adaptive version of DE which modifies it in three aspects.

1. JADE utilizes two mutation strategies, one with external archive, and the other without it. In DEELS, we utilize the former strategy.

2. For each individual $\mathbf{x}_i$, control parameter $F_i$ and the crossover probability, $CR_i$ are generated independently from Cauchy and normal distributions, respectively.
3. At each generation, the failed parents are sent to the archive. If the archive size exceeds the population size, $N_p$ some solutions are randomly deleted from it to keep its size equal to $N_p$. The archive inferior solutions play a roll in JADE's mutation strategy with archive. The archive not only provides information about direction, but improves the diversity as well.

**BFGS:** The BFGS method, which is also known as the quasi Newton algorithm, employs the gradient and Hessian in finding a suitable search direction. BFGS is considered as a good local search method due to its efficiency, but it is known as expensive due to its Hessian calculation. The details of BFGS are available in [2].

## 3 The Proposed Hybrid Algorithm: DEELS

In this section, we present the working of DEELS. DEELS begins with JADE and allows it to search for $\xi(=\frac{1}{3}maxFES)$ generations. It then selects the $q(=3)$ best individuals from the current population and applies BFGS to them for the first time to generate $q$ new solutions with $\gamma(=1)$ iterations. These $q$ new solutions are then introduced into the population and the worse $q$ solutions are removed from it. The objective of applying BFGS to the $q$ best individuals is to make them potential individuals to produce better offspring and to lead the search in promising directions.

The second purpose of calling BFGS after $\xi$ generations is to concentrate the population and to add local search ability to the overall scheme. This way, the algorithm will avoid getting trapped in the local optimal solutions. For these reasons, BFGS is invoked two more times in the evolution, with an interval of $\xi$ generations. If the function value of the current best solution is less than a threshold *error*, then it is in the neighborhood of the value to reach and the current best solution might lead the search to the desired optimal solution. It is desirable to apply the efficient LS, more than one iterations to this best solution. Therefore, BFGS is applied $\varsigma(=3)$ iterations when the best solution is in the vicinity of a local optimum. If the output solution of BFGS is the best known solution, then the algorithm stops; otherwise, it continues until the allowed maximum number of function evaluations is met. The pseudo-code of DEELS is given in Algorithm 1.

## 4 Experimental Results

To study the performance of the proposed algorithm, 15 test instances are chosen from CEC2005 competition [9]. The population size, $N_p$ is set to 75 and the problem

**Algorithm 1** The Algorithmic framework of DEELS

**Control Parameters:**
$q$: the number of points selected for LS;
$\gamma$: the number of iterations of LS for concentration;
$\varsigma$: the number of iterations of LS for refining solution;
$N_p$: population size;
$FES$: number of function evaluations;
$G$: generation counter;
$\xi$: interval between the LS calls;
**Step 0 Initialization:** Uniformly and randomly sample $N_p$ solutions, $\mathbf{x}_1, \mathbf{x}_2, \ldots, \mathbf{x}_{N_p}$ from the search space to form the initial population $P$. Evaluate each member of the population $P$, set $FES = N_p$ and set $G = 0$.
**Step 1**: **Global Search:** Run JADE to improve $P$ for $\xi$ generations.
**Step 2**: **LS:** Apply BFGS for $\gamma$ iterations on each of the top $q$ solutions in $P$ to generate $q$ new solutions.
**Step 3**: **Update Population:** Replace the $q$ worse solutions in $P$ with the $q$ new solutions generated in **Step 2**.
**Step 4**: If the current best solution is in the vicinity of the desired solution, go to **Step 5**; otherwise, go to **Step 6**.
**Step 5 Refinement:** Apply BFGS $\varsigma$ iterations to the current best solution and replace the worse solution of $P$ with this newly generated solution and go to **Step 6**.
**Step 6 Stopping Condition:** If a present stopping condition is met, stop and output the best solution obtained. Otherwise, set $G = G + 1$ and go to **Step 1**

dimension, $n$ is set to 30 for all the test instances in both JADE and DEELS. The other two parameters $F$ and $CR$ are set to 0.5 initially and then the parameter values used in JADE are adopted.

For fair comparison, 30 independent runs in MATLAB were conducted with both algorithms on each test instance. The mean and standard deviation of the function error $f(\mathbf{x}) - f(\mathbf{x}^*)$ values are recorded for each run.

## *4.1 Comparison with JADE*

The average and standard deviation of the function error values obtained in each run are used to compare the performance of both algorithms. They are summarized in Table 1.

### 4.1.1 Unimodal Test Instances (F1 - F5)

As shown in Table 1, the performance of DEELS on F1 and F3 is better than that of JADE in terms of solution quality. This performance improvement on F1 and F3 is due to the fact that BFGS has a greater exploitation ability. On test instances F2 and F4, both algorithms performed equally good, while on test instance F5 JADE showed superior performance. Thus, the overall performance of the proposed algo-

**Table 1** Experimental results of JADE and DEELS on 15 test instances of 30 variables with $3 \times 10^5$ FES. mean error and Std Dev of the function error values obtained in 30 runs.

| Test Instances | JADE Mean Error ± Std Dev | DEELS Mean Error ± Std Dev |
|---|---|---|
| F1 | $3.11E-11 \pm 4.02E-10-$ | $0.00E+00 \pm 0.00E+00$ |
| F2 | $1.02E-10 \pm 7.15E-10 \approx$ | $9.65E-009 \pm 2.93E-10$ |
| F3 | $9.99E+03 \pm 4.60E+03-$ | $7.31E+03 \pm 5.50E+03$ |
| F4 | $7.99E-10 \pm 1.81E-09 \approx$ | $9.58E-09 \pm 4.21E-10$ |
| F5 | $1.15E-06 \pm 6.90E-05+$ | $1.35E-05 \pm 4.57E-05$ |
| F6 | $0.00E+00 \pm 0.39E+00+$ | $1.29E+01 \pm 3.47E+01$ |
| F7 | $4.70E+03 \pm 2.64E-12-$ | $4.02E+02 \pm 4.13E+00$ |
| F8 | $2.09E+01 \pm 0.14E+00 \approx$ | $2.09E+01 \pm 0.24E+00$ |
| F9 | $1.20E-10 \pm 7.63E-10 \approx$ | $9.41E-09 \pm 6.74E-10$ |
| F10 | $2.29E+01 \pm 5.19E+00 \approx$ | $2.29E+01 \pm 5.05E+00$ |
| F11 | $2.58E+01 \pm 1.57E+00-$ | $2.53E+01 \pm 1.35E+00$ |
| F12 | $5.46E+03 \pm 3.968E+03+$ | $5.60E+03 \pm 5.27E+03$ |
| F13 | $1.27E+01 \pm 0.11E+01+$ | $1.35E+01 \pm 1.41E-01$ |
| F14 | $1.24E+01 \pm 2.74E-01-$ | $1.23E+01 \pm 3.15E-01$ |
| F15 | $3.53E+02 \pm 8.60E+01+$ | $3.67E+02 \pm 8.44E+01$ |
| − | 5 | |
| + | 5 | |
| ≈ | 5 | |

"-", "+"and "≈" denote that the performance of JADE is worse than, better than, or similar to that of DEELS, respectively.

rithm in this category of unimodal test instances is better on 40% test instances. On 40% it is comparable with JADE, while on the remaining 20% JADE is better.

### 4.1.2 Multimodal Test Instances (F6 - F14)

The performance of DEELS is better than JADE on 3 out of the 9 multimodal test instances, F7, F11 and F14. Although test instance F7 is considered as a hard instance, DEELS found a very good mean for it as compared to JADE. This success is due to the combination of BFGS with JADE, which helped the search in finding a local minimum which JADE could not find. On a further 3 out of the 9 functions, i.e, F8, F9 and F10 both JADE and DEELS showed comparable performances. However, on the remaining test functions, F6, F12 and F13 DEELS could not improve the mean error of them over JADE. The failure of DEELS on these three multimodal functions could be due to the reason that BFGS, being expensive in terms of function evaluations, may have spent the maximum number of function evaluations earlier than JADE, and as a result, could not get the desired solution.

Hence, it can be concluded that both JADE and DEELS showed comparable performance on this category of multimodal functions.

### 4.1.3 Hybrid Composition Test Instance (F15)

This test instance is considered among the hardest functions, because it is designed by combining several other test instances. The proposed algorithm did not improve its mean over the mean of JADE on F15, maybe due the reason as mentioned in the above section.

## 5 Conclusion

In this paper, we proposed DEELS, a new hybrid algorithm of JADE and BFGS. Its performance in terms of function error values is compared with that of JADE on fifteen test instances from CEC2005. The experimental results show that DEELS surpassed JADE on five test instances, F1, F3, F7, F11, F14, while the situation is vice versa for other five test instances, F5, F6, F12, F13, F15. However, both algorithms showed comparable performance on the rest of five test instances, F2, F4, F8, F9, F10.

## References

1. J. Zhang and A. C. Sanderson, "JADE: adaptive differential evolution with optional external archive," *Evolutionary Computation, IEEE Transactions on*, vol. 13, no. 5, pp. 945–958, 2009.
2. P. Venkataraman, *Applied Optimization with MATLAB programming*. John Wiley and sons, New York, 2002, pp. 238–240.
3. A. Skajaa, "Limited memory BFGS for nonsmooth optimization," *Master's thesis*, 2010.
4. R. Storn and K. Price, "Home page of differential evolution," 2003.
5. S. Das and P. N. Suganthan, "Tutorial:differential evolution, foundations, prospectives and applications," in *IEEE Symposium Series on Computational Intelligence (SSCI)*, Paris, France, April, 2011, pp. 1–59.
6. P. N. Suganthan and Swagatam, "Tutorial: Differential evolution," in *IEEE Symposium Series on Computational Intelligence (SSCI)*, Paris, France, April, 2011, pp. 1–76.
7. F. Neri and V. Tirronen, "Recent advances in differential evolution: a survey and experimental analysis," *Artificial Intelligence Review*, vol. 33, no. 1, pp. 61–106, 2010.
8. S. Das and P. N. Suganthan, "Differential evolution: A survey of the state-of-the-art," *Evolutionary Computation, IEEE Transactions on*, no. 99, pp. 1–28, 2011.
9. A. K. Qin and P. N. Suganthan, "Self adaptive differential evolution algorithm for numerical optimization," in *IEEE Congress on Evolutionary Computation*, vol. 2, 2005, pp. 1785–1791.